V International Conference on Problems of Mathematical and Theoretical Physics and Mathematical Modelling 2016

Journal of Physics: Conference Series Volume 788

Moscow, Russia
5 – 7 April 2016

Editors:

Nikolay A. Kudryashov
Oleg V. Nagomov
Sergey V. Popruzhenko

ISBN: 978-1-5108-3605-1
ISSN: 1742-6588

Printed from e-media with permission by:

Curran Associates, Inc.
57 Morehouse Lane
Red Hook, NY 12571

Some format issues inherent in the e-media version may also appear in this print version.

Printed by Curran Associates, Inc. (2017)

Institute of Physics
Dirac House, Temple Back
Bristol BS1 6BE UK

Phone: 44 1 17 929 7481
Fax: 44 1 17 920 0979

techtracking@iop.org

Additional copies of this publication are available from:

Curran Associates, Inc.
57 Morehouse Lane
Red Hook, NY 12571 USA
Phone: 845-758-0400
Fax: 845-758-2633
Email: curran@proceedings.com
Web: www.proceedings.com

Table of contents

Volume 788

V International Conference on "Problems of Mathematical and Theoretical Physics and Mathematical Modelling"

5–7 April 2016, Moscow, Russian Federation

Accepted papers received: 7 December 2016
Published online: 2 February 2017

Preface

Papers

Methods of mathematical physics

Problems of theoretical physics

Mathematical modelling

IOP Conf. Series: Journal of Physics: Conf. Series **788** (2017) 011001 doi:10.1088/1742-6596/788/1/011001

Preface

This annual conference was established at the National Nuclear Research University – Moscow Engineering Institute in the year 2012. The scope was initially on mathematical physics and mathematical modelling, while since this year theoretical physics was also included. The Conference was organized by three leading departments of the MEPhI in the field of theoretical and mathematical physics.

Each year, more than 100 participants from different countries including Belorussia, Czech Republic, France, Great Britain, Japan, Moldova, Russia, Slovakia, Spain, Sweden and the USA take part in the Conference with oral talks.

The aim of this conference is to bring together researchers with interests in: Analytical Theory of Differential Equations, Interaction of Radiation With Matter and Physics of Extreme Light, Nuclear Physics and Theory of Elementary Particles, Optimization Problems, Solid Mechanics, Soft Matter and Interdisciplinary Physics, Fluid Mechanics, Mathematical Modelling, Time Series.

Nikolay A. Kudryashov
Oleg V. Nagornov
Sergey V. Popruzhenko

Peer review statement

All papers published in this volume of *Journal of Physics: Conference Series* have been peer reviewed through processes administered by the proceedings Editors. Reviews were conducted by expert referees to the professional and scientific standards expected of a proceedings journal published by IOP Publishing.

The H-theorem for the chemical kinetic equations with discrete time and for their generalizations

S Adzhiev[1], I Melikhov[1] and V Vedenyapin[2]

[1] Lomonosov Moscow State University, Moscow, Russia
[2] Keldysh Institute of Applied Mathematics of Russian Academy of Sciences, Moscow, Russia; RUDN-University, Moscow, Russia

adzhiev@nm.ru, vicveden@yachoo.com

Abstract. In this paper the generalizations of equations of chemical kinetics, including classical and quantum chemical kinetics, is considered. We make time discrete in these equations and prove the H-theorem.

1. Introduction

The H-theorem was considered for the first time by Boltzmann in the paper "Weitere Studien über das Wärmegleichgewicht unter Gasmolekülen" (1872) [1]. This theorem, substantiated the convergence of solutions of the Boltzmann equations to the Maxwell distribution, Boltzmann associated with the law of entropy increase [2].

We consider the problem of the transition to discrete time. The discrete model should inherit the properties of the original equations. Therefore, in the equations of chemical kinetics necessary condition for the adequacy of the time discretization is fulfillment of the requirement of exact conservativeness (i.e., in the discrete model those and only those conservation laws that are in the original equation are present), and the implementation of the H-theorem: For which discretizations of the time is the H-theorem valid? It turns out that for the implicit time discretization the H-theorem is fulfilled. It ensures the stability of the stationary solution. In the present work we investigate the question of the physicochemical adequacy of models with discrete time in respect to the H-theorem.

Let we have a mixture of chemically interacting substances with spatially homogeneous concentrations. We denote by $f_i(t)$ the number of molecules of the i-th substance ($i = 1,2,\ldots,n$) at the time t.

In general, the equations for complex chemical reactions are written as [3]–[5]:

$$\frac{df_i}{dt} = \frac{1}{2}\sum_{(\boldsymbol{\alpha},\boldsymbol{\beta})\in\Im}(\beta_i - \alpha_i)\left(K_{\boldsymbol{\beta}}^{\boldsymbol{\alpha}}\mathbf{f}^{\boldsymbol{\alpha}} - K_{\boldsymbol{\alpha}}^{\boldsymbol{\beta}}\mathbf{f}^{\boldsymbol{\beta}}\right),\ i = 1,2,\ldots,n. \tag{1.1}$$

Here $\mathbf{f}^{\boldsymbol{\alpha}}$ denotes the product $\mathbf{f}^{\boldsymbol{\alpha}} = f_1^{\alpha_1} f_2^{\alpha_2} \ldots f_n^{\alpha_n}$, the summation is conducted on some finite set $\Im$ of multiindices $(\boldsymbol{\alpha},\boldsymbol{\beta})$, $\boldsymbol{\alpha} = (\alpha_1,\alpha_2,\ldots,\alpha_n)$ and $\boldsymbol{\beta} = (\beta_1,\beta_2,\ldots,\beta_n)$ – vectors with nonnegative integer components. $(\boldsymbol{\alpha},\boldsymbol{\beta})$ corresponds to the elementary reaction

$$\alpha_1 S_1 + \alpha_2 S_2 + \ldots + \alpha_n S_n \xrightarrow{K_{\boldsymbol{\beta}}^{\boldsymbol{\alpha}}} \beta_1 S_1 + \beta_2 S_2 + \ldots + \beta_n S_n,\ (\boldsymbol{\alpha},\boldsymbol{\beta}) \in \Im, \tag{1.2}$$

S_i – the chemical symbols of the reacting substances, $K_{\beta}^{\alpha} \geq 0$ – the coefficients of reaction rates (reaction constants). The coefficients α_i, β_i are stoichiometric coefficients. Without loss of generality we can assume that the set $\Im$ is symmetric with respect to permutations $\boldsymbol{\alpha}$ and $\boldsymbol{\beta}$. While the zero coefficients of the reaction rates may correspond to some couples $(\boldsymbol{\alpha}, \boldsymbol{\beta})$: $K_{\beta}^{\alpha} = 0$, it is possible $K_{\alpha}^{\beta} > 0$ (allowed irreversibility of reactions).

The next lemma is valid.

Lemma 1.1. Linear invariants of a system of equations of chemical kinetics: $I_{\mu}(t) = \sum_{i=1}^{n} \mu_i f_i(t) = (\boldsymbol{\mu}, \mathbf{f}(t))$, are the conservation laws of number of atoms of each kind entering into the composition of at least one of the substances S_i participating in the reaction, or a linear combination of these linear conservation laws. More subtle and wider, instead of atoms of each type should be considered different complexes of atoms, which do not change themselves during the reaction.

The H-theorem is just the proof of the inequality: $dH(\mathbf{f})/dt \leq 0$, the proof of decreasing some functional H on the solutions of a system of equations. The H-function for the equations of chemical kinetics has the form:

$$H(\mathbf{f}) = \sum_{i=1}^{n} f_i \left(\ln \frac{f_i}{\xi_i} - 1 \right). \tag{1.3}$$

Here the vector $\boldsymbol{\xi} = (\xi_1, \xi_2, \ldots, \xi_n)$, where $\xi_i > 0$, is a positive stationary solution of the system of chemical kinetics equations. If we take the function (1.3) with the minus sign, it is called the Kullback–Leibler information entropy [6].

A classification of conditions on the coefficients of the chemical kinetics equations (1.1), when the H-theorem is valid, is presented in [4], [5]. It is valid in the case of the symmetric constants of the reactions: $K_{\beta}^{\alpha} = K_{\alpha}^{\beta}$. A wider class of equations is when the condition of detailed balance is fulfilled. And the most general condition, when the H-theorem is valid, is a condition of dynamic equilibrium or the Stuckelberg–Batisheva–Pirogov condition [4], [5], [7], [8], [9], [10].

2. The H-theorem for generalizations of the chemical kinetics equations

Here we will study systems of more general type than (1.1), and will make the transition to discrete time.

Let consider the system of equations:

$$\frac{df_i}{dt} = \frac{1}{2} \sum_{(\boldsymbol{\alpha},\boldsymbol{\beta}) \in \Im} (\beta_i - \alpha_i) \sigma_{\beta}^{\alpha}(\mathbf{f}) \left(K_{\beta}^{\alpha} e^{(\boldsymbol{\alpha}, \nabla G(\mathbf{f}))} - K_{\alpha}^{\beta} e^{(\boldsymbol{\beta}, \nabla G(\mathbf{f}))} \right), \; i = 1, 2, \ldots, n, \tag{2.1}$$

where $\sigma_{\beta}^{\alpha}(\mathbf{f})$, $G(\mathbf{f})$ – given functions from $\mathbf{f} = (f_1, f_2, \ldots, f_n)$. Let $\sigma_{\beta}^{\alpha}(\mathbf{f}) = \sigma_{\alpha}^{\beta}(\mathbf{f}) > 0$; $K_{\beta}^{\alpha} \geq 0$, and $K_{\beta}^{\alpha} \neq 0$ or $K_{\alpha}^{\beta} \neq 0$ for $(\boldsymbol{\alpha}, \boldsymbol{\beta}) \in \Im$.

If $\sigma_{\beta}^{\alpha}(\mathbf{f}) = 1$, then we'll have the system:

$$\frac{df_i}{dt} = \frac{1}{2} \sum_{(\boldsymbol{\alpha},\boldsymbol{\beta}) \in \Im} (\beta_i - \alpha_i) \left(K_{\beta}^{\alpha} e^{(\boldsymbol{\alpha}, \nabla G(\mathbf{f}))} - K_{\alpha}^{\beta} e^{(\boldsymbol{\beta}, \nabla G(\mathbf{f}))} \right), \; i = 1, 2, \ldots, n. \tag{2.2}$$

If in (3.2) $\dfrac{\partial G(\mathbf{f})}{\partial f_i} = \ln f_i$, then we'll obtain the system (1.1).

Let's make the transition to discrete time $t = 0, \tau, 2\tau, 3\tau, \ldots$, where $\tau > 0$, in the systems (2.1) and (2.2), replacing the derivatives $\frac{df_i}{dt}$ by difference analogues $\frac{f_i(t+\tau) - f_i(t)}{\tau}$, and in the right parts of these systems taking time $t+\tau$. I.e. having an implicit discretization in time for (2.1) and (2.2), we obtain respectively (2.3) and (2.4):

$$f_i(t+\tau) = f_i(t) + \tau \frac{1}{2} \sum_{(\boldsymbol{\alpha},\boldsymbol{\beta}) \in \Im} (\beta_i - \alpha_i) \sigma_{\boldsymbol{\beta}}^{\boldsymbol{\alpha}}(\mathbf{f}(t+\tau)) \left(K_{\boldsymbol{\beta}}^{\boldsymbol{\alpha}} e^{(\boldsymbol{\alpha}, \nabla G(\mathbf{f}(t+\tau)))} - K_{\boldsymbol{\alpha}}^{\boldsymbol{\beta}} e^{(\boldsymbol{\beta}, \nabla G(\mathbf{f}(t+\tau)))} \right), \qquad (2.3)$$

where $i = 1, 2, \ldots, n$.

$$f_i(t+\tau) = f_i(t) + \tau \frac{1}{2} \sum_{(\boldsymbol{\alpha},\boldsymbol{\beta}) \in \Im} (\beta_i - \alpha_i) \left(K_{\boldsymbol{\beta}}^{\boldsymbol{\alpha}} e^{(\boldsymbol{\alpha}, \nabla G(\mathbf{f}(t+\tau)))} - K_{\boldsymbol{\alpha}}^{\boldsymbol{\beta}} e^{(\boldsymbol{\beta}, \nabla G(\mathbf{f}(t+\tau)))} \right), \; i = 1, 2, \ldots, n. \qquad (2.4)$$

Just for implicit time discretization of the generalization of the chemical kinetics equations (2.1) and (2.2) we can prove the H-theorem.

Let the system (2.3) (the system (2.4)) is solved for initial data $\mathbf{f}(0) \in M \subseteq \mathbb{R}^n$, where M is a convex domain, on which for all $(\boldsymbol{\alpha}, \boldsymbol{\beta}) \in \Im$ is defined functions $\sigma_{\boldsymbol{\beta}}^{\boldsymbol{\alpha}}(\mathbf{f})$ and a function $G(\mathbf{f})$ that has continuous partial derivatives of first order. And let solution of the system with initial data $\mathbf{f}(0) \in M$ in any time belongs to the set M. It'll be a condition of conservation of positivity, if M is a positive cone in $\mathbb{R}^n$: $M = \{\mathbf{f} \in \mathbb{R}^n : \forall i = 1, 2, \ldots, n \; f_i > 0\}$. Just we have such set M in the case of chemical kinetics and quantum chemical kinetics in the case of bosons ($\theta > 0$). In the case of fermions (i.e., for $\theta < 0$) $M = \{\mathbf{f} \in \mathbb{R}^n : \forall i = 1, 2, \ldots, n \, 0 < f_i < 1/|\theta|\}$.

Let's focus on the following questions: when the systems (2.3) and (2.4) have the H-function – the functional of entropy type, decreasing on non-stationary solutions? what conclusions from this can be done?

The generalized condition of detailed balance for the case of systems (2.3). Let there exist a vector $\boldsymbol{\xi} \in M$ such that for all reactions $(\boldsymbol{\alpha}, \boldsymbol{\beta}) \in \Im$ the next condition is fulfilled:

$$K_{\boldsymbol{\beta}}^{\boldsymbol{\alpha}} e^{(\boldsymbol{\alpha}, \nabla G(\boldsymbol{\xi}))} = K_{\boldsymbol{\alpha}}^{\boldsymbol{\beta}} e^{(\boldsymbol{\beta}, \nabla G(\boldsymbol{\xi}))}. \qquad (2.5)$$

Then we say that for system (3.3) the generalized detailed balance condition is valid.

The generalized condition of dynamic equilibrium for the case of systems (2.4) (or a generalized condition of Stuckelberg–Batisheva–Pirogov for (2.4) [9; formula (2,9) on page 21], [5], [7]). Suppose there is a vector $\boldsymbol{\xi} \in M$ which is the solution of the system of equations:

$$\sum_{\boldsymbol{\beta}} K_{\boldsymbol{\beta}}^{\boldsymbol{\alpha}} e^{(\boldsymbol{\alpha}, \nabla G(\boldsymbol{\xi}))} = \sum_{\boldsymbol{\beta}} K_{\boldsymbol{\alpha}}^{\boldsymbol{\beta}} e^{(\boldsymbol{\beta}, \nabla G(\boldsymbol{\xi}))}, \qquad (2.6)$$

here $\boldsymbol{\alpha}$ is such that $(\boldsymbol{\alpha}, \boldsymbol{\beta}) \in \Im$ with some $\boldsymbol{\beta}$. Then we say that for system (2.4) the generalized condition of dynamic equilibrium is fulfilled.

Let consider the functional

$$H(\mathbf{f}) = G(\mathbf{f}) - G(\boldsymbol{\xi}) - (\nabla G(\boldsymbol{\xi}), \mathbf{f} - \boldsymbol{\xi}). \qquad (2.7)$$

Let $G(\mathbf{f})$ is a strictly convex function. And let if the set M is unrestricted, then $\lim_{\substack{\|\mathbf{f}\| \to +\infty \\ \mathbf{f} \in M}} H(\mathbf{f}) = +\infty$.

Then the next theorem is valid.

Theorem 2.1. (the *H*-theorem with an arbitrary function G in the generalized condition of detailed balance for the coefficients $K_{\boldsymbol{\beta}}^{\boldsymbol{\alpha}}$ and symmetric sigma). For the system (2.3) we assume that the coefficients $K_{\boldsymbol{\beta}}^{\boldsymbol{\alpha}}$ are such that there exists at least one solution $\xi \in M$ of the system (2.5). Then the following assertions hold.

a) the *H*-function (2.7) is not-increasing on solutions of the system (2.3): $\Delta H \equiv H(\mathbf{f}(t+\tau)) - H(\mathbf{f}(t)) \leq 0$, where equality achieves only at stationary points. All stationary solutions of system (2.3) satisfy the equalities (2.5).

b) the system (2.3) possesses $n-r$ conservation laws of the form $\sum_{i=1}^{n} \mu_i^k f_i$, $k = 1,\dots,n-r$: $\sum_{i=1}^{n} \mu_i^k f_i(t) = A^k = const$ on the solutions of system (2.3), where vectors $\boldsymbol{\mu}^k \equiv (\mu_1^k, \mu_2^k, \dots, \mu_n^k)$ are linearly independent, r is the dimension of the linear span of the vectors $\boldsymbol{\alpha} - \boldsymbol{\beta}$ (the Boltzmann–Orlov–Moser–Bruno vectors [11]). The vectors $\boldsymbol{\mu}^k$ are orthogonal to all the vectors $\boldsymbol{\alpha} - \boldsymbol{\beta}$: $\sum_{i=1}^{n} \mu_i^k (\alpha_i - \beta_i) = 0$, and form a basis in the space of vectors $\boldsymbol{\mu}$ that form the space of the invariants for (2.3) of the form $\sum_{i=1}^{n} \mu_i f_i$ – the space of the linear conservation laws for (2.3). The stationary solution of the system (2.3) will be unique if we fix all the constants A^k of these conservation laws, and it can be found as the point $\mathbf{f}_0$ where the minimum of H-function is achieved under the condition that all the constants A^k of the linear conservation laws are fixed, so it is given by the formula

$$\mathbf{f}_0 = \varphi\left(\nabla G(\xi) + \sum_{k=1}^{n-r} \boldsymbol{\mu}^k \lambda^k \right), \tag{2.8}$$

where $\mathbf{f} = \varphi(\mathbf{x})$ is determined by equation $\nabla G(\mathbf{f}) = \mathbf{x}$ as an implicit vector-function, and the λ^k – by the values of A^k.

c) Such stationary solution exists if A^k is defined in terms of the initial condition $\mathbf{f}(0) \in M$ as $A^k = \sum_{i=1}^{n} \mu_i^k f_i(0)$. The solution $\mathbf{f}(t)$ with this initial condition exists for all $t = 0, \tau, 2\tau, \dots$, is unique and tends to the stationary solution (2.8).

H-theorem with the sigma-free (when all the sigmas are equal to 1) generalized Stuckelberg–Batishcheva–Pirogov condition is formulated similarly.

3. The H-theorem for the explicit time discretization of the chemical kinetics equations

In the case of explicit time discretization of the chemical kinetics equations the H-theorem, generally speaking, is not valid. Let us give the example of such case.

Let us consider the reaction $2S_1 \to S_2$. The system of the chemical kinetics equations with explicit time discretization for this reaction in the case of symmetrical coefficients of reaction rates: $K_{(0,1)}^{(2,0)} = K_{(2,0)}^{(0,1)} = \sigma$, has the form:

$$\begin{cases} f_1(t+\tau) = f_1(t) + 2\tau\sigma\left(f_2(t) - f_1^2(t)\right), \\ f_2(t+\tau) = f_2(t) + \tau\sigma\left(f_1^2(t) - f_2(t)\right). \end{cases} \tag{3.1}$$

It is possible to prove that the Kullback–Leibler entropy with sign "minus" (1.3): $H = f_1(\ln f_1 - 1) + f_2(\ln f_2 - 1)$, can increase on the solutions of (3.1).

Let us consider cases of explicit time discretization when H -theorem is valid.

Let we have written the system of the chemical kinetics equations in the form:

$$f_i(t+\tau)=\sum_{j=1}^{n} B_i^j(\mathbf{f}(t))f_j(t),\ i=1,2,\ldots,n. \tag{3.2}$$

Let us consider H -function such as for Markov processes and chains:

$$H=\sum_{i=1}^{n}\xi_i\psi(f_i/\xi_i), \tag{3.3}$$

where $\psi(x)$ is an arbitrary strictly convex function. The following theorem is valid.

Theorem 3.1. Let in (3.2) matrix $B(\mathbf{f}(t))$ is stochastic:

$$B_i^j(\mathbf{f}(t))>0,\ \sum_{i=1}^{n}B_i^j(\mathbf{f}(t))=1,$$

and let vector $\boldsymbol{\xi}=(\xi_1,\xi_2,\ldots,\xi_n)$ with positive components exists such that $\sum_{j=1}^{n}B_i^j(\mathbf{f}(t))\xi_j=\xi_i$. Then the next statements are valid.

a) H -function (3.3) does not increase on the solutions of system (3.2): $\Delta H\le 0$, moreover, equality is achieved only in stationary points (like for Markov chains). All stationary points have the form: $\mathbf{f}_0=\lambda\boldsymbol{\xi}$.

b) The system (3.2) has the unique (up to a constant multiplier) conservation law of the form $\sum_{i=1}^{n}\mu_i f_i$, i.e. the only linear invariant on the solutions of the system (3.2), which is the conservation law of the number of all particles in the system $\sum_{i=1}^{n}f_i(t)$:

$$\sum_{i=1}^{n}f_i(t)=A=const.$$

Stationary solution of system (3.2) with positive components will be unique, if the constant of the linear conservation law A is fixed. It is found as a conditional minimum of the H-function under the condition that the constant of the linear conservation law A is fixed, and is given by the formula

$$\mathbf{f}_0=\left(A\Big/\sum_{i=1}^{n}\xi_i\right)\boldsymbol{\xi}. \tag{3.4}$$

c) Such stationary solution will exist if A is determined by the initial condition $\mathbf{f}(0)$ with positive components: $A=\sum_{i=1}^{n}f_i(0)$. The solution $\mathbf{f}(t)$ with this initial condition exists for all $t=0,\tau,2\tau,\ldots$, is unique and tends to the stationary solution of (3.4).

This theorem is analogical to the H -theorem for Markov chains.

4. Conclusion

A. Poincare [12] V. V. Kozlov [13] and D. V. Treshev [14] discussed a new form of the H -theorem. It holds for the Liouville equation and its generalizations [10], [12]–[16]. The concept of the Boltzmann extremal there also works in this setting [10], [15], [16]: it coincides with the time mean (the Cesáro mean), which illustrates its mathematical and fundamental value as a method for finding stationary solutions for a broad class of equations (both linear equations like the Liouville equation, and non-linear equations), and as a broad generalization of the concept of entropy.

In the present paper it was not required the Cesáro mean because the limit when time tends to infinity exists. But, generally speaking, in the case of discrete time already for Markov chain convergence doesn't always take place (we didn't consider these cases), but the Cesáro means coincide with the Boltzmann extremals [10], [15], [16].

This paper was financially supported by the Ministry of Education and Science of the Russian Federation on the program to improve the competitiveness of Peoples' Friendship University of Russia (RUDN-University) among the world's leading research and educational centers in 2016-2020. It was also supported by the department of Mathematics of Russian Academy of Sciences, program 1.3.1 "Problems of Mathematical Computational Physics".

References

[1] Boltzmann L 1872 Weitere studien über das wärmegleichgewicht unter gasmolekülen *Wiener Berichte* 66 pp 275–370

[2] Boltzmann L 1878 Über die beziehung zwischen dem zweiten hauptsatze der mechanischen wärmetheorie und der wahrscheinlichkeitsrechnung, respektive den satzen über das wärmegleichgewicht *Wiener Berichte* **76** pp 373–435

[3] Vol'pert A I and Hudjaev S I 1985 *Analysis in classes of discontinuous functions and equations of mathematical physics, Mech. Anal..* 8 (Dordrecht: Martinus Nijhoff Publishers)

[4] Sinitsyn A, Dulov E and Vedenyapin V 2011 *Kinetic Boltzmann, Vlasov and related equations* (Waltham, MA: Elsevier Science)

[5] Batishcheva Ya G and Vedenyapin V V 2005 The second law of thermodynamics for chemical kinetics *Matematicheskoe Modelirovanie* (Russian) **17** pp 106–10

[6] Kullback S and Leibler R A 1951 On information and sufficiency *Ann. Math. Statist.* **22** pp 79–86

[7] Malyshev V A and Pirogov S A 2008 Reversibility and irreversibility in stochastic chemical kinetics *Russian Mathematical Surveys* **63** pp 1–34

[8] Gasnikov A V and Gasnikova E V 2013 On entropy-type functionals arising in stochastic chemical kinetics related to the concentration of the invariant measure and playing the role of Lyapunov functions in the dynamics of quasiaverages *Math. Notes* **94** pp 854–61

[9] Lifschitz E M and Pitajewski L P 1983 *Lehrbuch der theoretischen Physik ("Landau–Lifshits"),* vol. X: *Physikalische Kinetik* (Berlin: Akademie-Verlag)

[10] Vedenyapin V V and Adzhiev S Z 2014 Entropy in the sense of Boltzmann and Poincaré *Russian Mathematical Surveys* **69** pp 995–1029

[11] Vedenyapin V V and Orlov Yu N 1999 Conservation laws for polynomial Hamiltonians and for discrete models of the Boltzmann equation *Theoret. and Math. Phys.* **121** pp 1516–23

[12] Poincaré H 1906 Réflections sur la théorie cinétique des gaz *J. Phys. Théoret.* **5** pp 369–403

[13] Kozlov V V 2002 *Thermal equilibrium in the sense of Gibbs and Poincaré* (Moscow–Izhevsk: Contemporary Mathematics. Institut Komp'yuternykh Issledovaniĭ)

[14] Kozlov V V and Treschev D V 2003 Weak convergence of solutions of the Liouville equation for nonlinear hamiltonian systems *Theoret. and Math. Phys.* **134** pp 339–50

[15] Vedenyapin V V 2008 Time averages and Boltzmann extremals *Doklady Mathematics* **78** pp 686–88

[16] Adzhiev S Z and Vedenyapin V V 2011 Time averages and Boltzmann extremals for Markov chains, discrete Liouville equations, and the Kac circular model *Computational Mathematics and Mathematical Physics* **51** pp 1942–52

IOP Conf. Series: Journal of Physics: Conf. Series **788** (2017) 012002 doi:10.1088/1742-6596/788/1/012002

Symmetries and reductions of Burgers–Huxley equation

A V Aksenov and K P Druzhkov

Lomonosov Moscow State University, Moscow, Russia

E-mail: aksenov.av@gmail.com, Konstantin.Druzhkov@gmail.com

Abstract. For Burgers–Huxley equation we solve the problem of group classification and find all reductions to ordinary differential equation. We show that there are no reductions, except symmetry reductions.

1. Introduction

Burgers–Huxley equation

$$u_t + uu_x = u_{xx} + \gamma u + \delta u^2 + \varepsilon u^3 \tag{1}$$

is widely used to describe many nonlinear phenomena [1].

The problem of finding particular solutions of a given nonlinear partial differential equation is important [2, 3], as well as the problem of constructing reductions of given equation. The method of finding reductions of partial differential equations with two independent variables was proposed in the paper [4].

The method of constructing of reductions of partial differential equations with two independent variables, based on the idea of invariance, was proposed in the paper [5]. This method is a further development of Clarkson–Kruscal direct method [4].

In this paper we solve the problem of group classification of Burgers–Huxley equation and find all its reductions. We show that considered equation has no reductions, except symmetry reductions.

2. Group classification

The task of group classification is to find Lie symmetries admitted by the considered equation depending on parameters. We seek symmetry operators of the equation (1) as

$$X = \xi^1(x,t,u)\frac{\partial}{\partial x} + \xi^2(x,t,u)\frac{\partial}{\partial t} + \eta^1(x,t,u)\frac{\partial}{\partial u}\,.$$

Applying the invariance criterion [6], we obtain the following overdetermined linear homogeneous system of determining equations,

$$\begin{aligned}
&{\xi^2}_x = 0,\ {\xi^2}_u = 0,\ {\xi^1}_u = 0\,,\\
&\eta_{uu} = 0,\ {\xi^2}_t = 2{\xi^1}_x,\ 2\eta_{xu} = u{\xi^1}_x - {\xi^1}_t + \eta\,,\\
&\eta_{xx} + (2{\xi^1}_x - \eta_u)(\gamma u + \delta u^2 + \varepsilon u^3) = \eta_t + u\eta_x - \gamma\eta - 2\delta u\eta - 3\varepsilon u^2\eta\,.
\end{aligned}$$

1. γ, δ, ε – arbitrary constants.

$$\frac{\partial}{\partial x}, \qquad \frac{\partial}{\partial t}. \tag{2}$$

Symmetry operators (2) generate symmetry operators kernel.

2. $\varepsilon \neq 0$, $\delta = \gamma = 0$. In this case, the symmetry operators kernel is extended by the symmetry operator

$$x\frac{\partial}{\partial x} + 2t\frac{\partial}{\partial t} - u\frac{\partial}{\partial u}.$$

3. $\gamma \neq 0$, $\delta = \varepsilon = 0$. In this case, the symmetry operators kernel is extended by the symmetry operator

$$e^{\gamma t}\left(\frac{\partial}{\partial x} + \gamma\frac{\partial}{\partial u}\right).$$

4. $\gamma = \delta = \varepsilon = 0$. In this case, considered equation is Burgers equation.

3. Reductions

We seek reductions of equation (1) as $u = U(x,t,w(z))$, where $z = z(x,t)$. After substitution of $u = U(x,t,w(z))$ into equation (1) we get

$$U_t + U_w w' z_t + U(U_x + U_w w' z_x) = U_{xx} + 2U_{xw} w' z_x + U_{ww} {w'}^2 {z_x}^2 + U_w w'' {z_x}^2 + \\ + U_w w' z_{xx} + \gamma U + \delta U^2 + \varepsilon U^3.$$

Setting $U_w z_x \neq 0$, we can get the above equation in the following view

$$w'' + \frac{U_{ww}}{U_w}{w'}^2 + w'\frac{2U_{xw}z_x + U_w z_{xx} - U_w z_t - UU_w z_x}{U_w {z_x}^2} + \\ + \frac{U_{xx} - U_t - UU_x + \gamma U + \delta U^2 + \varepsilon U^3}{U_w {z_x}^2} = 0. \tag{3}$$

In case of reduction, coefficients of ${w'}^2$, w', ${w'}^0$ in that equation must be functions of z and w.

Particularly, $U_{ww}/U_w = \Gamma(z,w)$, then $U = \alpha(x,t)\widetilde{\Gamma}(z,w) + \beta(x,t)$. This means that all reductions of the equation (1) can be found in form of $U = \alpha(x,t)w + \beta(x,t)$.

We substitute $U = \alpha(x,t)w + \beta(x,t)$ into (3) and seek every coefficient of ${w'}^2, w'$, ${w'}^0$ as functions of $z(x,t)$. We get A-system for $\alpha(x,t)$, $\beta(x,t)$ and $z(x,t)$ (it's equations are vanishing Jacobians conditions).

To simplify the solution process of A-system we use invariants $\mu_1(x,t)$, $\mu_2(x,t)$, $\mu_3(x,t)$ as

$$\begin{aligned} z_t &= -\mu_1 \cdot z_x, \\ \alpha_t &= -\alpha_x \cdot \mu_1 + \alpha \cdot \mu_2, \\ \beta_t &= -\beta_x \cdot \mu_1 + \beta \cdot \mu_2 + \mu_3. \end{aligned}$$

In terms of invariants A-system:

$$\begin{aligned} \mu_2 &= -\mu_{1x}, \\ \mu_{1xx} &= -\delta \cdot \mu_{1x} - 3\varepsilon \cdot \mu_3, \\ \mu_{1xt} &= -2 \cdot {\mu_{1x}}^2 + (\delta^2 - 2\gamma)\mu_{1x} + (1 - 3\varepsilon)\mu_{3x} + \delta(3\varepsilon - 2)\mu_3, \\ \mu_{1t} &= -(2\mu_1 + 3\delta)\mu_{1x} + (1 - 9\varepsilon)\mu_3, \\ \mu_{3xx} &= 2\mu_3 \cdot \mu_{1x} + \mu_{3t} - \gamma\mu_3. \end{aligned} \tag{4}$$

IOP Conf. Series: Journal of Physics: Conf. Series **788** (2017) 012002 doi:10.1088/1742-6596/788/1/012002

1. $\varepsilon = 0$**.** Second equation of the system (4) is

$$\mu_{1_{xx}} = -\delta \cdot \mu_{1_x}.$$

1.1. $\delta \neq 0$**.** In this case $\mu_1 = f_1(t)e^{-\delta x} + f_2(t)$, then third and fourth equations of the system (4) are

$$\begin{aligned} &-\delta \dot{f}_1 e^{-\delta x} = -2\delta^2 e^{-2\delta x} f_1{}^2 - \delta(\delta^2 - 2\gamma)e^{-\delta x} f_1 + \mu_{3_x} - 2\delta\mu_3\,, \\ &\dot{f}_1 e^{-\delta x} + \dot{f}_2 = \delta(2f_1 e^{-\delta x} + 2f_2 + 3\delta)e^{-\delta x} f_1 + \mu_3\,. \end{aligned} \tag{5}$$

We express μ_3 in second equation of the system (5) and substitute it into first equation, then $f_1 = 0$, $f_2 = C_1$, and $\mu_1 = C_1$, $\mu_3 = 0$, $\mu_2 = 0$.

It is easy to verify that $\mu_1 = C_1$, $\mu_2 = 0$, $\mu_3 = 0$ give invariant solutions of considered equation, where $\varepsilon = 0$, $\delta \neq 0$.

1.2. $\delta = 0$**.** In this case $\mu_1 = f_1(t)x + f_2(t)$. We substitute it into third and fourth equations of the system (4), then $\gamma \cdot f_1 = 0$.

1.2.1. $\gamma \neq 0$**.** In this case $f_1 = 0$, then $\mu_1 = f_2(t)$, $\mu_3 = \dot{f}_2$. From fifth equation of the system (4): $f_2 = C_1 e^{\gamma t} + C_2$. Then $\mu_1 = C_1 e^{\gamma t} + C_2$, $\mu_2 = 0, \mu_3 = \gamma C_1 e^{\gamma t}$.

It is easy to verify that $\mu_1 = C_1 e^{\gamma t} + C_2$, $\mu_2 = 0, \mu_3 = \gamma C_1 e^{\gamma t}$ give invariant solutions of considered equation, where $\varepsilon = 0, \delta = 0, \gamma \neq 0$.

1.2.2. $\gamma = 0$**.** In this case equation (1) is Burgers equation. In paper [4] showed that Burgers equation has no reductions, except invariant reductions.

2. $\varepsilon \neq 0$**.** From second equation of the system (4):

$$\mu_3 = -\frac{1}{3\varepsilon}(\mu_{1_{xx}} + \delta \cdot \mu_{1_x}).$$

We substitute this into third, fourth and fifth equations of the system (4):

$$\begin{aligned} &\mu_{1_{xt}} = -2 \cdot \mu_{1_x}{}^2 + 2\left(\frac{\delta^2}{3\varepsilon} - \gamma\right)\mu_{1_x} + \left(1 - \frac{1}{3\varepsilon}\right)\mu_{1_{xxx}} + \frac{\delta}{3\varepsilon} \cdot \mu_{1_{xx}}\,, \\ &\mu_{1_t} = -\left(2\mu_1 + \frac{\delta}{3\varepsilon}\right)\mu_{1_x} + \left(3 - \frac{1}{3\varepsilon}\right)\mu_{1_{xx}}\,, \\ &-\frac{1}{3\varepsilon}(\mu_{1_{xxxx}} + \delta \cdot \mu_{1_{xxx}}) = -\frac{2}{3\varepsilon}(\mu_{1_{xx}} + \delta \cdot \mu_{1_x}) \cdot \mu_{1_x} - \\ &-\frac{1}{3\varepsilon}(\mu_{1_{xxt}} + \delta \cdot \mu_{1_{xt}}) + \frac{\gamma}{3\varepsilon}(\mu_{1_{xx}} + \delta \cdot \mu_{1_x})\,. \end{aligned} \tag{6}$$

Then we substitute second equation of the system (6) into its first equation:

$$\mu_{1_{xxx}} = \left(\mu_1 + \frac{\delta}{3\varepsilon}\right)\mu_{1_{xx}} + \left(\frac{\delta^2}{3\varepsilon} - \gamma\right)\mu_{1_x}\,. \tag{7}$$

Then from the equation (7), second equation of the system (6) and third equation of the system (6) we get:

$$\begin{aligned} &(-9\varepsilon^2(6\varepsilon + 1)\mu_{1_x} - 9\varepsilon^2\mu_1{}^2 - 6\delta\varepsilon\mu_1 - 81\gamma\varepsilon^3 + 27\delta^2\varepsilon^2 + 9\gamma\varepsilon^2 - 3\delta^2\varepsilon - \delta^2)\mu_{1_{xx}} + \\ &+ \left(3\varepsilon \cdot (3\gamma\varepsilon - \delta^2)\mu_1 + 18\delta^3\varepsilon^2 + 3\delta\gamma\varepsilon - 81\delta\gamma\varepsilon^3 - \delta^3\right)\mu_{1_x} = 0\,. \end{aligned}$$

Thereby, system (6) is equivalent to the system

$$\begin{aligned}
&\mu_{1t} = -\left(2\mu_1 + \frac{\delta}{3\varepsilon}\right)\mu_{1x} + \left(3 - \frac{1}{3\varepsilon}\right)\mu_{1xx}\,,\\
&\mu_{1xxx} = \left(\mu_1 + \frac{\delta}{3\varepsilon}\right)\mu_{1xx} + \left(\frac{\delta^2}{3\varepsilon} - \gamma\right)\mu_{1x}\,,\\
&(-9\varepsilon^2(6\varepsilon+1)\mu_{1x} - 9\varepsilon^2{\mu_1}^2 - 6\delta\varepsilon\mu_1 - 81\gamma\varepsilon^3 + 27\delta^2\varepsilon^2 + 9\gamma\varepsilon^2 - 3\delta^2\varepsilon - \delta^2)\mu_{1xx} +\\
&+ \big(3\varepsilon\cdot(3\gamma\varepsilon - \delta^2)\mu_1 + 18\delta^3\varepsilon^2 + 3\delta\gamma\varepsilon - 81\delta\gamma\varepsilon^3 - \delta^3\big)\mu_{1x} = 0\,.
\end{aligned} \tag{8}$$

Compatibility condition for first and second equations of the system (8) is

$$3\cdot{\mu_{1xx}}^2 - \left(2 - \frac{1}{3\varepsilon}\right)\cdot\left(\mu_1 + \frac{\delta}{3\varepsilon}\right)\mu_{1x}\cdot\mu_{1xx} - \frac{1}{3\varepsilon}\left(\gamma - \frac{\delta^2}{3\varepsilon}\right)\cdot{\mu_{1x}}^2 = 0\,, \tag{9}$$

then

$$\mu_{1xx} = \frac{1}{6}\cdot\mu_{1x}\left(\left(2 - \frac{1}{3\varepsilon}\right)\cdot\left(\mu_1 + \frac{\delta}{3\varepsilon}\right) \pm \sqrt{\left(2 - \frac{1}{3\varepsilon}\right)^2\left(\mu_1 + \frac{\delta}{3\varepsilon}\right)^2 + \frac{4}{\varepsilon}\left(\gamma - \frac{\delta^2}{3\varepsilon}\right)}\right). \tag{10}$$

Considering $\mu_{1x} = 0$, we get $\mu_{1t} = 0$. Then $\mu_1 = C_1$, $\mu_2 = 0$, $\mu_3 = 0$.

It is easy to verify that $\mu_1 = C_1$, $\mu_2 = 0$, $\mu_3 = 0$ give invariant solutions of considered equation in the form of a traveling wave, where γ, δ, ε are arbitrary constants.

Now we want to find such solutions of the system (8) that $\mu_{1x} \neq 0$. We substitute (10) into third equation of the system (8) and divide result by μ_{1x}:

$$\begin{aligned}
&-9\varepsilon^2(6\varepsilon+1)\left(\left(2 - \frac{1}{3\varepsilon}\right)\left(\mu_1 + \frac{\delta}{3\varepsilon}\right) \pm \sqrt{\left(2 - \frac{1}{3\varepsilon}\right)^2\left(\mu_1 + \frac{\delta}{3\varepsilon}\right)^2 + \frac{4}{\varepsilon}\left(\gamma - \frac{\delta^2}{3\varepsilon}\right)}\right)\mu_{1x} +\\
&+ \left(-9\varepsilon^2{\mu_1}^2 - 6\delta\varepsilon\mu_1 - 81\gamma\varepsilon^3 + 27\delta^2\varepsilon^2 + 9\gamma\varepsilon^2 - 3\delta^2\varepsilon - \delta^2\right)\left(\left(2 - \frac{1}{3\varepsilon}\right)\cdot\left(\mu_1 + \frac{\delta}{3\varepsilon}\right) \pm\right.\\
&\left.\pm\sqrt{\left(2 - \frac{1}{3\varepsilon}\right)^2\left(\mu_1 + \frac{\delta}{3\varepsilon}\right)^2 + \frac{4}{\varepsilon}\left(\gamma - \frac{\delta^2}{3\varepsilon}\right)}\right) +\\
&+ 6\big(3\varepsilon\cdot(3\gamma\varepsilon - \delta^2)\mu_1 + 18\delta^3\varepsilon^2 + 3\delta\gamma\varepsilon - 81\delta\gamma\varepsilon^3 - \delta^3\big) = 0\,.
\end{aligned} \tag{11}$$

2.1. $\gamma = \delta^2/(3\varepsilon)$, $\delta \neq 0$. If $\mu_{1xx} = 0$, then from third equation of the system (8): $\mu_{1x} = 0$. Thus, $\mu_{1xx} \neq 0$. Then from (9):

$$\mu_{1x} = f_1(t) + \frac{1}{3}\left(2 - \frac{1}{3\varepsilon}\right)\cdot\left(\frac{{\mu_1}^2}{2} + \frac{\delta}{3\varepsilon}\mu_1\right).$$

Above expression for μ_{1x} we substitute into second equation of the system (8). Then μ_1 is a root of polynomial, which coefficients don't depend on x. If $\varepsilon \neq 1/6$ then coefficient of ${\mu_1}^4$ is non zero, then μ_1 doesn't depend on x. Thus, $\varepsilon = 1/6$.

If $\varepsilon = 1/6$ then: $\mu_{1x} = f_1(t)$, but $\mu_{1xx} \neq 0$, therefore there are no reductions in the case of $\gamma = \delta^2/(3\varepsilon)$, $\delta \neq 0$, $\mu_{1x} \neq 0$.

2.2. $\gamma = \delta^2/(3\varepsilon)$, $\delta = 0$. Then $\gamma = 0$. From third equation of the system (8) we have:

$$\mu_{1xx} = 0$$

or

$$(6\varepsilon+1){\mu_1}_x+{\mu_1}^2=0\,.$$

2.2.1. $\mu_{1xx}=0$. In this case $\mu_1=f_1(t)x+f_2(t)$. It gives us the following equations from the system (8):

$$\dot{f_1}=-2{f_1}^2,\qquad \dot{f_2}=-2f_1\cdot f_2\,.$$

Then

$$\mu_1=\frac{x+C_2}{2t+C_1}\,,$$
$$\mu_2=-\frac{1}{2t+C_1}\,,$$
$$\mu_3=0\,.$$

It is easy to verify that this values of μ_i give invariant solutions of considered equation, where $\varepsilon\neq 0,\gamma=0,\delta=0$.

2.2.2. $\mu_{1xx}\neq 0$. In this case

$$(6\varepsilon+1){\mu_1}_x+{\mu_1}^2=0\,,$$

where $6\varepsilon+1\neq 0$. Then $\mu_1=(6\varepsilon+1)/(x+f_1(t))$. We substitute μ_1 to the system (8):

$$f_1=C_1\,,\qquad 6\varepsilon+4-\frac{1}{3\varepsilon}=0\,,\quad -3=6\varepsilon+1\,.$$

This system of equations is inconsistent. Thus, there are no reductions in case of $\varepsilon\neq 0$, $\delta=\gamma=0$, $\mu_{1xx}\neq 0$.

2.3. $\gamma\neq\delta^2/(3\varepsilon)$. In this case $\sqrt{\left(2-\frac{1}{3\varepsilon}\right)^2\left(\mu_1+\frac{\delta}{3\varepsilon}\right)^2+\frac{4}{\varepsilon}\left(\gamma-\frac{\delta^2}{3\varepsilon}\right)}$ is nonlinear with respect to μ_1. If $6\varepsilon+1=0$, equation (11) is irrational equation for μ_1 (without derivatives), which coefficient of ${\mu_1}^3$ is non zero. But x is not explicitly included in this equation. Then ${\mu_1}_x=0$. Thus, $6\varepsilon+1\neq 0$ and from equation (11) we have:

$${\mu_1}_x=F_1(\mu_1)\,.$$

Then from first equation of the system (8) we get:

$${\mu_1}_t=F_2(\mu_1)\,.$$

Suppose that the system (8) has such a solution that ${\mu_1}_x\neq 0$. If this system is consistent, we have $F_2/F_1=Const$. We substitute (10) to first equation of the system (8) and divide it by ${\mu_1}_x$

$$\frac{{\mu_1}_t}{{\mu_1}_x}=-\left(2\mu_1+\frac{\delta}{3\varepsilon}\right)+\frac{1}{6}\cdot\left(3-\frac{1}{3\varepsilon}\right)\cdot\left(2-\frac{1}{3\varepsilon}\right)\cdot\left(\mu_1+\frac{\delta}{3\varepsilon}\right)\pm$$
$$\pm\frac{1}{6}\cdot\left(3-\frac{1}{3\varepsilon}\right)\sqrt{\left(2-\frac{1}{3\varepsilon}\right)^2\left(\mu_1+\frac{\delta}{3\varepsilon}\right)^2+\frac{4}{\varepsilon}\left(\gamma-\frac{\delta^2}{3\varepsilon}\right)}\,.$$

Since the root is nonlinear with respect to μ_1, coefficients of μ_1 and root have to be equal to zero. Thus we get the system for ε:

$$\varepsilon=\frac{1}{9},\qquad 54\varepsilon^2+15\varepsilon-1=0.$$

This system is inconsistent. Therefore there are no reductions in the case of $\varepsilon\neq 0$, $\gamma\neq\delta^2/(3\varepsilon)$, ${\mu_1}_x\neq 0$.

4. Nonexistence of differential substitution

One can show that there is no analogue of Florin–Cole–Hopf substitution in the case of $\gamma^2+\delta^2+\varepsilon^2 \neq 0$. This follows from the fact that substitution $u = \phi(x,t,v,v_x,v_t)$ leads to inconsistent system in the case of $v_{xx} = A(x,t)v_{xt} + B(x,t)v_{tt} + a(x,t)v_x + b(x,t)v_t + c(x,t)v + f(x,t)$.

5. Conclusion

In paper it's shown that all reductions of Burgers-Huxley equation are symmetry reductions. This result corresponds to the result of the paper [4] on the reductions of the Burgers equation. Viewing method of construction of reductions can be effectively applied to other nonlinear equations.

Acknowledgments

This work was supported by RFBR grants 15-01-00361, 15-01-04677.

References

[1] Yefimova O Yu and Kudryashov N A 2004 Exact solutions of the Burgers–Huxley equation *J. Appl. Maths Mechs* **68** 413–20
[2] Polyanin A D and Zaitsev V F 2012 *Handbook of Nonlinear Partial Differential Equations* (New York: CRC Press, 2 Edition)
[3] Kudryashov N A 2010 *Methods of nonlinear mathematical physics* (Moscow: Intellekt) (in Russian)
[4] Clarkson P A and Kruskal M D 1989 New similarity reductions of the Boussinesq equation *J. Math. Phys.* **30** 2201–13
[5] Aksenov A V and Kozyrev A A 2013 Reductions of the stationary boundary layer equation with a pressure gradient *Doklady Mathematics* **87** 236-9
[6] Ovsiannikov L V 1982 *Group Analysis of Differential Equations* (New York: Academic Press)

The surface tension effect on viscous liquid spreading along a superhydrophobic surface

A V Aksenov, A D Sudarikova and I S Chicherin
Lomonosov Moscow State University, Moscow, Russia
E-mail: aksenov.av@gmail.com, 0650646@gmail.com, chicherin-ivan@rambler.ru

Abstract. Within the Stokes film approximation, unsteady plane-parallel spreading of a thin layer of a heavy viscous fluid along a horizontal superhydrophobic surface is studied. The forced spreading regimes induced by the mass supply are considered. Plane-parallel flow along the principal direction of the slip tensor of the superhydrophobic surface is studied in case that the corresponding slip tensor component is a power function of the spatial coordinate. An evolution equation for the film thickness is derived taking into account surface tension that is dependent on the spatial coordinate. The group classification problem is solved. Self-similar and invariant solutions are constructed for power and exponent time dependences on mass supply respectively at a special form of the surface tension coefficient. Surface tension is shown to have a significant influence on the character of the liquid spreading.

1. Introduction

For the analysis of mathematical models describing various physical processes the problem of finding exact solutions of the equations governing them is important [1]–[4]. The problem of Stokes film spreading from a localized source in the gravity field with the no-slip condition and its self-similar solutions were studied in many works. Steady flows along horizontal [5] and inclined [6] planes and also on conic surfaces [7] were considered. Unsteady Stokes film spreading was studied in the papers [8, 9]. In [10] within the Stokes film approximation unsteady spreading of a thin layer of a heavy viscous fluid along a horizontal superhydrophobic surface with the Navier slip condition [11, 12] was investigated in the presence of a given localized mass supply. Also, the evolution equation for the film thickness was derived, the forced spreading regimes were considered and the self-similar solutions analyzed in [10]. The surface tension effect was neglected in the works stated above.

The problem of a thin layer spreading of a heavy viscous fluid along a horizontal superhydrophobic surface has been studied in this work taking into account the surface tension. The surface tension was shown to have a significant effect on the film spreading character.

2. Statement of the problem

Similar to [10], it is possible to receive the film thickness equation for a heavy viscous fluid moving along a horizontal superhydrophobic surface on the assumption that surface tension is varying. It has the form

$$\frac{\partial h}{\partial t}-\frac{\partial}{\partial x}\left[\left(\frac{h^3}{3}+bh^2\right)\left(\frac{\partial h}{\partial x}-\frac{\partial}{\partial x}\left(D\frac{\partial^2 h}{\partial x^2}\right)\right)\right]=0. \tag{1}$$

Equation (1) is written in terms of dimensionless variables. Here, $h = h(x,t)$ is the film thickness, $b = b(x)$ is the effective slip length in the direction of spreading, $D = D(x)$ is the surface tension coefficient. Note that the no-slip condition corresponds to the case $b = 0$ and boundary frictionless condition corresponds to the case $b \to \infty$ [12]. As in [10] we will consider superhydrophobic surfaces with $b(x) = Bx^{\delta}$, $B > 0$, $\delta > 0$, where x is a dimensionless distance from the source.

The statement of the problem for finding the surface shape $h(x,t)$ is completed by the formulation of an integral condition following from the mass supply law and the requirement of zero thickness of the film at the leading front of the wetted area

$$\int_0^{x_f} h(x,t)dx = Q(t), \qquad h(x_f,t) = 0. \tag{2}$$

Here, $Q(t)$ is a known volume of the fluid in the film, $x_f = x_f(t)$ is a dimensionless coordinate of the leading front.

3. Group classification

We seek symmetry operators of equation (1) as

$$X = \xi^1 \frac{\partial}{\partial x} + \xi^2 \frac{\partial}{\partial t} + \eta \frac{\partial}{\partial h},$$

where $\xi^1 = \xi^1(x,t,h)$, $\xi^2 = \xi^2(x,t,h)$, $\eta = \eta(x,t,h)$. Applying the criterion of invariance [2] we obtain the system of the determining equations. The analysis of the system solution shows that the kernel of the symmetry operators consists of the symmetry operator

$$X_1 = \frac{\partial}{\partial t}.$$

The kernel of the symmetry operators has an expansion at a special type of the function $D(x) = Kx^2$, $K = \text{const}$

$$X_2 = x\frac{\partial}{\partial x} + (2-3\delta)t\frac{\partial}{\partial t} + \delta h \frac{\partial}{\partial h}.$$

4. Types of invariant solutions

We will consider the symmetry operator as $X = aX_1 + X_2$, $a = \text{const}$. The invariants of a transformation group corresponding to the symmetry operator X have the following appearance in accordance with the parameters a and δ

$$I_1 = x(t+t_0)^{-\frac{1}{2-3\delta}}, \qquad I_2 = h(t+t_0)^{-\frac{\delta}{2-3\delta}}, \qquad t_0 = \frac{a}{2-3\delta}, \qquad \text{if } \delta \neq 2/3;$$

$$I_1 = xe^{-\frac{t}{a}}, \qquad I_2 = he^{-\frac{2t}{3a}}, \qquad \text{if } \delta = 2/3, \quad a \neq 0;$$

$$I_1 = t, \qquad I_2 = hx^{-\frac{2}{3}}, \qquad \text{if } \delta = 2/3, \quad a = 0.$$

The relevant invariant solutions have a form

$$h = (t+t_0)^{\alpha} F(\eta), \quad \eta = \frac{x}{C(t+t_0)^{\beta}}, \quad \alpha = \frac{\delta}{2-3\delta}, \quad \beta = \frac{1}{2-3\delta}, \qquad \delta \neq 2/3; \tag{3}$$

$$h = e^{\alpha t} F(\eta), \quad \eta = \frac{x}{Ce^{\beta t}}, \quad \alpha = \frac{2}{3a}, \quad \beta = \frac{1}{a}, \qquad \delta = 2/3, \quad a \neq 0; \tag{4}$$

$$h = x^{\frac{2}{3}} F(t), \qquad \delta = 2/3, \quad a = 0. \tag{5}$$

We do not consider invariant solution (5) further since it does not satisfy condition (2) at the leading edge.

5. Self-similar solution

We consider invariant solution (3) at $t_0 = 0$. It is in accord with the one-parameter group of non-homogeneous dilations. Invariant solution (3) is self-similar since the symmetry operator corresponding to him specify one-parameter group of non-homogeneous dilations [2].

Since $F(\eta_f) = 0$ η_f is constant, hence the law of mass supply in the film is a power function of time and it can be calculated as follows

$$Q(t) = At^{\gamma}, \qquad A = C\int\limits_0^{\eta_f} F(\eta)d\eta = \text{const} > 0, \qquad \gamma = \alpha + \beta. \tag{6}$$

The value of the constant C is chosen so that the self-similar coordinate of the leading edge η_f is equal to unity. After the substitution of the self-similar form of the solution in equation (1) we obtain the ordinary differential equation

$$\begin{aligned}(K\eta^2F^3 + 3BK\eta^{\delta+2}C^{\delta}F^2)F'''' + (6\eta^{\delta+2}C^{\delta}BKFF' + 3K\eta^2F^2F' + \\ + 4K\eta F^3 + 3BK\delta\eta^{\delta+1}C^{\delta}F^2 + 12BK\eta^{\delta+1}C^{\delta}F^2)F''' + \\ + (6K\eta F^2 + 12BK\eta^{\delta+1}C^{\delta}F)F'F'' + (2KF^3 - F^3 + \\ + 6\eta^{\delta}C^{\delta}BKF^2 + 6BK\delta\eta^{\delta}C^{\delta}F^2 - 3B\eta^{\delta}C^{\delta}F^2)F'' - (3F^2 + \\ + 6B\eta^{\delta}C^{\delta}F)F'^2 - (3C^2\beta\eta - 3B\delta\eta^{\delta-1}C^{\delta}F^2)F' + 3C^2\delta\beta F = 0.\end{aligned} \tag{7}$$

After the substitution of the expression for α and β from (3) in the last ratio in (6) we obtain

$$\alpha = \frac{2\gamma - 1}{5}, \qquad \beta = \frac{3\gamma + 1}{5}, \qquad \delta = \frac{2\gamma - 1}{3\gamma + 1}. \tag{8}$$

Note that the following inequations are valid due to (8) and $\delta > 0$

$$\gamma > \frac{1}{2}, \qquad 0 < \delta < \frac{2}{3}, \qquad \alpha > 0.$$

Ordinary differential equation (7) has the fourth order. There is an singularity at the point $\eta = 1$ as $F(1) = 0$ (at the leading edge) at this point. We expand the required solution $F(\eta)$ in the asymptotic series in the neighborhood of $\eta = 1$ and take the first term of the series. After the substitution of the asymptotic expansion of the solution in equation (7) we obtain

$$F = \sqrt{\frac{8\beta C^{2-\delta}}{3BK}}(1-\eta)^{3/2} + o\left((1-\eta)^{3/2}\right). \tag{9}$$

The study of equation (7) solution behavior at $\eta \ll 1$ on the assumption that an order of magnitude $F(\eta)$ is larger than an order of magnitude η^{δ} shows self-similar solution does not exist when $0 < K \leqslant 1/2$.

Equation (7) was solved numerically using the fourth-order Runge-Kutta method. Asymptotics (9) has been used in the numerical calculations to shift the boundary condition from the singular point by a small distance to the point $1 - \eta_0$.The value of the constant C in the law of motion of the leading front was found by the shooting method with iterations, which continued until the integral mass conservation law was satisfied at a required accuracy.

Figure 1 a) shows the results of the numerical solution of equation (7) for $A = 1$, $\gamma = 1,5$, $B = 0,5$ and several values of the parameter K.

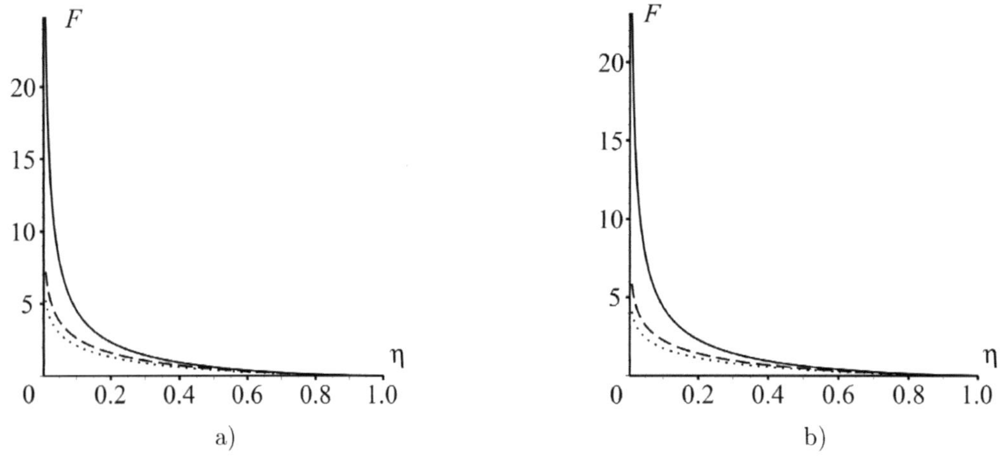

Figure 1. Function $F(\eta)$: $B = 0.5$, continuous lines correspond to $K = 1$, dashed lines to $K = 5$, dotted lines to $K = 10$, $\gamma = 1.5$, a) self-similar solution, b) invariant solution

From Figure 1 a) it follows that the curves of the film shape obtained with fixed value of the parameter γ lie closer to the horizontal axis η at the parameter K increasing. Calculations have shown the value of the constant C is increasing at the parameter K rising, thus for the same time the leading edge of the film passes a greater distance at a larger value of the parameter K than at smaller.

6. Invariant solution

In this section we consider invariant solution (4). As before the value of the constant C is chosen so that the self-similar coordinate of the leading edge $\eta = \eta_f$ is equal to unity. The variation of the fluid volume in the film is described by the formula

$$Q(t) = Ae^{\gamma t}, \qquad A = C\int_0^1 F(\eta)d\eta, \qquad \gamma = \alpha + \beta. \tag{10}$$

After the substitution of expression (4) in equation (1) we obtain the following ordinary differential equation

$$\begin{aligned}
&\frac{1}{3}K\eta^2F^3F'''' + BC^{2/3}K\eta^{8/3}F^2F'''' + \left(F + 2BC^{2/3}\eta^{2/3}\right)K\eta^2FF'F''' + \\
&+\frac{4}{3}K\eta F^3F''' + \frac{14}{3}BC^{2/3}\eta^{5/3}KF^2F''' + 2K\eta F^2F'F'' + \\
&+4BC^{2/3}K\eta^{5/3}FF'F'' + \frac{2}{3}KF^3F'' - \frac{1}{3}F^3F'' + \\
&+\left(\frac{10}{3}K - 1\right)BC^{2/3}\eta^{2/3}F^2F'' - 2BC^{2/3}\eta^{2/3}FF'^2 - \\
&-F^2F'^2 - \frac{3}{5}\gamma C^2\eta F' - \frac{2}{3}BC^{2/3}\eta^{-1/3}F^2F' + \frac{2}{5}\gamma C^2F = 0.
\end{aligned} \tag{11}$$

Analogously to the case of the self-similar solution formula (10) gives

$$\alpha = \frac{2\gamma}{5}, \qquad \beta = \frac{3\gamma}{5}.$$

Ordinary differential equation (11) is a forth-order differential equation. The point $\eta = 1$ is singular as $F(1) = 0$ (at the leading edge). In the neighborhood of $\eta = 1$ the approximate solution has the form

$$F = \sqrt{\frac{8\gamma C^{4/3}}{5BK}}(1-\eta)^{3/2} + o\left((1-\eta)^{3/2}\right).$$

The analysis of equation (11) solution behavior at $\eta \ll 1$ on the assumption that an order of magnitude $F(\eta)$ is larger than an order of magnitude η^{δ} shows invariant solution does not exist when $0 < K \leqslant 1/2$.

Equation (7) was solved numerically analogously to the procedure described in the previous section. Figure 1 b) shows the results of numerical solution of equation (11) for $A = 1$. For the same time the leading edge of the film passes a greater distance at a larger value of the parameter K than at smaller as in the case of the power law of mass supply.

7. Comparison of solutions taking into account surface tension and without it

Figure 2 shows the plots of self-similar solutions of the equation (1) taking into account surface tension and without it (for more detail, see [10]) at the given parameters of the problem.

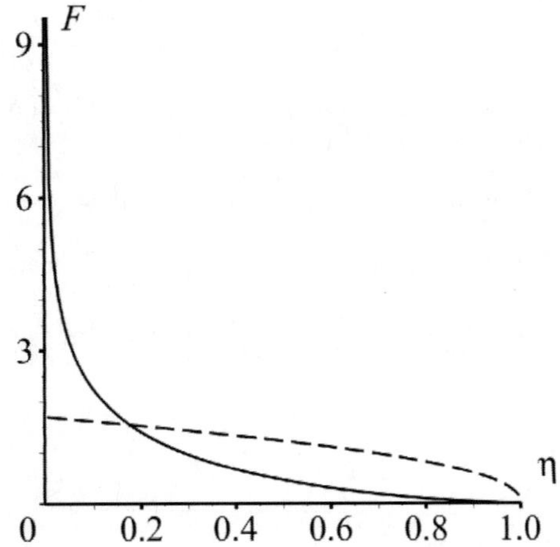

Figure 2. $B = 0.5$, $\gamma = 2.0$, continuous lines correspond to $K = 10$, dashed lines to $K = 0$

Structures of the related flows are differ significantly. It is caused by the fact that the surface tension influences on the film spreading.

For the same time the leading front of a film passes various distances in cases when capillary effects are considered and not. It is necessary to know values of a constant C to compare these distances. In case of the account of the surface tension the constant $C = 1.146$ at $K = 10$ and $C = 0.862$ in case of the surface tension absence. Hence for the case presented in Figure 2 the leading front of the film spreading along a superhydrophobic surface taking into account the surface tension moves quicker than the leading front of the film spreading along a superhydrophobic surface without surface tension.

8. Conclusion

The equation for a film thickness of a heavy viscous liquid, which spreading is plane-parallel, is derived in the present work. The symmetries of the investigated equation are found. The surface tension coefficient form of the dependence on the coordinate is stated. There is the maximum

Lie algebra expansion of the symmetry operators under the specified surface tension coefficient. Invariant solutions are found and investigated. the accounting of surface tension at the special selection of the surface tension coefficient is shown to cause a change of the current mode as opposed to a case with no surface tension. The surface tension coefficient has been chosen assuming that self-similarity of the problem is reserved. It allowed to carry out comparison with already known self-similar solution for the surface tension absence case. It is necessary to solve a nonisothermal problem for detailed research of the surface tension influence on a spreading mode of a viscous liquid. It will allow to determine the surface tension coefficient when it depends on temperature.

Acknowledgments

This work was supported by RFBR grants 15-01-00361, 15-01-04677.

References

[1] Sedov L I 1993 *Similarity and Dimensional Methods in Mechanics* (New York: CRC Press, Tenth Edition)
[2] Ovsiannikov L V 1982 *Group Analysis of Differential Equations* (New York: Academic Press)
[3] Polyanin A D and Zaitsev V F 2012 *Handbook of Nonlinear Partial Differential Equations* (New York: CRC Press, 2 Edition)
[4] Kudryashov N A 2010 *Methods of nonlinear mathematical physics* (Moscow: Intellekt) (in Russian)
[5] Huppert H E 1982 The propagation of two-dimensional and axisymmetric viscous gravity currents over a rigid horizontal surface *J. Fluid Mech.* **121** 43–58
[6] Smith P C 1973 A similarity solution for slow viscous flow down an inclined plane *J. Fluid Mech.* **58** 275–88
[7] Osiptsov A A 2006 Three-dimensional isothermal lava flows over a non-axisymmetric conical surface *Fluid Dynamics* **2** 31–45
[8] Lister J R 1992 Viscous flows down an inclined plane from point and line sources *J. Fluid Mech.* **242** 631–53
[9] Huppert H E 1982 Flow and instability of a viscous current down a slope *Nature* **300** 427–29
[10] Ageev A I and Osiptsov A N 2014 Self-similar regimes of liquid-layer spreading along a superhydrophobic surface *Fluid Dynamics* **49** 330–342
[11] Rothstein J P 2010 Slip on superhydrophobic surfaces *Annual Rev. Fluid Mech.* **42** 89–102
[12] Bazant M Z and Vinogradova O I 2008 Tensorial hydrodynamic slip *J. Fluid Mech.* **613** 125–34

Comparison of the model EBSD experiment results with the PF and ODF obtained experimentally

A O Antonova, T I Savyolova

Department of Applied Mathematics, Moscow Engineering Physics Institute (State University), Kashirskoe sh. 31, Moscow, 115409 Russia

aoantonova@mail.ru

Abstract. For study of EBSD parameters impact on the evaluation of the polycrystalline texture characteristics the mathematical model of EBSD experiment was developed. The specimen and experiment simulation allows us to compare texture entities calculated directly from specimen with corresponding ones obtained from model measurements. By the EBSD parameters a scanning step and a threshold disorientation angle are meant. In the given work the comparison of Pole Figures and Orientation Distribution Function calculated from model EBSD measurements and corresponding entities, obtained from real EBSD data, is presented. The most significant results are formulated.

1. Introduction

The problem of choosing measurement parameters in electron back scattering diffraction (EBSD) [1] remains insufficiently investigated, while these parameters have a large effect on measurement results. It was indicated than an unsuitable choice of measurement parameters leads to incorrect results. For example, an insufficient number of orientation measurements results in low accuracy of computed Pole Figures (PFs) [2]; disorientation angle measurements may lead to inconsistent average grain size values [3]; a small step of measurements leads to a considerable increase in time required for conducting an experiment (see [4]). To study the problem of EBSD parameters choosing we need the opportunity to compare the specimen data and obtained EBSD data. Accordingly, it is of interest to study this issue with the help of the mathematical simulation of an experiment.

Below we develop a two-dimensional mathematical model of a specimen and a subsequent EBSD experiment conducted for various sets of EBSD parameters. To study the impact of the measurement parameters PF and Orientation Distribution Function (ODF) of a model specimen and corresponding ones, calculated from model EBSD measurements, are compared. The real EBSD experiment was also performed. The results of the model experiment are correlated with such detected in the real EBSD data.

2. Specimen and experiment simulation

A specimen is a set of grains

$$\{(x_i, g_i, \omega_i)\ i = 1,2, \dots N-1, (x_N, g_N)\}, \tag{1.1}$$

where x_i is the size of the ith grain, which is equal to the diameter of a circle whose area is equal to that of the grain's cross section; $g_i = (\varphi_i, \theta_i, \psi_i)$ is the orientation of the ith grain corresponding to the Euler angles $-\pi \leq \varphi_i, \psi_i < \pi, 0 \leq \theta_i \leq \pi$; $\omega_i = (g_i, g_{i+1})$ is the disorientation angle between neighboring grains; and N is the number of grains in the specimen. Grain sizes are modeled by Neumann's method as random variables distributed satisfying a gamma distribution [5].

A set of grain orientations is generated by the special Monte-Carlo method with a distribution function in the form of central distribution function (CND) [6] with a given parameter $\varepsilon > 0$. Grains in equation (1.1) are separated by boundaries, which are variables uniformly distributed in $[0, 2\gamma/10)$, where $\gamma > 0$ is such that the area of the boundaries is 10% of the grains' area.

To simulate the EBSD experiment the following measurement parameters are specified: h is the scanning step, and ω_0 is a threshold disorientation angle for determining two different neighboring grains g_i and g_{i+1} from the disorientation angle between them.

Measurements are made at the lattice points $\Delta_k = kh, k = 1,2, \dots, N_1$ on the surface of the simulated specimen. If the point hits the boundary, it is assigned an orientation corresponding to the Euler angles equal to the arithmetic mean of the angles of two nearest determined orientations [6].

The resulting orientations are used to calculate the disorientation angles $\hat{\omega}_k = (\hat{g}_k, \hat{g}_{k+1})$, $k = 1, \dots, N_1 - 1$. If $\hat{\omega}_k \leq \omega_0$, then the kth and $(k + 1)$th steps belong to a single grain. If n measurements are inside the jth grain, then the grain size is set equal to $\tilde{x}_j = nh$ and the Euler angles are calculated as the arithmetic mean of the corresponding n measurements. If $\hat{\omega}_k > \omega_0$, then a new grain begins at the point $(k + 1)h$. As a result of the measurements, we obtain a set of grains

$$\{(\tilde{x}_i, \tilde{g}_i, \tilde{\omega}_i), i = 1,2, \dots \tilde{N} - 1, (\tilde{x}_{\tilde{N}}, \tilde{g}_{\tilde{N}})\}. \tag{1.2}$$

For calculation of ODF the following kernel ODF estimator for a sample of orientations in equations (1.1) and (1.2) was used

$$f_N(g) = \frac{1}{(\alpha)^3} \sum_{i=1}^{N} C_i q_1 \left(\frac{\varphi - \varphi_i}{\alpha}\right) q_3 \left(\frac{\psi - \psi_i}{\alpha}\right) q_2 \left(\frac{\cos\theta - \cos\theta_i}{\alpha}\right), C_i = \frac{x_i}{\sum_{i=1}^{N} x_i}; \tag{1.3}$$

where $q_i(t), i = 1,2,3$ are are smoothing kernels (see [6]).

A statistical estimate of PF with a kernel in the form of CND has the form [7]

$$\hat{P}_{\vec{h}}(\vec{y}) = \sum_{l(2)=0}^{\infty} (2l + 1) \exp\{-l(l + 1)\alpha^2\} \sum_{i=1}^{N} C_i P^l \left(g_{\vec{y}}^{-1} g_i^{-1} g_{\vec{h}}(3,3)\right), C_i = \frac{x_i}{\sum_{i=1}^{N} x_i};$$

$$\vec{y} = \{\eta, \chi\}, \quad -\pi \leq \eta \leq \pi, \quad 0 \leq \chi \leq \pi, \tag{1.4}$$

$$\vec{h} = \{\varphi, \theta\}, \quad -\pi \leq \varphi, \psi \leq \pi, \quad 0 \leq \theta \leq \pi,$$

where α is the kernel smoothing parameter (playing the role of regularization parameter); P^l are Legendre polynomials; $g_{\vec{y}}^{-1} g_i^{-1} g_{\vec{h}}(3,3)$ is the element with index 3×3 of the rotation matrix $g_{\vec{y}}^{-1} g_i^{-1} g_{\vec{h}}$, $g_{\vec{y}} = \{\vec{y}, 0\}, g_{\vec{h}} = \{\vec{h}, 0\}$.

3. Comparison of the results for model and real experiments

In the real EBSD experiment material for the investigation was a ferritic-martensitic steel EP823 (12Cr-Mo-W-Si-V-Nb-B) treated by the nitrogen gas high-temperature pulsed plasma flows (HTPPF). HTPPF surface modification were carried out by means of the Z-pinch type experimental plasma installation "Desna-M" [8]. This treatment results in the formation of the microcrystalline surface layer with a mean grain size of 2 mm. This layer has a single phase bcc type (a-Fe) equiaxed grain structure with a homogeneous distribution of alloying elements. An EBSD data were acquired using the HKL CHANNEL5 software (Oxford Instruments, UK) coupled to an EVO 50 XVP scanning electron microscope (ZEISS, Germany). EBSD measurement were performed with scanning step sizes h =

0.5 μm, 1.0 μm, 2.0 μm and then for each step size meaning the different threshold disorientation angle meanings were considered: $\omega_0 = 5°, 10°$ and $20°$.

Following the scheme described in Section 2, we modeled specimen, which is similar in the main properties with the sample in the real experiment. It is single-phased material with average grain size equal to 2.0μm. As the grain orientations were modeled by the special Monte-Carlo method for CND on $SO(3)$, shown in equation (1.2), with texture sharpness $\varepsilon = 1/8$ (see [6]), the simulated sample has strong texture. The number N of grains in a specimen was set equal to 1000. The following EBSD measurement parameters were chosen for model measurements: $h = 0.5$ μm, 1.0 μm, 2.0 μm ; $\omega_0 = 5°, 10°$ and $20°$. Below, the results of observations concerning the ODF and PF changing according to EBSD measurement parameters are given.

3.1. Influence of the EBSD parameters on PFs calculation

According to the results of specimen PF and PF of model measurements comparing it was obtained that for strongly textured specimens threshold disorientation angle is appeared to be more significant than scanning step; for weakly textured specimens, conversely, measurements are strongly affected by the scanning step [9]. The calculated RP-factor for measured PF, imaged on figure 1, verifies statement mentioned.

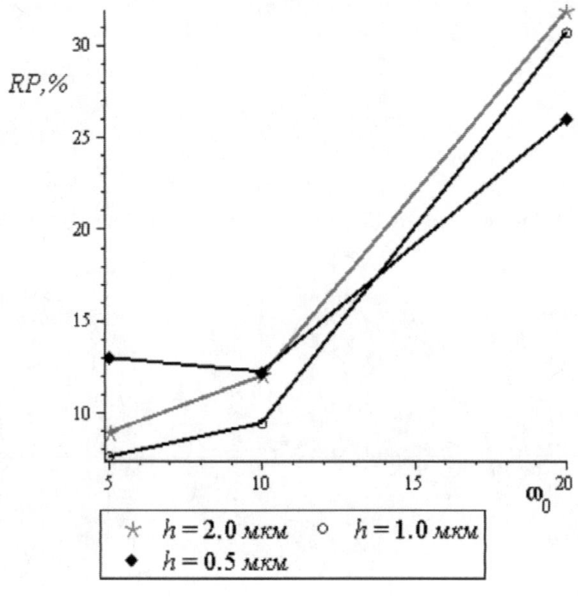

Figure 1. Dependences of the *RP* factor of the pole figures on the threshold misorientation angle.

There for calculation of the RP-factor the following formula is used

$$RP = \frac{1}{I}\sum_{i,j=1}^{n_i,n_j} \frac{\left|\hat{\hat{P}}_N(\eta_i,\chi_j) - \hat{P}_N(\eta_i,\chi_j)\right|}{\hat{P}_N(\eta_i,\chi_j)} \cdot 100\%, \tag{2.1}$$

where $\hat{P}_N(\eta,\chi)$ is a kernel estimate of PF for the model specimen, $\hat{\hat{P}}_N(\eta,\chi)$ is a kernel estimate of PF for result of measurements of the model specimen, I is a total number of partition cells.

According to the results of the real EBSD experiment PFs are strongly affected by the threshold disorientation angle (see figure 2), the influence of step size on PF evaluation is not so significant (see figure 3). The given statement is in a good agreement with the model experiment results.

IOP Conf. Series: Journal of Physics: Conf. Series **788** (2017) 012004 doi:10.1088/1742-6596/788/1/012004

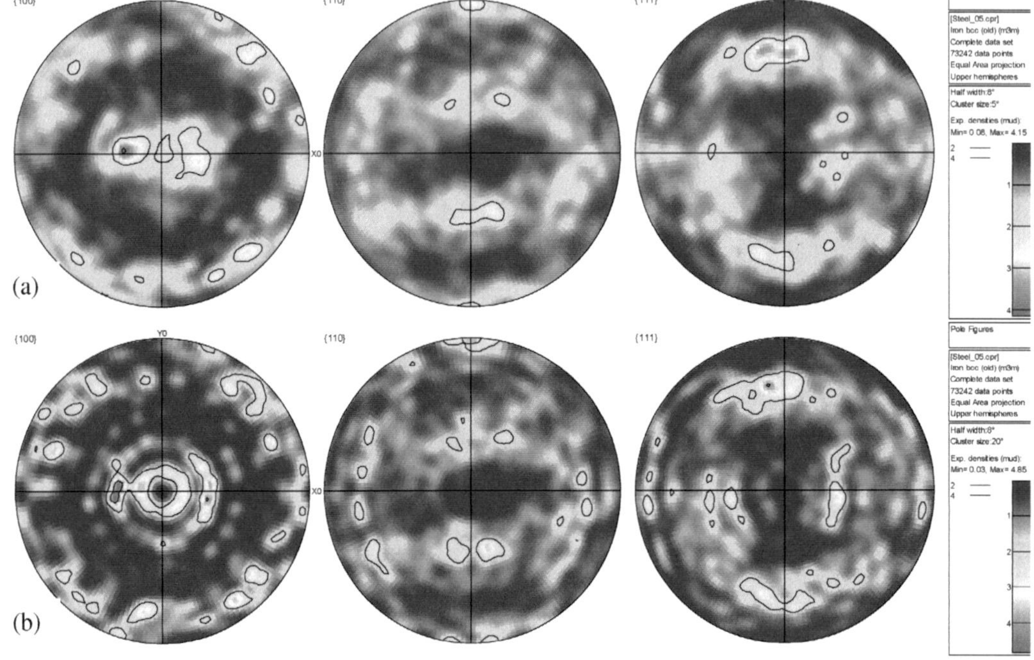

Figure 2. PF for fixed $h = 0.5$ µm, (a) for $\omega_0 = 5°$ and (b) for $\omega_0 = 20°$.

Figure 3 showes that PF for essentially different step sizes 2.0µm and 0.5µm the quite similar results were obtained.

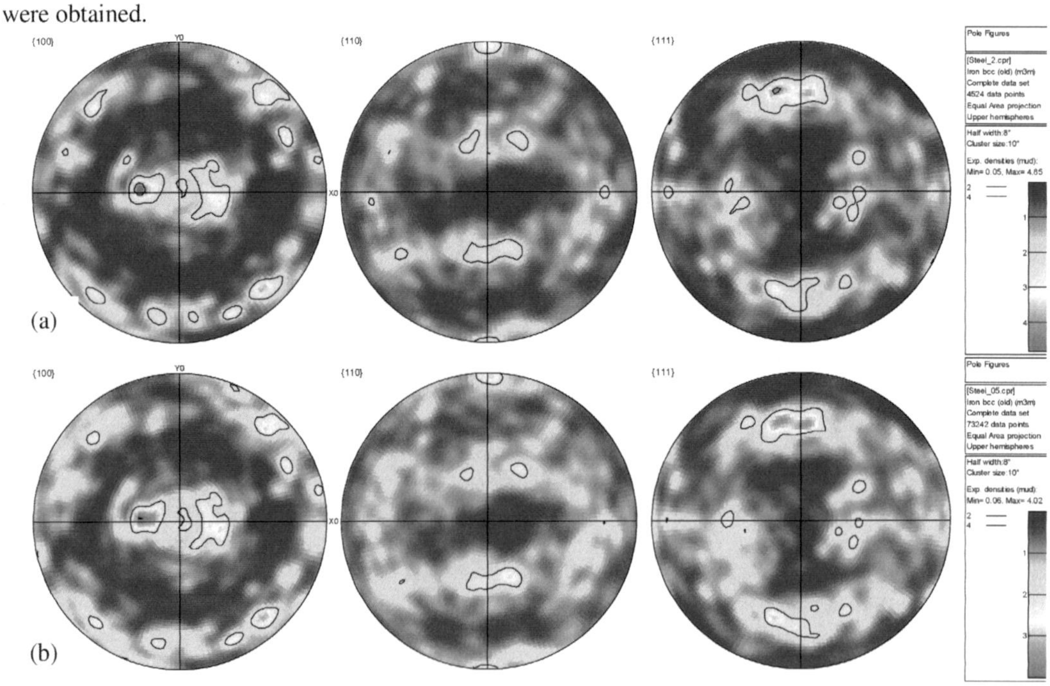

Figure 3. PF for fixed $\omega_0 = 10°$, (a) for $h = 2.0$ µm and (b) for $h = 0.5$ µm.

It is worth to mention at that the time expended in EBSD measurement for step sizes 2.0μm and 0.5μm increases by a factor of several tens under step size growth (from 10 minutes up to 2 hours 34 minutes).

We should also notice that according the results of simulation the PF maximum and its position are unstable to EBSD measurements parameters. Instability grows with weakening of specimen texture sharpness. It shows up in the shift of PF maximum position or second maximum appearing. The more grain size distribution approaches the uniform one, the stronger instability and the less accurate PF calculation [9].

In the real experiment the behavior of PF maximums was also considered. Table 2 presents PF maximums according to different EBSD parameters.

Table 2. Meanings of PF maximums.

h (μm)	ω_0 (°)		
	20°	10°	5°
2.0	5.56	4.85	4.82
1.0	4.93	4.11	4.23
0.5	4.85	4.02	4.15

Let us notice that PF maximum increases under step size growth for each fixed ω_0, but dependence of the PF maximum on the threshold disorientation angle is appeared to be not so clear: the maximum can increase or decrease with ω_0 growth. Such a behavior points out an instability of the PF calculation to EBSD parameters that is obtained in the model experiment mentioned above.

3.2. Influence of the EBSD parameters on ODF calculation

Below, the comparison of the ODF obtained from EBSD orientation data with ones calculated in the simulated experiment according to EBSD measurement parameters are given.

As well as for PFs the impact of a scanning step is appeared to be not such significant for ODF recovering from individual orientations. At the same time, the statistical sample is reduced substantially. For more detail on the effect of a sample on ODFs recovered from individual orientations, see [10].

At the same time in both experiments it was found out that ODF is also strongly affected by the threshold disorientation angle. For ODF at $\omega_0 = 20°$ and, in some cases, for $\omega_0 = 10°$, the maximum is observed to increase [5]. This behavior of the ODF is similar to that in the presence of dependent orientations [6], when the ODF maximums arise at $\omega_0 = 20°$ because of merging a large number of grains. What about the coordinates of the maximum we can not reasonably confirm whether they remain unchanged or not while the variable ψ is fixed. In the cased of strongly textured specimens, the ODF should be recovered with a small threshold disorientation angle. For the specimens under consideration, as such an angle, we can use $\omega_0 = 5°$.

Relying on the contour lines of ODF cross section (see figure 4) we receive the sufficient maximums growing (almost twice) for $\omega_0 = 20°$ versus ones for $\omega_0 = 5°$ and 10°. Angles $\phi1, \Phi, \phi2$ on figure 4 correspond to Euler angles φ, θ, ψ respectively. Besides unsuitable meaning of ω_0 may changed ODF character. This observation was also found out during the processing of model experiment results [5].

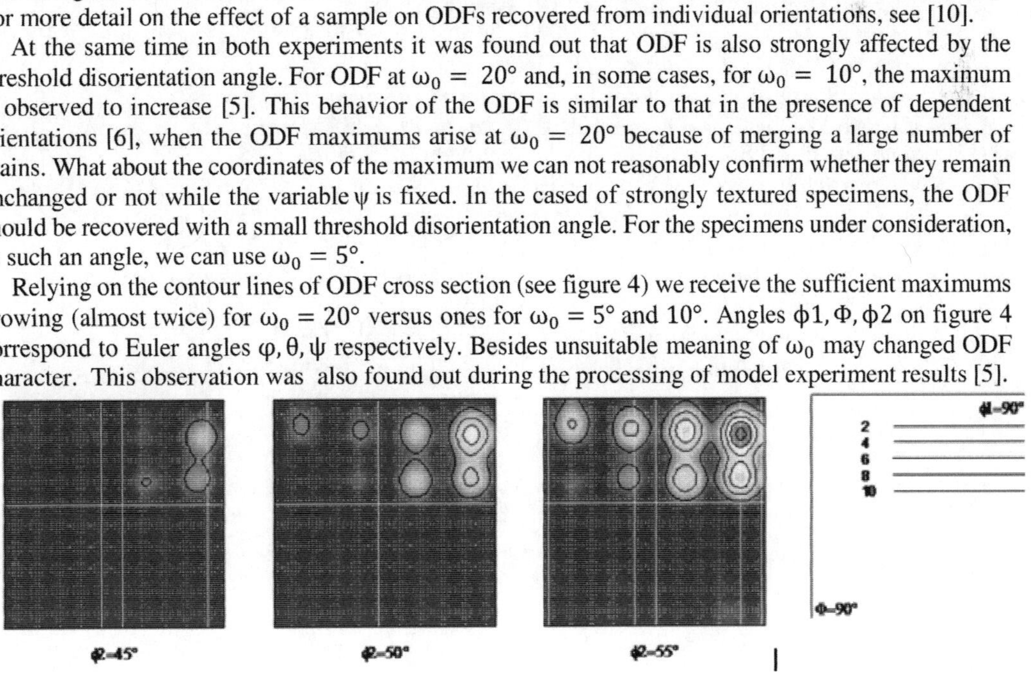

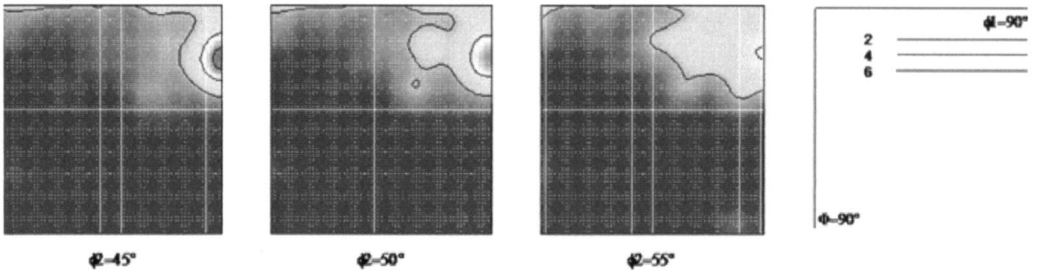

Figure 4. Contour lines of ODF cross section for fixed $h = 0.5$ μm, (a) for $\omega_0 = 20°$ and (b) for $\omega_0 = 10°$.

4. Conclusions

The processing of the model and real experiments observations has pointed out the correlation between behavior of texture entities obtained from simulated and really performed experiments.

This fact give us the ability to use our model for study of choosing measurement parameters. For strong textured material a step size changing does not lead to substantial changing in PFs and ODF calculation. The step size being less than a quarter of average grain size can result in unjustifiably high time of measuring in view of inessential result improvement. Thus, for the single-phase strong textured specimen with small amount of boundaries the threshold disorientation angle should not be more than 10° and the step size equal to half an average grain size is suitable.

To choose parameters for measurement of a specimen with other properties the corresponding modeling with new characteristics and further model experiment with parameters varying are needed to be conducted.

References

[1] Schwartz A J, Kumar M, Adams B L and Field D P 2009 *Electron Backscatter Diffraction in Materials Science*, 2nd. ed. (New York: Springer)

[2] Guilmeau E, Henrist C, Suzuki T S, Sakka Y, Chateigner D, Grossin D, and Ouladiaf B 2005 *Materials Science Forum* **495** 1395–1400

[3] Piazolo S, Sursaeva V G and Prior D J 2005 *Materials Science Forum* **495** 213–18

[4] Sypchenko M V and Savyolova T I 2010 *Cristallogr. Rep.* **55** 546–50

[5] Antonova A O and Savyolova T I 2015 *Comput. Math. Math. Phys.* **55** 317–29

[6] Savyolova T I, Ivanova T M and Sypchenko M V 2012 *Methods for Solving Ill-Posed Problems in Texture Analysis and Their Applications* (Moscow: NRNU MEPhI) [in Russian]

[7] Roginskii K N and Savyolova T I 2010 *Comput. Math. Math. Phys.* **50** 900–16

[8] Gribkov V A, F.I. Grigoriev F I, Kalin B A and Yakushin V L 2001 *Advanced Radiation-Beam Technologies of Materials Treatment* (Moscow: Krugly God) [in Russian]

[9] Antonova A O and Savyolova T I 2016 *Cristallogr. Rep.***61** 523–31

[10] Bozzolo N, Gerspach F, Sawina G and Wagner F 2007 *J. Microscopy* **227** 275–83

Mathematical modelling of risk reduction in reinsurance

R B Balashov, A V Kryanev, D E Sliva

National Research Nuclear University MEPhI, Kashirskoe highway 31, 115409, Moscow, Russia

E-mail: avkryanev@mephi.ru

Abstract. The paper presents a mathematical model of efficient portfolio formation in the reinsurance markets. The presented approach provides the optimal ratio between the expected value of return and the risk of yield values below a certain level. The uncertainty in the return values is conditioned by use of expert evaluations and preliminary calculations, which result in expected return values and the corresponding risk levels. The proposed method allows for implementation of computationally simple schemes and algorithms for numerical calculation of the numerical structure of the efficient portfolios of reinsurance contracts of a given insurance company.

1. Introduction

Reinsurance is a system of financial relations, during which the insurer, while forming the insurance portfolio, delegates part of the responsibility for the risk on certain agreed terms to other insurers to create a balanced portfolio of own insurance, ensuring financial stability and profitability of insurance operations within the generated portfolio. Consequently, the financial reinsurance principle is redistribution of the primary insurance fund among insurance companies [1, 2]. Reinsurance agreements differ in the ways risk is shared among insurers and reinsurers. There are three risk transfer systems for reinsurance: quota share, surplus and combined. In the case of quota share reinsurance, the reinsurer is involved in all the risks of the insurer in a certain proportion. Consequently, the quota share reinsurance system provides a great opportunity to create an insurance portfolio consisting of shares of the aggregate sums insured. In the case of the surplus reinsurance system the insurer passes to the reinsurer the insurance amount in excess of the limit value specified in the contract. If the limit value is, for example, 10 million rubles, the implied risks of an insurance amount lower than the limit value are not transferred to reinsurer and fully retained by the insurer. According to the insurance liabilities, when the sum insured exceeds the limit, the insurer reduces their risk to 10 million. rubles, while the remainder transfers to the reinsurer. Usually in practice surplus reinsurance agreements are concluded for large objects, such as power lines, gas pipelines, oil pipelines, nuclear power plants i.e. when the insurance value is high. A mixed system resides in establishing quota share insurance up to a certain limit of the insurance value beyond which the transfer of risk is accomplished by means of the surplus system [1–6]. Reinsurance market contributes to balancing insurance portfolio and the financial stability of the effective performance of insurance companies and their operations. The development of the reinsurance market, and mutual confidence of the insurance companies allows the reinsurance market to ensure the efficient functioning of insurance markets and, as a result, has a positive effect on the insurance market and the industries involved in the insurance process. International economic relations allow Russian insurance companies to work in international markets. Therefore, the reinsurance market is a regulator of the market of insurance services in Russian Federation [1–6]. Since the main purpose of the effective portfolio formation of an insurance company is the redistribution of risk in order to reduce its aggregate value, while maintaining an acceptable level of profitability, the task of forming an effective composition of portfolios is one of the main tasks of analytical management in insurance companies.

2. The setting of the problem

The problem of efficient portfolio formation in the reinsurance market, considered in this paper is states as follows (see also [7]):

$$\begin{gathered}\sum_{i=1}^{n} x_i w_i R_i - \max,\ \sum_{i=1}^{n} x_i w_i r_i - \min \\ 0 \le x_i \le \beta_i \le 1,\ i = 1, \dots, n \\ \sum_{i=1}^{n} x_i V_i = S,\ \sum_{i=1}^{n} \beta_i V_i > S,\ w_i = \frac{V_i}{V},\ V = \sum_{i=1}^{n} V_i,\end{gathered} \tag{1}$$

where $x_i, i = 1, \dots, n$ - share of the insurance coverage in the i - th contract of reinsurance included in the portfolio; $R_i, i = 1, \dots, n$ - the expected value of return in the i - th contract of reinsurance (for example, profitability or normalized return); ri, i=1,,n - the risk value of an insured event in the i - th contract of reinsurance (for example, the possibility of income deficiency); $V_i, i = 1, \dots, n$ the total insured amount in the i - th contract of reinsurance; S - the total insured amount of the portfolio $(0 < S < V)$, n - the number of potential contracts for reinsurance, which are tested for their possible inclusion in the contract portfolio of the insurance company.

The values of $R_i, i = 1, \dots, n$ $r_i, i = 1, \dots, n$ are determined according to expert evaluations of the uncertainties in return of insurance contracts or based on the processing of statistical data [8]. This problem has two criteria; Paretos solutions correspond to efficient portfolios of contracts of the insurance company.

To calculate the Paretos solutions of this problem it is sufficient to find the solutions of the two linear programming problems with linear objective functions

$$\sum_{i=1}^{n} x_i w_i R_i$$

and

$$\sum_{i=1}^{n} x_i w_i r_i,$$

respectively, and then to take their linear combination with an acceptable level of compromise between the expected values of profitability and risk generated by the portfolio.

Thus, finding efficient portfolios for reinsurance contracts in the statement (1) is reduced to single criterion solutions of two problems:

the problem of maximizing the efficiency of the portfolio

$$\begin{gathered}\sum_{i=1}^{n} x_i w_i R_i - \max \\ 0 \le x_i \le \beta_i \le 1,\ i = 1, \dots, n \\ \sum_{i=1}^{n} x_i V_i - S,\ \sum_{i=1}^{n} \beta_i V_i > S,\ w_i = \frac{V_i}{V}, V = \sum_{i=1}^{n} V_i,\end{gathered} \tag{2}$$

and the problem of minimizing portfolio risk

$$\begin{gathered}\sum_{i=1}^{n} x_i w_i r_i - \min \\ 0 \le x_i \le \beta_i \le 1,\ i = 1, \dots, n \\ \sum_{i=1}^{n} x_i V_i = S,\ \sum_{i=1}^{n} \beta_i V_i > S,\ w_i = \frac{V_i}{V}, V = \sum_{i=1}^{n} V_i,\end{gathered} \tag{3}$$

3. Solutions

Problems (2),(3) are linear programming problems, each of which has a unique solution $\vec{x}_1^*, \vec{x}_2^*$, respectively

$$\vec{x}_1^* = (x_{11}^*, \dots, x_{1n}^*)^T$$
$$\vec{x}_2^* = (x_{21}^*, \dots, x_{2n}^*)^T$$

The entire set of Paretos solutions of (1) defined by the equality

$$\vec{x}^*(\alpha) = \alpha \vec{x}_1^* + (1-\alpha)\vec{x}_2^* \tag{4}$$

where $\alpha \in [0, 1]$.

Choosing a numerical value priority α, we obtain the specific structure of the effective reinsurance portfolio. In particular, in the case where $\alpha = 1$, we obtain the structure of the effective portfolio of reinsurance company, that corresponds to the maximum expected value of return, and in the case where $\alpha = 0$, we obtain the structure of the effective reinsurance portfolio, that corresponds to the minimum value of risk.

The numerical solution of linear programming problems (2),(3) can be realized with the help of, the two-stage method for solving linear programming problems under group constraints, developed by one of the authors of this paper [9].

Table 1 shows the results of numerical solution of problems (2),(3) for the following initial data: $V = 750$ million rubles, $S = 150$ million rubles;

$$\vec{W} = (0.05, 0.15, 0.1, 0.03, 0.07, 0.12, 0.03, 0.01, 0.04, 0.11, 0.06, 0.23)^T;$$

$$\vec{R} = (0.13, 0.05, 0.07, 0.1, 0.11, 0.12, 0.07, 0.09, 0.08, 0.13, 0.14, 0.04)^T;$$

$$\vec{r} = (0.05, 0.03, 0.07, 0.09, 0.15, 0.1, 0.06, 0.15, 0.25, 0.12, 0.08, 0.14)^T;$$

$$\vec{\beta} = (0.25, 0.5, 0.35, 0.25, 0.3, 0.2, 0.25, 0.1, 0.15, 0.1, 0.5, 0.3)^T.$$

The problem of efficient portfolio formation in the reinsurance market (1) can be considered in the case of the more severe group restrictions under which the mathematical model of efficient portfolios is stated as (5).

Table 1. Results of numerical solution.

	$x^*_{1_i}$	$x^*_{2_i}$	$x^*_{1_i} \cdot V_i$, **mln. RUB**	$x^*_{2_i} \cdot V_i$, **mln. RUB**
1	0	0.25	0	9.375
2	0.29(3)	0.5	33	56.25
3	0	0.35	0	26.25
4	0	0.25	0	5.625
5	0.3	0.3	15.75	15.75
6	0.2	0.0375	18	3.375
7	0	0.25	0	5.625
8	0.1	0.1	0.75	0.75
9	0	0.15	0	4.5
10	0.1	0	8.25	0
11	0.5	0.5	22.5	22.5
12	0.3	0	51.75	0
Total			$S = 150$	$S = 150$

$$\begin{gathered}\sum_{i=1}^{n} x_i w_i R_i - \max, \ \sum_{i=1}^{n} x_i w_i r_i - \min, \\ 0 \le \alpha_j \le x_{j_1} + x_{j_2} + \ldots + x_{j_{n_j}} \le \beta_j \le 1, \ j = 1, \ldots, m, \\ \sum_{j=1}^{m} \beta_j \ge 1, \ \sum_{j=1}^{m} \alpha_j \le 1, \ \sum_{j=1}^{m} n_j = n, \end{gathered} \tag{5}$$

where m is the number of groups.

Then, instead of problems (2),(3) we approach, respectively, problems (6),(7):

$$\begin{gathered}\sum_{i=1}^{n} x_i w_i R_i - \max, \\ 0 \le \alpha_j \le x_{j_1} + x_{j_2} + \ldots + x_{j_{n_j}} \le \beta_j \le 1, \ j = 1, \ldots, m, \\ \sum_{j=1}^{m} \beta_j \ge 1, \ \sum_{j=1}^{m} \alpha_j \le 1, \ \sum_{j=1}^{m} n_j = n, \\ \sum_{i=1}^{n} x_i V_i = S, \ \sum_{i=1}^{n} \beta_i V_i \ge S, \ w_i = \frac{V_i}{V}, \ V = \sum_{i=1}^{n} V_i, \end{gathered} \tag{6}$$

$$\begin{gathered}\sum_{i=1}^{n} x_i w_i r_i - \min, \\ 0 \le \alpha_j \le x_{j_1} + x_{j_2} + \ldots + x_{j_{n_j}} \le \beta_j \le 1, \ j = 1, \ldots, m, \\ \sum_{j=1}^{m} \beta_j \ge 1, \ \sum_{j=1}^{m} \alpha_j \le 1, \ \sum_{j=1}^{m} n_j = n, \\ \sum_{i=1}^{n} x_i V_i = S, \ \sum_{i=1}^{n} \beta_i V_i \ge S, \ w_i = \frac{V_i}{V}, \ V = \sum_{i=1}^{n} V_i, \end{gathered} \tag{7}$$

Numerical solutions of specific problems of linear programming type (6),(7) can also be computed using the effective two-stage method [9].

IOP Conf. Series: Journal of Physics: Conf. Series **788** (2017) 012005 doi:10.1088/1742-6596/788/1/012005

4. Conclusion

The balance of the insurance portfolio leads to financial sustainability and efficient operation of insurance companies. The development of the reinsurance market, and mutual confidence of insurance companies ensure reinsurance market its efficient functioning, which will positively impact the insurance market. The internationalization of economic relations makes it possible for Russian insurance companies to operate not only in the domestic market, but also on the international reinsurance markets, creating efficient portfolios of reinsurance contracts with the inclusion of foreign market participants.

References

[1] Pass C, Lowes B and Davies L 1998 *COLLINS Dictionary of Economics* (Saint Petersburg: Higher School of Economics)
[2] Golubin A Y 2003 *Mathematical models in the theory of insurance: construction and optimization* (Moscow: Ankil)
[3] Zhuravlev Y 1994 *Dictionary of terms in insurance and reinsurance* (Moscow: Ankil)
[4] Kornilov I A 2004 *Basics of Insurance Mathematics* (Moscow: FIZMATLIT)
[5] Tulinov V V and Gorin V S 2000 *Insurance and risk management. Terminological dictionary* (Moscow: Nauka)
[6] Fedorova T A (ed) 2009 *The essence of the form and legal basis of reinsurance. Insurance: a textbook* (Moscow: Magistr)
[7] Balashov R B and Kryanev A V 2016 *Vestnik Natsionalnogo Issledovatelskogo Yadernogo Universiteta MIFI* **5** 174–6
[8] Kryanev A V, Lukin G V and Udumyan D K 2010 *Metric analysis and data processing* (Moscow: FIZMATLIT)
[9] Kryanev A V and Cherny A I 1996 *Mathematical modeling* **8** 97–103

IOP Conf. Series: Journal of Physics: Conf. Series **788** (2017) 012006 doi:10.1088/1742-6596/788/1/012006

Automatic analysis of D-partition

V G Bogaevskaya

PhD student, Yaroslavl State University, Yaroslavl, Russia

BogaevskayaVG@yandex.ru

Abstract. The paper is dedicated to automatization of D-partition analysis. D-partition is one of the most common methods for determination of solution stability in systems with time-delayed feedback control and its dependency on values of control parameters. A transition from analytical form of D-partition to plain graph has been investigated. An algorithm of graph faces determination and calculation of count of characteristic equation roots with positive real part for appropriate area of D-partition has been developed. The algorithm keeps an information about analytical formulas for edges of faces. It allows to make further analytical research based on the results of computer analysis.

1. Introduction

Mathematical modeling in spheres such as physics, biology, neurodynamic and many others often tends to necessity to research the solution dynamic for differential equations with parameters. Among thousands of examples a special place belongs to differential equation with time-delayed feedback control. Starting with 1992 when K. Pyragas published his article about the control of chaos with help of time-delayed feedback [1], this field became more and more popular due to easy physical implementation and high effectivity [2, 3]. For this type of problems it is important to determine solution stability. Usually for this aim researchers use the method of D-partition suggested by Yu. I. Neimark [4]. In accordance with Lyapunov theorem, if we linearize given system on the periodic solution or equilibrium state and construct characteristic equation, the count of its roots with positive real part will give us an information about the solution stability in most cases. The method of D-partition reflects imaginary axis of roots plane to the plane of two chosen parameters, dividing it to areas with constant number of roots from right complex half-plane and invariable solution stability. An analysis of D-partition consists in a calculation of these numbers for all areas in special order. Usually researchers do this process manually, but there are systems for which such analysis is very complicated and time-consuming because of complex structure of the D-partition [5]. Also for systems with big number of parameters it may be needed to analyze huge amount of significantly different D-partitions. All of this is the reason why the issue of process automatization is an actual problem.

2. Statement of the problem

2.1. General statement

Consider some differential or difference-differential equation with two real parameters α and β. Let linearize it on the periodic solution or equilibrium state. For linearized system it is possible to construct characteristic equation

$$h(\mu, \alpha, \beta) = 0, \tag{1}$$

To investigate the issue of solution stability we need to know the number of roots of the equation (1) that have positive real part. We suppose that μ continuously depends on parameters, so the only possible situation, when the number of roots from right complex half plane can change, is crossing of imaginary axis during parameters changing.

We suppose that from analytical research of (1) we know:

1) Formulas for D-curves on plane of parameters (α, β) that give us
   a) a root of characteristic equation moving through the point of origin when β is changing
      $\alpha = l_i(\beta)$,
   b) a pair of roots moving through complex conjugate points ±iω on imaginary axis when β is changing
      $\alpha = a_i(\omega), \qquad \beta = b_i(\omega)$.
2) Formulas of conditions that determine the direction of roots movement through imaginary axis
3) Number of roots with positive real part for the case when α = β = 0.

Also we define a rectangle on the plane of parameters (α, β) that contains the area for computer analysis. We will name it plot area.

In this paper we pose the problem to analyze automatically the D-partition defined by information given above and to calculate the count of roots with positive real parts for each area inside given plot area.

2.2. The statement of problem in terms of graphs

The problem can be reformulated in terms of graphs. Let consider following points as vertexes of graph:

1) Points of crossing axes and borders of the plot area with each other
2) Points of crossing D-curves with axes and borders
3) Points of crossing D-curves with each other and itself
4) Points on D-curves where the condition of movement direction changes its state

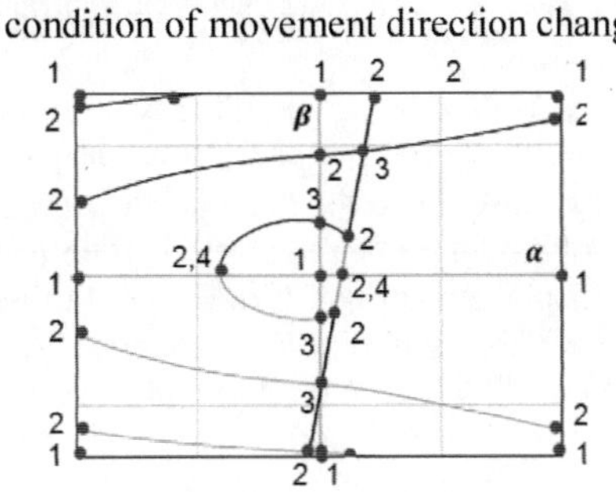

Figure 1. Types of vertexes

Notice that one point can satisfy to several criteria.

Let consider parts of D-curves, axes and borders connecting two vertex as edges. Due to choice of vertexes, it is obvious that

1) Edges cross each other only in vertexes and these vertexes are ends for the given edges,
2) Roots movement direction is the same for all points of each edge.

As a result the graph faces will have following properties:

1) Each face belongs to only one area of given D-partition,
2) Each area of D-partition may include several graph faces, but if two faces are adjacent, their common edge always will be part of axis. So, the whole area can be constructed by union of faces that have such type of common edge.

3) Due to using axes as graph edges, it's always possible to determine two edges for each of four faces around the point of origin. It's very important because the analysis of D-partition should be started with one of them.

So, in these terms, the main problem is to construct the graph, based on analytically gained formulas, to find all its faces and to calculate the count of characteristic equation roots with positive real part for each of them in special order.

3. The algorithm of graph's elements determination

3.1. Search of vertexes

Let approximate each D-curve with polygon. For this aim we set value of step h. For the case of movement through the point of origin we will change β from its minimum to its maximum value on the plot area. For the case of pair of roots movement through points ±iω we will use h to change ω from zero to some given value. As a result we will be able to operate by analytically gained D-curves using a computer.

3.1.1. Search of points of crossing for two curves. By definition the search of most vertexes types requires a method of finding crossing points for two curves. We use the standard formulas for crossing of two segments to determine an existence of the point of curves crossing. There are examples demonstrating that crossing of segments isn't equivalent to curves crossing, but we suppose that h is small enough to give the segment that approximates this part of curve with high accuracy on every step. If the point of crossing segments exists, we divide the increment (of β for the movement through the point of origin case and ω for the other one) by two and calculate appropriate points on curves. As a result we have four segments and we check crossing for them. If we find the pair of crossing segments, we use for them the same algorithm as described above. The process stops when the length of segments becomes less than admissible error. Furthermore, we keep an information which formula was used for each edge, therefore, it gives us an opportunity to precise points of crossing analytically.

3.1.2. Crossing of borders and axes. There are three types of border crossing: self-crossing of borders, crossing with axes and crossing with D-curves. For the first two cases we know coordinates of crossing points and can include them to the list of vertexes at the beginning without any calculations. To find points of crossing D-curves with borders, on each step we check whether a new point on D-curve is inside the plot area or not. If the state of this condition changes, we start the process of crossing point search for each of borders and the current part of the D-curve. Noteworthy that every D-curve should have its start and end outside of the plot area, on the border or in crossing point with some other D-curve, otherwise we will have dangling vertex but by the definition of D-partition, it can't have any of them. The D-curve of movement through the point of origin starts and ends with values of β equal to the minimum and maximum values for β inside the plot area accordingly, so there will not be such problems here. The D-curve of movement through ±iω starts with ω =0. It is easy to see that this point will lay on D-curve of movement through the point of origin. To make computer analysis finite, we need to restrict the maximum value of ω, but in this case we can't guarantee where this point will be. We use the indicator of being inside the plot area to avoid the situation with dangling vertex. If maximum value of ω is reached but the current point is inside plot area we will continue increasing ω until new point leaves the plot area or stay at the same place during huge amount of steps. In the first case we find the point of crossing with border, in the second one the point should lay on some other D-curve. It's possible to prove that then it coincides with the point of direction of movement changing on the second D-curve or the edge based on it and previous vertex can be excluded from consideration. So, we mark such points as vertex to analyze it later.

If points of border crossing have been found, we create an array of subsegments that lay inside the plot area and use them for further analysis. Points of axes crossing could be found similarly. We check, in which coordinate quarter current point lays, and if the value changes, start the process of crossing point search. When the point of crossing is found we check if we already have a vertex distance, to which from this point is smaller than admissible error. If we find such vertex, we take it as the current crossing point and add an information about the current D-curve and value of changing variable to it. If we haven't such vertex, we add a new one to the list of them.

3.1.3. Self-crossing of D-curves. Let divide plot area by a grid with relatively small cell width. As it was mentioned above increasing changing variable we move in the D-curve. After completion of each step we keep an information which cells of the grid are crossed by the current part of D-curve. So, when we process a new segment (or subsegment, if we have found points of crossing with borders and/or axes) we determine in which cells it lays and check crossing only with previously processed segments that also are there. If we find a point of crossing, we check whether this point was already added as a vertex and either add a new vertex, or update an information about D-curves in the old one. Also we replace stored segment by two with ends in new crossing point. We split current segment to two subsegments too and continue analysis with them separately. As a result all stored segments will have vertexes only as ends. It will be useful for edges search.

3.1.4. Direction of movement changing. At the end of analysis of the segment or subsegment we check the state of movement direction condition for its end points. If states are different, we precise the point by dividing the changing variable by two and check which of new subsegments has different states of the condition at its ends. We repeat the process until the length of subsegment becomes smaller than admissible error and then we add a new vertex to the list.

3.2. Edges searching
When all vertexes are found we can define edges of the graph. For each D-curve we store a list of vertex sorted by changing variable. We go through this list and set edges as parts of D-curve that connect two neighboring vertexes. However some edges can lay outside of the plot area and we should exclude them from analysis. Sufficient condition of exclusion is that an arbitrary point from the edge is out of the plot area because we already found all border crossing points and they are vertexes in our graph. So, before adding edge to the list we check if it is inside the plot area or not. Also for each edge we calculate a polar angle in each of its vertexes. We take this value as the angle of the small segment approximating the curve near the vertex. We also add this information to the vertex, where we sort list of its edges by the angle.

4. An algorithm of graph faces searching and classification

In accordance with the rules of D-partition analysis our algorithm should work in following order:

1) The first face that we find should have the point of origin as its vertex. Only for such faces we know the count of roots with positive real part initially.
2) When some faces are already found and the number of roots is calculated for them, we should choose as the next face the one that is adjacent with already processed face. To make it easier to detect such faces, we will count for each edge how many processed faces it belongs. So, if the face has an edge with the number equal to 1 and this edge isn't a border (borders can be "used" in faces only once), we can process it.
3) We will repeat step 2 until there will be no edge with the index "1". But we still can have edges with the index "0" if we have nested faces. In this case we need to determine inside which face nested component located, find its outer contour, mark them by "1" and then go to step 2.

Let consider the process in more details.

4.1. Faces around the point of origin

Due to using axes as edges we can easily get four vertexes nearest to the point of the origin and belonging to one of them. Let choose two such vertex: one from positive part of β-axis and one from positive part of α-axis. Let name the edge connecting the vertex on β-axis with the vertex in the point of the origin "start edge". In the list of edges of the vertex from the point of the origin, sorted by polar angle, the second edge is a neighbor of the start edge. Besides we know that it is on anticlockwise direction. We keep this information and start following process: among all edges from the list of the vertex on α-axis we choose a neighbor of given edge that lay on the same direction as we store (anticlockwise in this case). This new edge also belongs to the face we determine. Then we do the same thing but for the second end of new edge and continue this process until we get the start edge as the next one. As a result we have full list of edges defining the face. By the problem statement the number of roots of characteristic equation with positive real part is already known. We increase the index for all these edges to 1 and store the information about found face to the each of edges. We can do the same process for other three faces, but it's not necessary because they will be processed anyway as adjacent faces.

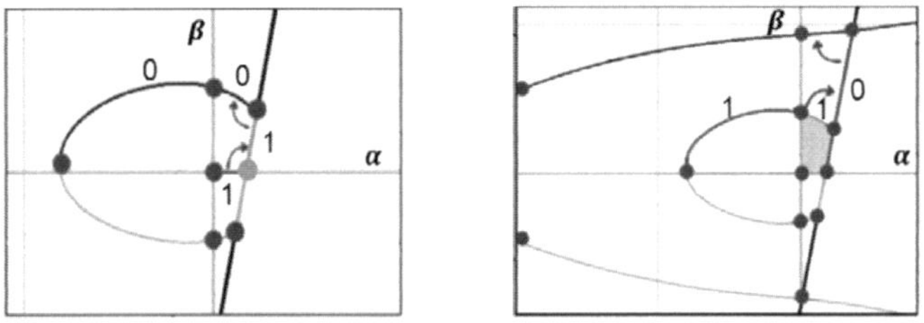

Figure 2. Search of faces. Left picture is the search of the face around the point of origin. Right one is the search of adjacent face. Numbers show the index of edge. Arrows are the direction of movement. Grey field is already processed face.

4.2. Search of adjacent faces

Let take one of the edges with index "1" as the start edge. Consider one of its two vertexes and its list of edges. If there is only one other edge, we take it as the next one and look at its second vertex. If there are several other edges, we choose the neighbor by polar angle, that doesn't belong to the processed face which contains start edge. We keep an information about the direction of movement to the neighbor from the current edge (clockwise or anticlockwise) and continue round of the face according to it until we get the start edge as the next one. As before we increase the index for all these edges to 1 and store the information about found face to the each of edges.

4.3. Calculation of root count for the adjacent face

From the statement of the problem we know how the count of roots will change if we cross an edge in the direction of β increasing. If β decreases, the count will change in opposite way. Therefore, to calculate how many roots have positive real part if parameters α and β are taken from the inside of processing face we need to detect which face is above of common edge and which one is below. Let hold a vertical line through an arbitrary point on the start edge and count the number of its crossing with edges of the processing face. It has been proved that if this number is even, the processing face is located below the start edge, but if the number is odd, the processing face is located above. In the first case we need to deduct the value, got from the condition, from the count of roots for adjacent face processed earlier. In the second case we will add this value.

4.4. The case of nested components

The process described above allows to find faces only for one coherent component of the graph. If we processed all edges with index "1" but we still have edges with index "0", consider one of the edges never used before and start the search of nested component. Let take an arbitrary point on the edge and hold a vertical line through it. Consider all points of crossing this line with edges and find the nearest one that lays above the point and belongs to the edge with index "2". For this edge we know its faces, their numbers of roots and the condition of changing roots count when β grows up. So it's possible to get which face is located below. It is proved that this face contains given nested component. Consider the last edge with index "0" crossed by the line and located above the point. This edge belongs to outer contour of the nested component. Using it find all edges of outer contour using standard algorithm, set their indexes to "1" and start the same process as before.

5. Summary

The algorithm described in this paper allows us to construct D-partition based on analytically gained formulas inside specified rectangle, find all its areas as lists of edges and calculate the count of characteristic equation roots with positive real part for each of them. It also keeps an information about formulas used for each edge. It gives more accurate calculations and can be useful, if we want to continue analytical research after computer processing. Furthermore, algorithm can automatically determine a possibility of stabilization by the check whether face with the count of roots equal to zero exists or not.

6. References

[1] Pyragas K 1992 Continuous control of chaos by self-controlling feedback. *Phys. Lett.* A **170** 421–28
[2] Flunkert V 2011 *Delay-Coupled Complex Systems and Applications to Lasers* (Berlin: Springer)
[3] Bogaevskaya V and Kashchenko I 2014 Influence of delayed feedback control on the stability of periodic orbits *Automatic Control and Computer Sciences* **48** 53–65
[4] Neimark Yu I 1992 Robust stability and D-partition *Automation and Remote Control* **53** 957–65
[5] Bogaevskaya V and Kashchenko I 2015 Cycle stabilization by one and two delay feedback control. *Nonlinear Phenomena in Complex Systems* **18** 175–80

IOP Conf. Series: Journal of Physics: Conf. Series **788** (2017) 012007 doi:10.1088/1742-6596/788/1/012007

Impact of waves on the circulation flow in the Iguasu gas centrifuge

S Bogovalov, V Kislov, I Tronin
National Research Nuclear University MEPhI (Moscow Engineering Physics Institute)
E-mail: vakislov@mephi.ru

Abstract. 2D axisymmetric transient flow induced by a pulsed braking force in the Iguasu gas centrifuge (GC) is simulated numerically. The simulation is performed for two cases: transient and stationary. The braking forces averaged over the period of rotation are equal to each other in both cases. The transient case is compared with the stationary case where the flow is excited by the stationary braking force.Two models of the gas cenrifuge is simulated. There are two cameras in the first model and three cameras in the second one. In the transient case for the two cameras model pulsations almost doubles the axial circulation flux in the working camera. In the transient case for the three cameras model the gas flux through the gap in the bottom baffle exceeds on 15 % the same flux in the stationary case for the same gas content and temperature at the walls of the rotor. We argue that the waves can reduce the gas content in the GC on the same 15 %.

1. Introduction

Separation of heavy isotopes in strong centrifugal fields is used for industrial production of enriched uranium. Rotors of gas centrifuges (GC) rotate with linear velocity a few hundreds meters per second [1]. The centrifugal acceleration can reach $\sim 10^6 g$ at the radius of the rotor of a few centimeters. Nevertheless, strong centrifugal field is not the only factor which results into efficient isotope separation in GC. A range of physical processes is explored to produce an additional axial circulation, which essentially increases the efficiency of separation of the GC [2]. This secondary circulation is one of the key feature of the industrial GC.

A pair of scoops located near the end caps of the GC are used to remove enriched and depleted gas mixture from the GC. Simultaneously, they provide the additional axial circulation due to the mechanical brake of the gas. Mach number of the gas in GC is of the order 7 because the sound velocity of the working gas UF_6 is of the order 86 m/s at room temperatures. Interaction of the gas with the scoop is accompanied by formation of strong shock wave which propagates along the rotational axis. Figure 1 schematically shows that the shock wave forms a spiral wave propagating from one end of the GC to another. It can reflect from the end caps of the GC forming waves running in both directions along the rotational axis. The amplitude of the shock waves is damped rather quickly. In the largest part of the rotor, they propagate as small amplitude waves.

It is well known that the waves can produce the flow of gas due to their absorption [4]. These are so-called acoustic flows described by Rayleigh [5, 6]. They are produced due to transfer of the energy and momentum from the waves to the gas due to the molecular viscosity. This way the waves generated by the scoops can provide an additional mechanism of generation of the

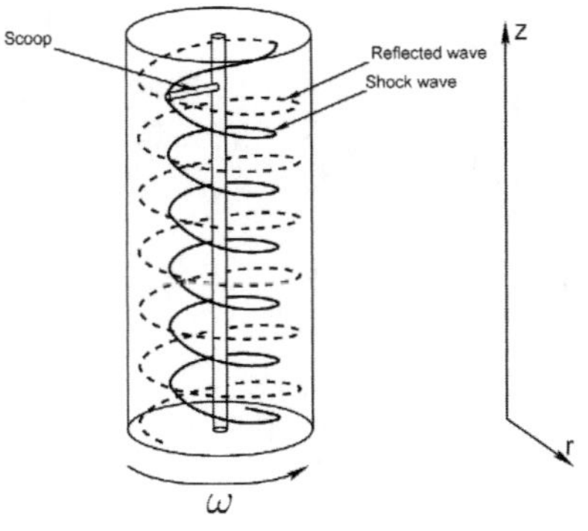

Figure 1. Scheme of the gas centrifuge with scoops. Solid line shock wave produced by the scoop, dashed line the wave reflected from the upper end cap

secondary axial circulation in GC. This mechanism can essentially differ from the conventional mechanism of the circulation generation. The waves can propagate and therefore transfer the breaking torque at larger distance from the scoops than it happens at the axisymmetric brake. Recently, authors have shown that conventional acoustic waves are essentially modified in strong centrifugal fields [3]. Exploration of this mechanism can change efficiency of the isotope separation and working parameter of the GC.

2. Formulation of the problem

The waves produced by the scoops are stationary in the laboratory frame system. Computer modelling of the gas flow in the GC is performed in rotating frame system [7, 8, 9, 10, 11]. 2-D axisymmetric model of source-sinks is applied [9]. In this work we use axisymmetric model of the sources and sinks in the rotating frame system as well. The waves are not stationary in this model. Special computer code has been developed in MEPhI for numerical simulation of the non stationary gas dynamics in strong centrifugal fields . The results of verification of the code were presented in [12, 13].

The schemes of the computational domain is presented in Figure 2. The first domain consists of two cameras: top and working. The model is closed.We do not consider here the problems connected with feed, product and waste fluxes. This allows us to investigate the role of the wave in the simplest model. In the upper camera a small region is selected where the source of the momentum and energy is introduced in to the flow. The circulation flow is induced by this source.

The second domain consists of three cameras: top, working and bottom. Product (enriched) and waste (depleted) fluxes are removed in the bottom and top cameras, respectively by the sinks shown by solid points 6 and 2. Feed flow (F) is injected at the middle of the working camera. Circulation flow (C) is induced due to two mechanisms: temperature gradient ΔT at the rotor wall and brake of the gas in the top camera. The braking force is located in the point source 2.

We consider two types of the sources in both cases. The first one is the conventional steady state source.The second one is the pulsed source of energy and momentum with the same as in the steady state case integral of the breaking force over the period of rotation.The shape of the pulse is determined by the following equation:

$$A(t) = A'e^{-15\sin(0,25t/\tau_0)}, \quad (1)$$

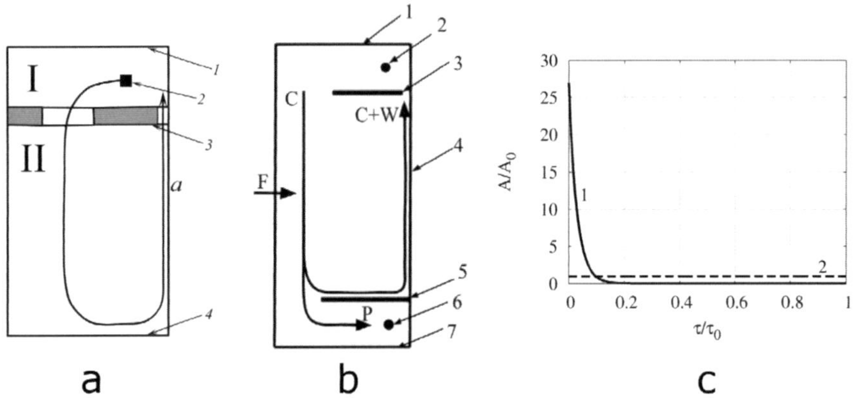

Figure 2. (a,b) – Schemes of the computational domain. (a) – Scheme of the two cameras domain. 1 – upper end cap, 2 – the source of the energy and momentum, 3 – the baffle, 4 – the lower end cap. I – upper and II – working cameras. (b) – Scheme of the three cameras domain. F,P,W – feed, product and waste fluxes, respectively, and C– circulation flow. 1 – the top cap, 2 – source of the momentum and energy, sink of the waste flux, 3 – the top baffle, 4 – the rotor wall, 5 – the bottom baffle, 6 – sink of the product flux, 7 – the bottom cap. (c) – The dimensionless braking force in the top source versus dimensionless period of rotation. 1 – non-stationary case, 2 – stationary case. A_0 is the force in the stationary case, τ_0 is the period of rotation.

where τ_0 is the period of the rotation.

The dependence on time of the pulsating braking force in the time interval (0 - τ_0) is shown in Figure 2. In the transient case the duration of the braking force is about $\frac{\tau_0}{10}$. Amplitude is equal $A' = 27A_0$, where A_0 is the amplitude of the force in the stationary case. This reproduces the impact of the rotating scoop on the gas.

Rotation of the gas results into high pressure gradient in radial direction about 10^5 Pa per 1 cm at the wall of the rotor. At a certain radius the pressure becomes so small that hydrodynamical approximation fails. At this radius free slip and adiabatic boundary conditions are specified according to [11]. No slip boundary condition is specified at the rotor wall. The wall is rotating with angular velocity ω.

Parameters used for the computer simulation of the Iguasu centrifuge are presented in Table 1 [14].

3. Results

3.1. Two cameras model

The braking force produces unbalanced radial gradient of pressure. The gas moves to the axis of rotation after the slow down increasing the pressure near the axis. This results into expansion of the gas and motion of its part through the hole of the baffle. This way the perturbation generated in the upper camera propagates into the working camera. Here the perturbation takes a form of the wave which can propagate to the bottom end cap of the GC. After reflection from the bottom end cap of the GC the wave propagates into reversed direction. Multiple reflections of the wave from the end caps result into rather complicated time dependent flow. Snapshot of this flow is shown in Fig. 3 . It follows from this figure that the wave satisfies to the dispersion equation of the form $\omega = kc$, where c is the conventional sound velocity, k is the wave vector and ω is the angular frequency of the wave.

Table 1. Parameters of the Iguasu centrifuge

Parameter	Value
Molar mass, M	352 g/mol
Specific heat at constant pressure, c_p	385 J/kg$\cdot$ K
Viscosity, μ	$1.83\cdot 10^{-5}$ Pa$\cdot$ s
Thermal conductivity, κ	0.0061 J/(m$\cdot$ s$\cdot$ K)
Angular frequency, ω	$1700 * 2\pi$ s^{-1}
Radius of the rotor, a	0.065 m
Inner radius of the computational domain, r_in	0.052 m
Temperature of the bottom baffle, T_{low}	300 K
Temperature gradient on the rotor wall between top and bottom baffles, ΔT	10 K
Force in the stationary case, A_0	$17.8\cdot 10^{-6}$ N

Comparison of the velocity field obtained with the stationary breaking force (see Fig. 3) with the snapshot of the velocity field with the pulsating breaking force shows that they are well different. In the nonstationary problem there are velocity perturbations which exceed the velocity in the stationary case. For the practical applications it is important to estimate the impact of the waves on the separation and efficiency of the GC. However, the snapshot of the flow field does not allow us to do that. It is necessary to consider the average flow field.

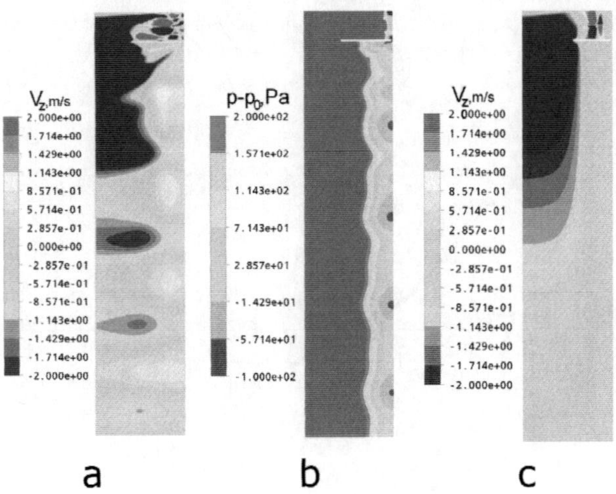

Figure 3. (a, b) – The waves in the working camera of the GC: (a) axial velocity; (b) $(p-p_0)e^{-\frac{M\omega^2 r^2}{2RT_0}}$, where $p_0 = p_w e^{\frac{M\omega^2(r^2-a^2)}{2RT_0}}$. (c) – Axial velocity in the working camera of the GC in the stationary case.

The velocity fields for stationary breaking force and the velocity field for pulsating breaking force averaged over last 6 periods of rotation are shown in Figs.5 and 6. It is seen that the averaged velocity field well coincides with the stationary one. The flow pattern is quite identical. Nevertheless, the quantitative analysis shows that the pulsating breaking force essentially changes the axial circulation of the gas.

The axial circulation flow in the rotor can be described by the integral functions showing the mass flux $\dot{m}_-$ down the rotor and mass flux $\dot{m}^+$ up the rotor. In the stationary case they should

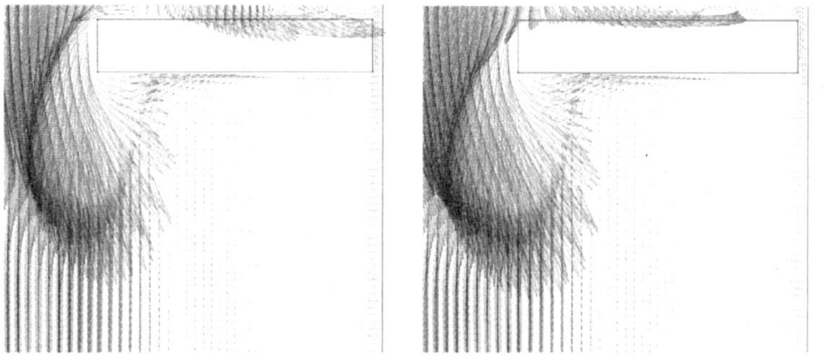

Figure 4. (a) – Velocity under the baffle in the stationary case. (b) – Averaged velocity under the baffle in the case of pulsating breaking force.

coincide due to the equation of discontinuity. In the nonstationary case the same functions after averaging also should coincide with each other. By the way, this coincidence is an additional test of the correctness of the simulation. These mass fluxes are shown in Fig. 5. It is most important that the averaged mass flux twice exceeds the mass flux of the axial circulation produced by the steady state breaking force. The nature of this effect can be understood in terms of the acoustic flow [5, 6]. The scoop generates all possible types of waves including ones with fast damping due to the molecular viscosity. These waves transfer their momentum to the gas increasing the mass flux.

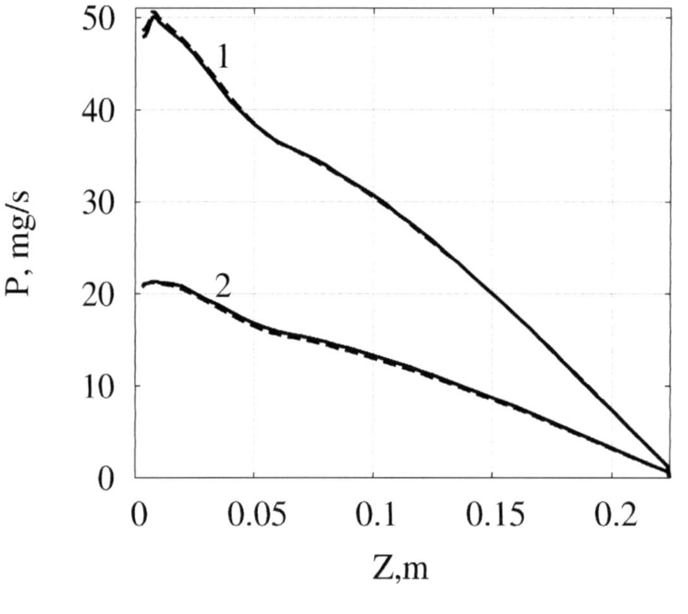

Figure 5. Dependence of $\dot{m}^+$ (solid lines) and $\dot{m}_-$ (dashed lines) on axial coordinate z. Zero of this coordinate corresponds to the point right under the baffle. 1 – average case, 2 – stationary case.

3.2. Three cameras model

For the three cameras model for the steady state case the waste flux is specified in the point 2 equal to 0.55 of the feed flux. The pressure in the working camera is obtained in the process of the solution. It equals to 79 mm Hg. This pressure provides the product flux satisfying the mass conservation. It equals to 7.2 mg/s.

In the transient case all the fluxes and temperature at the walls of the rotor are taken as in the stationary case. The stationary solution is taken as the initial state for the transient case. In the transient case the flux through the gap in the bottom baffle appears on 15% higher than the same flux in the stationary case. Correspondingly, the product flux increases on 15% as well while the feed and waste fluxes remain fixed. This happens because the decrease of the gas content due to the leakage of the gas through the product sink is insignificant during the simulation. The gas fluxes are not balanced. The mass conservation will take place when gas content in the working camera will fall down to reduce the product mass flux. But duration of simulation of such process exceeds duration of our simulation many times. We do not reach the equilibrium state where all the fluxes are balanced. Nevertheless, we can conclude that the waves have to change the gas content on the same 15% to balance all the gas fluxes in the GC.

4. Conclusion

The numerical simulation allows us to investigate the axial circulation of the gas under impact of the axisymmetric pulsating breaking force in the upper camera of the GC. The axisymmetric waves with dispersion law $\omega = kc$ are excited due to this force in accordance with the theoretical predictions.

In two cameras model the averaged field of velocity qualitatively coincides with the stationary one. However, the pulsating breaking force almost twice increases the mass flux in the axial circulation flow in comparison with the mass flux at the stationary breaking force.

In three cameras model the product flux in the transient case 15 % exceeds the product flux of the stationary case for the same gas content and temperature of the rotor walls. This means that in the steady-state regime, when all the gas fluxes are balanced, the gas content in the transient case is expected to be lower than the gas content in the stationary case approximately on the same 15%.Conventional axisymmetric models of the gas flow in the GC do not take into account this effect. This could result into in correct estimate of the optimal breaking force.

References

[1] Glaser A 2008 *Sci. Glob. Secur.* **16(12)** pp. 126
[2] Borisevich V, Borman V, Sulaberidze G et al. 2011 *Fizicheskie osnovy razdeleniya isotopov v gazovoi centrifuge (Physical basics of the isotope separation in gas centrifuge),in Russian* (Moscow:Izdatelskii dom MEI)
[3] Bogovalov S, Kislov V and Tronin I 2015 *Theoretical and Computational Fluid Dynamics* **29(1-2)** 111–25
[4] Broman G I and Rudenko O V 2010 *Phys. Usp.* **53** 91–8
[5] Rayleigh L 1884 *Philos. Trans. R.Soc. Lond. Ser. A* **175** 1–21
[6] Rayleigh L 1896 *The Theory of Sound* (London:MacMillan)
[7] Wood H and Morton J 1980 *Journal of Fluid Mechanics* **101(1)** 1–31
[8] Doneddu F, Roblin P and Wood H 2000 *Separation Science and Technology* **35(8)** 1207–21
[9] Matsuda T, Sakurai T and Takeda H 1975 *Journal of Fluid Mechanics* **69(1)** 197–208
[10] Aisen E, Borisevich V and Levin E 1997 *Mathematical models and computer experiment* **9(4)** 27–38
[11] Von Halle E, Wood H and Lowry R 1983 *Nuclear Technology* **62 (6)** 325–34
[12] Abramov V, Bogovalov S, Borisevich V, Borman V D, Kislov V A, Tronin I V, Tronin V N and Yupatov S V 2013 *Computational Mathematics and Mathematical Physics* **53(6)** 789–97
[13] Bogovalov S V, Borisevich V D, Borman V D, Kislov V A, Tronin I V and Tronin V N 2013 *Computers and Fluids* **86** 177–84
[14] Borisevich V D, Borman V D, Bogovalov S V, Kislov V A, Tronin V N, Tronin I V, Abramov V I and Yupatov S V 2014 *Separation Science and Technology (Philadelphia)* **49(3)** 329–34

Plasmastatic model of toroidal trap "Galatea-belt"

K V Brushlinskii[1], **A S Goldich**[2]

[1] Keldysh Institute of Applied Mathematics RAS, Miusskaya pl., 4, 125047, Moscow, Russia
[2] Department of Applied Mathematics, National Nuclear Research University MEPHI, Kashirskoe highway, 31, 115409, Moscow, Russia

E-mail: `dephmaster@gmail.com`

Abstract. Magnetic galatea-traps for thermonuclear plasma confinement with current carrying conductors immersed into the plasma volume, are represented by an example of the toroidal trap "The Belt" with two circular conductors. Numerical models of equilibrium plasma and field configurations are investigated in straightened into cylinder analogues of some toroidal galateas in a series of works by the authors. This paper presents a plasmastatic model of configurations in the toroidal variant of "The Belt" in terms of a boundary problem with the Grad-Shafranov equation. Distinctions of their geometry and quantitative characteristics from the cylindrical analogues and their dependence of parameters are determined in computation.

1. Introduction

The work deals with mathematical simulations and computation of physical processes in the magnetic trap for plasma confinement research. The traps are a basic element of plasma systems in many programs aimed at controlled nuclear fusion realization, because the magnetic field is the only accessible materials for keep the hot thermonuclear plasma. Because of some established reasons a considerable part of elaborations and researches treat with the toroidal traps, for example the well-known tokamak and stellarator. In the first one the magnetic field is induced mainly by the toroidal electric current in the plasma, the field of the second one owes its origin to a system of conductors outside the plasma volume. It is of interest a promising class of traps where the current carrying conductors are immersed into the plasma, but don't immediately contact with the hot one. The magnetic field geometry is more complicated and varied in this case, that allows to expect a more effective plasma confinement. A special attention to such traps was concentrated in a series of works by A.I. Morozov, who named them "Galateas" (see [1,2] with references). The galatea-traps are supposed also to be toroidal, but some researches of general or conceptual problems in theoretical and computational plasma physics admit to replace the torus by its straightened into the cylinder analogue. The mathematical means become more simple in this case.

Investigation of magnetic traps include as an essential part the mathematical modeling and computation using the highly efficient computer technique up to day. Plasma, treated there, may be considered dense enough for dealing with the models in continuous media mechanics terms, i.e., magnetogasdynamics. The plasma confinement time, needed for nuclear fusion reaction expected, is significantly greater than characteristic ones of low-scaled plasma processes. Therefore equilibrium plasma and field configurations in the trap, more exactly, their plasmastatic models form our investigation object.

IOP Conf. Series: Journal of Physics: Conf. Series **788** (2017) 012008 doi:10.1088/1742-6596/788/1/012008

"The Belt" is a simple and visual example of the galatea-traps. It is a plasma torus with two circular conductors, suggested by analogy with a cylindrical device in the General Physics Institute Russian Academy of Sciences for investigation of current sheets by the group of S.I. Syrovatskii followers [3]. Numerical simulation of "The Belt" and its differences from the current sheets are also considered in the cylindrical geometry in the series of papers by the authors (see. [4, 5] with references).

A plasmastatic model of equilibrium configurations in the toroidal variant of "The Belt" is developed in this paper. We put a question of the differences between toroidal and cylindrical configurations and their quantitative characteristics, depending on the trap parameters. Some responses are obtained in computation.

2. Plasmastatic model. Setting of the problem and means of its solving.

The mathematical apparatus of plasmastatics is based on the magnetogasdynamics equations. In our case (the velocity $\mathbf{v} \equiv 0$ and the time dependence $t\ \partial/\partial t \equiv 0$ absent) only the equilibrium one of them is nontrivial

$$\nabla p = \frac{1}{c}\mathbf{j} \times \mathbf{H}, \tag{1}$$

where $p, \mathbf{j}, \mathbf{H}$ are respectively the plasma pressure, electric current density and magnetic field strength. Together with the Maxwell's equations

$$\mathbf{j} = \frac{c}{4\pi}\mathrm{rot}\mathbf{H}; \quad \mathrm{div}\mathbf{H} = 0, \tag{2}$$

they form a closed system for three unknown functions, that determine the equilibrium configuration required in the giving space region with the given boundary conditions.

Boundary value problems with the equations (1)-(2) are generally rather complicated. But they may be essentially simplified, given a symmetry, in particular in axially symmetric problems on "The Belt" configurations in torus, where $\partial/\partial\varphi \equiv 0$ in the cylindrical coordinates (r, φ, z). It is convenient to express the the (r, z)- components of solenoidal vectors $\mathbf{H}$ and $\mathbf{j}$ by their flux functions Ψ and I

$$rH_r = -\frac{\partial\Psi}{\partial z}; \quad rH_z = \frac{\partial\Psi}{\partial r}; \quad rj_r = -\frac{\partial I}{\partial z}; \quad rj_z = \frac{\partial I}{\partial r}, \quad I = \frac{c}{4\pi}rH_\varphi \tag{3}$$

It follows from (1),(3), firstly, that the magnetic force lines $\Psi(r, z) = \mathrm{const}$ and electric current stream lines $I(r, z) = \mathrm{const}$ are isobars ($p = \mathrm{const}$), i.e., the three function p, Ψ, I are interdependent:

$$p = p(\Psi); \quad I = I(\Psi). \tag{4}$$

Secondly, the function Ψ satisfies the Grad-Shafranov equation [6,7]

$$\triangle^*\Psi + 4\pi r^2\frac{dp}{d\Psi} + \left(\frac{4\pi}{c}\right)^2 I\frac{dI}{d\Psi} = 0, \tag{5}$$

where

$$\triangle^*\Psi \equiv r\frac{\partial}{\partial r}\left(\frac{1}{r}\frac{\partial\Psi}{\partial r}\right) + \frac{\partial^2\Psi}{\partial z^2}.$$

Its low minor order terms include two functions (4), that must be given from additional requirements, relating with trap configurations.

Boundary value problems on the galatea-trap configurations must be considered, strictly speaking, in the torus cross section without conductors immersed. But we don't want to operate with a multiply connected regions and admit one more simplification: the conductor current is given by a function $j_\varphi^{ex}(r,z)$ localized in their cross sections, the equation (1) contains only the plasma current $\mathbf{j}^{\mathbf{pl}}$, and the whole current $\mathbf{j} = \mathbf{j}^{\mathbf{pl}} + \mathbf{j}^{\mathbf{ex}}$ takes part in the Maxwell's equations (2). The equation (5) obtains in this case the additional term $4\pi r j^{ex}/c$ in the left hand side.

In the specific case of a toroidal trap "The Belt" with two circular conductors, those cross-section centers are situated at $z = \pm z_0, r = r_0$ the current j^{ex} is given in the form

$$j^{ex} = \sum_{k=1}^{2} j_0 \exp\left(-\frac{(r-r_0)^2 + (z-z_k)^2}{r_c^2}\right), \tag{6}$$

where $z_1 = z_0$, $z_2 = -z_0$, and r_c is the conventional conductor radius. The factor j_0 is chosen by means of the condition

$$\int\int j^{ex}(r,z)drdz = J_c,$$

where J_c is the given current value in each conductor, and the integral is taken around its vicinity.

There is no "toroidal" magnetic field H_φ in the trap, thus $I(\Psi) \equiv 0$ in the equation (5). The relation $p(\Psi)$ is chosen, starting from the disire to dispose plasms in the trap, in particular, to insulate the conductor from it. We can gain such effect, concentrating plasma around the magnetic field singular point and hence along the separatrix $\Psi = \Psi_0$ passing through this point (the magnetic field topology in cylindrical and toroidal "Belt" variants are presented at the Fig. 1a and 2a). The function $p(\Psi)$ must be nonmonotone, have the maximum in this point and decrease rapidly in the both sides from the separatrix. For example,

$$p = p_0 \exp\left(-\frac{(\Psi - \Psi_0)^2}{q^2}\right) \tag{7}$$

Boundary value problem with equation (5), including refinement made above, are considered in the rectangular (for simplicity sake) space region

$$R_1 < r < R_2; \quad -Z < z < Z. \tag{8}$$

The simplest boundary conditions correspond to the impermeable for the magnetic field boundary $H_n = 0$, where $\mathbf{n}$ is normal direction, i.e. $\Psi = \text{const}$. For definiteness put

$$\Psi = 0; \quad \text{at} \quad (r,z) \in \Gamma, \tag{9}$$

where Γ is region (8) boundary. More complex conditions, taking into account magnetic transparent boundaries, are considered in [8].

The problem is considered in dimensionless variables, using units, composed from the given constants. The units of length, field strength, flux function, current density and pressure are respectively

$$r_u = z_0;\ H_u = \frac{2J_c}{cz_0};\ \Psi_u = H_u r_u^2;\ j_u = \frac{cH_u}{4\pi r_u};\ p_u = \frac{H_u^2}{4\pi}. \tag{10}$$

where z_0 is the half-distance between the conductors, J_c - the current value in the each one. In these units, the Grad-Shafranov equation(5) is

$$\triangle^*\Psi + r^2\frac{dp}{d\Psi} + rj^{ex}(r,z) = 0 \tag{11}$$

The conductor coordinates and the constant j_0 in Eq. (7) are

$$z_1 = 1,\ z_2 = -1,\ j_0 = \frac{2}{r_c^2},$$

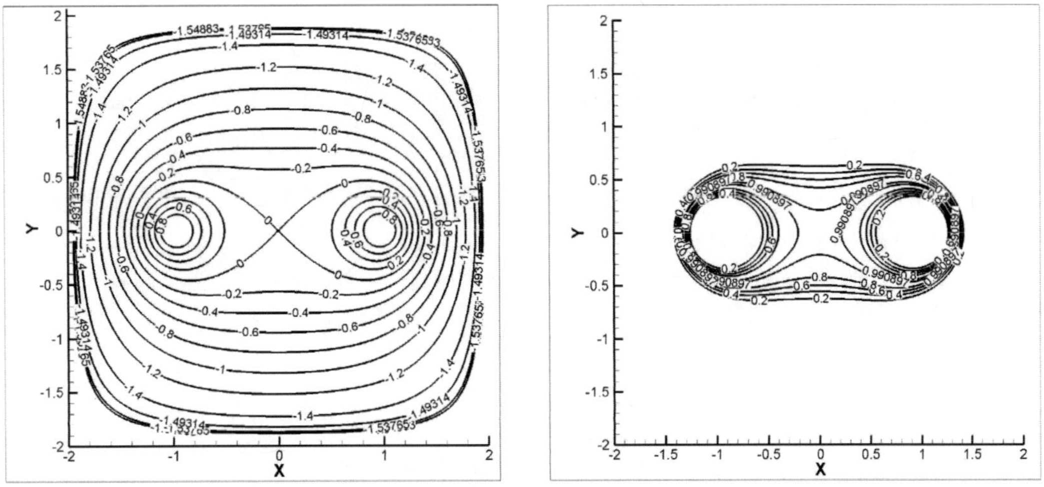

a b

Figure 1. Equilibrium configuration in the cylindrical "Belt" variant a) magnetic field ($\Psi(r,z) = \text{const}$), b) plasma pressure ($p = \text{const}$).

Formula (7) and the region (8) boundary are the same, but with parameters, related to the units (10).

The problem posed is solved numerically by means of iterative relaxation method. We have to do with the initial- boundary value problem for the parabolic equation

$$\frac{\partial \Psi}{\partial t} = \triangle^* \Psi + r^2 \frac{dp}{d\Psi} + r j^{ex}(r,z) \tag{12}$$

in assumption that its solution would approache to the required one with the time. Its difference analogue is solved by the well-known alternating direction method [9]. In this case the computation of the next ($n+1$-th) "time"-step or iteration uses the value of nonlinear function $p(\Psi)$ from the preceding (n-th) one.

A particular feature of the problem is the requirement: the parameter Ψ_0 in eq. (7) must coincide with the value of the solution Ψ in the field singular point. We use the following suitable way to determine it: the value Ψ_0 in computation of the $n+1$-th step is the one of Ψ in the singular point on the n-th step, that is its maximum of the vertical line $z = 0$. Thus

$$(\Psi^0)^{n+1} = \max_r (\Psi(r,0))^n \tag{13}$$

The model of equilibrium configurations in a straight cylinder for comparison with our results below is constructed in a similar manner under the plane symmetry assumption: $\partial/\partial z \equiv 0$ in the Cartesian coordinates. The dimensionless Grad-Shafranov type equation is

$$\triangle \Psi + \frac{dp}{d\Psi} + j^{ex}(x,y) = 0 \tag{14}$$

The magnetic field is presented by the flux function Ψ ,

$$H_x = \frac{\partial \Psi}{\partial y}; H_y = -\frac{\partial \Psi}{\partial x}$$

the independent variable space region is a rectangle $|x| < X, |y| < Y$ (or a circle $x^2+y^2 < R^2$[4]), the current in the conductors is distributed as (6), the plasma pressure is given in the form (7)with the maximum in the center and at the separatrix, passing through it.

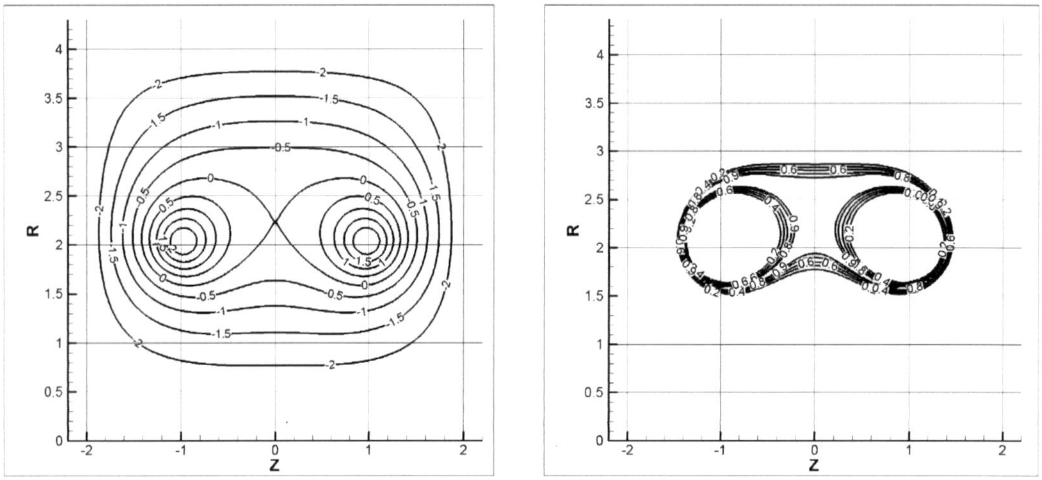

a b

Figure 2. Equilibrium configuration in the toroidal "Belt" variant a) magnetic field ($\Psi(r, z) =$ const), b) plasma pressure ($p =$ const)

3. Computation results.

In numerical solving of the problems in the torus as in the cylinder equilibrium configurations of plasma, magnetic field and electric current are obtained under some conditions. The examples of them in cylindrical and toroidal traps are presented at the fig 1, 2 by the magnetic force lines $\Psi =$ const and isobars $p =$ const. The plasma is insulated from the conductors and concentrated around the field singular point. It has a form of curvilinear quadrangle with concave boundaries and thin branches along the separatrix surrounding the conductors. The electric current in plasma, orthogonal to the figure plane, vanishes in the configuration center and is concentrated at its boundaries, where the pressure gradient is maximal. It is positive (of conductor current direction) outside the separatrix and negative inside it, i.e. around the conductors. Interacting with the magnetic field of anti-clockwise direction, it promotes to confine plasma near the separatrix. The configuration in the straight cylinder are symmetric relating the center $x = y = 0$. The ones is the torus are deformed: the singular point at the fig.2 is situated farther than the conductors from the z- axis at the distance δr. Its value decreases with the growth of the current radius r_0 because the configuration geometry tends to the cylindrical one, when $r_0 \to \infty$.

The problem includes the dimensionless parameter p_0 in (7). Its physical meaning is the maximum plasma pressure, referred to the characteristic magnetic one p_u (10). The equilibrium mentioned above is established as usual under the restriction

$$p_0 < p_0^{cr}, \qquad (15)$$

where the critical value p_0^{cr} depends on the parameter q in (7) and on the torus radius r_0. It restricts the capability of a trap with fixed field and current magnitude to confine the plasma with pressure desired. The values p_0^{cr} in the toroidal "Belt" exceed the ones in the cylinder and increase, when the radius r_0 takes away. The total current in plasma and its positive (outside the separatrix) and negative (inside it) parts are obtained in computation under the condition (15). They increase proportional the pressure p_0. The singular point shift δr increases with p_0 about twice, when $0 < p_0 < p_0^{cr}$. The parameter Ψ_0 in (7) that means the magnetic flux between the separatrix and outer boundary, also increases with p_0 but weekly.

The restriction of the type (15) takes place in a large class of the mathematical models of reaction and diffusion interaction processes. Their mathematical nature is connected with the solution existence and uniqueness in boundary problems with semilinear elliptic type equations and with spectral analysis of their linearized operators(see [10] with detailed references).

The research was supported by the Russian Scientific Foundation (project 16-11-10278)

References

[1] Morozov A I 1992 *Sov. J. Plasma Phys.* B **18** 159
[2] Morozov A I and Savelyev V V 1998 *Phys.-Uspekhi.* **11** 11 1049
[3] Morozov A I and Frank A G 1994 *Plasma Phys. Report* **20** 879
[4] Brushlinskii K V and Ignatov P A 2010 *Comp. Math. and Mathematical Phys.* **50** 12 2071
[5] Brushlinskii K V, Goldich A S and Davydova N A 2016 *Math. Models and Comp. Simulation.* **7**
[6] Shafranov V 1958 *Sov. Phys. JETP* **6** 545
[7] Grad H and Rubin H 1959 *Proc. 2nd United Nations Int. Conf. on the Peaceful Use of Atomic Energy. Geneva* **31** 190
[8] Brushlinskii K V and Goldich A S 2013 *Vestnik National Nuclear Research University MEPHI* **2** 3 292
[9] Pieceman D W and Rachford H H 1955 *J. Soc. Industr. Appl. Math.* **3** 1 28
[10] Brushlinskii K V 2009 *Mathematical and computational problem in magnetogasdynamics* (Moscow: Binom)

On the magnetohydrodynamics flows in curved coaxial channels

K V Brushlinskii[1,4] and E V Styopin[2,3,5]

[1] Keldysh Institute of Applied Mathematics, Russian Academy of Sciences, Miusskaya sq. 4, 125047 Moscow, Russia
[2] National Research Nuclear University MEPHI, Kashirskoe sh. 31, 115409 Moscow, Russia
[3] Author to whom any correspondence should be addressed.

[4] brush@keldysh.ru
[5] styopin.evgeniy@gmail.com

Abstract. Stationary plasma flows obtained by using numerical solutions of corresponding stationary problems in curved coaxial channels are considered. It is supposed that plasma interacts with own azimuthal magnetic field. The main attention of the paper is focused on accelerating transonic flows, which attract a considerable interest in the theory of plasma accelerators. It was demonstrated that a curvilinear configuration of a channel makes the significant contribution to distributions of basic plasma parameters.

1. Introduction

1.1. Coaxial plasma accelerators
Research and development of plasma plants attract a considerable interest due to wide range of its applications to solve different technical problems. One of the most promising directions in the theory of plasma plants is the design of plasma accelerators due to high velocities and energies of working medium flows emerging from them. In terms of technical applications these devices can be used to design spaceship engines for cosmic flights and for its orbit adjustment.

Mathematical modeling and numerical simulation of magnetohydrodynamics flows in channels play a considerable role in the successful development of several generations of plasma accelerators [1]. It allows to obtain main regularities in plasma flow properties and thereby to reduce costs of expensive experiments.

For an analysis of a plasma motion in coaxial accelerators mathematical models in which plasma flow parameters depends only on cross section of the channel have been used up to the present moment [2]. Further development of the theory and expansion of perspectives of its applications is the construction and calculation of mathematical model of much less studied plasma flows in curved coaxial channels (figure 1). Some issues of the problem were discussed in [3], where, particularly, the first integrals of the stationary system of differential equations describing the process were obtained. In [4] plasma and gas motions in the channel in the form of a nozzle with curved upper and lower electrodes were considered. Particularly, a significant impact of the curvilinear configuration of the plant on the working fluid was noticed in this paper.

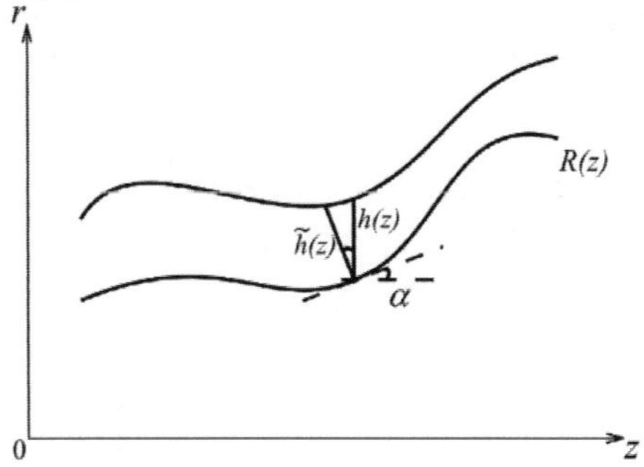

Figure 1. Scheme of the curved coaxial channel.

1.2. Objective of the work
The objective of the present work is the modeling and calculating of the accelerated dense plasma flows in curved coaxial channels in its interaction with own azimuthal magnetic field.

The basic regularities can be obtained by using a quasi-one-dimensional (hydraulic) approximation. The analysis is based on the solving of the corresponding stationary MHD problems subject to the boundary conditions. Neglecting, as usual in this class of physical problems, plasma thermal conductivity and viscosity, and assuming its electric conductivity is infinite, the mathematics of the MHD model is reduced to the algebraic equation – consequence of the first integrals of the stationary system of the differential equations which represent energy conservation laws.

2. Mathematical model

2.1. Quasi-one-dimensional (hydraulic) approximation
Plasma with specified density ρ_0, temperature T_0, and hence, with the pressure $P_0 = C_V(\gamma-1)\rho_0 T_0$, where γ – adiabatic index, C_V – thermal capacity at constant volume, enters the channel through its input section. It is natural to assume that the plasma enters the channel without rotation, i.e. its azimuthal velocity υ_φ at the channel entrance equals zero. The value of full electric current J in the system which flows through the left end of the accelerator is also assumed to be known. It means that the azimuthal magnetic field $H_{\varphi 0}$ is set in the channel input section.

In the present paper, the problem is considered in the quasi-one-dimensional (hydraulic) approximation. It is assumed that distance $h(z)$ between the electrodes is small, and all sought-for quantities are averaged over the cross section of the channel $S(z) = \pi(R(z)+h(z))^2 - \pi R(z)^2 \approx 2\pi R(z)h(z)$, where $R(z)$ – lower electrode configuration. It will be convenient to introduce $\tilde{S}(z) \approx 2\pi R(z)\tilde{h}(z)$, where $\tilde{h}(z) = h(z)\cos\alpha = h(z)\left(1+tg^2\alpha\right)^{-1/2}$.

The tangent of angle α is determined as derivative $R'(z)$ at a given place of the lower electrode (figure 1).

The electrodes forming the side walls of the channel are assumed to be impervious and equipotential, and the channel is formed by the trajectories $\upsilon_n = 0$. In mathematical language it is expressed in the following form:
$\upsilon_r / \upsilon_z = r_z' = tg\alpha$, where υ_r – radial component of the plasma velocity vector, υ_z – its longitudinal component.

Plasma velocity along the channel attracts physical interest for the analysis, so it also will be convenient to introduce $\tilde{\upsilon} = \left(\upsilon_z^2 + \upsilon_r^2\right)^{1/2} = \upsilon_z\left(1 + R'^2\right)^{1/2}$.

2.2. Stationary equations

In the quasi-one-dimensional approximation in the dimensionless form, neglecting aforementioned dissipative effects, stationary plasma flows in curved coaxial channels in its interaction with own azimuthal magnetic field and in the absence of the rotation at the channel entrance are described by the following ordinary differential equations resolved relatively to derivatives:

$$\begin{cases} \dfrac{d\rho}{dz} = -\dfrac{\rho}{\tilde{\upsilon}^2 - \tilde{C}_f^2}\left[\tilde{\upsilon}^2\dfrac{\tilde{S}'}{\tilde{S}} - 2C_\varphi^2\dfrac{R'}{R}\right] = -\dfrac{\rho}{\tilde{\upsilon}^2 - \tilde{C}_f^2}\left[\tilde{\upsilon}^2\dfrac{\tilde{h}'}{\tilde{h}} + \left(\tilde{\upsilon}^2 - 2C_\varphi^2\right)\dfrac{R'}{R}\right] \\ \dfrac{dH_\varphi}{dz} = -\dfrac{H_\varphi}{\tilde{\upsilon}^2 - \tilde{C}_f^2}\left[\tilde{\upsilon}^2\dfrac{\tilde{S}'}{\tilde{S}} + \left(C_T^2 - C_\varphi^2 - \tilde{\upsilon}^2\right)\dfrac{R'}{R}\right] = -\dfrac{H_\varphi}{\tilde{\upsilon}^2 - \tilde{C}_f^2}\left[\tilde{\upsilon}^2\dfrac{\tilde{h}'}{\tilde{h}} + \left(C_T^2 - C_\varphi^2\right)\dfrac{R'}{R}\right] \\ \dfrac{d\tilde{\upsilon}}{dz} = \dfrac{\tilde{\upsilon}}{\tilde{\upsilon}^2 - \tilde{C}_f^2}\left[\left(C_T^2 + C_\varphi^2\right)\dfrac{\tilde{S}'}{\tilde{S}} - 2C_\varphi^2\dfrac{R'}{R}\right] = \dfrac{\tilde{\upsilon}}{\tilde{\upsilon}^2 - \tilde{C}_f^2}\left[\left(C_T^2 + C_\varphi^2\right)\dfrac{\tilde{h}'}{\tilde{h}} + \left(C_T^2 - C_\varphi^2\right)\dfrac{R'}{R}\right] \end{cases} \quad (1)$$

where

- $C_\varphi = H_\varphi \rho^{-1/2}$ – Alfven velocity calculated by an azimuthal magnetic field;
- $C_T = \left(\gamma P/\rho\right)^{1/2}$ – gas-dynamic (thermal) sound velocity;
- $\tilde{C}_f^2 = C_T^2 + C_{A\varphi}^2$ – fast magnetosonic velocity square.

The units of measurement are the dimensional quantities participating in the formulation of the problem:

$$\rho_U = \rho_0,\ z_U = L,\ S_U = S(0), T_U = T_0, H_U = H_{\varphi 0},\ \upsilon_U^2 = \frac{H_{\varphi 0}^2}{4\pi\rho_0},\ P_U = \frac{H_{\varphi 0}^2}{4\pi}, \quad (2)$$

where L – the length of the channel.

The problem also includes the dimensionless parameter $\beta = 8\pi P_0 / H_{\varphi 0}^2$ traditionally participating in MHD. It shows, how many times a magnetic pressure differs from a gas-dynamic pressure.

The analysis of the equations (1) allows to make the conclusion that the curvilinear configuration of the channel enables the influence of the correlation between gas-dynamic (defined by C_T) and electromagnetic (defined by C_φ) flow properties. In terms of mathematics this assertion is manifested in the existence of the additional summands in the nominators of the basic plasma flow parameters.

2.3. The first integrals

Stationary system of MHD equations (1) has full set of the first integrals – laws of conservation:

$$C_1 = \rho\upsilon_z S; C_2 = \frac{P}{\rho^\gamma}; C_3 = C_1\upsilon_\varphi R; C_4 = \frac{C_1 H_\varphi}{R\rho}; C_5 = \frac{\tilde{\upsilon}^2 + \upsilon_\varphi^2}{2} + \frac{\gamma}{\gamma - 1}\frac{P}{\rho} + \frac{C_4}{C_1}H_\varphi R \quad (3)$$

Constant C_1 is proportional to the longitudinal plasma velocity in the input section of the channel υ_{z0}, which is defined based on the condition of a transonic accelerated flow regime receiving. Values of other constants are determined by the boundary conditions in the input section of the channel.

We can express from the equations (3) all plasma parameters via its density ρ and obtain one algebraic equation which defines ρ as a function of z and C_1:

$$\Phi(\rho,z,C_1)=\rho^2\left(C_5-\frac{\gamma}{\gamma-1}C_2\rho^{\gamma-1}-\frac{C_4^2\rho R^2(z)}{C_1^2}\right)-\frac{C_1^2}{2\tilde{S}^2(z)}=0 \quad (4)$$

The solution of the equation (4) completely describes a quasi-one-dimensional stationary MHD flow in case of infinite plasma conductivity.

2.4. *Method of calculation*

The following algorithm is used as a numerical method of calculation of stationary transonic accelerated plasma flows in curved coaxial channels:

1. In order to determine plasma density in the transition point in the center of the channel and the plasma stream inflow velocity in which the studied transonic flow regime is obtained, the following system of equations is numerically solved:
$$\begin{cases}\Phi\left(\rho_{0.5},0.5,\upsilon_{z0}\right)=0\\ \dfrac{\partial\Phi}{\partial\rho}\left(\rho_{0.5},0.5,\upsilon_{z0}\right)=0\end{cases} \quad (5)$$
2. By defining inflow velocity, the problem of forming the plasma flow density distribution $\rho(z)$ along the channel is numerically solved by using the equation $\Phi(\rho,z,\upsilon_{z0})=0$.
3. The other sought-for plasma parameters distributions along the channel, namely, $\tilde{\upsilon}(z)$ and $H_\varphi(z)$, are built using the first integrals equations (3) and plasma density distribution $\rho(z)$ obtained.

3. Results

In order to analyze plasma flows in real plasma plants without using two-dimensional calculations, it is expedient to subdivide the channel area into thin tubes. The quasi-one-dimensional mathematical model of the plasma flow motion in curved thin tubes was created in the present paper. The tubes configuration into which the channel is subdivided is completely defined by the configurations of upper and lower electrodes of the plant.

It is important to note that the boundary condition on the azimuthal magnetic field and the value of the dimensionless parameter β are individual for each tube because of the condition $H_\varphi r=const$ in the two-dimensional problem. At the time boundary conditions in the input section of each tube have the following form:

$$\rho=1,T=1,\upsilon_{\varphi0}=0,H_{\varphi0}=\frac{r_0}{R(0)}, \quad (6)$$

where r_0 – plant characteristic radius and $R(0)$ – configuration of the lower boundary of the tube at the entrance of the channel. At the same time we choose $\beta=0.1\left(R(0)/r_0\right)^2$, because electromagnetic properties exceed gas-dynamic properties in real plasma plants.

3.1. *Plasma accelerators with curved upper electrode*

We consider plasma accelerator with the following configuration:

$$electrode_{upper}=1.2(z-0.5)^2+0.4;electrode_{lower}=0.3;r_0=0.5 \quad (7)$$

In order to analyze transonic accelerated plasma flows, the accelerator area is subdivided into thin tubes corresponding to the flow streams (figure 2). Parameter a defines a tube curvature and its location in the channel area.

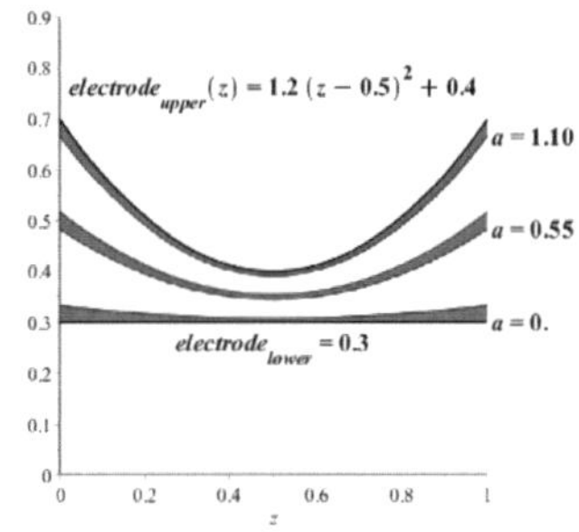

Figure 2. Subdivision of plasma accelerator with curved upper electrode into thin tubes.

In figure 3, the distributions of the basic plasma parameters in the transonic accelerated flow regime in the thin tubes corresponding to the different parts of plasma accelerator with curved upper electrode are represented.

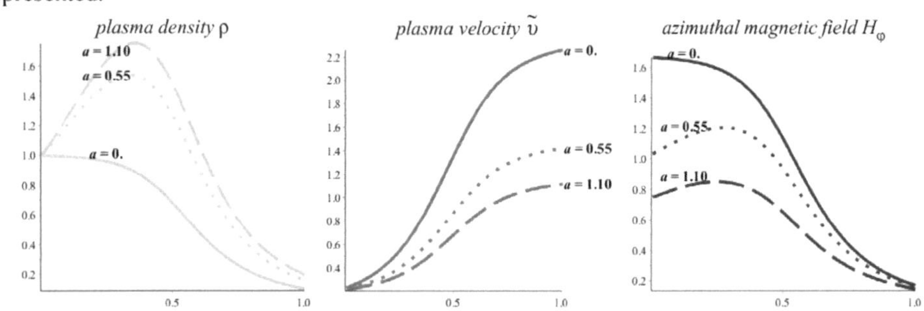

Figure 3. Distributions of basic plasma parameters in the set of thin tubes in plasma accelerator with curved upper electrode.

Plasma compression zone is formed and then is amplified during the transfer from lower electrode to curved upper electrode: the maximum of the plasma density distribution occurs and then increases moving to the channel center. The outflow plasma density is also increases. Accelerating characteristics of the channel get worse: outflow plasma velocity decreases. During the transfer from lower electrode to curved upper electrode inverse currents in plasma are generated because of forming regions of intensity rise. At the same time the full electric current quantity defined by the difference between output and input azimuthal magnetic field intensity values decreases.

3.2. Plasma accelerators with curved lower electrode

We consider plasma accelerator with the following configuration (figure 4):

$$electrode_{upper} = 0.5, electrode_{lower} = -0.8(z-0.5)^2 + 0.3, r_0 = 0.3 \tag{8}$$

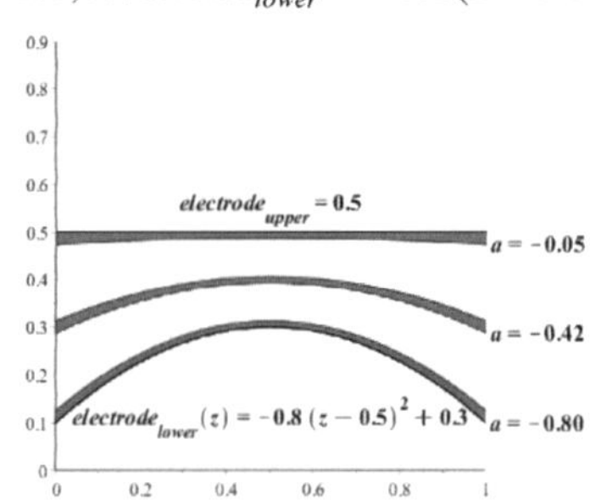

Figure 4. Subdivision of plasma accelerator with curved lower electrode into thin tubes.

In figure 5, the distributions of the basic plasma parameters in the transonic accelerated flow regime in the thin tubes corresponding to the different parts of plasma accelerator with curved lower electrode are given.

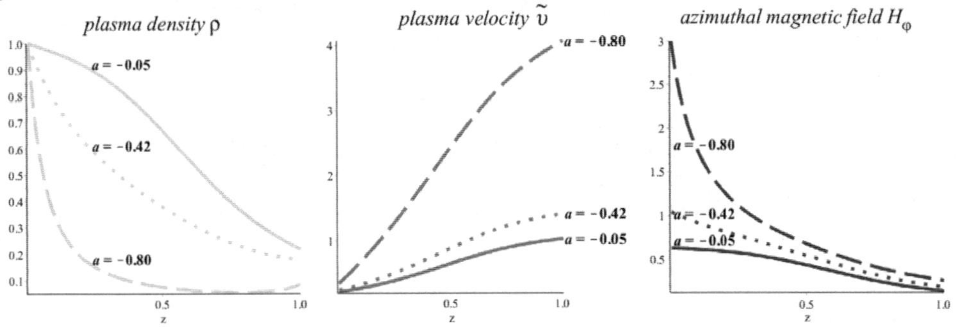

Figure 5. Distributions of basic plasma parameters in the set of thin tubes in plasma accelerator with curved lower electrode.

Plasma density decreasing, and hence, thermal energy consumption become more intensive during the transfer from upper electrode to curved lower electrode. At the same time ill-defined zone of plasma rarefaction occurs near the lower electrode. Azimuthal magnetic field intensity drop, and hence, electromagnetic energy consumption increase because of input magnetic field intensity augmentation. Along with the increasing of electromagnetic pressure influence, the considered regularities improve accelerating characteristics of the channel: plasma is accelerated significantly more intensive near the lower electrode.

4. Conclusion

- In this work, a quasi-one-dimensional stationary MHD model of a plasma flow in curved narrow coaxial channels in its interaction with own azimuthal magnetic field has been considered and realized.
- It has been established that a curvilinear geometry of a channel enables the influence of the correlation between gas-dynamic and electromagnetic properties of a process on the flow parameters.
- By using method of the subdivision of a plasma accelerator area into thin tubes, it has been demonstrated that plants with a curved lower electrode is preferable to plants with a curved upper electrode in terms of the acceleration effectiveness.
- It has been established that in curved channels in contrast to channels with constant average radius, depending on a channel geometry and a correlation between gas-dynamic and electromagnetic pressures, areas of plasma compression and rarefaction, areas of the electric current generation and inverse currents can occur in a transonic accelerated plasma flow regime.

Acknowledgments
The study was performed by the grant from the Russian Science Foundation (project № 16-11-10278).

References

[1] Morozov A I 1978 *Physical Principles of Space Electric Propulsion Thrusters* (Moscow: Atomzidat)
[2] Brushlinskii K V 2009 *Mathematical and Computational Problems in Magnetohydrodynamics* (Moscow: Binom)
[3] Morozov A I and Solov'ev L S 1974 *Reviews of Plasma Physics* (Moscow: Atomizdat) pp 3–87
[4] Styopin E V 2015 *Vestnik natsional'nogo issledovatel'skogo yadernogo universiteta «MIFI»* **4** 407-20

Analysis of the incomplete Galerkin method for modelling of smoothly-irregular transition between planar waveguides

D Divakov[1], L Sevastianov[1,2] and N Nikolaev[1]

[1] RUDN University, Moscow, Russia
[2] Joint Institute for Nuclear Research, Dubna, Moscow region, Russia

E-mail: dmitriy.divakov@gmail.com, leonid.sevast@gmail.com, nnikolaev@sci.pfu.edu.ru

Abstract. The paper deals with a numerical solution of the problem of waveguide propagation of polarized light in smoothly-irregular transition between closed regular waveguides using the incomplete Galerkin method. This method consists in replacement of variables in the problem of reduction of the Helmholtz equation to the system of differential equations by the Kantorovich method and in formulation of the boundary conditions for the resulting system. The formulation of the boundary problem for the ODE system is realized in computer algebra system Maple. The stated boundary problem is solved using Maples libraries of numerical methods.

1. Introduction

The paper deals with a numerical solution to the problem of waveguide propagation of polarized light in smoothly-irregular transition between closed regular waveguides using the incomplete Galerkin method [1, 2]. This method consists in replacement of variables in the problem of reduction of the Helmholtz equation to the system of differential equations by the Kantorovich method [3] and in formulation of the boundary conditions for the resulting system. The Kantorovich expansion is made in the system of functions that satisfy the reduced boundary conditions [3]. Coefficient functions of the Kantorovich expansion are the desired values of the boundary problem. The formulation of the boundary problem for the ODE system is realized by using computer algebra system Maple. The stated problem is solved using Maples libraries of numerical methods.

2. Statement of the problem

Consider smoothly-irregular along z closed waveguide transition between closed planar regular waveguides. The lower and upper boundaries of the transition are given by the equations $x = 0$ and $x = h(z)$. The condition of smooth irregularity of the waveguide transition is represented by continuity of $h(z)$ and $h'(z)$ for all z.

Function $h(z)$ describes the variable height of irregular waveguide transition, and in order to meet the conditions of smooth irregularity it is necessary that $h(0) = h_1$, $h(d) = h_2$ and $h'(0) = h'(d) = 0$. Let us consider the problem of propagation of polarized electromagnetic field in the described structure made of material with refractive index n_f.

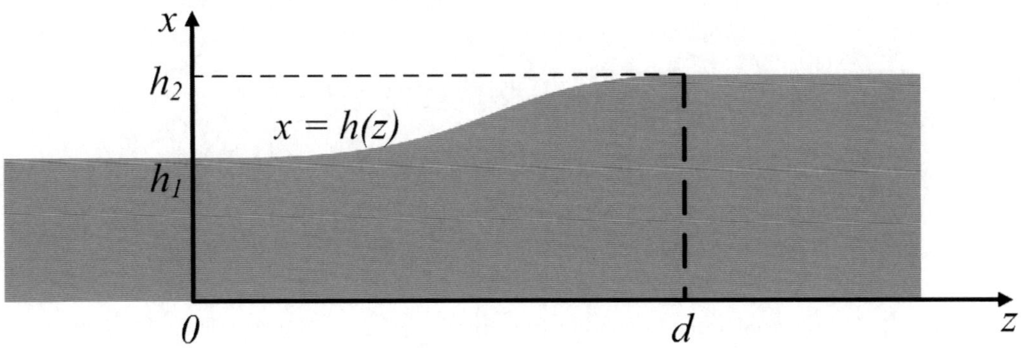

Figure 1. Waveguide transition between two closed planar waveguides of constant cross-section.

Electromagnetic field propagating in the waveguide satisfies the Maxwell's equations, constitutive relations and boundary conditions [4, 5]. For the described planar waveguide structure (Fig. 1) the Maxwells equations can be reduced to two independent subsystems with respect to the desired component of the electromagnetic field – subsystems for TE- and TM-modes. Subsystem for TE-modes can be formulated as the Helmholtz differential equation with respect to the component E_y and two differential equations relating the components H_x and H_z with E_y. Let us denote E_y by $u = u(x,\, z)$ and write down the Helmholtz differential equation[1]

$$\left(\frac{\partial^2}{\partial x^2} + \frac{\partial^2}{\partial z^2} + k^2\right) u = 0 \tag{1}$$

with coefficient $k^2 = (2\pi n_f/\lambda)^2$ (λ is wavelength), and the boundary conditions of the first kind[2]:

$$u|_{x=0} = u|_{x=h(z)} = 0 \tag{2}$$

The function $u = u(x,\, z)$ meets the conditions of excitation and emission at infinity[2]:

$$u|_{z\leq 0} = \sum_{n=1}^{\infty} R_n e^{-i\gamma_n z} \sin\left(\frac{\pi n x}{h_1}\right) + \; A e^{i\gamma_{n_0} z} \sin\left(\frac{\pi n_0 x}{h_1}\right) \tag{3}$$

$$u|_{z\geq d} = \sum_{n=1}^{\infty} T_n e^{i\Gamma_n z} \sin\left(\frac{\pi n x}{h_2}\right) \tag{4}$$

where $\gamma_n = k^2 - \left(\frac{\pi n}{h_1}\right)^2$, $\Gamma_n = k^2 - \left(\frac{\pi n}{h_2}\right)^2$, A is the given amplitude of the incident mode of number n_0, R_n and T_n are unknown amplitude coefficients.

According to the algorithm proposed in [1, 2], we introduce new variables which transform an irregular area in the strip:

$$\begin{aligned} \xi(x,\, z) &= x/h(z) \\ \eta(x,\, z) &= z. \end{aligned} \tag{5}$$

In the new coordinates the Helmholtz equation takes the form:

$$\left(\frac{1}{h^2(\eta)} + \xi^2 b^2(\eta)\right) \times \frac{\partial^2 u}{\partial \xi^2} - 2\xi b(\eta) \times \frac{\partial^2 u}{\partial \xi\, \partial \eta} + \frac{\partial^2 u}{\partial \eta^2} + \xi c(\eta) \times \frac{\partial u}{\partial \xi} + k^2 u = 0 \tag{6}$$

where $b(\eta) = h'(\eta)/h(\eta)$, $c(\eta) = b^2(\eta) - b'(\eta)$. The boundary conditions (2) and the conditions of excitation and emission at infinity (3), (4) in the new coordinates take the following form:

$$u|_{\xi=0} = u|_{\xi=1} = 0 \tag{7}$$

$$u|_{\eta\le 0} = \sum_{n=1}^{\infty} R_n e^{-i\gamma_n \eta} \sin(\pi n\xi) + A e^{i\gamma_{n_0}\eta} \sin(\pi n_0 \xi) \tag{8}$$

$$u|_{\eta\ge d} = \sum_{n=1}^{\infty} T_n e^{i\Gamma_n \eta} \sin(\pi n\xi) \tag{9}$$

Following [1, 2], we apply the Kantorovich method [3] for the formulation of a system of differential equations and boundary conditions for this system. We use the eigenfunctions of the waveguide problem in the zeroth approximation by $h'(\eta)$ as the basis functions of the Kantorovich method, and we justify such a choice for the problem in the zeroth approximation.

3. Zeroth approximation

Consider the case, when $h'(\eta)$ is a small quantity: $h'(\eta) \le \delta \ll 1$. Equation (6) in the zeroth approximation by the parameter δ takes the form:

$$-h^2(\eta)\left(\frac{\partial^2 u}{\partial \eta^2} + k^2 u\right) = \frac{\partial^2 u}{\partial \xi^2} \tag{10}$$

We seek a solution of (10) that can be represented as a product $u(\xi,\eta) = V(\eta)\varphi(\xi)$. We substitute the estimated form of solution to the equation (10) and divide the equation by $V\varphi$. The resulting equation must be satisfied for all values of the variables, for $u(\xi,\eta) = V(\eta)\varphi(\xi)$ to be a solution of (10), which means:

$$-h^2(\eta)\left(\frac{V''}{V}(\eta) + k^2\right) = \frac{\varphi''}{\varphi}(\xi) = -\lambda \tag{11}$$

where λ is an unknown constant. Relation (11) and conditions (7) formulate the Sturm-Liouville problem:

$$\begin{aligned} &\varphi'' + \lambda\varphi = 0, \\ &\varphi(0) = \varphi(1) = 0. \end{aligned} \tag{12}$$

Eigenvalues of the problem (12), given by $\lambda_n = (\pi n)^2$ where $n = 1, 2, 3 \ldots$, correspond to eigenfunctions (defined with an accuracy up to a factor), which take the form:

$$\varphi_n(\xi) = \sin\left(\sqrt{\lambda_n}\xi\right), \quad n = 1, 2, 3 \ldots \tag{13}$$

Functions (13) depend on z parametrically, and the system of functions (13) is orthogonal for all z from $[0,\ d]$ and coincides with the system of eigenfunctions of the left regular waveguide at $z = 0$ and with the system of eigenfunctions of the right regular waveguide at $z = d$.

We use the system of functions (13) as the basis in the Kantorovich method [3], because the functions (13) satisfy the exact boundary conditions (7).

4. Reduction to the boundary problem for ODE system

We seek an approximate solution to the equation (6) in the form of a partial sum of the series in the system of functions (13) satisfying the boundary conditions (7):

$$u^N(\xi,\eta)=\sum_{n=1}^{N}V_n(\eta)\sin(\pi n\xi) \tag{14}$$

where N is the number of terms in the partial sum of the series.

We substitute the solution (14) into the equation (6) and the boundary conditions (8), (9) and apply projection scheme of the Galerkin method [6, 7]. As a result we get the boundary problem for the system of $N\times N$ ordinary differential equations of the second order:

$$\mathbf{A}(\eta)\vec{v}''+\mathbf{P}(\eta)\vec{v}'+\mathbf{Q}(\eta)\vec{v}=\vec{0} \tag{15}$$

$$\begin{aligned}&(\vec{v}'+i\mathbf{D}_\gamma\vec{v})|_{\eta=0}=2i\gamma_{n_0}\vec{\delta}_{n,n_0}A/h_1\\&(\vec{v}'-i\mathbf{D}_\Gamma\vec{v})|_{\eta=d}=0\end{aligned} \tag{16}$$

where $\vec{v}=(V_1(\eta),V_2(\eta),\ldots,V_N(\eta))^T$ is the vector of the desired amplitude (coefficient) functions of the Kantorovich expansion, $\mathbf{A}(\eta)$, $\mathbf{P}(\eta)$, $\mathbf{Q}(\eta)$ are matrix functions, elements of which are calculated using Maple, $\mathbf{D}_\gamma=diag\{\gamma_n\}_{n=1}^N$, $\mathbf{D}_\Gamma=diag\{\Gamma_n\}_{n=1}^N$, $\vec{\delta}_{n,n_0}$ is a vector with zero components, with the exception of one, standing on n_0-th position.

The described algorithm of stating the boundary problem (15), (16) by the incomplete Galerkin method is realized in computer algebra system Maple. The boundary problem (15), (16) is being solved numerically in Maple, the results of calculations are presented below.

5. Numerical calculation

We solve numerically the problem of waveguide propagation of monochromatic light of wavelength $0,55\,\mu$ in the waveguide transition made of material with refractive index $n_f=1,51$. Function $h(z)$ is third-degree polynomial and its value changes smoothly from $h_1=0,70\,\mu$ to $h_2=0,92\,\mu$ along the interval of length $d=2,00\,\mu$. In the waveguide of the thickness $h_1=0,70\,\mu$ there exist 3 non-evanescent modes, in the waveguide of the thickness $h_2=0,92\,\mu$ - 5 non-evanescent modes.

We consider two cases: in the first case the first non-evanescent mode ($n_0=1$) is incident on the described structure, in the second case the first evanescent mode ($n_0=4$) is incident on the structure. The total number of modes in the expansion $N=16$. The boundary problem is solved by finite difference scheme with partition into 2048 points using Maple function `dsolve`.

The numerical calculation (see Fig. 2) shows the evolution of the amplitude functions of all allowable modes when the mode with the number $n_0=1$ is incident on the irregular transition. We can observe that the second and the third non-evanescent (or guided) modes are excited. Amplitude of incident mode ($n_0=1$) decreases because of the energy redistribution between guided modes of irregular transition.

The second example (see Fig. 3) shows the evolution of the amplitude functions in the case when the first evanescent mode ($n_0=4$) is incident on the irregular transition. Amplitude of the incident evanescent mode decreases with the thickness of the waveguide decreasing till the critical value. When the thickness of the waveguide becomes less than the critical thickness, the evanescent mode becomes the guided mode with decreased amplitude. And as in the first example, other guided modes are excited in the waveguide as well.

The results of numerical experiments in both examples are in good agreement with the theoretical expectation of the behavior of waveguide modes.

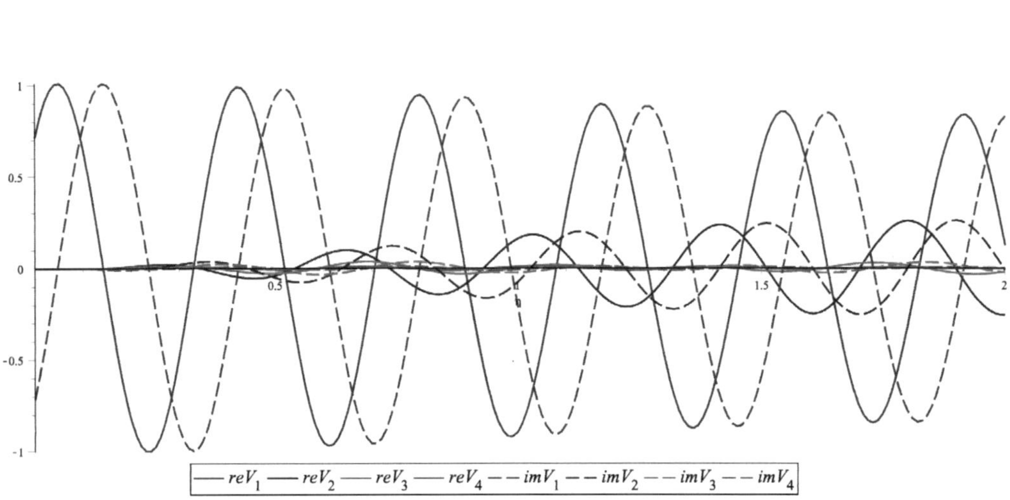

Figure 2. First four amplitude (coefficient) functions: real parts are shown by solid lines, imaginary parts – by dashed lines. Incident mode has the number $n_0 = 1$ and amplitude $A = (i+1)/2$.

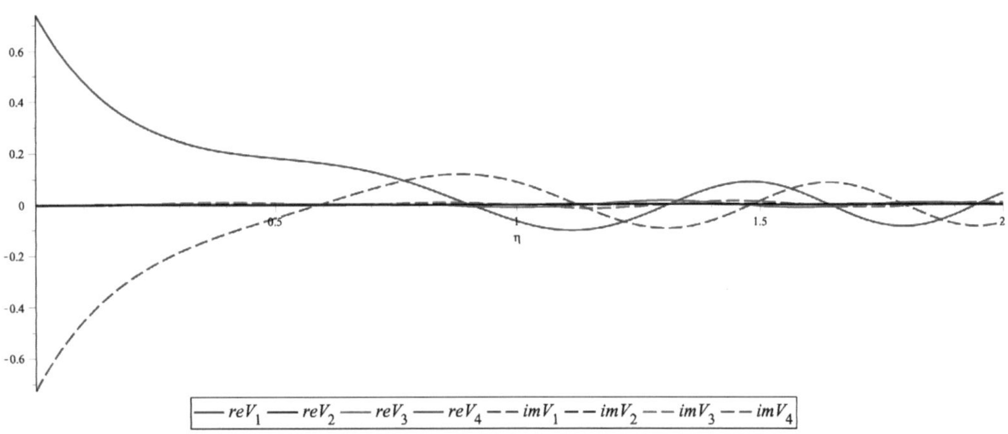

Figure 3. First four amplitude (coefficient) functions: real parts are shown by solid lines, imaginary parts - by dashed lines. Incident mode has the number $n_0 = 4$ and amplitude $A = (i+1)/2$.

6. Conclusion

The paper considered using of the incomplete Galerkin method to calculate the waveguide propagation of monochromatic polarized light in closed waveguide structures with irregularity along the propagation direction. Algorithms for symbolic transformations of the original problem for the Helmholtz equation to a boundary problem for ODE system and numerical solution of the reduced boundary problem are implemented using computer algebra system Maple. The numerical experiments show that using of eigenfunctions of waveguide problem in zeroth approximation as the basis in the Kantorovich method is adequate to the problem. The numerical results show qualitative agreement with the theoretically expected behavior of waveguide modes and evanescent modes.

Authors considered the waveguide propagation of monochromatic polarized light in an open

IOP Conf. Series: Journal of Physics: Conf. Series **788** (2017) 012010 doi:10.1088/1742-6596/788/1/012010

irregular transition between two regular planar open waveguides in [8, 9]. Comparing the results of current research and the results from [8, 9] shows qualitative agreement and the ability to generalize the incomplete Galerkin method from the closed to the open irregular waveguide transitions.

Acknowledgments

The work was partially supported by RFBF grants No 14-01-00628, No 15-07- 08795, No 16-07-00556. The reported study was funded within the Agreement No 02.a03.21.0008 dated 24.04.2016 between the Ministry of Education and Science of the Russian Federation and RUDN University.

References

[1] Sveshnikov A G 1963 On the proof of a method of calculation for irregular waveguides *Zh. Vychisl. Mat. Mat. Fiz.* **3**:1 170–79
[2] Sveshnikov A G 1962 Calculations for matching of plane waveguides *Zh. Vychisl. Mat. Mat. Fiz.* **2**:1 175–79
[3] Kantorovich L V and Krylov V I 1964 *Approximate Methods of Higher Analysis* (Interscience, New York)
[4] Tamir T 1990 *Guided-Wave Optoelectronics* (Springer-Verlag, Berlin)
[5] Adams M J 1981 *An Introduction to Optical Waveguides* (New York, Wiley)
[6] Fletcher C 1984 *Computational Galerkin Methods* (Springer, N.-Y.)
[7] Mikhlin S G 1971 *The numerical performance of variational methods* (Wolters-Noordhoff Publ. Groningen)
[8] Divakov D V and Sevastianov L A 2015 Application of incomplete Galerkin method to irregular junction in open planar waveguides *Matem. Mod.* **27**:7 44–50
[9] Divakov D V, Sevastianov L A and Nikolaev N E 2016 Modeling Open Transitions of the Horn Type between Open Planar Waveguides *EPJ Web of Conf.* V **108** 02020

Asymptotic solutions of a fourth–order analogue for the Painlevé equations

I Yu Gaiur, N A Kudryashov
National Research Nuclear University MEPhI, Kashirskoe sh. 31, Moscow, 115409 Russia
E-mail: iygayur@gmail.com

Abstract. Asymptotic solutions of a fourth–order analogue for the Painlevé equations that is self–similar reduction of the modified Sawada–Kotera and Kaup–Kupershmidt equation is considered. The Boutroux variables of two types have been found which allows us to find asymptotic solutions of the equation in the neighbourhood of the infinity. It was shown that asymptotic of self–similar solution for the modified Sawada–Kotera and Kaup–Kupershmidt equations can be determined as solutions of autonomous differential equations. Asymptotic solutions expressed by elementary functions have been found too. Besides asymptotic solutions expressed by logarithmic derivative of two elliptic Weierstrass functions have been found. Connection between obtained asymptotic solutions and asymptotic solutions of the Sawada–Kotera and Kaup–Kupershmidt equations has been discussed.

1. Introduction

In this paper we consider a fourth–order analogue of Painlevé equations in the form [1]:

$$w_{zzzz} + 5w_z w_{zz} - 5w^2 w_{zz} - 5ww_z^2 + w^5 - zw - \beta = 0. \tag{1}$$

This equation can be obtained taking into account the self–similar variables

$$u = \frac{1}{(5t)^{\frac{1}{5}}} w(z), \quad z = \frac{x}{(5t)^{\frac{1}{5}}}, \tag{2}$$

from the following nonlinear partial differential equation

$$\frac{\partial u}{\partial t} - \frac{\partial}{\partial x}\left(u_{xxxx} - 5u_x u_{xx} + 5uu_x^2 + 5u^2 u_{xx} - u^5\right) = 0. \tag{3}$$

Equation (3) is the modified Kaup–Kupershmidt and Sawada–Kotera equations. It is known that solutions of the Kaup–Kupershmidt and Sawada–Kotera equations can be found via the solutions of equation (3) by means of the Miura transformations. These equations plays significant role in the description of non–linear waves in a liquid with gas bubbles and in other application [2, 3].

Equation (1) is one of the fourth–order analogues of the Painlevé equations [4, 5]. It is known that solutions of equation (1) can determine new special functions [1, 7, 8]. It was found that solutions of equation (1) can be obtained using the Bäcklund transformations at some values of parameters [9]. The Cauchy problem for equation (1) can be solved using the Lax pair [10].

There are some rational solutions of equation (1) expressed via special polynomials at some values of parameters too [11, 12]. The Lagrangian description of equation (1) was obtained in work [13]

Asymptotic solutions of the first and the second Painlevé equations [14, 15] in the complex domain were obtained by Piere Boutroux in his classical work [16]. At the present time there are a lot of works where asymptotic properties of the solutions for Painlevé equations were studied. One of approaches for finding the Boutroux variables of the Painleve equations was proposed in work by A. Bruno [17], that is also known as the three dimensional power geometry method. Using this method the author has found asymptotic solutions determined in terms of elliptic and periodic functions in neighborhood of infinity. Investigation of the asymptotic properties of solutions for the Painlevé equations in the complex domain using the method of isomonodromy deformation is presented in works s [18, 19, 20, 21].

Rational solutions of equation (1) have been found in paper [22]. Asymptotic solutions of equation (1) determined as series by the powers of z have been obtained in paper [23]. However, asymptotic solutions determined in the Boutroux variables have not been found yet.

The aim of this paper is to find the Boutroux variables of equation (1) using power geometry methods and to determine Boutroux asymptotic solutions in the neighborhood of infinity.

2. Boutroux variables for asymptotic solutions of equation (1)

To determine asymptotic solutions of equation (1) we are going to look for solutions in the following form

$$w(z) = z^{\alpha} f\,(t)\,, \quad t = \frac{z^{\gamma}}{\gamma}. \tag{4}$$

The main idea of the following approach to choose real parameters α and γ thus substitution (4) transform equation (1) to the next form

$$F(f,\, f_t,\, f_{tt},\, f_{ttt},\, f_{tttt}) + \sum_{i\in\mathbb{Z}_+} \frac{G_i(f,\, f_t,\, f_{tt},\, f_{ttt},\, f_{tttt})}{t^i} = 0. \tag{5}$$

Thus we can propose that damping asymptotic solutions can be determined as solution of

$$F(f,\, f_t,\, f_{tt},\, f_{ttt},\, f_{tttt}) = 0, \tag{6}$$

when $|t| \to \infty$. It has been found that there are two types of the Boutroux variables for equation (1). The first one is

$$w(z) = \frac{f(t)}{z}, \quad t = \frac{4}{5} z^{\frac{5}{4}} \tag{7}$$

and the second one is

$$w(z) = z^{\frac{1}{4}} v\,(t)\,, \quad t = \frac{4}{5} z^{\frac{5}{4}}. \tag{8}$$

Substitutions (7) and (8) transforms equation (1) to the form (5).

3. Trigonometric asymptotic solutions of equation (1)

Substituting (7) into equation (1) we have the following equality

$$\begin{gathered} w_{zzzz} + 5w_z w_{zz} - 5w^2 w_{zz} - 5ww_z^2 + w^5 - zw - \beta = \\ = f_{tttt} - f - \beta - 2\frac{f_{ttt} - f_t f_{tt}}{t} - \frac{16ff_{tt} - 27f_{tt} + 28f_t^2 + 16f^2 f_{tt} + 16ff_t^2}{5t^2}. \end{gathered} \tag{9}$$

At $|t| \to \infty$, we obtain that the asymptotic solution of equation (1) can be determined from the following equation

$$f_{tttt} - f - \beta = 0. \tag{10}$$

General solution of equation (10) can be written in form

$$f(t) = -\beta + C_1 e^{it} + C_2 e^{-it} + C_3 e^{-t} + C_4 e^{t}. \tag{11}$$

To obtain asymptotic solution of equation (1) we should choose constants in such that we obtain solution with grow behaviour less than the first power of t

For example, to obtain asymptotic solutions of equation (1) on the real axes we should choose constants in the following way

$$C_4 = 0, \quad t \to \infty. \tag{12}$$

Thus, we obtain asymptotic solution on real axis of equation (1) in the following form

$$w_1(z) = \frac{-\beta + C_1 e^{i\frac{4}{5}z^{5/4}} + C_2 e^{-i\frac{4}{5}z^{5/4}} + C_3 e^{-\frac{4}{5}z^{5/4}}}{z}, \quad z \to +\infty, \tag{13}$$

The asymptotic solution (13) is illustrated in Fig. 1.

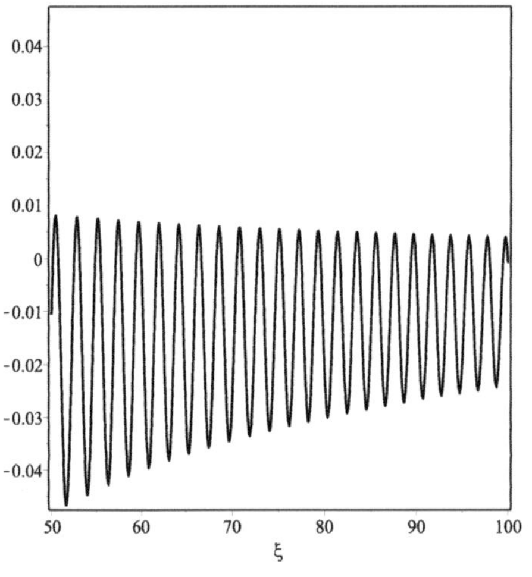

Figure 1. Trigonometric asymptotic solution (13) of equation (1) at $\beta = 1$, $C_1 = C_2 = C_3 = 1$, $z \to \infty$.

4. Elliptic asymptotic solutions of equation (1)

Taking into account transformation (8) we obtain the following equality

$$w_{zzzz} + 5w_z w_{zz} - 5w^2 w_{zz} - 5ww_z^2 + w^5 - zw - \beta =$$

$$= v_{tttt} + 5v_t v_{tt} - 5vv_t^2 - 5v^2 v_{tt} + v^5 - v - \frac{5v^2 v_t - vv_{tt} + \frac{4}{5}\beta - 3v_t^2 - 2v_{ttt}}{t} - \tag{14}$$

$$-\frac{3v_{tt} - 2v^3}{5t^2} - \frac{3v^2 - 15v_t}{25t^3} - \frac{231}{625}\frac{v}{t^4} = 0$$

At $|t| \to \infty$, we obtain asymptotic solution of (1) determined by the following equation

$$v_{tttt} + 5v_t v_{tt} - 5vv_t^2 - 5v^2 v_{tt} + v^5 - v = 0. \tag{15}$$

Equation (15) has solution in form of elliptic function. This asymptotic solution of (14) can be written as [24]

$$v(t) = \frac{d}{dt} \ln \left[\Theta(t)\right], \quad \Theta(t) = \eta(t) - \mu(t) \tag{16}$$

where η and μ are any two solutions of the equation

$$y_{tt} = 6y^2 - \frac{1}{24}. \tag{17}$$

The solution of equation (17) is the Weierstrass elliptic function $\wp\left(t, \frac{1}{12}, g_3\right)$, where g_3 is an arbitrary constant. The asymptotic solution of equation (1) takes the form

$$w_3(z) = z^{\frac{1}{4}} \frac{\wp'\left(\frac{4}{5}z^{\frac{5}{4}} - z_1, \frac{1}{12}, g_3\right) - \wp'\left(\frac{4}{5}z^{\frac{5}{4}} - z_2, \frac{1}{12}, \tilde{g}_3\right)}{\wp\left(\frac{4}{5}z^{\frac{5}{4}} - z_1, \frac{1}{12}, g_3\right) - \wp\left(\frac{4}{5}z^{\frac{5}{4}} - z_2, \frac{1}{12}, \tilde{g}_3\right)} \tag{18}$$

where z_1, z_2, g_3, $\tilde{g}_3$ - are arbitrary constants. The asymptotic solution (18) is illustrated in Fig. 2.

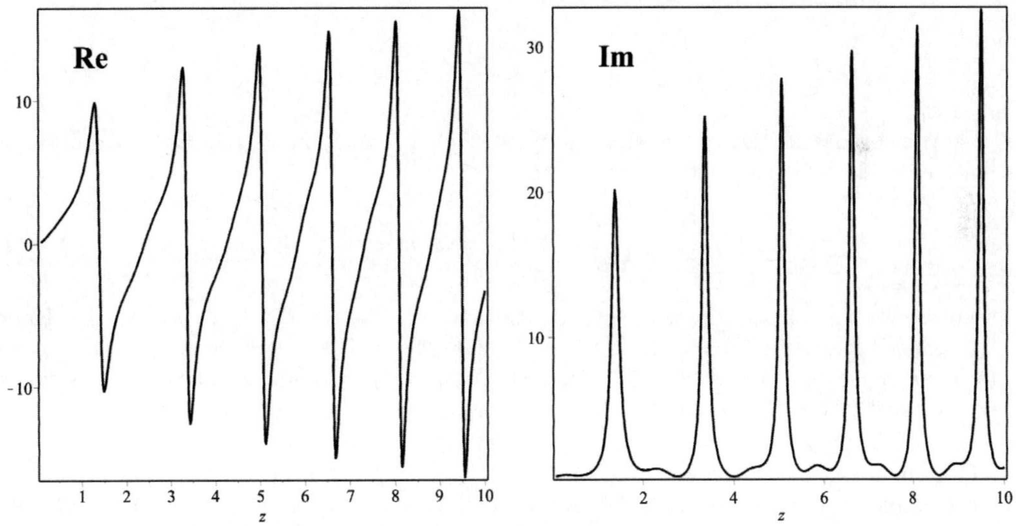

Figure 2. Real and imaginary part of solution (18) at $g_3 = 4$, $\tilde{g}_3 = 0$, $z_1 = 2i$, $z_2 = 2i + 4$.

5. The relationship with the Savada–Kotera equation and the Kaup–Kupershmidt equation.

It is also known that solutions of equation (3) are connected with the Savada–Kotera and the Kaup–Kupershmidt equations by the Miura transformation. Let us demonstrate the connection between equation (1) and self—similar solutions of the Savada–Kotera equation and the Kaup–Kupershmidt equation.

The Savada–Kotera equation and the Kaup–Kupershmidt equation can be written in form [25]

$$u_t + u_{xxxxx} + 10uu_{xxx} + 5\nu u_x u_{xx} + 20u^2 u_x = 0, \tag{19}$$

where we take $\nu = 2$ for the Kaup–Kupershmidt equation [26], and $\nu = 5$ for the Savada–Kotera equation [27, 28]. Using self–similar variables

$$u = \frac{1}{(5t)^{\frac{2}{5}}} g(z), \quad z = \frac{x}{(5t)^{\frac{1}{5}}}, \tag{20}$$

we obtain the following differential equation

$$g_{zzzzz} + 10gg_{zzz} + 5\nu g_z g_{zz} + 20g^2 g_z - 2g - zg_z = 0. \tag{21}$$

Using the Miura transformation in the case $\nu = 5$ for the Sawada–Kotera equation in the form

$$g = w_z - \frac{w^2}{2} \tag{22}$$

we obtain

$$\left(\frac{d}{dz} - w\right) \frac{d}{dz} f(z, w) = 0. \tag{23}$$

where $f(z, w)$ is differential sum of equation (1).

In the case of the Kaup–Kupershmidt equation the Miura transformation takes form

$$g = -\frac{w_z + w^2}{2}. \tag{24}$$

Using transformation (24) we obtain that self–similar reduction of the Kaup–Kupershmidt equation can be written in the following form

$$\frac{1}{2}\left(\frac{d}{dz} + 2w\right) \frac{d}{dz} f(z, w) = 0. \tag{25}$$

Thus, we can obtain the asymptotic solutions of the Savada–Kotera and the Kaup–Kupershmidt equations (19) using formulae (22) and (24). This relationship gives us opportunity to obtain asymptotic solutions of the SK and KK equations by using Miura on the obtained asymptotic solutions (18) and (13).

6. Conclusion

In this work we have considered a fourth–order analogue for the Painlevé equations. This equation can be also obtained as a self–similar reduction of the modified Sawada–Kotera and Kaup–Kupershmidt equations. We have found two types of the Boutroux variables for equation (1). We have constructed asymptotic solutions of equation (1) via trigonometric functions (13) and Weierstrass elliptic functions (18). We have also demonstrated that asymptotic self–similar solutions of the Sawada–Kotera and Kaup–Kupershmidt equations can be obtained via asymptotic solutions (13) and (18) by means of the Miura transformation.

7. Acknowledgement

This research was supported by Russian Science Foundation grant No. 14–11–00258.

References

[1] Kudryashov N A 2002 *J. Phys.* A: Math. Gen. **35** 4617
[2] Kudryashov N A and Sinelshchikov D I 2014 *Regular and Chaotic Dynamics* **19** 576
[3] Kudryashov N A and Sinelshchikov D I 2014 *International Journal of Non-Linear Mechanics* **63** 31
[4] Kudryashov N A 1999 *Phys. Lett.* A **252** 173
[5] Kudryashov N A 2003 *J. Math. Phys.* **44** 6160
[6] Gordoa P R and Pickering A 2011 *Applied Mathematics and Computation* **218** 3942
[7] Kudryashov N A and Soukharev M B 1998 *Physics Letters* A **237** 206
[8] Kudryashov N A 1997 *Physics Letters* A **224** 353
[9] Kudryashov N A 2000 *Physics Letters* A **273** 194
[10] Kaup D J 1980 *Studies in Applied Mathematics* **62** 189
[11] Kudryashov N A and Pickering A 1998 *J. Phys.* A: Math. Gen. **31** 9505
[12] Kudryashov N A 2008 *Physics Letters* A **372** 1945
[13] Choudhury A G, Guha P and Kudryashov N A 2012 *Applied Mathematics and Computation* **218** 6612
[14] Painlevé P 1902 *Acta Mathematica* **25** 1
[15] Polyanin A D and Zaitsev V F 2012 *Handbook of Nonlinear Partial Differential Equations, Second ed.* (Chapman and Hall/CRC, Boca Raton)
[16] Boutroux P 1913 *Annales scientifiques de l'École Normale Supérieure* **30** 255
[17] Bruno A D 2015 *International Journal of Differential Equations Volume* **2015** 13
[18] Kitaev A V 1994 *Russian Mathematical Surveys* **49** 81
[19] Novokshenov V Y 1984 *Functional Analysis and Its Applications* **18** 260
[20] Its A R and Kapaev A A 1988 *Mathematics of the USSR-Izvestiya* **31** 193
[21] Its A R, Fokas A S and Kapaev A A 1994 *Nonlinearity* **7** 1291
[22] Kudryashov N A and Demina M V 2007 *Phys. Let.* A **363** 346
[23] Efimova O Y and Kudryashov N A 2007 *Regular and Chaotic Dynamics* **12** 198
[24] Cosgrove C 2000 *Studies in Applied Mathematics* **104** 1
[25] Foursov M V and Moreno Maza M 2001 *Technical report, LIFL, Universite de Lille-I, France*
[26] Kupershmidt B A 1984 *Phys. Lett.* A **102** 25
[27] Sawada K and Kotera T 1974 *Prog. Theor. Phys.* **51** 1355
[28] Caudrey P J, Dodd R K and Gibbon J D 1976 . *R. Soc. London, Ser.* A **351** 407

Isomonodromic problem for K_2^2 analogue of the Painlevé equations

I Yu Gaiur, N A Kudryashov

National Research Nuclear University MEPhI, Kashirskoe sh. 31, Moscow, 115409 Russia

E-mail: iygayur@gmail.com

Abstract. A fourth–order analogue for the Painlevé equations that is self–similar reduction of the modified Sawada–Kotera and Kaup–Kupershmidt equation is considered. Zero curvature representation of Lax pair for the modified Sawada–Kotera equation is presented. Linear problem for a fourth–order analogue for the Painlevé equations is obtained from Lax pair for the modified Sawada–Kotera equation using self-similar variables.

1. Introduction

Nonlinear ordinary differential equations play significant role in the description of different physical phenomena as exact mathematical models or arise as symmetry reductions of nonlinear partial differential equations [1, 2, 3]. Properties of the second–order differential equations of the form

$$w_{zz} = R(z, w, w_z) \tag{1}$$

have been studied by Painlevé [4] and Gambier [5]. They proved that there are fifty canonical equations of the form (1) with general solutions without critical movable points. Since that ordinary differential equations having solutions without critical movable points are called equations of the P–type. Thus all second-order differential equations of the P–type can be transformed to the integrating equations or to the one of the six non–autonomous equations which also called the Painlevé equations. It was proved that these equations have no movable critical points and determines new special functions which called Painlevé transcendents.

It is well known that equations of the P-type are closely connected with solvable by inverse scattering transform nonlinear evolution equations. According to Painlevé conjecture all exact reductions of the nonlinear partial differential equations solvable by inverse scattering transform are equations of the P–type. So investigation of nonlinear ordinary differential equations of P–type which can determine new special functions is important issue in modern mathematical physics.

Generalizations of the Painlevé equations were obtained by Kudryashov (K and P hierarchies) [6], Flashka and Newell (FN–hierarchy) [7], Jimbo and Miwa (JM–hierarchy) [8]. The following generalizations of the Painlevé give us higher order equations of P–type which can determine new higher transcendents. FN and JM hierarchies of the Painlevé equations were founded as the compatibility of the linear problem given by 2×2 matrix systems of the form

$$\begin{aligned} \partial_\mu \Psi &= F\Psi \\ \partial_z \Psi &= G\Psi \end{aligned} \tag{2}$$

So the existence of linear isomonodromic problem for FN and JM hierarchies is completely closed problem. A different situation arises with the K–hierarchies which were obtained as generalization of fourth–order equations which pass Painlevé test and define new special functions [6]. Lately in works [9] the linear problem representation was obtained for P and K_1 hierarchies as the 2×2 matrix systems. Exact form of the linear problem representation which also called isomonodromic problem for K_2 hierarchy is still open problem. To get more closer to solution of this problem we are going to find the linear problem representation for the fourth—order member of the K_2–hierarchy which can be written in the following form

$$w_{zzzz} + 5w_z w_{zz} - 5w^2 w_{zz} - 5ww_z^2 + w^5 - zw - \beta = 0. \tag{3}$$

It is well known that equation (3) can be obtained taking into account self–similar variables

$$v = \frac{1}{(5t)^{\frac{1}{5}}} w(z), \quad z = \frac{x}{(5t)^{\frac{1}{5}}}, \tag{4}$$

from the following nonlinear partial differential equation

$$v_t - v_{xxxxx} + 5v_{xx}^2 + 5v_x v_{xxx} + 5v_x^3 + 20vv_x v_{xx} + 5v^2 v_{xxx} - 5v^4 v_x = 0 \tag{5}$$

Equation (5) is the modified Kaup–Kupershmidt and Sawada–Kotera equations. It is known that solutions of the Kaup–Kupershmidt and Sawada–Kotera equations can be found via the solutions of equation (5) by means of the Miura transformations. These equations play significant role in the description of non–linear waves in a liquid with gas bubbles and in other application [2, 3].

The Lax pair representation for equation (5) was obtained in work [10]. The authors of this work show that equation (5) can be obtained as the compatibility of the following linear system

$$L\psi = \lambda\psi, \quad \psi_t = B\psi \tag{6}$$

where L and B are differential operators given by

$$\begin{aligned} L &= \partial_x^3 - 3v\partial_x^2 + 2(v^2 - v_x)\partial_x \\ B &= -\partial_x + 45v\partial_x^4 + 75(v_x - v^2)\partial_x^3 + (60v_{xxx} - 165vv_x + 45v^3)\partial_x^2 \\ &+ 20(v_{xxx} - 45v_x^2 - 70vv_{xx} + 70v^2 v_x - 5v^4)\partial_x \end{aligned} \tag{7}$$

The connection between equation (3) and equation (5) gives us opportunity to find isomonodromic problem for equation (3) using self–similar variables (4) for the Lax pair (6)–(7).

2. Main results

Let us rewrite the Lax pair (6)–(7) in zero curvature representation form

$$\begin{aligned} &L_1\mathbf{\Psi} = 0, \quad L_2\mathbf{\Psi} = 0 \\ &L_1 = \partial_x - A, \quad L_2 = \partial_t - M \\ &[L_1, L_2] = [\partial_x - A, \partial_t - M] = A_t - M_x + [A, M] = 0 \end{aligned} \tag{8}$$

where $\mathbf{\Psi}$ is N-vector, A and M are $N \times N$ matrices, N is order of the Lax pair. The order of the Lax pair (6)-(7) is equal 3. Let us choose the following 3-vector

$$\mathbf{\Psi} = \begin{pmatrix} \psi_{xx} \\ \psi_x \\ \psi \end{pmatrix} \tag{9}$$

Obtaining all derivatives from the second order by x and all mixed derivatives of ψ from equations (6)-(7) we find the form of matrices A and B in the following form

$$A = \begin{pmatrix} 3v & -2(v^2 - v_x) & \lambda \\ 1 & 0 & 0 \\ 0 & 1 & 0 \end{pmatrix}, \quad M = \begin{pmatrix} m_{11} & m_{12} & m_{13} \\ m_{21} & m_{22} & m_{23} \\ m_{31} & m_{32} & m_{33} \end{pmatrix} \tag{10}$$

where

$$\begin{aligned}
m_{11} &= 3v^5 + 8v^3 v_x - 7v^2 v_{xx} - 12v^2\lambda + vv_x^2 - 2vv_{xxx} - 5v_x v_{xx} - 6v_x\lambda + v_{xxxx} \\
m_{12} &= -2v^6 - 6v^4 v_x - 6v^3 v_{xx} + 18v^3\lambda - 22v^2 v_x^2 - 6v^2 v_{xxx} - 50vv_x v_{xx} + 12v\lambda v_x \\
&- 10v_x^3 + 2vv_{4x} - 10v_x v_{xxx} - 10v_{xx}^2 + 3v_{xx}\lambda - 9\lambda^2 + 2v_{5x} \\
m_{13} &= \lambda m_{21}, \quad m_{21} = v^4 + 4v^2 v_x - vv_{xx} - 9v\lambda + 3v_x^2 - v_{xxx} \\
m_{22} &= -8v^3 v_x - 8v^2 v_{xx} + 15v^2\lambda - 16vv_x^2 + 2vv_{xxx} - 10v_x v_{xx} + 3v_x\lambda + 2v_{xxxx} \\
m_{23} &= -9\lambda^2, \quad m_{31} = 6vv_x - 3v_{xx} - 9\lambda, \quad m_{32} = v^4 - 14v^2 v_x + 2vv_{xx} + 18v\lambda - 3v_x^2 + 2v_{xxx} \\
m_{33} &= -3\lambda(v^2 - v_x).
\end{aligned}$$

Obtained zero curvature representation can be rewritten as the following one–form

$$d\mathbf{\Psi} = A(v, x, t)\mathbf{\Psi} dx + M(v, x, t)\mathbf{\Psi} dt. \tag{11}$$

The goal is to find one-form representation corresponding to (3) using transformation (4)

$$v = \frac{1}{(5t)^{\frac{1}{5}}} g(z), \quad z = \frac{x}{(5t)^{\frac{1}{5}}}. \tag{12}$$

Using transformation (4) we obtain that

$$dx = (5t)^{\frac{1}{5}} dz + \frac{z}{(5t)^{\frac{4}{5}}} dt. \tag{13}$$

So equation (11) can be rewritten in the following form

$$\begin{aligned}
d\mathbf{\Psi} &= \hat{A}(g, z, t)dz + \hat{B}(g, z, t)dt \\
\hat{A}(g, z, t) &= (5t)^{\frac{1}{5}} A\left(\tfrac{g(z)}{(5t)^{\frac{1}{5}}}, z(5t)^{\frac{1}{5}}, t\right) \\
\hat{B}(g, z, t) &= \tfrac{z}{(5t)^{\frac{4}{5}}} A\left(\tfrac{g(z)}{(5t)^{\frac{1}{5}}}, z(5t)^{\frac{1}{5}}, t\right) + B\left(\tfrac{g(z)}{(5t)^{\frac{1}{5}}}, z(5t)^{\frac{1}{5}}, t\right)
\end{aligned} \tag{14}$$

Choosing new variable μ in the following way

$$\mu = \left(\frac{t}{5}\right)^{\frac{1}{5}}, \quad \frac{d\mu}{\mu} = \frac{dt}{t} \tag{15}$$

we get rid of variable t and obtain the following form of matrices

$$d\mathbf{\Psi} = N(g, z, \mu)dz + F(g, z, \mu)d\mu \tag{16}$$

where

$$N = \begin{pmatrix} 3g & -\frac{2(g^2-g_z)}{\mu} & \mu\lambda \\ \mu & 0 & 0 \\ 0 & \mu & 0 \end{pmatrix}, \tag{17}$$

and

$$5^{-\frac{1}{5}}F = 45\lambda\mu^4 \begin{pmatrix} 0 & \lambda & 0 \\ 0 & 0 & \lambda \\ 1 & 0 & 0 \end{pmatrix} - 45\lambda g\mu^3 \begin{pmatrix} 0 & 0 & \lambda \\ 1 & 0 & 0 \\ 0 & -2 & 0 \end{pmatrix} +$$

$$+ 15\lambda\mu^2 \begin{pmatrix} 2(2g^2-g_z) & 0 & 0 \\ 0 & -5g^2+g_z & 0 \\ 0 & 0 & g^2+g_z \end{pmatrix} + 15\mu \begin{pmatrix} 0 & \lambda m_1(g) & 0 \\ 0 & 0 & 0 \\ -(m_1(g)+6(gg_z-g^3)) & 0 & 0 \end{pmatrix} -$$

$$- \begin{pmatrix} 0 & 0 & \lambda m_{0,0}(g) \\ m_{0,0}(g) & 0 & 0 \\ 0 & m_{0,0}(g)+m_{0,1}(g) & 0 \end{pmatrix} + \frac{1}{\mu} \begin{pmatrix} m_{-1,0}(g) & 0 & 0 \\ 0 & 2(m_{-1,0}(g)+3gm_{-1,1}(g)) & 0 \\ 0 & 0 & 0 \end{pmatrix} +$$

$$+ \frac{1}{\mu^2} \begin{pmatrix} 0 & m_{-2}(g) & 0 \\ 0 & 0 & 0 \\ 0 & 0 & 0 \end{pmatrix}$$

$$\begin{aligned} & m_1(g) = g_{zz} - 4gg_z + 6g^3, \quad m_{0,0}(g) = 5g_{zzz} + 5g^4 - 20g^2g_z - 5gg_{zz} + 15g_z^2 - z \\ & m_{0,1}(g) = -15g_{zzz} + 90g^2g_z + 15gg_{zz} - 30g_z^2 \\ & m_{-1,0}(g) = 5g_{zzzz} + 15g^5 - 40g^3g_z - 35g_{zz}g^2 + 5gg_z^2 + 25g_zzg_z + 10g_{zzz}g - 3zg \\ & m_{-1,1}(g) = 5(g_{zzz} + g^4) - 20g^2g_z + 15g_z^2 - 5gg_{zz} - z \\ & m_{-2}(g) = 5g_{zzzzz} + 5g^6 - 15g^4g_z + 55g^2g_z^2 + 15g^3g_{zz} - 25g_z^3 - \\ & - 125gg_zg_{zz} - 15g^2g_{zzz} - g^2z + 25g_zg_{zzz} - g_zz + 25g_{zz}^2 - 5gg_{zzzz}. \end{aligned} \tag{18}$$

The isomonodromic deformations equations of system (16)-(18) reduce to the following equation

$$g_{zzzz} - 5g_zg_{zz} - 5gg_z^2 - 5g^2g_{zz} + g^5 - \frac{1}{5}zg + \alpha = 0, \tag{19}$$

which can be transformed to (3) by the following transformation

$$w(z') = -5^{\frac{1}{5}}g(z), \quad z' = 5^{-\frac{1}{5}}z, \quad \beta = 5\alpha. \tag{20}$$

Using scaling deformations (20) for the following linear system

$$\begin{aligned} \mathbf{\Psi}_z &= N\mathbf{\Psi} \\ \mathbf{\Psi}_\mu &= F\mathbf{\Psi} \end{aligned} \tag{21}$$

we obtain isomonodromic problem representation for K_2^2 equation.

3. Conclusion

In this work we have considered a fourth order analogue of the Painlevé equations (3). We have shown that equation (3) can be obtained from (5) using self–similar variables. Using Lax pair representation (6)-(7) for equation (5) we have constructed linear system which isomonodromic deformations equation gives equation (19) which can be transformed to equation (3). Obtained linear representation gives us chance to look for asymptotic solutions of (3) using isomonodromic deformation technic and find corresponding Riemann–Hilbert problem.

4. Acknowledgement

This research was supported by Russian Science Foundation grant No. 14–11–00258.

References

[1] Moore G 1990 *Prog. Theor. Phys. Supplement* **102** 255
[2] Kudryashov N A and Sinelshchikov D I 2014 *Regular and Chaotic Dynamics* **19** 576
[3] Kudryashov N A and Sinelshchikov D I 2014 *International Journal of Non-Linear Mechanics* **63** 31
[4] Painlevé P 1902 *Acta Mathematica* **25** 1
[5] Gambier B 1910 *Acta Mathematica* **33** 1
[6] Kudryashov N A 1997 *Physics Letters* A **224** 353
[7] Flaschka H and Newell A C 1980 *Comm. Math. Phys.* **76** 65
[8] Jimbo M and Miwa T 1981 *Phys.* D **2** 407
[9] Kudryashov N A 2003 *J. Math. Phys.* **44** 6160
[10] Guo-Liang H, Xian-Guo G 2012 *Chinese Physics* B **21** 070205

On the analytical properties and some exact solutions of the Glukhovsky-Dolzhansky system

I R Garashchuk [1], **N A Kudryashov** [2] **and D I Sinelshchikov** [3]
Department of Applied Maths, NRNU MEPhI, 31 Kashirskoe shosse, Moscow, Russia, 115409
E-mail: [1] ivan.mail4work@yandex.ru, [2] nakudr@gmail.com, [3] disine@gmail.com

Abstract. Non-linear dynamical systems describe many physical processes. In this work we investigate a three-dimensional Lorenz-like system - the Glukhovsky-Dolzhansky system. We consider analytical properties of the studied system. The problem of existence of meromorphic solution is discussed. We perform the Painlevè test and find conditions imposed on parameters of the system for which meromorphic solutions exist. Laurent series locally representing solutions are built. First integrals are obtained. We also find simply periodic solutions with one and two poles in a stripe of periods.

1. Introduction

Dynamical systems are used to describe various physical processes [1, 2]. In this work we study the Glukhovsky–Dolzhansky system. Comparing to well-known Lorenz system it has an additional non-linear term, which leads to essential differences in analytical structure and dynamics of the system. The Glukhovsky–Dolzhansky system describes following physical processes: convective fluid motion in an ellipsoidal rotating cavity, a rigid body rotation in a resisting medium, the forced motion of a gyrostat, a convective motion in harmonically oscillating horizontal fluid layer [1, 2]. Initially, this system was obtained by Glukhovsky and Dolzhansky as a three-mode model of convection for a fluid in an ellipsoidal rotating cavity, which can be interpreted as one of the models of ocean flows. Note that from this point of view system under investigation is significantly different from the Lorenz system. In the Lorenz system the flow of the two-dimensional convection is considered only. In the Glukhovsky–Dolzhansky system, the flow of the three-dimensional convection can be considered.

The Glukhovsky–Dolzhansky system was numerically investigated in detail (see, e.g. [1, 2]). However it has not been considered from the analytical point of view yet. Consequently in this work we apply analytical methods for studying the Glukhovsky-Dolzhansky system.

2. Painlevè analysis

The Glukhovsky–Dolzhanksy system takes the following form:

$$\begin{aligned} x_t &= -\sigma(x-y) - ayz, \\ y_t &= rx - y - xz, \\ z_t &= -bz + xy, \end{aligned} \tag{2.1}$$

where σ, a, r, b are physical parameters. Detailed discussion of physical meaning of the parameters can be found in work [2]. Note that system (2.1) is invariant under translations in

t. Consequently, to simplify the notation, further we will skip the arbitrary constant t_0 keeping in mind that in every solution we can always make the substitution $t \to t - t_0$. Generally, system (2.1) cannot be reduced to a third order ordinary differential equation. We begin our investigation with applying the Painlev'e test to system (2.1) (see, e.g. [3]).

Leading terms of system (2.1) are $x_t = -ayz,\ y_t = -xz,\ z_t = xy$. Substituting $x = \alpha/t^k, y = \beta/t^l, z = \gamma/t^m$ into the leading terms we find that $k = l = m = 1$ and four branches of solution:

$$\begin{array}{ll} \alpha_1 = i, \beta_1 = \frac{i}{\sqrt{a}}, \gamma_1 = \frac{1}{\sqrt{a}}, & \alpha_2 = -i, \beta_2 = -\frac{i}{\sqrt{a}}, \gamma_2 = \frac{1}{\sqrt{a}}, \\ \alpha_3 = i, \beta_3 = -\frac{i}{\sqrt{a}}, \gamma_3 = -\frac{1}{\sqrt{a}}, & \alpha_4 = -i, \beta_4 = \frac{i}{\sqrt{a}}, \gamma_4 = -\frac{1}{\sqrt{a}}. \end{array} \quad (2.2)$$

Then, substuting the following series

$$x = \alpha \sum_{j=0}^{\infty} \alpha_j t^{j-1}, \quad y = \beta \sum_{j=0}^{\infty} \beta_j t^{j-1}, \quad z = \gamma \sum_{j=0}^{\infty} \gamma_j t^{j-1}, \quad (2.3)$$

into (2.1), we get $\alpha_0 = \beta_0 = \gamma_0 = 1$ and $\alpha_1 = -\left(\sqrt{a}\,(b - \sigma + 1) \pm (\sigma - ra)\right) / \left(2\sqrt{a}\right)$, $\beta_1 = -\left(\sqrt{a}\,(\sigma + b - 1) \mp (\sigma - ra)\right) / \left(2\sqrt{a}\right)$, $\gamma_1 = -\left(\sqrt{a}\,(b - \sigma - 1) \pm (\sigma + ra)\right) / \left(2\sqrt{a}\right)$. For further coefficients we obtain the following recurrent relation

$$\begin{pmatrix} j-1 & 1 & 1 \\ 1 & j-1 & 1 \\ 1 & 1 & j-1 \end{pmatrix} \begin{pmatrix} \alpha_j \\ \beta_j \\ \gamma_j \end{pmatrix} = \begin{pmatrix} -\sum_{k=1}^{j-1} \beta_{j-k}\gamma_k - \sigma\alpha_{j-1} \pm \frac{\sigma}{\sqrt{a}}\beta_{j-1} \\ -\sum_{k=1}^{j-1} \alpha_{j-k}\gamma_k - b_{j-1} \pm r\sqrt{a}\alpha_j - 1 \\ -\sum_{k=1}^{j-1} \alpha_{j-k}\beta_k - b\gamma_{j-1} \end{pmatrix} \quad (2.4)$$

Accordingly, the Fuchs indicies are $j = -1, 2, 2$. For $j = 2$ consistency conditions must be imposed. Restrictions inflicted on the parameters of the system are shown in the table below.

Table 1. Parameters for which consistency conditions for $j = 2$ are satisfied.

	First pair of branches	Second pair of branches	Intersection
1.	$b = 0, r = \frac{\sigma - \sqrt{a}(\sigma - 1)}{a}$	$b = 0, r = \frac{\sigma + \sqrt{a}(\sigma - 1)}{a}$	$b = 0, r = \frac{1}{a}, \sigma = 1$
2.	$b = \sigma + 1, r = \frac{\sigma(-\sigma - 2\sqrt{a})}{a}$	$b = \sigma + 1, r = \frac{\sigma(-\sigma + 2\sqrt{a})}{a}$	$b = 1, r = 0, \sigma = 0$
3.	$b = \frac{\sigma(\sqrt{a} - 1)}{\sqrt{a}}, r = \frac{\sigma - \sqrt{a}(\sigma - 1)}{a}$	$b = \frac{\sigma(\sqrt{a} + 1)}{\sqrt{a}}, r = \frac{\sigma + \sqrt{a}(\sigma - 1)}{a}$	$\emptyset$

Now we consider the results given in Tab. 1 in more details. Case 1 does not correspond to any known physical processes because $b = 0$. In case 2 for second pair of branches positive sets of the parameters are possible. They may describe at least some extreme cases of real physical systems (for example a fluid with huge heat-conduction coefficient, very low viscosity coefficient inside a slowly rotating ellipsoidal cavity). In case 3 there are some sets of positive parameters for both pairs of roots. We can suppose that they may contain some cases that have physical meaning. One can note that if necessary conditions for the existence of Laurent series for all four solutions branches are held simultaneously (see column 3 of Tab. 1) the corresponding value of the parameters have no physical sense. Further coefficients of the Laurent series are not shown in this work, but they can be easily obtained by means of relation (2.4).

3. First integrals

Dynamical system (2.1) is dissipative, if we suppose that physical parameters are positive:

$$\frac{\partial x_t}{\partial x} + \frac{\partial y_t}{\partial y} + \frac{\partial z_t}{\partial z} = -\sigma - 1 - b < 0$$

Consequently we have reasons to expect that first integrals will be time-dependent. Now we consider two types of restrictions on the parameters of the system that are given in the last column of Tab. 1.

If $b = 0, r = \frac{1}{a}, \sigma = 1$ we can obtain first integral in the form of $e^{2t}(y^2 + z^2) = C_1$. Using this integral we can reduce our system to a second order differential equation that has the form

$$\left(C_1 e^{-2t} - z^2\right) z_{tt} + z z_t^2 + 2C_1 e^{-2t} z_t + az\left(C_1 e^{-2t} - z^2\right)^2 + C_1 e^{-2t} z = 0 \tag{3.1}$$

In the case of $b = 0,\ r = \frac{1}{a},\ \sigma = 1$, we get first integral in the form of $e^{2t}(x^2 - ay^2) = C_1$. By means of this integral initial system can be reduced to the following equation

$$\left(C_1 e^{-2t} + ay^2\right) y_{tt} - ayy_t^2 - y\left(\left(C_1 e^{-2t} + ay\right)^2 + C_1 e^{-2t}\right) = 0 \tag{3.2}$$

Necessary condition to possess the Painlevè property is satisfied for both equations (3.1) (3.2). However, in general, we can't transform them to one of the canonical equations of the Painlevè type [4] using the following transformation (similar for $z(t)$)

$$y(t) = \frac{k(t)w(s) + l(t)}{m(t)w(s) + n(t)},\ s = h(t).$$

Consequently, we think that equations (3.1), (3.2) don't have the Painlevè property in general, but can have it in some particular cases (for example if $C_1 = 0$).

4. Simply periodic solutions with one pole in a stripe of periods

It is convenient to look for simply periodic solutions with one pole in a stripe of periods in the following form: $x(t) = A_{1x} + A_{2x}w(t),\ y(t) = A_{1y} + A_{2y}w(t),\ z(t) = A_{1z} + A_{2z}w(t)$, where $w(t)$ is the solution of the Riccati equation looking like $w_t = -w^2 + B$. If $B \neq 0$ then $w(t) = \sqrt{B}\tanh\left(\sqrt{B}t\right)$ otherwise $w(t) = \frac{1}{t}$. Substituting x, y, z in this form into (2.1) for A_{2x}, A_{2y}, A_{2z} we obtain values (2.2) and three possible cases

$$1) B = 0,\ b = 0,\ r = \frac{\sigma \mp (\sigma\sqrt{a} - \sqrt{a})}{a}. \tag{4.1}$$

$$2) \begin{array}{l} B = \pm\dfrac{\sigma(\sigma+1)\left(4\sigma\sqrt{a} \pm (\sigma^2 + \sigma + 4a)\right)}{4a}, \\ b = 1 + \sigma,\ r = \mp\dfrac{(\pm\sigma + 2\sqrt{a})\,\sigma}{a}. \end{array} \tag{4.2}$$

$$3) \begin{array}{l} B = \mp\dfrac{\sigma^2}{4}\dfrac{- a^{3/2} - \sqrt{a} + 2\sigma a^{3/2} + 4\sigma\sqrt{a} \mp [\sigma + 5\sigma a - 2a]}{a^{3/2}\left(\frac{\sigma}{\sqrt{a}} \pm [1 - 2\sigma]\right)}, \\ b = \dfrac{\sigma\left(\sqrt{a} \mp 1\right)}{\sqrt{a}},\ r = \dfrac{\sigma \mp \sqrt{a}(\sigma - 1)}{a}. \end{array} \tag{4.3}$$

If conditions (4.1) are satisfied, we get $A_{1x} = 0,\ A_{1y} = 0,\ A_{1z} = \mp\frac{\sigma(\mp 1+\sqrt{a})}{a}$.
In case (4.2) coefficients take the following form for the first pair of branches of solution (according to (2.2)):

$$A_{1x} = \mp\left(\tfrac{1}{2}i\sigma^2\sqrt{\tfrac{1}{a}} + i\sigma + i + \tfrac{i\sigma}{2}\sqrt{\tfrac{1}{a}}\right), A_{1y} = \pm\tfrac{i\sigma(\sigma+1)}{2a},\ A_{1z} = -\tfrac{\sigma\left(\sigma+2\sqrt{a}-1\right)}{2a},$$

and for the second pair of branches:

$$A_{1x} = \pm\left(\tfrac{1}{2}i\sigma^2\sqrt{\tfrac{1}{a}} - i\sigma - i + \tfrac{i\sigma}{2}\sqrt{\tfrac{1}{a}}\right), A_{1y} = \pm\tfrac{i\sigma(\sigma+1)}{2a},\ A_{1z} = \tfrac{\sigma\left(-\sigma+2\sqrt{a}+1\right)}{2a}.$$

In this case $B = 0$ if $\sigma = 0$, $\sigma = -1$ or $a = \mp\frac{1}{4}\dfrac{\sigma(\sigma+1)^2}{\pm(\sigma+1) + 2(\mp\sigma + \sqrt{-\sigma})}$.
Considering case (4.3) coefficients for the first pair of branches of solution look like $A_{1x} = \mp\frac{i\sigma\left(\sqrt{a}-1\right)}{2\sqrt{a}}$, $A_{1y} = \mp\frac{i\sigma\left(\sqrt{a}-1\right)}{2a}$, $A_{1z} = -\frac{\sigma\left(\sqrt{a}-1\right)}{2a}$, and for the second pair of branches they take form $A_{1x} = \mp\frac{i\sigma\left(\sqrt{a}+1\right)}{2\sqrt{a}}$, $A_{1y} = \pm\frac{i\sigma\left(\sqrt{a}+1\right)}{2a}$, $A_{1z} = \frac{\sigma\left(\sqrt{a}+1\right)}{2a}$. B turns into zero if $a = 1$ only for the first pair of branches of solution and if $\sigma = 0$ for all four branches of solution.

5. Simply periodic solutions with two poles in a stripe of periods

According to [5, 6] we look for simply periodic solutions with two poles in a stripe of periods in the following form

$$x(t) = \sqrt{L}C_{-1}^{1x}\frac{A_{1x}\cot\left(\sqrt{L}t\right) + \sqrt{L}}{A_{1x} - \sqrt{L}\cot\left(\sqrt{L}t\right)} + \sqrt{L}C_{-1}^{0x}\cot\left(\sqrt{L}t\right) + h_x, \tag{5.1}$$

and similarly for $y(t)$ and $z(t)$. Now we consider situation where C_{-1}^0 and C_{-1}^1 are coefficients of the first and the second branches of solution in (2.2) respectively. To obtain parameters of solutions we use the Laurent series for (5.1) in a neighborhood of $t = 0$. After that we convert cot functions into tanh and put the phase offset of $\pi/2$ into the arbitrary constant t_0 (again we will skip it in further notation). As a result, the corresponding solutions can be represented by two following expressions:

$$1)\quad \begin{aligned} &b = 0,\ r = -\tfrac{\sigma\sqrt{a}-\sigma-\sqrt{a}}{a},\\ &x = -i\sqrt{-L}\frac{\left(\tanh(\sqrt{-L}t) - 1\right)\left(\tanh(\sqrt{-L}t) + 1\right)}{\tanh(\sqrt{-L}t)},\\ &y = -i\frac{\sqrt{-L}}{\sqrt{a}}\frac{\left(\tanh(\sqrt{-L}t) - 1\right)\left(\tanh(\sqrt{-L}t) + 1\right)}{\tanh(\sqrt{-L}t)},\\ &z = -\frac{\sqrt{-L}}{\sqrt{a}}\frac{\left(1 + \tanh^2(\sqrt{-L}t)\right)}{\tanh(\sqrt{-L}t)} - \frac{\sigma\left(\sqrt{a}-1\right)}{a}. \end{aligned} \tag{5.2}$$

$$2)\quad \begin{aligned} &b = \frac{\sigma\left(\sqrt{a}-1\right)}{\sqrt{a}},\ r = -\frac{\sigma\sqrt{a}-\sigma-\sqrt{a}}{a},\ L = -\tfrac{1}{4}\tfrac{\sigma^2\left(\sqrt{a}-1\right)^2}{a},\\ &x = -i\tfrac{b}{2}\left(1 + \tanh(\tfrac{b}{2}t)\right),\ y = -i\tfrac{b}{2\sqrt{a}}\left(1 + \tanh(\tfrac{b}{2}t)\right),\ z = -\tfrac{b}{2\sqrt{a}}\left(\tanh(\tfrac{b}{2}t) - 1 \mp 2\right). \end{aligned} \tag{5.3}$$

Solutions in form of (5.3) degenerate to functions with one pole in a stripe of periods. Solutions (5.2) have no physical sense. The parameters of system (2.1) in case (5.3) may describe at least some extreme cases of physical systems. Methods given in [5, 6] allow us to seek out simply periodic solutions with three and four poles in a stripe of periods. However, for system (2.1) we can't find such solutions.

In accordance with [5, 6, 7] it is reasonable to look for elliptic solutions of the system under discussion in the form of

$$x(t) = \sum_{i \subset I} \frac{C_{-1}^{(i),x}\left(\wp_t(t) + B_{ix}\right)}{2\left(\wp(t) - A_{ix}\right)} + h_x, \tag{5.4}$$

and similarly for $y(t)$ and $z(t)$. We tried this method for functions with two and four poles in a parallelogram of periods. But unfortunately this type of solutions can't be obtained for system (2.1).

6. Conclusion

In this work we studied the Glukhovsky-Dolzhansky system (2.1). The Painlevè test for this system of equations has been carried out. Coniditions imposed on parameters of the Glukhovsky–Dolzhansky system for meromorphic solutions to exist have been obtained. They are shown in Tab. 1. Solution in the form of Laurent series has been built. We have also found first integrals of system (2.1). We have shown that by means of this integrals the system can be reduced to second order ordinary differential equations (see (3.1), (3.2)). It seems that these equations do not possess the Painlevè property in general, but can have it in some particular cases.

We have obtained some meromorphic partial solutions of the system. We have found simply periodic solutions with one and two poles in a stripe of periods. Solutions with one pole in a stripe of periods are expressed via solution of the Riccati equation. They are given by formulae (4.1), (4.2), (4.3). Solutions with two poles in a stripe of periods are shown in (5.2), (5.3). However solutions (5.3) degenerate to functions with one pole in a stripe of periods. We have not found either periodic solutions with more poles in a stripe of periods or elliptic solutions by the method applied.

References

[1] Leonov G A, Kuznetsov N V and Mokaev T N 2015 *Eur. Phys. J.* Special Topics **224**
[2] Leonov G A, Kuznetsov N V and Mokaev T N 2015 *Physics Letters* A **379**
[3] Tabor M and Weiss J 1981 *Physical Review* A **24**
[4] Ince E L 1944 *Ordinary differential equations* (Dover Publications)
[5] Demina M V and Kudryashov N A 2010 *Physics Letters* A **374**
[6] Demina M V and Kudryashov N A 2011 *Communications in Nonlinear Science and Numerical Simulation* **16**
[7] Kudryashov N A, Soukharev M B and Demina M V 2012 *Communications in Nonlinear Science and Numerical Simulation* **17**

Interaction of solitary waves in longitudinal magnetic field in two-fluid MHD

M B Gavrikov[1] **and V V Savelyev**[1,2]

[1] Keldysh Institute of Applied Mathematics, Russian Academy of Sciences, Miusskaya sq., 4, Moscow, 125047, Russia
[2] National Research Nuclear University (MEPhI), Kashirskoe highway, 31, Moscow, 115409, Russia

Email: nadya_p@cognitive.ru, ssvvvv@rambler.ru

Abstract. The interaction of solitary waves in a model of two-fluid MHD is studied analytically and numerically in the most general case of waves in cold plasma in longitudinal magnetic field. The distinctive feature of this work is the use of "exact" equations rather than an approximate approach (a model equation). Numerical analysis of the solutions of this system of eight partial differential equations shows that the the interaction of solitary waves found in this case is the same (with great accuracy) as that of solitons, i.e., solitary waves that are solutions of various model equations. The solitary waves considered here transport plasmoids with velocities of the order of the Alfven velocity. The main finite-difference method used here for solving the said equations is a natural generalization of the classical two-step Lax-Wendorff scheme.

1. Introduction

The study of solitons is traditionally based on certain model equations [1], starting with the Korteweg de Vries equation, the nonlinear Schrödinger equation, the sine-Gordon equation. For plasma environments, this list has been considerably extended due to Zakharov equations, Kadomtsev-Petviashvili equation, the equations of nonlinear Alfven waves and magnetic-sound waves [2], etc.

In this paper, we study numerically the interaction of solitary waves of a speical type: the so-called wave packets, each having the form of nonlinear vibrations whose amplitude is modulated by a solitary wave. It is the longitudinal magnetic field that (being responsible for the rotation of plasma particles in the transverse plane) makes the transverse velocity components, as well as the magnetic and the electric fields, experience nonlinear vibrations which, under certain conditions, form a wave packet running along the magnetic field. In the case of cold plasma, these waves are presented in [3]. The principal distinction of the present work from other similar studies consists in avoiding the ideology of model equations, thereby making it possible to use the exact equations of two-fluid plasma hydrodynamics [4] (i.e., the equations expressing the fundamental laws of conservation of mass, energy, and momentum of electrons and ions, as well as the laws of electrodynamics) for both finding the solitary waves and studying their interaction.

It is shown here that, when colliding, the wave packets are similar to material particles: they preserve their shape, velocity, amplitude, etc., and the collision process has a finite duration.

2. Basic Equations

In the absence of dissipation, for fully ionized two-component quasi-neutral cold plasma the equations of two-fluid hydrodynamics of plasma [4] can be written in the following one-fluid form [5]:

$$\frac{\partial\rho}{\partial t}+\mathrm{div}\rho\mathbf{U}=0,\quad \frac{\partial\rho\mathbf{U}}{\partial t}+\mathrm{Div}(\rho\mathbf{UU}+\Pi^p+\Pi^c)=0$$
$$\mathbf{E}+\frac{c^2\lambda_i\lambda_e}{4\pi\rho}\mathrm{rotrot}\mathbf{E}=-\frac{1}{c}[\mathbf{U},\mathbf{H}]+\frac{1}{\rho}\mathrm{Div}\mathrm{W} \tag{1}$$
$$\frac{1}{c}\frac{\partial\mathbf{H}}{dt}+\mathrm{rot}\mathbf{E}=0,\quad \mathrm{div}\mathbf{H}=0\ ,\ \mathrm{rot}\mathbf{H}=\frac{4\pi}{c}\mathbf{j}$$

where the tensors Π^p, Π^c, W have the form

$$\Pi^p=\frac{H^2}{8\pi}\mathrm{I}_3-\frac{\mathbf{HH}}{4\pi},\quad \Pi^c=\lambda_i\lambda_e\frac{\mathbf{jj}}{\rho}\ ,\quad \mathrm{W}=(\lambda_e-\lambda_i)(\Pi^p+\Pi^c)+\lambda_i\lambda_e(\mathbf{Uj}+\mathbf{jU}) \tag{2}$$

In (1)-(2) I_3 is the three-dimensional identity tensor, $\rho=\rho_e+\rho_i$ is the total density, $\mathbf{U}=(\rho_i\mathbf{v}_i+\rho_e\mathbf{v}_e)/\rho$ is the mass hydrodynamic plasma velocity, $\lambda_e=m_e/e$, $\lambda_i=m_i/(Ze)$, where Z is the ion charge multiplicity. Thus, (1)-(2) is a closed system of equations for the unknown quantities ρ , $\mathbf{U}$, $\mathbf{H}$, $\mathbf{E}$.

Consider the solutions of equations (1)-(2) as functions of t, $\mathbf{r}$ in terms of the combined argument $\theta=\mathbf{rk}-at$, where $\mathbf{k}$ is a vector of unit length and a is a constant. Such solutions are usually referred to as *planar traveling waves*, a is the phase velocity of the wave, $\mathbf{k}$ is the wave propagation vector. We are interested in the interaction of solitary traveling waves such that all their respective parameters have finite and equal limit values as $\theta\to\pm\infty$. The parameters of an arbitrary traveling wave $\mathbf{U}(\theta)$, $\rho(\theta)$, $\mathbf{H}(\theta)$, $\mathbf{E}(\theta)$, for fixed $\mathbf{k}$ and a, are found by substituting these functions into system (1)-(2). After simple transformations, we obtain the following equations for the transverse magnetic field $\mathbf{H}_\perp(\theta)$ and the longitudinal velocity $u(\theta)=U_{\|}(\theta)-a$ (in the reference frame of the traveling wave):

$$Ju+\frac{H_\perp^2}{8\pi}=D$$
$$\left(u-\frac{H_{\|}^2}{4\pi J}\right)\mathbf{H}_\perp-\frac{c^2\lambda_i\lambda_e}{4\pi J}u\frac{d}{d\theta}\left(u\frac{d\,\mathbf{H}_\perp}{d\,\theta}\right)+\frac{cH_{\|}}{4\pi J}(\lambda_i-\lambda_e)u\left[\mathbf{k},\frac{d\,\mathbf{H}_\perp}{d\,\theta}\right]+\mathbf{q}=0 \tag{3}$$

Where each vector is expanded along (||) and across ($\perp$) the direction of wave vector $\mathbf{k}$, and $J\neq 0$, $D>0$, $\mathbf{q}\perp\mathbf{k}$ are arbitrary constants of integration. Moreover, we always have $H_{\|}(\theta)=\mathrm{const}$ in the traveling wave. In a special case, the two-fluid form of system (3) is obtained in [6], and in the general case, in [3].

In [7], the authors study collisions of solitary waves represented by the solutions of system (3) of the form $\mathbf{H}_\perp(\theta)=H(\theta)\mathbf{e}_0$ with $H_{\|}=0$, where $\mathbf{e}_0\perp\mathbf{k}$ is an arbitrary unit vector, $\mathbf{q}=q\mathbf{e}_0$, in which case the magnetic field $\mathbf{H}$ varies only in magnitude and has a constant direction in the transverse plane. The case $\mathbf{q}=0$ corresponds to system (3) being invariant with respect to rotations in the transverse plane.

2. Solving Equations (3) for Solitary Waves in the Case $\mathbf{q}=0$

System (3) with $\mathbf{q}=0$ is completely integrable. Let us write this system in the dimensionless form

$$u+H_{\perp}^2=2$$
$$(\frac{u}{2}-b^2)\mathbf{H}_{\perp}-\xi^2 u\frac{d}{d\theta}(u\frac{d\,\mathbf{H}_{\perp}}{d\theta})+\xi\Lambda bu\,[\mathbf{k},\frac{d\,\mathbf{H}_{\perp}}{d\theta}]=0 \tag{4}$$

With the dimensionless parameters ξ, b, Λ expressed by $\xi^2=c^2\lambda_i\lambda_e/(4\pi\rho_0 L_0^2)$, $b=H_{\parallel}/H_0$, $\Lambda=(\lambda_i/\lambda_e)^{1/2}-(\lambda_e/\lambda_i)^{1/2}$

Here, L_0, H_0, V_0, ρ_0 are the characteristic scales of length, magnetic field strength, velocity, and density, respectively; and H_0, V_0, ρ_0 are chosen related to the traveling wave constants J and D by $H_0=\sqrt{4\pi D}$, $\rho_0=4J^2/D$, $V_0=D/(2J)$, $V_0=H_0/(\sqrt{4\pi\rho_0})$. Thus, we seek a solution of system (5) in the fixed domain $|\mathbf{H}_{\perp}|<\sqrt{2}$ and this solution depends on the parameters ξ, b.

In order to find a solution of system (4), we set $\mathbf{H}_{\perp}=(0, H_y, H_z)$ and pass to the polar coordinates $H_y=H\cos\varphi$, $H_z=H\sin\varphi$ on the transverse plane. From system (4), we obtain the solution in explicit form

$$\theta(H)=\pm\xi\int\frac{(2-H^2)dH}{\sqrt{2(E-\Phi(H))}} \;,\; \varphi(H)=\pm\int\frac{2\Omega+\Lambda bH^2}{2H^2\sqrt{2(E-\Phi(H))}}dH \tag{5}$$

Here, E, Ω are arbitrary constants, and the potential function $\Phi(H)$ has the form

$$\Phi(H)=\frac{H^2}{8}-\frac{1}{2}(1-\eta^2b^2)H^2+\frac{\Omega^2}{2H^2} \;,\; \eta^2=1+\frac{\Lambda^2}{4}$$

The desired solutions of the solitary wave type correspond to the degenerate cases and exist only if

$$\Omega=0 \;,\; E=0 \;,\; 0<\alpha^2=1-\eta^2b^2<1/2. \tag{6}$$

Then formulas (5) define two semibounded arcs whose analytic expressions are obtained by calculating the integrals $(0<H\le H_m=2\alpha)$

$$\theta(H)=\pm\int\frac{(2-H^2)}{\sqrt{-2\Phi(H)}}dH=\mp 2\xi(\frac{1}{H_m}\ln\frac{H_m+\sqrt{H_m^2-H^2}}{H_m-\sqrt{H_m^2-H^2}}-\sqrt{H_m^2-H^2})$$
$$\varphi(H)=\pm\frac{\Lambda b}{2}\int\frac{dH}{\sqrt{-2\Phi(H)}}=\mp\frac{\Lambda b}{2H_m}\ln\frac{H_m+\sqrt{H_m^2-H^2}}{H_m-\sqrt{H_m^2-H^2}} \tag{7}$$

Consider the boundary value problem of exciting a solitary wave (in resting plasma of density ρ_∞ with the magnetic field strength $H_\infty=H_{\parallel}$) traveling along the magnetic field. For the characteristic scales we then have

$u_\infty=-a$, $J=\rho_\infty u_\infty=-a\rho_\infty$, $D=\rho_\infty u_\infty^2=\rho_\infty a^2$, $V_0=|a|/2$, $\rho_0=4\rho_\infty$, $H_0=\sqrt{4\pi\rho_\infty}\,|a|$

Hence, we get $b=H_{\perp}/H_0=V_A/|a|$, where $V_A=H_{\perp}/\sqrt{4\pi\rho_\infty}$. Then, from (6) we obtain the following constraints on the plase velocity a

$$|V_A|\eta<|a|<\sqrt{2}|V_A|\eta \tag{8}$$

Moreover, it is not difficult to obtain formulas for the relative amplitude $A(a)$ of the solitary wave

$$A(a)=\frac{H_m H_0}{|H_{\parallel}|}=2\left|\frac{a}{V_A}\right|\sqrt{1-\eta^2(\frac{V_A}{a})^2}=2\sqrt{(\frac{a}{V_A})^2-\eta^2} \tag{9}$$

The function $A(a)$ is monotonically increasing in $|a|$ from 0 to $2\eta^2$ In particular, for electron-ion plasma, the maximal amplitude is $\approx \Lambda$, which, for instance, in the case of hydrogen plasma is ≈ 42.

3. Numerical Modeling Method

Consider three problems regarding the interaction of the above waves. We are interested in three cases: A) motion of two solitary waves of the same amplitude towards each other; B) motion of two solitary waves of different amplitudes towards each other; C) a solitary wave of greater amplitude catching up with a solitary wave of a smaller amplitude moving in the same direction.

In this connection, we are going to obtain numerical solutions of problem (1)-(2) in the plane case, $\partial/\partial y = \partial/\partial z = 0$, on the line $-\infty < x < +\infty$ for $t \geq 0$, with the initial conditions at $t = 0$ (to be specified below) and the boundary conditions at infinity

$$\rho(\pm\infty) = \rho_\infty \ , \ U_x(\pm\infty) = 0 \ , \ U(\pm\infty) = 0 \ , \ H(\pm\infty) = 0 \ , \ E(\pm\infty) = 0 \qquad (10)$$

Where the x axis is directed along the vector $\mathbf{k}$ and complex notation is used for the components of the vectors $\mathbf{U}$, $\mathbf{H}$, $\mathbf{E}$: $U = U_y + iU_z$, $H = H_y + iH_z$, $E = E_y + iE_z$.

In dimensionless form, system (1)-(2) can be writes as

$$\frac{\partial \mathbf{u}}{\partial t} + \frac{\partial \mathbf{f}(\mathbf{u}, H)}{\partial x} = 0 \ , \ \frac{\partial H}{\partial t} + i\frac{\partial E}{\partial x} = 0$$
$$E - g(\mathbf{u})\frac{\partial^2 E}{\partial x^2} = D(\mathbf{u}, H) \qquad (11)$$

With the boundary conditions

$$u_1(\pm\infty) = 1 \ , \ u_2(\pm\infty) = 0 , \ u_3(\pm\infty) = 0 \ , \ H(\pm\infty) = 0 \ , \ E(\pm\infty) = 0 \qquad (12)$$

Here

$$\mathbf{u} = (u_1, u_2, u_3) = (\rho, \rho U_x, \rho U)$$

$$\mathbf{f}(\mathbf{u}, H) = (u_2, \frac{u_2^2}{u_1} + \frac{|H|}{2}, \frac{u_2 u_3}{u_1} - H_{\|} H) = (\rho U_x, \rho U_x^2 + \frac{|H|^2}{2}, \rho U_x U - H_{\|} H)$$

$$g(\mathbf{u}) = \frac{\xi^2}{u_1} \ , \ D(\mathbf{u}, H) = \frac{i}{u_1}(H_{\|} u_3 - H u_2 - i\Lambda\xi H_{\|}\frac{\partial H}{\partial x} + \xi^2 \frac{\partial}{\partial x}(\frac{u_2}{u_1}\frac{\partial H}{\partial x}))$$

Where i is the imaginary unit, ξ is the similarity number from Sect. 2, and the characteristic scales satisfy the relations $E_0 = V_0 H_0 / c$, $V_0 = H_0 / \sqrt{4\pi\rho_0}$, $t_0 = L_0 / V_0$, $\rho_0 = \rho_\infty$.

Our numerical method for solving system (11)-(12) is based on the two-step Lax-Wendorff [8] scheme for hydrodynamic equations.

4. Calculation Results

We start with rewriting (7) for the solitary wave in dimensionless variables adopted in (11):

$$\theta(H) = \pm\frac{\xi}{a}[\sqrt{H_m^2 - H^2} - \frac{a^2}{H_m}\ln\frac{H_m + \sqrt{H_m^2 - H^2}}{H_m - \sqrt{H_m^2 - H^2}}]$$
$$\varphi(H) = \pm\frac{\Lambda}{2H_m}\ln\frac{H_m + \sqrt{H_m^2 - H^2}}{H_m - \sqrt{H_m^2 - H^2}} \ , u = -\frac{2a^2 - H^2}{2a} \ , \rho = \frac{2a^2}{2a^2 - H^2} \ , E = iaHe^{i\varphi} \qquad (13)$$

Here, $0 < H \le H_m = (4a^2 - 4 - \Lambda^2)^{1/2}$, $\eta < |a| < \sqrt{2}\eta$, a is the dimensionless phase velocity (the characteristic velocity scale is $V_A = H_{\|} / \sqrt{4\pi\rho_\infty}$). The function $H(\theta) = H_a(\theta)$ is obtained by inverting $\theta(H)$ in (13).

Consider the calculation results in the case of electron-positron plasma, for $\Lambda = 0$, $\varphi(H) = const$. Assuming this constant to be equal to zero, we get $H_y = H_a(\theta)$, $H_z = 0$, $E_y = 0$, $E_z = aH_y$, while the remaining parameters are determined by (13). Thus, in the case of electron-positron plasma, the wave packet degenerates into a linearly polarized solitary wave. It is assumed below that $\xi = 1$, $H_\perp = 1$.

A) Consider the evolution of two waves of the same amplitude $A = 0.9$ moving towards each other with phase velocity $a = 1.1$. In this case, system (11)- (12) is solved numerically with the initial condition for the magnetic field

$$H(0,x) = H_{1.1}(\mathrm{x}-25) \ , \quad 0 \le \mathrm{x} \le 50$$

$$H(0,x) = H_{-1.1}(\mathrm{x}-75) \ , \quad 50 \le \mathrm{x} \le 100$$

on the interval $[0,100]$. The initial conditions for the other plasma parameters are calculated from $H(0,x)$ with the help of (13). Figure 1 represents the calculation results for the profiles $H(t,x)$ at different time instants.

B) Consider two solitary waves of different amplitudes moving towards each other with different phase velocities: the wave moving from left to right has amplitude $A = 0.9$ and phase velocity $a = 1.1$, while $A = 0.7$ and $a = -1.05$ for the wave moving from right to left. In this case, the initial magnetic field has the form

$$H(0,x) = H_{1.1}(\mathrm{x}-25) \ , \quad 0 \le \mathrm{x} \le 50$$

$$H(0,x) = H_{-1.05}(\mathrm{x}-75) \ , \quad 50 \le \mathrm{x} \le 100$$

Figure 2 shows the calculated profiles of $H(t,x)$ at different time instants.

C) Finally, consider the case of a wave with greater amplitude $A = 1.3$ and a greater phase velocity $a = 1.2$ catching up with a wave with smaller amplitude $A = 0.7$ and a smaller phase velocity $a = 1.05$. The initial condition for the magnetic field has the form

$$H(0,x) = H_{1.2}(x-60) \ , \quad 0 \le x \le 80$$

$$H(0,x) = H_{1.05}(x-115) \ , \quad 80 \le x \le 140$$

The solution is sought on the interval $[0, 700]$ and $H(0,x) = 0$ for $140 \le x \le 700$. Figure 3 shows the profiles $H(t,x)$ at different time instants. Time goes on all the figures from top to bottom.

5. Concluding Remarks

The general qualitative conclusion of this investigation is that the solitary waves in plasma, considered here in the framework of the two-fluid MHD, exhibit the behavior of genuine solitons, in the sense that after their interaction they preserve their characteristics (velocity, amplitude, etc). Our calculations show that in the case of waves moving towards each other, the tail of each wave after the interaction exhibits oscillations of small amplitude filling the space between the diverging waves. These oscillations can be interpreted as radiation produced upon the collision of waves. This question, however, requires further investigation.

All profiles shown above were obtained for the magnetic field, but similar distributions along the x axis take place for the other plasma parameters. The properties of interacting solitary waves described in this paper are typical mainly for the solutions of model equations and are usually explained by the complete integrability of the corresponding equations [1,2] and an infinite number of their first integrals.

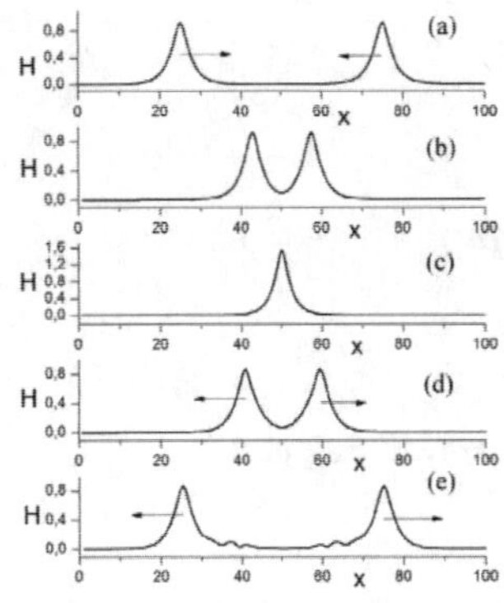

Figure 1. A=0.9, a=1.1,-1.1

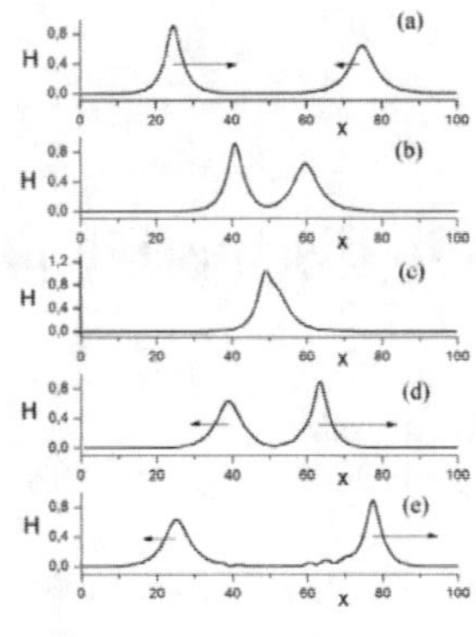

Figure 2. A=0.9, 0.7, a=1.1,-1.05

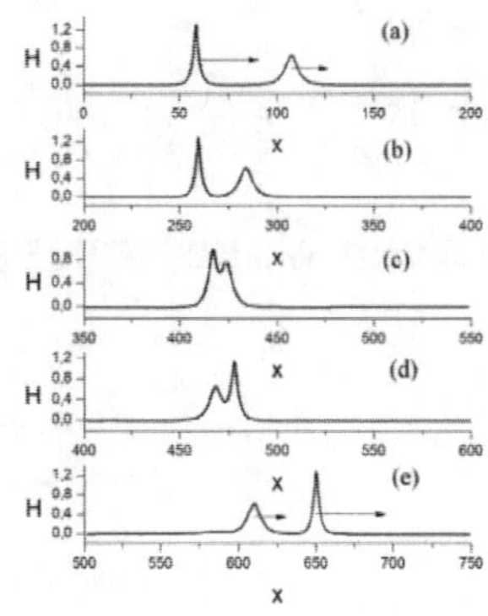

Figure 3. A=1.3, 0.7, a=1.2, 1.05

In this paper, these effects have been discovered numerically on the basis of a complete system of equations of two-fluid hydrodynamics of cold plasma, i.e., the laws of conservation of mass and momentum, together with the Maxwell equations. It should be mentioned that some mathematical properties (the Painleve property) of traveling wave equations (8) were considered in [9].

The present work continues the attempts to construct new models of plasma for the investigation of processes in plasma accelerators and new types of magnetic traps.

This work has been financially supported by the Russian Scientific Foundation (Project 16-11-10278).

6. References

[1] Manakov S V, Novikov S P, Pitaevskii L P and Zakharov V E 1984 Theory *of Solitons. The Inverse Scattering Method* (New York and London: Consultants Bureau)

[2] B.B. Kadomtsev, V.I. Petviashvili 1970 On the stability of solitary waves in media with weak dispersion *Dokl. Akad. Nauk SSSR* **102** 753-6

[3] Gavrikov M B 1991 Aperiodic Vibrations of Cold Plasma *Preprint 33 Keldysh Inst. Appl.Math. Academy of Sciences of the USSR, Moscow*

[4] Braginskii S I 1965 Transport processes in Plasma In *Reviews of Plasma Physics* **1**, ed. M A Leontovich (New York: Consultants Bureau) p 205

[5] Gavrikov M B and Sorokin R V 2009 Homogeneous deformation of a two-fluid plasma with allowance for electron inertia *Fluid Dynamics* **43** 977–89

[6] Saffman P G 1961 Propagating of a solitary wave along a magnetic field in a cold collision-free plasma *J. of Fluid Mech.* **11** 16–20

[7] Gavrikov M B, Savelyev V V and Tayurskii A A 2010 Solitons in two-fluid MHD with allowance for electron inertia. *Izv. Vuzov. Applied Nonlinear Dynamics* **18** 132–47

[8] Roache P J 1982 *Computational Fluid Dynamics* (Hermosa Publs, Albuquerque, NM)

[9] Gavrikov M B, Kudryashov N A, Petrov V A, Savelyev V V and Sinelshchikov D I 2016 Solitary and periodic waves in two-fluid magnetohydrodynamics *CNSNS* **38** 1–7

Alfven wave absorption in dissipative plasma

M B Gavrikov[1] **and A A Taiurskii**[1]
[1]Keldysh Institute of Applied Mathematics of Russian Academy of Sciences

E-mail: nadya_p@cognitive.ru, tayurskiy2001@mail.ru

Abstract. We consider nonlinear absorption of Alfven waves due to dissipative effects in plasma and relaxation of temperatures of electrons and ions. This study is based on an exact solution of the equations of two-fluid electromagnetic hydrodynamics (EMHD) of plasma. It is shown that in order to study the decay of Alfven waves, it suffices to examine the behavior of their amplitudes whose evolution is described by a system of ordinary differential equations (ODEs) obtained in this paper. On finite time intervals, the system of equations on the amplitudes is studied numerically, while asymptotic integration (the Hartman-Grobman theorem) is used to examine its large-time behavior.

1. Introduction

We examine time decay of Alfven EMHD waves due to dissipative factors (magnetic and hydrodynamic viscosities of electrons and ions, as well as relaxation of their temperatures) under the assumption that the Alfven wave has been initially excited in plasma occupying the entire space. This study is based on the the equations of electromagnetic hydrodynamics (EMHD) of plasma [1,2] that take into account the electron-ion structure of plasma and are written out in Section 2. In Section 3, it is shown that the nonlinear absorption of an Alfven wave due to dissipation is described by a system of ordinary differential equations (ODEs) for the amplitudes of the Alfven wave parameters. Solutions of the ODEs for the amplitudes on finite time intervals are studied numerically in Section 5, while large time solutions are obtained in Section 4 by asymptotic integration with the help of the Hartman-Grobman theorem [3]. This investigation allows us to find some important relationships characterising the conversion of magnetic and kinetic energies of an Alfven wave into thermal energy of electrons and ions. These relationships are of principal value for explaining abnormal heating of plasma.

2. EMHD Equations of Plasma

In view of the electron-ion structure of plasma, in particular, taking full account of electron inertia, we can write the equations of hydrodynamics in the form

$$\begin{gathered}\frac{\partial\rho}{\partial t}+\mathrm{div}\rho\mathbf{U}=0,\quad \frac{\partial\rho\mathbf{U}}{\partial t}+\mathrm{Div}\Pi=\mathrm{DivP}\\ \frac{\partial T_\pm}{\partial t}+\mathbf{U}\cdot\nabla T_\pm+T_\pm(\gamma-1)\mathrm{div}\mathbf{U}\pm\lambda_\mp\rho^{\gamma-2}\mathbf{j}\cdot\nabla\left(\frac{T_\pm}{\rho^{\gamma-1}}\right)=\\ =\frac{\lambda_\Sigma e_\pm(\gamma-1)}{k\rho}\{\mathrm{div}(\chi_\pm\nabla T_\pm)+\mathrm{tr}(\Pi_\pm \mathrm{D}_\pm)+\frac{m_\mp}{m_\Sigma}\frac{j^2}{\sigma}\pm b(T_- - T_+)\}\\ \frac{1}{c}\frac{\partial\mathbf{H}}{\partial t}+\mathrm{rot}\mathbf{E}=0,\quad \mathrm{div}\mathbf{H}=0,\quad \mathbf{j}=\frac{c}{4\pi}\mathrm{rot}\mathbf{H}\end{gathered}\tag{1}$$

$$\mathbf{E}+\frac{c^2\lambda_+\lambda_-}{4\pi\rho}\mathrm{rotrot}\mathbf{E}=\frac{\mathbf{j}}{\sigma}-\frac{1}{c}[\mathbf{U},\mathbf{H}]+\frac{1}{\rho}\mathrm{Div}\mathrm{W}$$

Here, k is the Botzmann constant; $\lambda_\pm=m_\pm/e_\pm$, $\lambda_\Sigma=\lambda_++\lambda_-$, $m_\Sigma=m_++m_-$, $\rho=\rho_++\rho_-$, $\mathbf{U}=(\rho_+\mathbf{v}_++\rho_-\mathbf{v}_-)/\rho$, where $\rho_\pm$, $\mathbf{v}_\pm$, $m_\pm$, $e_\pm$ are, respectively, the densities, hydrodynamic velocities, masses, and absolute values of charges of electrons and ions, which are assumed to be ideal polytropic gasses with the common adiabatic exponent γ. Thus, we have a closed system of equations for the unknown functions ρ, $\mathbf{U}$, $T_\pm$, $\mathbf{H}$, $\mathbf{E}$. The momentum flux density tensor $\Pi=\Pi^{(h)}+\Pi^{(p)}+\Pi^{(c)}$, the viscous stress tensor $\mathrm{P}=\Pi_*^c+\Pi^U$, and the Hall stress tensor W have the form

$$\Pi^{(h)}=\rho\mathbf{U}\mathbf{U}+p_\Sigma\mathrm{I}_3,\ \Pi^{(p)}=\frac{H^2}{8\pi}\mathrm{I}_3-\frac{\mathbf{H}\mathbf{H}}{4\pi},\ \Pi^{(c)}=\lambda_+\lambda_-\frac{\mathbf{jj}}{\rho},\ p_\Sigma=p_++p_-$$

$$\Pi^U=2\mu_\Sigma\mathrm{D}^U+(\nu_\Sigma-2\mu_\Sigma/3)\mathrm{trD}^U\mathrm{I}_3,\ \Pi^c=2\mu^*\mathrm{D}^c+(\nu^*-2\mu^*/3)\mathrm{trD}^c\mathrm{I}_3$$

$$\mathrm{W}=(\lambda_--\lambda_+)(\Pi^{(p)}+\Pi^{(c)})+(\lambda_-p_+-\lambda_+p_-)\mathrm{I}_3+\lambda_+\lambda_-(\mathbf{jU}+\mathbf{Uj})-\Pi_*^U-\Pi^c$$

$$\Pi_*^U=2\mu_*\mathrm{D}^U+(\nu_*-2\mu_*/3)\mathrm{trD}^U\mathrm{I}_3,\ \Pi_*^c=2\mu_*\mathrm{D}^c+(\nu_*-2\mu_*/3)\mathrm{trD}^c\mathrm{I}_3$$

where $\mathrm{D}^U=\mathrm{def}\mathbf{U}$, $\mathrm{D}^c=\mathrm{def}(\mathbf{j}/\rho)$ are strain tensors. The viscous stress tensors of electrons and ions are assumed to have the form $\Pi_\pm=2\mu_\pm\mathrm{D}_\pm+(\nu_\pm-2\mu_\pm/3)\mathrm{trD}_\pm\mathrm{I}_3$, where $\mathrm{D}_\pm=\mathrm{def}\mathbf{v}_\pm$; $\mu_\pm$, $\nu_\pm$ are the first and the second hydrodynamic viscosities of electrons and ions, respectively; $\mu_\Sigma=\mu_++\mu_-$, $\mu_*=\lambda_-\mu_+-\lambda_+\mu_-$, $\mu^*=\lambda_-^2\mu_++\lambda_+^2\mu_-$, and ν_Σ, ν_*, ν^* are expressed similarly. Finally, b and $\chi_\pm$ are the coefficients of thermal relaxation and heat conductivity of electrons and ions, respectively. Below, it is assumed that $\sigma=RT_-^{3/2}$, $\mu_\pm=T_\pm^{5/2}R_\pm$, $b=R_0\rho^2T_-^{-3/2}$, $\nu_\pm=0$, where the constants R, $R_\pm$, R_0 have the form [4,5]

$$R_+=\frac{4\pi^{1/2}e^4Z^4L}{0.96\cdot 3m_i^{1/2}k^{5/2}},\quad R_-=\frac{4(2\pi)^{1/2}e^4ZL}{0.733\cdot 3m_e^{1/2}k^{5/2}}$$

$$R_0=\frac{5m_e^{1/2}e^4Z^3L}{m_i^3k^{1/2}},\quad \sigma=\frac{3k^{3/2}}{4(2\pi m_e)^{1/2}e^2ZL\cdot 0.5129}$$

Here, $e_-=e$ is the electron charge, $Z=e_+/e_-$ is the ion charge multiplicity, $L=15$ is the Coulomb logarithm, $m_i=m_+$, $m_e=m_-$. We omit the expressions of $\chi_\pm$, since the results obtained below do not depend on $\chi_\pm$.

The law of conservation of total energy holds on the solution of system (1):

$$\frac{\partial}{\partial t}\left[\rho\left(\frac{U^2}{2}+\varepsilon+\frac{\lambda_+\lambda_-j^2}{2\rho^2}\right)+\frac{H^2}{8\pi}\right]+\mathrm{div}\left[\rho\mathbf{U}\left(\frac{U^2}{2}+\varepsilon+\frac{p_\Sigma}{\rho}+\frac{\lambda_+\lambda_-j^2}{2\rho^2}\right)+\frac{c}{4\pi}[\mathbf{E},\mathbf{H}]+A\mathbf{j}\right]= \tag{2}$$

$$=\mathrm{div}\{\chi_+\nabla\mathrm{T}_++\chi_-\nabla\mathrm{T}_-+\Pi_+\mathbf{v}_++\Pi_-\mathbf{v}_-\}$$

where $A=\lambda_+\lambda_-\langle\mathbf{U},\mathbf{j}\rangle/\rho+\lambda_+\lambda_-(\lambda_--\lambda_+)j^2/(2\rho^2)+\gamma(\lambda_-p_+-\lambda_+p_-)/(\rho(\gamma-1))$ and $\varepsilon=(\lambda_+\varepsilon_++\lambda_-\varepsilon_-)/\lambda_\Sigma$ is plasma internal energy density.

3. Decay of Alfven Waves in EMHD

Consider plane flows ($\partial/\partial y=\partial/\partial z=0$) of uniform plasma ($\rho=\mathrm{const}$) described by system (1). For such flows, we have $H_x=\mathrm{const}$, $\mathrm{U}_x=0$. In the absence of dissipation ($\nu_\pm=\mu_\pm=0$, $b=0$, $\chi_\pm=0$, $\sigma=+\infty$), system (1) in the plane case admits solutions on the line $x\in R$ with the initial conditions

$$t=0:\quad U_\perp=u_0e^{i\kappa x},\ H_\perp=h_0e^{i\kappa x},\ E_\perp=e_0e^{i\kappa x},\ T_\pm=\mathrm{T}_\pm^0 \tag{3}$$

where $\kappa > 0$ is a constant (wave number) and complex notation $U_\perp = U_y + iU_z$, $H_\perp = H_y + iH_z$, etc., is used for the transverse components of the corresponding vector fields. This solution has the form

$$U_\perp = u(t)e^{i\kappa x},\ H_\perp = h(t)e^{i\kappa x},\ E_\perp = e(t)e^{i\kappa x},\ T_\pm = T_\pm^0 \tag{4}$$

where $u(t)$, $h(t)$, $e(t)$ are found by substituting the expressions (4) into system (1):

$$u(t) = C_1 e^{i\omega_+ t} + C_2 e^{i\omega_- t},\ h(t) = \frac{(4\pi\rho)^{1/2}}{\kappa v_A}\{C_1\omega_+ e^{i\omega_+ t} + C_2\omega_- e^{i\omega_- t}\}$$

$$e(t) = \frac{i}{1+r^2}\left\{\left(\frac{H_x}{c} + \Lambda\sqrt{\lambda_+\lambda_-}\,\omega_+\right)C_1 e^{i\omega_+ t} + \left(\frac{H_x}{c} + \Lambda\sqrt{\lambda_+\lambda_-}\,\omega_-\right)C_2 e^{i\omega_- t}\right\},\ r = \frac{\kappa c}{\omega_p},\ \omega_p = \left(\frac{4\pi\rho}{\lambda_+\lambda_-}\right)^{1/2} \tag{5}$$

$$\omega_\pm = \frac{\kappa v_A}{2}\left\{\frac{r\Lambda}{1+r^2} \pm \left[\frac{r^2\Lambda^2}{(1+r^2)^2} + \frac{4}{1+r^2}\right]^{1/2}\right\},\ \Lambda = \sqrt{\frac{\lambda_+}{\lambda_-}} - \sqrt{\frac{\lambda_-}{\lambda_+}},\ v_A = H_x/\sqrt{4\pi\rho}$$

$$C_1 = \left(\frac{\kappa v_A}{\sqrt{4\pi\rho}}h_0 - \omega_- u_0\right)(\omega_+ - \omega_-)^{-1},\ C_2 = \left(\frac{\kappa v_A}{\sqrt{4\pi\rho}}h_0 - \omega_+ u_0\right)(\omega_- - \omega_+)^{-1}$$

The solution (4), (5), called an *Alfven wave,* is a superposition of transverse waves moving in the direction of the magnetic field or in the opposite direction with different phase velocities depending on the wave length $2\pi/\kappa$. In the MHD-limit, $r << 1$, the solution (4), (5) turns into the classical Alfven wave.

For a plane flow, consider the solution of system (1) on the line with the initial conditions (3) and dissipation taken into account. This solution describes the decay of the Alfven wave (4), (5) due to dissipative effects and has the form

$$U_\perp = u(t)e^{i\kappa x},\ H_\perp = h(t)e^{i\kappa x},\ E_\perp = e(t)e^{i\kappa x},\ T_\pm = T_\pm(t) \tag{6}$$

where the amplitudes $u(t)$, $h(t)$, $T_\pm(t)$ satisfy the system of ODEs obtained by substituting the functions (6) into system (1):

$$\mathrm{d}u/\mathrm{d}t = a_{11}u + a_{12}h,\ \mathrm{d}h/\mathrm{d}t = a_{21}u + a_{22}h$$

$$\frac{\mathrm{d}T_\pm}{\mathrm{d}t} = Z_\pm a_*\left\{\mu_\pm\kappa^2\left|u - \frac{c\kappa\lambda_\mp}{4\pi\rho}h\right|^2 + \frac{m_\mp}{m_\Sigma}\frac{c^2\kappa^2}{16\pi^2\sigma}|h|^2 \pm b(T_- - T_+)\right\} \tag{7}$$

where $Z_+ = Z$, $Z_- = 1$, $a_{11} = -\kappa^2\mu_\Sigma\rho^{-1}$, $a_{12} = \kappa c(\mathrm{i}H_x c^{-1} + \kappa^2\mu_*\rho^{-1})(4\pi\rho)^{-1}$, $a_{21} = \lambda_+\lambda_-\omega_p^2 a_{12}(1+r^2)^{-1}$, $a_{22} = \kappa c\left[-c\kappa(4\pi\sigma)^{-1} + \mathrm{i}\kappa\Lambda v_A\omega_p^{-1} - c\kappa^3\mu^*(4\pi\rho^2)^{-1}\right](1+r^2)^{-1}$, $a_* = \lambda_\Sigma e(\gamma-1)/(k\rho)$. Here, $e(t) = (a_{21}u + a_{22}h)(\kappa c)^{-1}$. From (6), (7), it follows that: (i) the decay of plane waves is purely temporal, in the sense that only the amplitudes $u(t)$, $h(t)$, $e(t)$, $T_\pm(t)$ are varying in time, while the spatial sine distribution of the plasma parameters remains unchanged; (ii) the decay of Alfven waves does not depend on the thermal conductivity of electrons and ions. From (2), it follows that the conservation law

$$\frac{\rho|u(t)|^2}{2} + (1+r^2)\frac{|h(t)|^2}{8\pi} + \frac{T_+(t)}{Za_*} + \frac{T_-(t)}{a_*} = C_0 = \mathrm{const} \tag{8}$$

holds on the solution (6), where C_0 is determined by the initial condition (3).

4. Asymptotic Integration of Amplitude Equations for $t \to +\infty$

Let us write system (7) in dimensionless form, choosing the following characteristic scales of the density, the magnetic field strength, velocity, etc.: $\rho_0 = \rho$, $H_0 = H_x$, $U_0 = v_A$, $L_0 = c\omega_p^{-1}$, $t_0 = L_0U_0^{-1}$, $T_0 = v_A^2\lambda_\Sigma e(2k)^{-1}$. Thus, we obtain the system

$$\frac{du}{dt}=r\left(\mathrm{i}+\frac{r^2}{\zeta}\alpha_1\right)h+\frac{r^2}{\zeta}\beta_1 u,\quad \frac{dh}{dt}=\frac{r}{1+r^2}\left\{\left(\mathrm{i}+\frac{r^2}{\zeta}\alpha_1\right)u+r\left(-\frac{\xi}{T_-^{3/2}}+\mathrm{i}\Lambda+\frac{r^2}{\zeta}\beta\right)h\right\}$$

$$\frac{dT_\pm}{dt}=2Z_\pm(\gamma-1)\left\{\frac{m_\mp}{m_\Sigma}r^2\zeta\frac{|h|^2}{T_-^{3/2}}\pm\zeta\eta\frac{(T_- - T_+)}{T_-^{3/2}}+\alpha^\pm\frac{r^2}{\zeta}\left[|u|^2+r^2\frac{\lambda_\mp}{\lambda_\pm}|h|^2\mp r\left(\frac{\lambda_\mp}{\lambda_\pm}\right)^{1/2}(\bar{u}h+u\bar{h})\right]\right\} \tag{9}$$

where $\alpha_1=\alpha^+(\lambda_-/\lambda_+)^{1/2}-\alpha^-(\lambda_+/\lambda_-)^{1/2}$, $\beta_1=-(\alpha^+ +\alpha^-)$, $\beta=-\alpha^+(\lambda_-/\lambda_+)-\alpha^-(\lambda_+/\lambda_-)$, $\alpha^\pm=T_\pm^{5/2}R_\pm^{-1}$, and $R_\pm$, η are universal constants; r, ζ are similarity numbers,

$$\zeta=0.386LZ^3\frac{ce^3}{m_+^2}\frac{(4\pi\rho)^{5/2}}{H_x^4}\left(\frac{\lambda_+}{\lambda_\Sigma}\right)^{3/2},\quad r=\frac{\kappa c}{\omega_p}=\frac{c\sqrt{\lambda_+\lambda_-}}{\sqrt{4\pi\rho}}$$

$$R_+=2.87\left(\frac{m_-}{m_+}\right)^{1/2}\frac{\lambda_+}{\lambda_\Sigma},\ R_-=5.313\frac{\lambda_+}{\lambda_\Sigma},\ \eta=1.46\frac{m_-}{m_+}\frac{\lambda_\Sigma}{\lambda_+} \tag{10}$$

Thus, if $\ell=2\pi/\kappa$ is the length of an Alfven wave, then the problem of wave decay has two determining parameters: $\rho^{5/2}/H_x^4$ and $\ell\rho^{1/2}$. Moreover, the energy integral (8) can be rewritten in the dimensionless form

$$|u|^2+(1+r^2)|h|^2+\frac{T_+}{Z(\gamma-1)}+\frac{T_-}{(\gamma-1)}=C_0 \tag{11}$$

Separating the real and the imaginary parts in system (9), we pass to the real unknown functions u_1, u_2, h_1, h_2, with $u_1+\mathrm{i}u_2=u$, $h_1+\mathrm{i}h_2=h$, and exclude T_+ from the unknown quantities with the help of the energy integral (11). Thus we obtain a modification of system (9) that consists of five ODEs for five real unknown functions (T,u_1,u_2,h_1,h_2). It is not difficult to verify the following statements:

1) The modified system (9) has a unique singular point $(T^0,0,0,0,0)$, where $T^0=Z(\gamma-1)C_0(1+Z)^{-1}$, with $C_0=|u_0|^2+(1+r^2)|h_0|^2+T_+^0(Z(\gamma-1))^{-1}+T_-^0(\gamma-1)^{-1}$ being the value of the energy integral calculated on the basis of the initial values.

2) The eigenvalues of the Jacobi matrix J coincide with $\lambda_0=-2(\gamma-1)\eta\zeta(1+Z)(T^0)^{-3/2}$ and the roots of two quadratic equations

$$\lambda^2(1+r^2)-\lambda r^2[\beta_0+\beta_1\zeta^{-1}(1+r^2)\pm\mathrm{i}\Lambda]+r^4\zeta^{-1}[\beta_1\beta_0-r^2\zeta^{-1}\alpha_1^2\pm\mathrm{i}(\Lambda\beta_1-2\alpha_1)]+r^2=0. \tag{12}$$

3) If $\lambda_1\neq\lambda_2$ are the roots of (12) with the upper sign, then $\bar{\lambda}_1\neq\bar{\lambda}_2$ are the roots of (12) with the lower sign.

4) Each equation in (12) has neither multiple, nor conjugate, nor real roots.

5) All roots of equations (12) have negative real parts.

6) All eigenvalues $\{\lambda_0,\lambda_1,\lambda_2,\bar{\lambda}_1,\bar{\lambda}_2\}$ of the matrix J are single and there is a basis of the space C^5 that consists of eigenvectors of J.

It follows that the only singular point of the modified system (9) is an attractive stable multidimensional focus and, by the Hartman-Grobman theorem, the topology of the integral curves in a neighborhood of the singular point of this system coincides with that of its linearization at this singular point. Thus, for $t\to+\infty$, the decay of the Alfven wave is correctly described by the linearization of the modified system (9) at the singular point $(T^0,0,0,0,0)$. The solutions of the linearized system

$$(\dot{T},\dot{u}_1,\dot{u}_2,\dot{h}_1,\dot{h}_2)^*=\mathrm{J}(T,u_1,u_2,h_1,h_2)^* \tag{13}$$

(here the dot and the asterisk indicate differentiation in t and transposition, respectively) can be easily obtained in explicit form. Let $\lambda_1\neq\lambda_2$ be the roots of the characteristic equation (12) with the upper sign and let $x_j + iy_j \neq 0$ be the eigenvector of J corresponding to λ_j, $j = 1,2$. If $\lambda_j = a_j + ib_j$, $j = 1,2$, $x_0 = (1,0,0,0,0)$, then $\{x_0, x_1, y_1, x_2, y_2\}$ is a basis of R^5 in which the Jacobi matrix J has the form

$$\mathrm{J} = \mathrm{diag}\left\{1, \begin{pmatrix} a_1 & b_1 \\ -b_1 & a_1 \end{pmatrix}, \begin{pmatrix} a_2 & b_2 \\ -b_2 & a_2 \end{pmatrix}\right\}$$

Therefore, if $(z_0, z_1, z_2, z_3, z_4)$ are the coordinates of a vector in R^5 in the basis $\{x_0, x_1, y_1, x_2, y_2\}$, then system (13), in that basis, splits into three independent subsystems,

$$\dot z_0 = \lambda_0 z_0, \quad \begin{cases} \dot z_1 = a_1 z_1 + b_1 z_2 \\ \dot z_2 = -b_1 z_1 + a_1 z_2 \end{cases}, \quad \begin{cases} \dot z_3 = a_2 z_3 + b_2 z_4 \\ \dot z_4 = -b_2 z_3 + a_2 z_4 \end{cases}$$

whose solutions can be easily written out, which gives us the solution of system (13). This solution represents a two-frequency spiral (with frequencies b_1, b_2) in five-dimensional space. The spiral winds around the origin with the decrements of distance from the origin being equal to $|a_j|$, $j = 1,2$, $|\lambda_0|$. In particular, we have

$$(u_1, u_2, h_1, h_2)^* = \sum_{j=1}^{2} D_j e^{a_j t} (\cos(\varphi_j - b_j t) x_j + \sin(\varphi_j - b_j t) y_j)$$

where the constants D_0, D_1, D_2, φ_1, φ_2 are found by expanding the initial vector $(u_1^0, u_2^0, h_1^0, h_2^0)$ with respect to the basis $\{x_1, y_1, x_2, y_2\}$. The explicit expressions for x_j, y_j, $j = 1,2$, are the following:

$$x_j = \left(0, \zeta r \cdot \frac{(r^2\beta_1 - \zeta a_j) - r^2\alpha_1 b_j}{(r^2\beta_1 - \zeta a_j)^2 + \zeta^2 b_j^2}, -r \cdot \frac{b_j\zeta^2 + r^2\alpha_1(r^2\beta_1 - \zeta a_j)}{(r^2\beta_1 - \zeta a_j)^2 + \zeta^2 b_j^2}, 0, 1\right)$$

$$y_j = \left(0, r \cdot \frac{b_j\zeta^2 + r^2\alpha_1(r^2\beta_1 - \zeta a_j)}{(r^2\beta_1 - \zeta a_j)^2 + \zeta^2 b_j^2}, \zeta r \cdot \frac{(r^2\beta_1 - \zeta a_j) - r^2\alpha_1 b_j}{(r^2\beta_1 - \zeta a_j)^2 + \zeta^2 b_j^2}, -1, 0\right)$$

The constants a_j, b_j can be easily calculated by the formulas for the roots of quadratic equations and square roots of complex numbers. The resulting expressions are rather lengthy, but can be simplified in some special or limit cases. Thus, for $r >> 1$ (short waves), $h_0 \neq 0$, $\mu_\pm \neq 0$, we have the asymptotic formulas

$$a_{1,2} \sim \left(\frac{Z(\gamma-1)}{1+Z}\right)^{5/2} |h_0|^5 r^7 \frac{\lambda_\Sigma}{\lambda_\pm} \frac{1}{\zeta R_\pm}, \quad b_{1,2} \sim \frac{1}{2}\left\{\Lambda \pm \frac{\Lambda R^* - \Lambda R_\Sigma - 4R_*}{\lambda_\Sigma(\lambda_+^{-1}R_+^{-1} - \lambda_-^{-1}R_-^{-1})}\right\} \tag{14}$$

where $R^* = (\lambda_-/\lambda_+)R_+^{-1} + (\lambda_+/\lambda_-)R_-^{-1}$, $R_\Sigma = R_+^{-1} + R_-^{-1}$, $R^* = (\lambda_-/\lambda_+)^{1/2}R_+^{-1} + (\lambda_+/\lambda_-)^{1/2}R_-^{-1}$, with the upper and the lower signs in (14) being in agreement. For $r >> 1$ (long waves), we have

$$a_{1,2} \sim -\frac{r^2}{2\zeta A_0^{3/2}}[\zeta^2 + R_\Sigma A_0^4], \quad b_{1,2} \sim \pm r, \quad A_0 = \frac{Z(\gamma-1)}{Z+1}\left[|h_0|^2 + |u_0|^2 + \frac{T_+^0}{Z(\gamma-1)} + \frac{T_-^0}{\gamma-1}\right]$$

The coordinate T^0 of the singular point is equal to the equilibrium temperature established in plasma after the complete absorption of the Alfven wave and the relaxation of electron and ion temperatures:

$$T^0 = \frac{T_0^+ + ZT_0^-}{1+Z} + \frac{Z}{1+Z}\frac{\lambda_\Sigma e(\gamma-1)}{k}\left\{\frac{|u_0|^2}{2} + \frac{|h_0|^2}{8\pi\rho}\left[1 + \left(\frac{\kappa c}{\omega_p}\right)^2\right]\right\} \tag{15}$$

It follows from (15) that the equilibrium temperature does not depend on the plasma magnetization H_x, but depends on the wave length $2\pi/\kappa$. Theoretically, (15) indicates that fairly short Alfven waves, even of small amplitudes u_0, h_0, may heat up plasma to arbitrarily high temperatures.

5. Results of Numerical Analysis

The absorption of an Alfven wave amounts to the conversion of its kinetic energy $\varepsilon_{\text{kin}} = \rho|u(t)|^2/2$ and its total (with the kinetic energy of the relative motion of electrons taken into account) magnetic energy $\varepsilon_{\text{m}} = (1+r^2)|h(t)|^2/(8\pi)$ into the thermal energy of electrons and ions $\varepsilon_- = T_- a_*^{-1}$, $\varepsilon_+ = T_+(Za_*)^{-1}$. This process is superimposed on the relaxation of the electron and ion temperatures determined by the coefficient b. Numerical solutions of the Cauchy problem for system (14) show that the absorption of an Alfven wave splits into two stages: (i) first, there is a rapid conversion of its magnetic energy and a considerable part of its kinetic energy into the thermal energy of (mostly) electrons; (ii) then, slow (for the most part) relaxation of temperatures occurs, which is approximated by the solution of system (9) with $u = 0$, $h = 0$; here, the remainder of the kinetic energy is converted into heat. The curves in Figure 1 represent typical values of thermal energies of electrons and ions, as well as the magnetic and the kinetic energies, versus time in the case of $r = 0.1$, $\zeta = 300$, $T_+^0 = 0.1$, $T_-^0 = 1$, $h_0 = 5$, $u_0 = 1.5$, $\mu_\pm = 0$. If the hydrodynamic viscosity of ions is taken into account, then the absorption process becomes much faster. For instance, the time of magnetic energy absorption becomes equal to $\sim(\omega_c^+\omega_c^-)^{-1/2}$, where $\omega_c^\pm$ are cyclotron frequencies of electrons and ions. If we additionally take into account electron viscosity, then the absorption process becomes even faster, occurring in a fraction of $(\omega_c^+\omega_c^-)^{-1/2}$, and the absorption time for the magnetic energy becomes $\sim 10^{-2}(\omega_c^+\omega_c^-)^{-1/2}$.

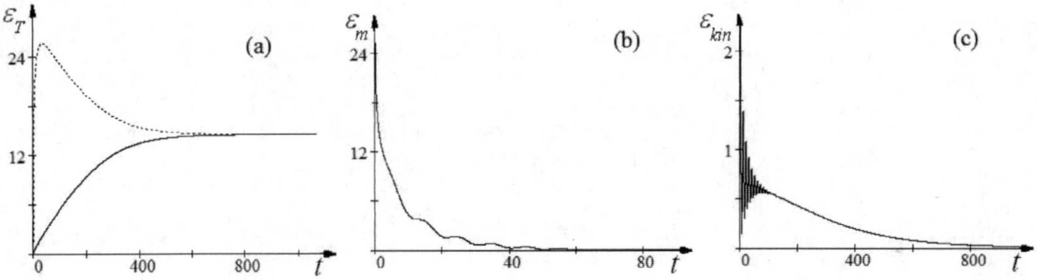

Figure 1. Time dependence of the thermal energy of electrons ("···") and ions ("—") in the Alfven wave (a), the magnetic energy of the Alfven wave (b), the kinetic energy of the Alfven wave (c)

Acknowledgments

The study was performed by the grant from the Russian Foundation for Basic Research (projects №15-01-03085, №16-31-00302)

Bibliography

[1] Braginskii S I 1965 *Reviews of Plasma Physics* vol 1 ed M A Leontovich (New York: Consultants Bureau) pp 205-311
[2] Gavrikov M B and Sorokin R V 2008 *Fluid Dynamics* **43** 6 pp 977-989
[3] Hartman P 2002 *Ordinary Differential Equations* (SIAM)
[4] Spitzer L 1962 *Physics of Fully Ionized Gases* (New York: 2nd ed Interscience)
[5] Chapman S and Cowling T G 1952 *The Mathematical Theory of Non-uniform Gases* (Cambridge University Press)

Existence and stability of periodic solutions of quasi-linear Korteweg – de Vries equation

S D Glyzin, A Yu Kolesov, M M Preobrazhenskaia
P.G. Demidov Yaroslavl State University, Sovetskaya str., 14, Yaroslavl, 150000, Russia
E-mail: rita.preo@gmail.com

Abstract. We consider the scalar nonlinear differential-difference equation with two delays, which models electrical activity of a neuron. Under some additional suppositions for this equation well known method of quasi-normal forms can be applied. Its essence lies in the formal normalization of the Poincare – Dulac obtaining quasi-normal form and the subsequent application of the theorems of conformity. In this case, the result of the application of quasi-normal forms is a countable system of differential-difference equations, which can be turned into a boundary value problem of the Korteweg – de Vries equation. The investigation of this boundary value problem allows us to draw a conclusion about the behaviour of the original equation. Namely, for a suitable choice of parameters in the framework of this equation is implemented buffer phenomenon consisting in the presence of the bifurcation mechanism for the birth of an arbitrarily large number of stable cycles.

1. General formulation of the problem

We consider the scalar non-linear differential equation with two delays

$$\dot{u} = \lambda\Big((a+1)f(u(t-h_1)) - a - bg(u(t-h_2))\Big)u \tag{1}$$

which is used to model the impulse activity of a neuron (see [1, 2, 3, 4]). Here $u(t) > 0$ is the normalized membrane potential of the neuron, the sufficiently large parameter λ characterizes the rate of the electrical processes in the system, the parameters h_1, h_2, a, b are positive. The functions $f(u)$, $g(u)$ are from $C^\infty(\mathrm{R}_+)$, where $\mathrm{R}_+ = \{u \in \mathrm{R} : u \geq 0\}$, and satisfy (see [2]) the following conditions.

Condition 1. We assume that $f'(u) < 0$, $g'(u) > 0$ $\forall u \in \mathrm{R}_+$ and $f(0) = 1$, $g(0) = 0$, $\lim\limits_{u\to+\infty} f(u) = 0$, $\lim\limits_{u\to+\infty} g(u) = 1$.

The monotony of functions $f(u)$ and $g(u)$ ensures the existence and uniqueness of solution $u = u_0(a,b) > 0$ of the equation $(a+1)f(u) - a - bg(u) = 0$.

Condition 2. Let the parameter $a > 0$ be fixed. The equation

$$\psi(b) \equiv (a+1)f'(u)|_{u=u_0(a,b)} + bg'(u)|_{u=u_0(a,b)} = 0$$

has the unique solution $b = b_* > 0$ on the half-line $b \in \mathrm{R}_+$ such that $\psi'(b_*) > 0$.

For the last condition we need the Taylor expansions of the following functions at the point $v = 0$: $V_1(v,b) = (f(u(1+v)) - f(u) - f'(u)uv)/(2f'(u)u)|_{u=u_0(a,b)}$ and $V_2(v,b) =$

$-b(g(u(1+v))-g(u)-g'(u)uv)/(2(a+1)f'(u)u)|_{u=u_0(a,b)}$. The expansions begin with quadratic terms:

$$V_j(v,b_*) = a_{j2}v^2 + a_{j3}v^3 + O(v^4),\ j=1,2. \tag{2}$$

Condition 3. We assume that $d \equiv 4(a_{12}^2 - a_{22}^2) + 2(a_{23} - a_{13}) < 0$.

Under the above restrictions we consider the question about the self-oscillating modes of the equation (1), which bifurcates from the equilibrium state $u = u_0(a,b)$ as the parameters $b,\ h_1,\ h_2$ are changed.

First of all, choose the parameters. As in [2] we are interested in the singular perturbed case: $h_2/h_1 = \varepsilon h,\ h = const > 0,\ \varepsilon = \lambda^{-1},\ 0 < \varepsilon \ll 1$. We choose the parameter h later and assume that b is close to $b = b_*$, from condition 2. Then we make the problem suitable for use in the method of quasi-normal forms. Assume

$$b = b_* + \alpha(\mu),\ |\mu| \ll 1, \tag{3}$$

where μ is small enough parameter, the function $\alpha(\mu),\ \alpha(0) = 0$, is defined from equation $\psi(\alpha,\mu) \equiv (1+a)f'(u)/((b_*+\alpha)g'(u))^{-1}|_{u=u_0(a,b_*+\alpha)} + (1-2\mu)(1+2\mu)^{-1} = 0$. Put (3) to (1) and introduce a new variable v: let $u = u_*(\mu)(1+v)$ where $u_*(\mu) = u_0(a, b_* + \alpha(\mu))$. As a result of normalization $\varepsilon(\chi_*(\mu))^{-1} \mapsto \varepsilon,\ h\chi_*(\mu) \mapsto h$, where $\chi_*(\mu) \equiv -2(1+a)f'(u_*(\mu))u_*(\mu)(1-2\mu)^{-1} > 0$, the equation (1) is transformed to

$$\varepsilon\dot v = -\Big((\tfrac{1}{2}-\mu)v(t-\varepsilon h) + (\tfrac{1}{2}+\mu)v(t-1) + \Delta_1(v(t-\varepsilon h),\mu) + \Delta_2(v(t-1),\mu)\Big)(1+v) \tag{4}$$

where $\Delta_j(v,\mu) = (1-2\mu)V_j(v, b_* + \alpha(\mu)),\ j=1,2$.

Now choose the parameter h. Consider the characteristic equation

$$2\varepsilon\lambda + (1-2\mu)e^{-\varepsilon h\lambda} + (1+2\mu)e^{-\lambda} = 0 \tag{5}$$

corresponding to zero equilibrium state of the equation (4). The equation (5) is analysed in [5] where in case $h < \pi$ all its roots are divided into two groups. The first group contains the roots from the left complex half plane $\{\lambda : \mathrm{Re}\,\lambda < 0\}$. Besides, there are no roots from the group close to the imaginary axis as $\varepsilon,\ \mu \to 0$. All other roots $\lambda_n(\varepsilon,\mu),\ \overline{\lambda}_n(\varepsilon,\mu),\ n \in \mathrm{N}$, belong to the second group. As $\varepsilon = \mu = 0$ the second group roots become the roots $\lambda_n = i\omega_n,\ \omega_n = \pi(2n-1),\ n \in \mathrm{N}$, of the equation $e^{-\lambda} = -1$. In the article [5] the roots are defined by equations

$$\lambda_n(\varepsilon,\mu) = i\omega_n(1 + \varepsilon(h-2) + \varepsilon^2(h-2)^2) - 2\varepsilon^2\omega_n^2(1-h) + 4\mu + O(\varepsilon^3 + \varepsilon\mu),\ \varepsilon,\mu \to 0. \tag{6}$$

The last equations allow to define the smallness order of μ and ε. Let below

$$h = 1 - h_0\varepsilon,\ \ \mu = \mu_0\varepsilon^3,\ \ h_0, \mu_0 = const > 0. \tag{7}$$

Note that with this choice of parameters the order of $\mathrm{Re}\,\lambda_n(\varepsilon,\mu)$ is ε^3.

2. Application of the quasi-normal form method

In this case, the algorithmic part of the quasi-normal form method is as follows. Put (7) to (4) and make change of time

$$\tau = (1+\tilde\sigma)t,\ s = \varepsilon^3 t,\ \ \tilde\sigma = \varepsilon\sigma_1 + \varepsilon^2\sigma_2 + \varepsilon^3\sigma_3,\ \ \sigma_1 = -1,\ \sigma_2 = 1 - h_0,\ \sigma_3 = 2h_0 - 1 \tag{8}$$

where $\sigma_1,\ \sigma_2,\ \sigma_3$ are from (6). We find the main asymptotic behavior of self-oscillating mode as the series

$$v = \sum_{k=0}^{\infty} v_k(s,\tau)\varepsilon^{(k+3)/2}. \tag{9}$$

Here $v_k(s,\tau)$ is 2-periodic by τ, the initial approximation $v_0(s,\tau)=\xi(s,\tau)$ satisfies anti-periodic condition $\xi(s,\tau+1)\equiv-\xi(s,\tau)$ where ξ is unknown. The reason why ξ is anti-periodic is clear from (6): ξ has to be expandable into Fourier series by $e^{i\omega_n\tau}$, $\omega_n=\pi(2n-1)$, $n\in\mathrm{N}$.

Then we put (9) to (4) with condition (8), transform $v(s-\tilde{s},\tau-\tilde{\tau})$, $v(s-\varepsilon^3,\tau-1-\tilde{\sigma})$ to Taylor series by delays $\tilde{s}$, $\tilde{\tau}$, ε^3, $\tilde{\sigma}$ ($\tilde{s}=\varepsilon^4(1-h_0\varepsilon)$, $\tilde{\tau}=\varepsilon(1-h_0\varepsilon)(1+\tilde{\sigma})$), equate the coefficients of the same powers of ε in the left and right parts of the v_k, $k=0,1,2,\dots$ equation. As a result, we get equations

$$(v_k(s,\tau)+v_k(s,\tau-1))/2=F_k(s,\tau),\ k=0,1,2,\dots \tag{10}$$

where functions $F_k(s,\tau)$ are 2-periodic by τ. For the analysis (see [5]) of the equations (10) divide the function $F_k(s,\tau)$ to two summands $F_k^{(1)}(s,\tau)+F_k^{(2)}(s,\tau)$ such that $F_k^{(1)}(s,\tau+1)\equiv -F_k^{(1)}(s,\tau)$, $F_k^{(2)}(s,\tau+1)\equiv F_k^{(2)}(s,\tau)$ where $F_k^{(1)}(s,\tau)=(F_k(s,\tau)-F_k(s,\tau-1))/2$, $F_k^{(2)}(s,\tau)=(F_k(s,\tau)+F_k(s,\tau-1))/2$. The system (10) has 2-periodic solution if $F_k^{(1)}(s,\tau)\equiv 0$. In this case the solution of the system is $v_k(s,\tau)=F_k^{(2)}(s,\tau)$.

The choice of σ_1, σ_2, σ_3 ensures that the equation (10) has 2-periodic solution with numbers from $k=1$ to 5. So like above we find $v_1(s,\tau)\equiv 0$, $v_2(s,\tau)\equiv 0$, $v_3(s,\tau)=-(a_{12}+a_{22})\xi^2$, $v_4(s,\tau)\equiv 0$, $v_5(s,\tau)=(4a_{12}+1)\xi\xi_\tau$.

When $k=6$ from the solvability condition we have the boundary value problem for $\xi(s,\tau)$

$$\xi_s=2h_0\xi_{\tau\tau}+\frac{1}{3}\xi_{\tau\tau\tau}+4\mu_0\xi+(4(a_{12}^2-a_{22}^2)+2(a_{23}-a_{13}))\xi^3,\ \ \xi(s,\tau+1)\equiv-\xi(s,\tau). \tag{11}$$

We call the boundary value problem (11) quasi-normal form of (4).

Consider the statement of the correspondence between the self-oscillating modes of quasi-normal form (11) and equation (4). As the phase space of problem (11) we take the Sobolev space $W_2^3[0,1]$ of anti-periodic functions. And let the phase space of equation (4) be $C[-1,0]$.

Theorem 1. *Let μ and h satisfy (7) and quasi-normal form (11) has the periodic traveling-wave solution*

$$\xi=\xi_0(y),\ y=\alpha s+\tau,\ \xi_0(y+1)=-\xi_0(y),\ \alpha_0=const, \tag{12}$$

which is orbitally exponentially stable or dichotomous. Then, there exists $\varepsilon_0>0$ such that, for all $0<\varepsilon\le\varepsilon_0$ solution (12) corresponds to a cycle of equation (4) with the same stability properties. The main asymptotic of the cycle can be found from the part of above constructed series (9) using the formulas (12) and (8).

The proof of this fact is based on results from works [6], [7]. It is clear from the theorem 1 that we should be interested in periodic solutions of (11).

3. Secondary quasi-normalization

We suppose that $\mu_0=\epsilon/12$ $h_0=\epsilon\kappa/6$ as $0<\epsilon\ll 1$ is a new small parameter, $\kappa=const>0$, then after normalization $s/3\mapsto s$, $-\tau\mapsto\tau$, $\sqrt{-d}\xi\mapsto\xi$ as d is from condition 3, the problem (11) is transformed to

$$\xi_s+\xi_{\tau\tau\tau}=\epsilon\kappa\xi_{\tau\tau}+\epsilon\xi-\xi^3,\ \ \xi(s,\tau+1)=-\xi(s,\tau). \tag{13}$$

Here s serves as a time, τ — as space variable. The phase space is the Sobolev space W_2^3 of anti-periodic function.

Let us consider the question of the existence and the stability of invariant toruses of the system (13). We find self-oscillating solution of the boundary value problem (13) as the sum

$$\xi=\sum_{k=0}^{\infty}\xi_k(s,\theta,\tau)\epsilon^{(2k+1)/2}, \tag{14}$$

where $\xi_0(s,\theta,\tau) = \sum\limits_{k=1}^{\infty} (z_k(\theta)e^{iy_k} + \bar{z}_k(\theta)e^{-iy_k})$, $y_k = \nu_k^3 s + \nu_k\tau$, $\nu_k = (2k-1)\pi$, $\theta = \epsilon s$. $z_k(\theta)$ are unknown, and will be defined. The function $\xi_0(s,\theta,\tau)$ satisfies by s,τ linear homogeneous equation

$$\xi_s + \xi_{\tau\tau\tau} = 0.$$

We find the function ξ_1 as the trigonometrical series by y_k, $k \in \mathrm{N}$. Put (14) to (13) and equate the coefficients of the same powers of $\epsilon^{3/2}$. As a result, we get boundary value problem

$$\frac{\partial \xi_1}{\partial s} + \frac{\partial^3 \xi_1}{\partial \tau^3} = \kappa \frac{\partial^2 \xi_0}{\partial \tau^2} + \xi_0 - \frac{\partial \xi_0}{\partial \theta} - \xi_0^3,\ \xi_1(s,\theta,\tau+1) = -\xi_1(s,\theta,\tau) \tag{15}$$

for unknown function ξ_1 where θ is parameter. Using the explicit formula of ξ_0 we get that right part of (15) is the trigonometrical series by $e^{\pm i(y_k+y_l+y_m)}$ and $e^{\pm i(y_k+y_l-y_m)}$, $k,l,m \in \mathrm{N}$. We find the function ξ_1 in the same view. But it is possible if and only if right part of (15) does not contain e^{iy_k}, $k \in \mathrm{N}$. Therefore we equate the coefficients of e^{iy_k} to zero. As a result, for complex amplitudes $z_k(\theta)$ we obtain a countable system of ordinary differential equations

$$\dot{z}_k = (1-\kappa\nu_k^2)z_k - 6\sum_{l=1}^{\infty} z_k|z_l|^2 + 3z_k|z_k|^2,\ k \in \mathrm{N}, \tag{16}$$

which is quasi-normal form of the boundary value problem (13).

Note that it is possible to find the function ξ_1 from (15) as a trigonometrical series by $e^{\pm i(y_k+y_l+y_m)}$ and $e^{\pm i(y_k+y_l-y_m)}$, $k,l,m \in \mathrm{N}$, $l \neq m \neq k$. because here $(\nu_k^3+\nu_l^3+\nu_m^3) - (\nu_k+\nu_l+\nu_m)^3 \neq 0$, $(\nu_k^3+\nu_l^3-\nu_m^3) - (\nu_k+\nu_l-\nu_m)^3 \neq 0$, $k,l,m \in \mathrm{N}$, $l \neq m$.

For the analysis of the quasi-normal form (16) we make the replacements $z_k = \sqrt{\eta_k}e^{i\psi_k}$, $\tau \to \tau/2$ where η_k, ψ_k are real variables. So, for η_k we get the splitting system

$$\dot{\eta_k} = \Big(1 - \kappa\nu_k^2 - 6\sum_{l=1}^{\infty}\eta_l + 3\eta_k\Big)\eta_k,\ k \in \mathrm{N}. \tag{17}$$

4. Analysis of the amplitude system

Note that the system (17) has no equilibrium state with infinite count of non-zero coordinates. (The proof of the same fact can be found in the book [8].) Fix an arbitrary positive p, $n_1 < n_2 < \ldots < n_p$. For system (17) we investigate the question of the existence and stability of equilibrium states $\eta = (\eta_1, \eta_2, \ldots)$ with coordinates

$$\eta_{n_m} > 0,\ m = 1,2,\ldots,p;\ \eta_k = 0,\ \text{where } k \neq n_m,\ m = 1,\ldots,p. \tag{18}$$

Put (18) to (17) for finding η_{n_m} and obtain the linear inhomogeneous system

$$1 - \kappa\nu_{n_m}^2 - 6\sum_{l=1}^{p}\eta_{n_l} + 3\eta_{n_m} = 0,\ m = 1,\ldots,p. \tag{19}$$

The direct verification shows that this system has the unique solution $\hat{\eta} = (\hat{\eta}_1, \hat{\eta}_2, \ldots)$ where

$$\hat{\eta}_{n_m} = \frac{1}{3}\left(\frac{2}{2p-1}\Big(p - \kappa\sum_{l=1}^{p}\nu_{n_l}^2\Big) + \kappa\nu_{n_m}^2 - 1\right),\ \hat{\eta}_k = 0,\ \text{where } k \neq n_m,\ m = 1,\ldots,p. \tag{20}$$

Since variables η_k are modules we are interested in case when all $\hat{\eta}_{n_m}$, $m = 1,\ldots,p$, are positive. So, $\hat{\eta}_{n_m} > 0$ if and only if κ satisfies inequality $\kappa < \left(2\sum\limits_{l=1}^{p}\nu_{n_l}^2 - (2p-1)\min\limits_{1\leq m\leq p}\nu_{n_m}^2\right)^{-1}$.

Now we study the stability of the equilibrium state (20). To this end, we denote the right-hand side in (17) through $f(\eta) = (f_1(\eta), \ldots, f_k(\eta), \ldots)$, where $f_k(\eta) \equiv \Big(1 - \kappa\nu_k^2 - 6 \sum\limits_{l=1, l\neq k}^{\infty} \eta_l - 3\eta_k\Big)\eta_k$.

Partial derivatives of functions $f_k(\eta)$ at the point $\hat{\eta}$ are as follows:

$$\frac{\partial f_k}{\partial \hat{\eta}_j}\bigg|_{\eta=\hat{\eta}} = \begin{cases} -6\hat{\eta}_{n_m}, & k = n_m,\ m = 1, \ldots, p,\ \ j \neq k, \\ -3\hat{\eta}_{n_m}, & k = n_m,\ m = 1, \ldots, p,\ \ j = k, \\ 0, & k \neq n_m,\ m = 1, \ldots, p,\ \ j \neq k, \\ 1 - \kappa\nu_k^2 - 6\sum_{l=1}^{p} \hat{\eta}_{n_l}, & k \neq n_m,\ m = 1, \ldots, p,\ \ j = k. \end{cases}$$

Then the system in variations of the interesting us equilibrium state has the form $\dot{h} = J_{n_1,\ldots,n_p} h$ where $J_{n_1,\ldots,n_p}$ is the Jacobi matrix of $f(\eta)$.

It is convenient to investigate the state of equilibrium in two cases: $p = 1$ and $p \geq 2$. Firstly, let $p = 1$ and fix natural n. Then the solution (20) of the system (17) is transformed to

$$\eta_n = (1 - \kappa\nu_n^2)/3, \ \eta_k = 0 \ \ k \neq n. \tag{21}$$

Besides, it exists if $\kappa < \nu_n^{-2}$. The operator J_n has eigenvalues $\lambda_n = -3\eta_n$, $\lambda_k = 2\kappa\nu_n^2 - \kappa\nu_k^2 - 1$, $k \neq n$. By requiring that all eigenvalues are negative, we obtain the condition for the stability of the equilibrium state (21):

$$\kappa < (2\nu_n^2 - \nu_1^2)^{-1}. \tag{22}$$

Now let $p \geq 2$. Consider a finite-dimensional operator

$$\hat{J}_{n_1,\ldots,n_p} = \begin{pmatrix} -3\eta_{n_1} & -6\eta_{n_1} & \ldots & -6\eta_{n_1} \\ -6\eta_{n_2} & -3\eta_{n_2} & \ldots & -6\eta_{n_2} \\ \vdots & \vdots & \ddots & \vdots \\ -6\eta_{n_p} & -6\eta_{n_p} & \ldots & -3\eta_{n_p} \end{pmatrix}.$$

Note that the eigenvalues of this operator are the eigenvalues of the operator $J_{n_1,\ldots,n_p}$. The eigenvalues are found from the characteristic equation

$$\left(1 - \sum_{m=1}^{p} \frac{6\eta_{n_m}}{3\eta_{n_m} - \lambda}\right) \prod_{l=1}^{p} (3\eta_{n_l} - \lambda) = 0. \tag{23}$$

It follows from the analysis of (23) that the equilibrium state (20) of system (17) with the number of nonzero coordinates $p \geq 2$ is unstable. Thus, the equilibrium state (20) is stable with only one non-zero coordinate when κ satisfies (22). Besides, (22) implies that the number of stable equilibrium states increases without bound with decreasing κ.

Theorem 2 (see [1]). *Any equilibrium state of the amplitude system (17) with p non-zero positive coordinate corresponds to an invariant p-dimension torus of the problem (13). The stability of the equilibrium state and the torus is the same.*

The following statement implies from the above research and the theorem 2.

Theorem 3. *For each natural n there exists a sufficiently small $\epsilon_n > 0$ such that for all $0 < \epsilon \leq \epsilon_n$ and the parameter κ, satisfying the condition (22), the problem (13) has a stable travelling wave, defined by the equality $\xi = \sqrt{\epsilon}\xi_0(y) + O(\epsilon^{3/2})$ as $y = (2n-1)\pi x + \nu_n(\epsilon)t$, $\nu_n(\epsilon) = (2n-1)^3\pi^3 + O(\epsilon)$. If $p \geq 2$, then the p-dimensional toruses of the problem (13) are unstable.*

As follows from the condition (22) the number $K(\kappa)$ of stable cycles of the boundary value problem (13) satisfies

$$K(\kappa) \geq \left\lfloor \left(\sqrt{\pi^{-2}\kappa^{-1} + 1} + 1\right)/2 \right\rfloor. \tag{24}$$

(Here $\lfloor\cdot\rfloor$ denotes the integer part.) As mentioned above, $K(\kappa)$ grows without bound with decreasing κ.

The cycles from theorem 3 are stable not only as $0 < \epsilon \ll 1$, but also as $\epsilon \sim 1$. However, it is possible to check only by means of computer experiment.

5. Numerical experiment

For a description of the numerical experiment we fix an arbitrary integer N and replace the partial derivatives with respect to τ in (13) by difference approximations (see [2])

$$\xi_{\tau\tau}\Big|_{\tau=k/N} \approx N^2(\xi_{k+1} - 2\xi_k + \xi_{k-1}),\ \xi_{\tau\tau\tau}\Big|_{\tau=k/N} \approx N^3(\xi_{k+2} - 2\xi_{k+1} + 2\xi_{k-1} - \xi_{k-2})/2 \quad (25)$$

where $\xi_k = \xi(s,\tau)|_{\tau=k/N}$, $k = 0, 1, \ldots, N-1$. After time normalization $N^3/2 \mapsto N$ we get the system for $\xi_k(s)$:

$$\dot{\xi_k} = -(\xi_{k+2} - 2\xi_{k+1} + 2\xi_{k-1} - \xi_{k-2}) + 2\epsilon\kappa(\xi_{k+1} - 2\xi_k + \xi_{k-1})/N + 2\epsilon\xi_k/N^3 - 2\xi_k/N^3,$$
$$k = 0, 1, \ldots, N-1,\ \xi_N = -\xi_1,\ \xi_{-1} = -\xi_{N-1}.$$

Numerical analysis was performed using the software package *tracer 3.7* developed by D. S. Glyzin (the method of the fourth-order Runge-Kutta precision increments of 10^{-3}) for $N = 25$, $\kappa = 10^{-3}$, $\epsilon = 1$ and the initial conditions

$$\xi_{0k} = (2/\sqrt{3})\cos\left((2j+1)\pi k/N\right),\ k = 1, \ldots, N. \quad (26)$$

Note that the initial conditions (26) are taken from the theoretical analysis, namely the initial condition is determined as the solution of the homogeneous equation corresponding to (13) as $\epsilon = 1$, $s = 0$, $\tau = j/N$, $j = 1, \ldots, N$.

It follows from (24) that the amount $K(\kappa)$ of the coexisting stable cycles of problem (13) satisfies the inequality $K(\kappa) \geq 5$ as $\kappa = 10^{-3}$. The fact is confirmed by the numerical simulations. Indeed, as a result of the computer experiment we get five different stable cycles.

Thus, we are developing the theory that not only works for small values of $\epsilon \ll 1$, but also predicts good result for $\epsilon \sim 1$. In the last case the number of simultaneously existing stable cycles coincides with their number expected from the theory for a small ϵ. Thus, the results for $\epsilon \sim 1$ are qualitatively the same and the problem (13) has the buffering property also as $\epsilon \sim 1$. From the theorem 1 stable solutions of the original problem (4) correspond to exponentially stable solutions of the boundary value problem (13).

References

[1] Glyzin S D, Kolesov A Yu, Rozov N Kh 2013 Modeling the Bursting Effect in Neuron Systems *Math. Notes* **93** 676–90
[2] Glyzin S D, Kolesov A Yu, Rozov N Kh 2012 Buffer phenomenon in neurodynamics *Doklady Mathematics* **85** 297–300
[3] Glyzin S D, Kolesov A Yu, Rozov N Kh 2012 Discrete autowaves in neural systems *Comput. Math. and Math. Physics* **52** 702–19
[4] Kashchenko S 2015 Models of Wave Memory (Springer) 239
[5] Glyzin S D, Kolesov A Yu, Rozov N Kh 2009 Extremal dynamics of the generalized Hutchinson equation *Comput. Math. and Math. Physics* **49** 71–83
[6] Kolesov A Yu, Mishchenko E F, Rozov N Kh 1998 Asymptotic Methods of Investigation of Periodic Solutions of Nonlinear Hyperbolic Equations *Proc. of the Steklov Inst. of Math.* **222** 1–189
[7] Kolesov A Yu, Mishchenko E F, Rozov N Kh 2007 New Methods for Proving the Existence and Stability of Periodic Solutions in Singularly Perturbed Delay Systems *Proc. of the Steklov Inst. of Math.* **259** 101–27
[8] Kolesov A Yu, Rozov N Kh 2004 Invariantnye tory nelineynykh volnovykh uravneniy [in Russian] (Moscow: FIZMATLIT) 408

Comparison of abilities of two trend definition techniques for experimental data time series processing

I O Ivanov[1], V V Borog[1], A V Kryanev[1], V G Getmanov[2] and R V Sidorov[2]

[1]National Research Nuclear University MEPhI, Kashirskoe highway 31, 115409, Moscow, Russia

[2]Geophysical Center of the Russian Academy of Sciences, Molodezhnaya St. 3, 119296, Moscow, Russia

E-mail: varbright@gmail.com

Abstract. The abilities of two processing techniques for time series of physical data are studied. One of this techniques is based on the implementing of singular spectrum analysis, while the other is based on local approximation approach. The operation of these techniques is analyzed on a model time series. Both methods are compared in terms of accuracy of trend definition for different values of the noise level.

1. Introduction

The problem of trend separation is very important for experimental data time series analysis. There are many physical processes time series for which consist of some slow trend and high-frequency component superimposed on the first one. Since usually only one of them is of interest the problem of trend separation arises. In this paper the two methods are considered: the method of trend separation based on singular spectrum analysis(SSA)[1, 2] and the second one based on local sliding approximation piecewise linear models with weighted average(SLAM)[3, 4]. These methods were tested on the specially generated model time series. To test the methods on different time scales we constructed time series from three consecutive fragments of periodical functions: decaying, amplitude-modulated and long-period ones. After this the model signal was mixed with white noise. Since a noise can significantly affect result of trend definition, we need to check the dependence of quality of trend separation on its level. To achieve this we varied the value of standard deviation of the noise (σ) and run describing technics on the obtained time series. In the Fig.1 the model time series with noise level $\sigma = 0.01$ is shown.

2. Singular spectrum analysis

Consider experimental data $y_1, \ldots, y_n$. Construct the trajectory matrix of this time series:

$$Y = \begin{pmatrix} y_1 & y_2 & \cdots & y_M \\ y_2 & y_3 & \cdots & y_{M+1} \\ \cdots & \cdots & \cdots & \cdots \\ y_L & y_{L+1} & \cdots & y_{L+M-1} \end{pmatrix}$$

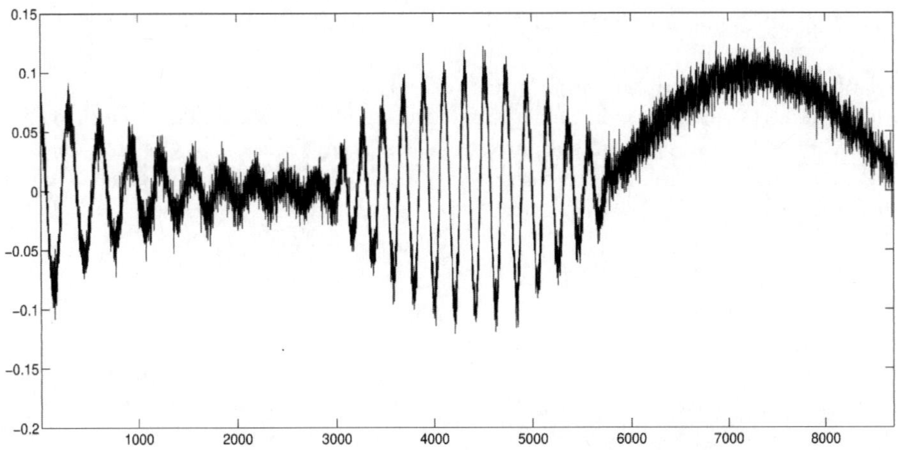

Figure 1. Model time series with noise level $\sigma = 0.01$.

M stands for the sliding window size, it is the parameter of this method and should be selected based on physical properties of the problem. Perform the singular value decomposition of trajectory matrix. Eigenvalues of matrix Y^TY are called singular values of matrix Y. Sort singular values in decreasing order of magnitude and denote $\psi_j = (\psi_{j1}, \ldots, \psi_{jM})^T, j = 1, \ldots, M$ is the corresponding orthonormal system of eigenvectors. Construct orthogonal matrix $\Omega = (\psi_1, \ldots, \psi_M)$ and define the matrix of principal components: $V = Y\Omega = (V_1, \ldots, V_M)$

So,

$$Y = V\Omega^T = \sum_{j=1}^{M}(V_j\psi_j^T)$$

is the decomposition of the original time series into principal components $V_j, j = 1, \ldots, M$.

The magnitude of singular values determines the level of "information contribution" of corresponding principal component to the source time series. Components with larger indices corresponds to more frequent fluctuations of the signal. Therefore, for trend separation we need to exclude the last few principal components, corresponding to high-frequency part of time series (noise). After this we get the next expression for time series decomposition.

$$Y' = \sum_{j=1}^{m}(V_j\psi_j^T)$$

where m is the number of the remaining components.

The resulting time series is reconstructed using one-to-one correspondence between trajectory matrix and time series. Applying diagonal averaging to result matrix Y' we get reconstructed series:

$$y'_1 = y'_{11}; y'_2 = (y'_12 + y'_21)/2; y'_3 = (y'_{13} + y'_{22} + y'_{31})/3$$

and so on.

Selecting of value for sliding window size is very important here, because it determines number of principal components i.e. "information resolution" of the method. If it is too small, the result trend will be mixed with noise. And vice versa, if the window size is too big, the real trend will be divided into several components. Because of these reasons, choosing of window size value is very difficult in case of a real problems due to lack of information about the real trend.

3. Method of local sliding approximation models

The second method of trend removal is based on using so-called local sliding approximation models with weighted averaging, and special correlation analysis for provision of consistency. Suppose that observations of original time series $y(T_i), i = 0, 1, \dots, N_f - 1$ are given. Construct the local sliding intervals:

$$N_{1j} = N_d(j-1); N_{2j} = N_{1j} + N - 1$$

where N_d is a sliding step, N is the size of local interval and $j = 1, \dots, m; N_d(m-1) + N = N_f$. Let the local sliding models will be represented by the functions $y_{Mj}(c_j, T_i)$; outside the local interval the models equal to zero. In a particular case local models can be piecewise linear - $y_{Mj}(c_j, Ti) = c_{1j} + c_{2j}Ti$. Let us take weighted sum of local models as a model of trend.

$$y_M(c, Ti) = \sum_{j=1}^{m} \alpha_j y_{Mj}(c_j, Ti)$$

α_j - weights, $c^T = (c_1^T, \dots, c_m^T)$ - parameters of model. The optimal model of trend $y_M(c^0, Ti)$ is a result of minimization of functional $S(c, y)$

$$S(c, y) = \sum_{i=0}^{N_f - 1} (y(Ti) - y_M(c, Ti))^2, c^0 = \arg\min_{c} S(c, y)$$

The model $y_M(c^0, Ti)$ and difference $\Delta y(Ti) = y(Ti) - y_M(c^0, Ti)$ depend on parameters N_d, N. This ones are being adjusted so that time series $\Delta y(Ti), (i = 0, 1, 2, \dots, N_f - 1)$ would be the most statistically similar to white noise. For this purpose we define local sliding autocorrelation functions and calculate ratios of their values for big and small time lags. Then we adjust parameters N_d, N to minimize the sum of calculated ratios.

4. Trend separation

Let us consider the work of described methods on model series with $\sigma = 0.01$. Since we are dealing with nonstationary series, which greatly changes its behavior on different intervals, we selected value for sliding window size as 32. The plot of singular values for model series is shown in Fig.2. As it is evident from the graph, the first singular value is a several tens of times larger than the others, so we can conclude that the first principal component corresponds to the trend. Result of trend definition by SSA method is shown in Fig.3

Let us consider the work of SLAM method with local interval size equal to 64 and sliding step equal to 4. The result of trend definition using method SLAM is shown in Fig.4.

As you can see, the result is almost the same as in the case of SSA method. Since we use artificial series, and know the real trend, the quality of considered methods can be compared using simple formula based on mean squared error [5]:

$$\delta = \sum_{i=1}^{N} (y_i^0 - \tilde{y}_i)^2,$$

where y^0 - source signal without noise and $\tilde{y}$ - resulting trend separated from the series. These calculations were performed for several values of noise level (i.e. standard deviation).

The plot shown in Fig.5 illustrates relationship between trend definition error and noise level for both techniques. Numerical values of the error are shown in Table 1.

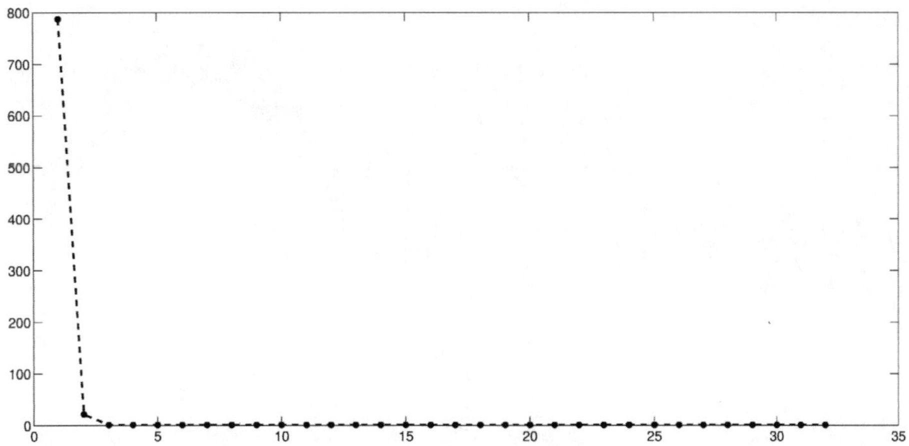

Figure 2. Singular values of model series with $\sigma = 0.01$.

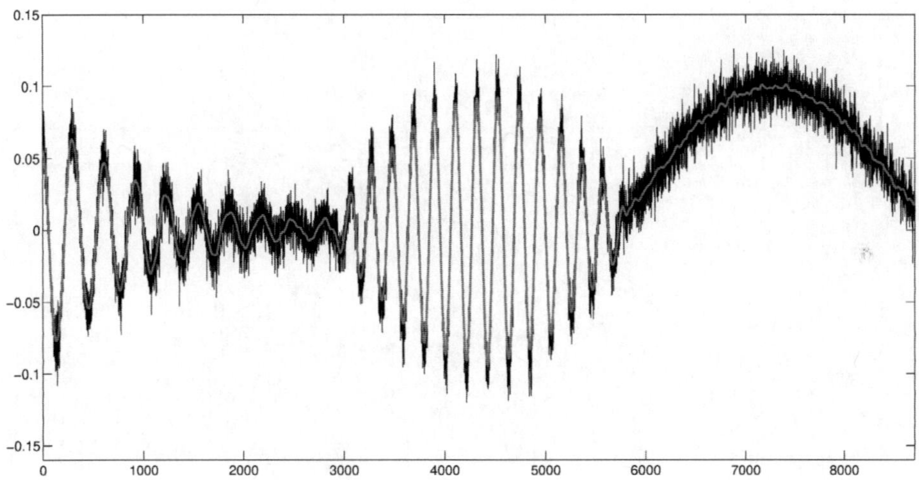

Figure 3. Result of trend definition using SSA method.

Table 1. Trend definition error for different values of noise level

σ	0.001	0.005	0.01	0.015	0.02	0.03	0.04	0.05	0.06
δ_{SSA}	0.2233	0.2322	0.2729	0.2666	0.3827	0.4859	0.5552	0.7021	0.8654
δ_{SLAM}	0.0565	0.0894	0.1489	0.1920	0.3168	0.4527	0.5239	0.6819	0.8768

5. Conclusions

In this paper two methods for trend definition are considered. These methods were tested on the specially generated model series. To compare two methods the mean squared error was

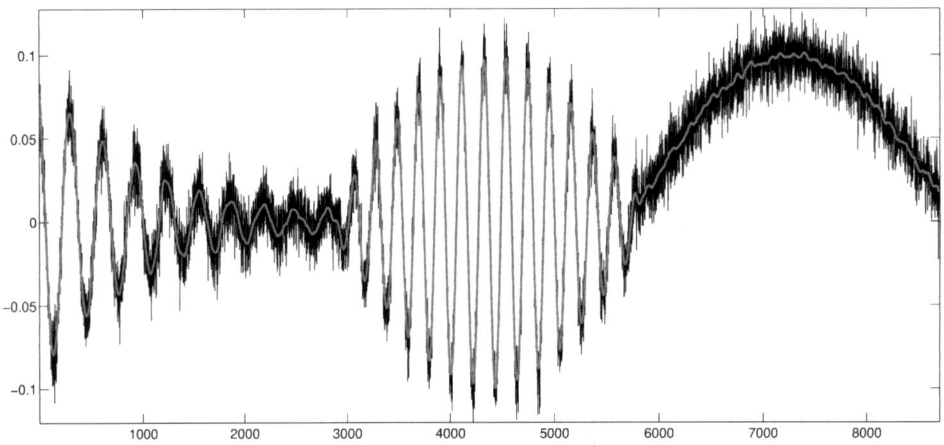

Figure 4. Result of trend definition using SLAM method.

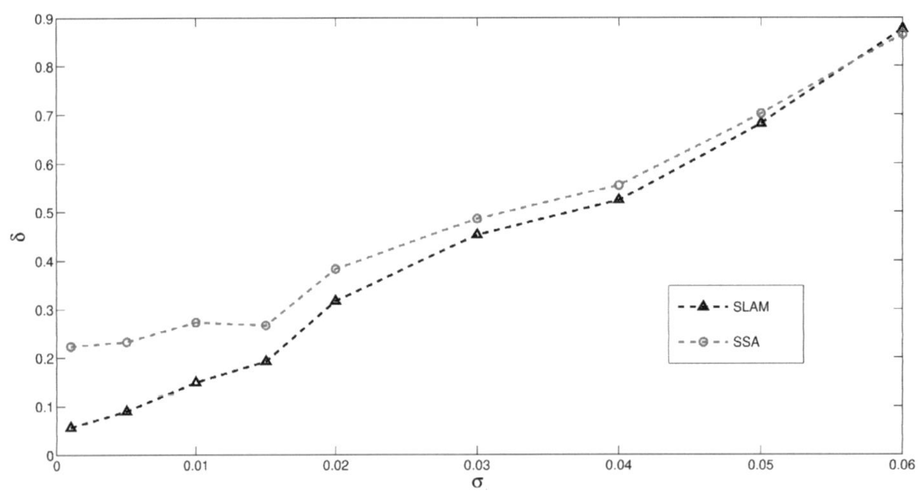

Figure 5. Trend definition error depending on noise level

calculated. As we can see from the plot in Fig.5, the method of local sliding intervals is about two times more accurate than SSA for the low noise levels. And for the high ones both methods shows almost equal accuracy.

References

[1] Golyandina N and Osipov E 2007 *Journal of Statistical Planning and Inference* **137** 2642
[2] Kryanev A, Lukin G and Udumyan D 2010 *Metric analysis and data processing* (Moscow: FIZMATLIT)
[3] Getmanov V G 2013 *Optoelectronics, Instrumentation and Data Processing* **49** 21
[4] Getmanov V G, Sidorov R V and Dabagyan R A 2015 *Measurement Techniques* **58** 1029
[5] Weisberg S 2005 *Applied Linear Regression* (Hoboken, New Jersey: Wiley-Interscience)

IOP Conf. Series: Journal of Physics: Conf. Series **788** (2017) 012018 doi:10.1088/1742-6596/788/1/012018

Model of attenuation of long waves under continuous ice layer

M B Kochanov, B A Petrov

Department of Applied Mathematics, National Research Nuclear University MEPhI, 31 Kashirskoe Shosse, 115409 Moscow, Russian Federation

E-mail: mbkochanov@mephi.ru

Abstract. In this work new mathematical model of long wave propagation on water surface with ice cover is proposed. The model of thin elastic plate is used to describe ice layer movement. Equation for ice cover contain additional term to takes into account dissipation effects in the ice cover to explain wave attenuation. Proposed model was reduced to one nonlinear evolution equation for water level perturbation. The expression for wave energy was obtained under assumption of long waves. Proposed model is numerically studied, energy of system is computed. Obtained results are compared with results of suggested before model that takes into account the flow law of Glen.

1. Introduction

The process of attenuation of surface wave under ice layer is well studied at the present time. There are some models for description energy loss in the system ice-water under wave propagation. Almost all proposed models consider system under linear theory approximation.

The one model was proposed by Wadham [1]. He takes into account the Glen law to describe energy lose in ice layer due to creep. He consider the propagation of a one-dimensional monochromatic wave propagating at right angle to the ice edge. As a result, he obtain that energy of wave is directly proportional to expression

$$\left(\frac{1}{(n-1)Sx + 1/A_i^{n-1}}\right)^2 \tag{1}$$

where S is some constant depending on system and wave properties, A_i is initial amplitude of wave and n is power in flow law of Glen. Wadham find that attenuation rate of wave is best fitted by exponent $n = 3$.

Another approach to problem was proposed by Weber [2]. He assume that ice floes very small and treat ice field as a viscous Newtonian fluid. Moreover he studies case of infinite rotating ocean. His model fit well field data from the marginal ice zone.

Considering only linear effects gives good approximation for wave propagation over few tens of wavelengths [3]. But waves in ocean can propagate over longer distance and nonlinear effects will accumulate and affect the wave form. So a nonlinear model for describing considered problem is needed. The model that takes into account Glen flow law proposed by Wadham can not be directly applied in this case.

We consider a problem of long wave propagation under ice layer. In this work the new phenomenological model is proposed. Thin plate model is used for ice layer. Proposed model takes into account creep effects in ice layer by adding new term in equation for its motion. The new term in equation is chosen in such a way that derived equation for water level perturbation is dissipative.

In nature the waves in open ocean usually are long, so perturbation method can be applied to reduce origin system of equations to simpler problem. In section II the nonlinear evolution equation that describe water level perturbation is derived. The expression for the total energy of wave in terms of water level perturbation written in non-dimensional variables is obtained. Section III contain description of numerical algorithm for solving boundary value problem for derived equation. In section IV boundary value problem is solved numerically. The behaviour of solitary wave is studied, time dependence of the wave energy is calculated.

2. Model

2.1. Equation derivation

Let us consider the potential incompressible flow of ideal fluid under thin elastic plate over flat bottom. This motion is described by the following equations [4, 5]

$$\begin{cases} \varphi_{xx} + \varphi_{yy} = 0, \quad -\infty < x < +\infty,\, 0 < y < h(x,t), \\ \dfrac{\partial \varphi}{\partial y} = 0, \quad y = 0, \\ \eta_t + \varphi_x \eta_x = \varphi_y, \quad y = h(x,t), \\ \varphi_t + \dfrac{\varphi_x^2 + \varphi_y^2}{2} + gy + \dfrac{P}{\rho_w} = gh_0 + \dfrac{P_0}{\rho_w}, \quad y = h(x,t), \\ \dfrac{Eh_i^3}{12(1-\nu^2)}\eta_{xxxx} - h_i \sigma_{xx}\eta_{xx} = P - P_0 - \rho_i h_i \eta_{tt} - \kappa \eta_{xxx}. \end{cases} \tag{2}$$

Here φ is the velocity potential of the liquid, h_0 is depth of liquid, η is the deviation of liquid from its equilibrium position ($h(x,t) = h_0 + \eta(x,t)$), ρ_i and h_i are the density and thickness of the ice, g is the gravitation acceleration, P_0 is the atmospheric pressure, ρ_w is the density of water, P is the external load, E is the elastic modulus and $\sigma_{xx} = \text{const}$ is the component of stress tensor. We introduce the dissipative term $-\kappa\eta_{xxx}$ to the right part of the equation of plate motion, so κ is the phenomenological coefficient describing dissipation of wave energy under ice cover due to bend.

After changing of variables in (2)

$$x = lx', \quad y = h_0 y', \quad \eta = a\eta', \quad t = \frac{l}{c_0}t', \quad \varphi = \frac{gla}{c_0}\varphi', \tag{3}$$

where l is a wavelength, a is a characteristic amplitude and $c_0 = \sqrt{gh_0}$ is a phase velocity in the linear approximation, we obtain the following problem

$$\begin{cases} \mu\varphi_{xx} + \varphi_{yy} = 0, \quad -\infty < x < +\infty,\, 0 < y < 1 + \varepsilon\eta(x,t), \\ \dfrac{\partial \varphi}{\partial y} = 0, \quad y = 0, \\ \eta_t + \varepsilon\varphi_x\eta_x = \dfrac{\varphi_y}{\mu}, \quad y = 1 + \varepsilon\eta, \\ \varphi_t + \dfrac{\varepsilon}{2}\varphi_x^2 + \dfrac{\varepsilon}{2\mu}\varphi_y^2 + \eta + \delta\eta_{tt} - \beta\eta_{xx} + \tau\eta_{xxx} + \gamma\eta_{4x} = 0, \quad y = 1 + \varepsilon\eta. \end{cases} \tag{4}$$

Here ε, μ, δ, β, τ, and γ are small parameters of the same orders defined by the following expressions

$$\varepsilon = \frac{a}{h_0}, \quad \mu = \frac{h_0^2}{l^2}, \quad \delta = \frac{\rho_i h_i h_0}{\rho_w l^2}, \quad \beta = \frac{h_i \sigma_{xx}}{\rho_w g l^2}, \quad \tau = \frac{\kappa}{\rho_w g l^3}, \quad \gamma = \frac{Eh_i^3}{12(1-\nu)^2 \rho_w g l^4}. \tag{5}$$

Here we suppose that all parameters have the same order.

Expanding $\varphi(x, y, t)$ in a Taylor series in powers of y we obtain, from the first equation and first boundary condition of problem (4)

$$\varphi(x, y, t) = \sum_{k=0}^{\infty}(-1)^k \frac{\mu^k}{(2k)!} \frac{\partial^{2k} f(x, t)}{\partial x^{2k}} y^{2k}, \tag{6}$$

Substitution of (6) into the second and third boundary conditions of (4) yields system of nonlinear partial differential equation, which with an accuracy up to terms of order $O(\varepsilon^2)$ looks

$$\begin{aligned} &w_t + \eta_x + \varepsilon w w_x - \frac{1}{2} w w_{xxt} + \gamma\eta_{5x} - \beta\eta_{xxx} + \tau\eta_{4x} + \delta\eta_{ttx} + \\ &+\varepsilon\mu\left(-(\eta w_{xt})_x - \frac{1}{2} w w_{xxx} + \frac{1}{2} w_x w_{xx}\right) + \frac{1}{24}\mu^2 w_{xxxxt} = 0, \\ &\eta_t + w_x + \varepsilon(w\eta)_x - \frac{1}{6}\mu w_{xxx} - \frac{1}{2}\varepsilon\mu(\eta w_{xx})_x + \frac{1}{120}\mu^2 w_{5x} = 0, \end{aligned} \tag{7}$$

where $w(x, t) = \dfrac{\partial}{\partial x} f(x, t)$.

In the zero order approximation system (7) takes a form

$$\begin{aligned} &w_t + \eta_x = 0, \\ &\eta_t + w_x = 0, \end{aligned}$$

and for compatibility of this system we have to assume

$$\begin{aligned} w = {} &\eta + A^1\varepsilon + A^2\mu + A^3\gamma + A^4\beta + A^5\delta + A^6\tau + A^{12}\varepsilon\mu + A^{13}\varepsilon\gamma + A^{14}\varepsilon\beta + A^{15}\varepsilon\delta + \\ &+ A^{16}\varepsilon\tau + A^{23}\mu\gamma + A^{24}\mu\beta + A^{25}\mu\delta + A^{26}\mu\tau + A^{34}\gamma\beta + A^{35}\gamma\delta + A^{36}\gamma\tau + \\ &+ A^{45}\beta\delta + A^{46}\beta\tau + A^{56}\delta\tau + A^{11}\varepsilon^2 + A^{22}\mu^2 + A^{33}\gamma^2 + A^{44}\beta^2 + A^{55}\delta^2 + A^{66}\tau^2, \end{aligned} \tag{8}$$

where $A^k = A^k(x, t)$.

Applying perturbation method in the form given in [5] to (7) we obtain differential equation for describing nonlinear waves of liquid under continuous ice layer

$$\begin{aligned} &\eta_t + c_3\eta_{xxx} + c_4\eta_{4x} + c_5\eta_{5x} + c_6\eta_{6x} + c_7\eta_{7x} + c_8\eta_{8x} + c_9\eta_{9x} + \\ &+ c_{01}\eta\eta_x + c_{03}\eta\eta_{xxx} + c_{12}\eta_x\eta_{xx} + c_{04}\eta\eta_{4x} + c_{22}\eta_{xx}^2 + c_{13}\eta_x\eta_{xxx} + \\ &+ c_{05}\eta\eta_{5x} + c_{14}\eta_x\eta_{4x} + c_{23}\eta_{xx}\eta_{xxx} + c_{001}\eta^2\eta_x = 0. \end{aligned} \tag{9}$$

Values of constant coefficients c_k are defined through small parameters (5) by following expressions

$$\begin{aligned} &c_3 = \frac{\mu}{6} - \frac{\beta}{2} + \frac{\delta}{2}, \quad c_4 = \frac{\tau}{2}, \quad c_5 = -\frac{\mu\beta}{12} + \frac{\mu\delta}{4} - \frac{\beta\delta}{4} + \frac{\gamma}{2} + \frac{19\mu^2}{360} + \frac{3\delta^2}{8} - \frac{\beta^2}{8}, \\ &c_6 = \frac{\mu\tau}{12} + \frac{\delta\tau}{4} + \frac{\beta\tau}{4}, \quad c_7 = \frac{\mu\gamma}{12} + \frac{\gamma\delta}{4} + \frac{\beta\gamma}{4} - \frac{\tau^2}{8}, \quad c_8 = -\frac{\tau\gamma}{4}, \quad c_9 = -\frac{\gamma^2}{8}, \\ &c_{01} = \frac{3\varepsilon}{2}, \quad c_{03} = \frac{7\varepsilon\delta}{4}, \quad c_{12} = \frac{5\beta\varepsilon}{8} + \frac{23\varepsilon\mu}{24} + \frac{31\varepsilon\delta}{8}, \quad c_{04} = \frac{\tau\varepsilon}{4}, \quad , c_{13} = -\tau\varepsilon \\ &c_{22} = -\frac{9\tau\varepsilon}{8}, \quad c_{05} = \frac{\gamma\varepsilon}{4}, \quad c_{14} = -\frac{11\gamma\varepsilon}{8}, \quad c_{23} = -\frac{15\gamma\varepsilon}{4}, \quad c_{001} = -\frac{3\varepsilon^2}{8}. \end{aligned} \tag{10}$$

2.2. Small parameters of equation

Values of coefficients (10) were analyzed for typical parameters of model: the length of wave is from 100 m to 1 km, the wave amplitude is of the order of 1 m, water depth is from 10m to 50m, the thickness of ice is of the order of 1 m. Direct calculation of all coefficients in this range shows that the biggest coefficient is c_{01}. The another coefficients in equation is usually two to three orders less than c_{01}.The coefficients c_3, c_{03}, c_{12}, c_{001} can be of the order 10^{-3}–10^{-4}. The remaining coefficients has the lower order. The order of coefficient c_4 depend on value of parameter κ. We want to take into account dissipative effects in the model, so we keep it in equation.

Waves in ocean can penetrate several hundreds of kilometers. It corresponds to time 10^3–10^4 in nondimensional variables for wave length from 100 m to 1 km. So we can omit all coefficients in equation having the order 10^{-5} and less: c_5, c_6, c_7, c_8, c_9, c_{04}, c_{22}, c_{13}, c_{05}, c_{14}, c_{23}.

2.3. Energy of long wave

The total energy of long wave consist of kinetic and potential energy. Elastic energy of ice layer we do not take into account. We suppose that potential energy of nonperturbed system is zero. The energy of unit-width wave can be calculated by the formula (potential energy can be negative for the wave with negative amplitude)

$$E = \frac{\rho_w}{2}\int_{-\infty}^{\infty}\int_{0}^{h(x,t)}\left(\varphi_x^2+\varphi_y^2\right)dxdy + \frac{\rho_w g}{2}\int_{-\infty}^{\infty}\operatorname{sgn}(h-h_0)\,(h-h_0)^2\,dx.$$

In nondimentional variables (3) energy has the form

$$E = E_0\varepsilon\left(\int_{-\infty}^{\infty}\int_{0}^{1+\epsilon\eta(x,t)}\left(\varphi_x^2+\frac{1}{\mu}\varphi_y^2\right)dxdy + \int_{-\infty}^{\infty}\operatorname{sgn}(\eta)\eta^2dx\right) \tag{11}$$

Here E_0 is potential energy of unit-width unperturbed water reservoir measured with respect to bottom $y=0$. Note, that $\varphi_x^2 = O(1)$, while $\varphi_y^2 = O(\mu^2)$. So the kinetic energy related to vertical motion has the higher order of smallness than the energy related to horizontal motion.

Taking into account expansion (6) for potential and the expansion (8) expression for the energy (11) can be written in terms of $\eta(x,t)$

$$E = E_0\varepsilon^2\int_{-\infty}^{\infty}\left((1+\operatorname{sgn}(\eta))\eta^2 + (\beta-\delta)\eta_x^2 + \gamma\eta_{xx}^2 + \frac{\varepsilon}{2}\eta^3\right)dx. \tag{12}$$

Therefore the first correction to the total energy of nonperturbed system has the order $O(\varepsilon^2)$.

3. Numerical modeling

As was said above, we can omit following coefficients in Eq.(9): c_5, c_6, c_7, c_8, c_9, c_{04}, c_{22}, c_{13}, c_{05}, c_{14}, c_{23}. As a result we have equation (hereafter we denote function $\eta(x,t)$ as $u(x,t)$)

$$u_t + c_3u_{xxx} + c_4u_{4x} + c_{01}uu_x + c_{03}uu_{xxx} + c_{12}u_xu_{xx} + c_{001}u^2u_x = 0. \tag{13}$$

For numerical studying of wave attenuation the boundary value problem for Eq.(13) is numerically solved with necessary number of boundary conditions:

$$u(0,t) = u(L,t) = u_x(0,t) = u_x(L,t) = u_{xx}(0,t) = u_{xx}(L,t) = 0. \tag{14}$$

Here L is big enough for wave was localized in domain $[0; L]$ for all $t < T$.

Finite difference method is used to solve problem (13), (14). Eq.(13) can be written in the form

$$\frac{\partial u}{\partial t} + L[u] = 0,$$
$$L[u] = c_{01}uu_x + c_3u_{3x} + c_4u_{4x} + c_{03}uu_{3x} + c_{12}u_xu_{xx} + c_{001}u^2u_x.$$

For numerical modeling we use implicit finite difference scheme, which leads to system

$$\boldsymbol{F}(\boldsymbol{u}^j) = 0,$$
$$F_i(\boldsymbol{u}^j) = \frac{u_i^j - u_i^{j-1}}{\tau} + \hat{L}\left[\frac{u_i^j + u_i^{j-1}}{2}\right].$$

Here $\hat{L}$ is finite difference operator. All differential operators in L are approximated with central differences to a second order of accuracy. The truncation error of this scheme is $O(\tau^2) + O(h^2)$. Numerical experiments shows that this scheme is unconditionally stable. A couple of numerical algorithms for solving KdV equations is presented in work [6]. The suggested methods can be generalized for solving derived is Section II equation.

We have a system of nonlinear algebraic equations on each time layer, which is solved by Newton's method. On the each iteration of Newton's method solution is computed by formula

$$\boldsymbol{u}^{j\ (n+1)} = \boldsymbol{u}^{j\ (n)} - J^{-1}\boldsymbol{F}(\boldsymbol{u}^{j\ (n)}), \quad J = \frac{\partial \boldsymbol{F}(\boldsymbol{u}^{j\ (n)})}{\partial \boldsymbol{u}^{j\ (n)}}. \tag{15}$$

Exact form of Jacobin matrix J can be found in Appendix. For the first iteration of algorithm the solution from the previous time layer is used

$$\boldsymbol{u}^{j\ (0)} \equiv \boldsymbol{u}^{j-1}.$$

L^∞-norm is used to control convergence of Newton's method

$$\frac{\|\boldsymbol{u}^{j\ (n+1)} - \boldsymbol{u}^{j\ (n)}\|}{\|\boldsymbol{u}^{j\ (n+1)}\|} < 1E - 16.$$

So, a five-tridiagonal system of equations is required to be solved at each step of Newton's method. We use DGBSV solver from LAPACK library to solve this system of equations.

4. Results

To verify the proposed mathematical model several cases of problem was numerically solved. In table 4 the values of model parameters and corresponding values of coeffcieints in Eq.(13) are presented. The following initial condition was used in numerical simulation

$$u(x, 0) = \frac{1}{\cosh^2(x - x_0)}.$$

Here x_0 is some constant. In the first two cases value $x_0 = 10$ was used. In the third case $x_0 = 50$. This condition correspond to inis-size wave written in nondimensional variables (3). The time and spatial steps is chosen to be 0.02. Numerical tests with exact solutions for special cases of equation give the error or order 10^{-3}–10^{-4} with chosen steps.

On the figure 4 the numeric solution of Eq.(13) with boundary conditions (14) is presented for all cases from table 4. As it can be seen, the wave shape is various for all cases and heavily

Table 1. Values of parameters (5) and coefficients (10) for $E = 3 \times 10^9\,\mathrm{N/m^2}$, $\sigma_{xx} = 1 \times 10^4\,\mathrm{N/m^2}$, $\nu = 0.3$, $\rho_i = 900\,\mathrm{kg/m^3}$, $\rho_w = 1025\,\mathrm{kg/m^3}$.

h_0, m	l, m	a, m	h_i	expansion parameters (5)		coefficients in Eq.(9)	
				$\varepsilon = 6.00 \times 10^{-2}$	$\mu = 6.94 \times 10^{-3}$	$c_{01} = 1.00 \times 10^{-1}$	$c_3 = 1.94 \times 10^{-3}$
15	150	1	1	$\gamma = 3.38 \times 10^{-6}$	$\beta = 1.11 \times 10^{-5}$	$c_{001} = 1.67 \times 10^{-3}$	$c_{12} = 7.92 \times 10^{-4}$
				$\delta = 2.44 \times 10^{-4}$	$\tau = 1.00 \times 10^{-2}$	$c_{03} = 3.45 \times 10^{-4}$	$c_4 = 2.50 \times 10^{-4}$
				$\varepsilon = 1.25 \times 10^{-2}$	$\mu = 4.00 \times 10^{-2}$	$c_{01} = 1.88 \times 10^{-2}$	$c_3 = 7.09 \times 10^{-3}$
40	200	0.5	1	$\gamma = 1.71 \times 10^{-5}$	$\beta = 2.49 \times 10^{-5}$	$c_{12} = 5.22 \times 10^{-4}$	$c_4 = 2.50 \times 10^{-4}$
				$\delta = 8.78 \times 10^{-4}$	$\tau = 1.00 \times 10^{-2}$	$c_{03} = 2.27 \times 10^{-4}$	$c_5 = 1.02 \times 10^{-4}$
				$\varepsilon = 2.67 \times 10^{-2}$	$\mu = 3.52 \times 10^{-2}$	$c_{01} = 4.00 \times 10^{-2}$	$c_3 = 6.26 \times 10^{-3}$
75	400	2	2	$\gamma = 8.30 \times 10^{-6}$	$\beta = 1.28 \times 10^{-5}$	$c_{12} = 9.84 \times 10^{-4}$	$c_{03} = 2.29 \times 10^{-4}$
				$\delta = 9.38 \times 10^{-4}$	$\tau = 1.00 \times 10^{-2}$	$c_{001} = 2.67 \times 10^{-4}$	$c_4 = 2.50 \times 10^{-4}$

depend on the parameters of the model. The attenuation of the wave energy was numerically calculated with taking into account expression (12). Results presented on the figure 4. The shape of plot for the energy is similar to results, obtained by Wadham [1]. The value of parameter τ was fixed in all three cases. The additional tests for other values of τ shows that the shape of energy curve does not crucially depend on it exact value. Additional investigation is required to explore the dependence of energy attenuation on it value.

However obtained curve of energy against time can not be properly fitted by the expression (1), which has good agreement with field data [1]. There some reasons for this. The proposed model is phenomenological and require additional argumentation. Furthermore proposed model does not take into account some processes. As it can be seen on figure 4, the curve of the solution in some points is very bended. In practice, this leads to the ice break up and subsequent energy release. The model must be changed to take into account such processes. The bottom surface of the ice layer is not plain for real ice floe and turbulent mixing must be considered.

Acknowledgments
This work was supported by Russian Foundation For Basic Research, grant no. 16-31-00498.

Appendix
The nonzero coefficients in Jacobian (15) have a form

$$J_{0,0} = 1,$$

$$J_{1,0} = -\frac{3}{2h}, \quad J_{1,1} = \frac{2}{h}, \quad J_{1,2} = -\frac{1}{2h},$$

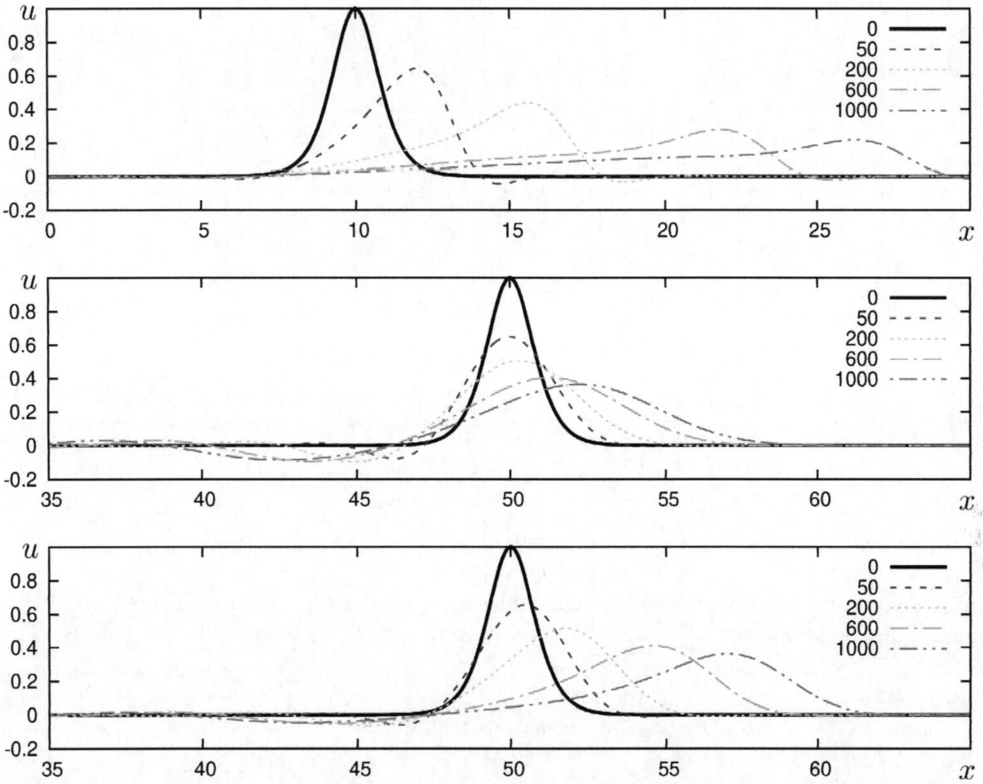

Figure 1. Numerical solution of the problem (13), (14) for three cases from table 4 for different time

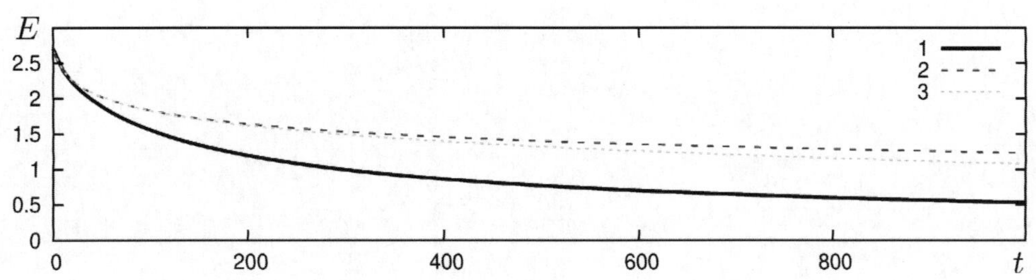

Figure 2. Energy of wave (12) for three cases from table 4

$$
\begin{aligned}
J_{i,i+2}(\boldsymbol{u}^j) &= \frac{c_3}{4h^3} + \frac{c_4}{2h^4} + \frac{c_{03}}{8h^3}\left(u_i^j + u_i^{j-1}\right),\\
J_{i,i+1}(\boldsymbol{u}^j) &= -\frac{c_3}{2h^3} - \frac{2c_4}{h^4} + \frac{c_{01}}{8h}\left(u_i^j + u_i^{j-1}\right) - \frac{c_{03}}{4h^3}\left(u_i^j + u_i^{j-1}\right) +\\
&+ \frac{c_{12}}{4h^3}\left(-u_i^j + u_{i+1}^j - u_i^{j-1} + u_{i+1}^{j-1}\right) + \frac{c_{001}}{16h}\left(u_i^j + u_i^{j-1}\right)^2,\\
J_{i,i}(\boldsymbol{u}^j) &= \frac{3c_4}{h^4} + \frac{c_{01}}{8h}\left(u_{i+1}^j - u_{i-1}^j + u_{i+1}^{j-1} - u_{i-1}^{j-1}\right) +\\
&+ \frac{c_{03}}{8h^3}\left(-u_{i-2}^j + 2u_{i-1}^j - 2u_{i+1}^j + u_{i+2}^j - u_{i-2}^{j-1} + 2u_{i-1}^{j-1} - 2u_{i+1}^{j-1} + u_{i+2}^{j-1}\right) -\\
&- \frac{c_{12}}{4h^3}\left(u_{i+1}^j - u_{i-1}^j + u_{i+1}^{j-1} - u_{i-1}^{j-1}\right) +\\
&+ \frac{c_{001}}{8h}\left(u_i^j + u_i^{j-1}\right)\left(u_{i+1}^j - u_{i-1}^j + u_{i+1}^{j-1} - u_{i-1}^{j-1}\right),\\
J_{i,i-1}(\boldsymbol{u}^j) &= \frac{c_3}{2h^3} - \frac{2c_4}{h^4} - \frac{c_{01}}{8h}\left(u_i^j + u_i^{j-1}\right) + \frac{c_{03}}{4h^3}\left(u_i^j + u_i^{j-1}\right) +\\
&+ \frac{c_{12}}{4h^3}\left(u_i^j - u_{i-1}^j + u_i^{j-1} - u_{i-1}^{j-1}\right) - \frac{c_{001}}{16h}\left(u_i^j + u_i^{j-1}\right)^2,\\
J_{j,i-2}(\boldsymbol{u}^j) &= -\frac{c_3}{4h^3} + \frac{c_4}{2h^4} - \frac{c_{03}}{8h^3}\left(u_i^j + u_i^{j-1}\right).
\end{aligned}
$$

Here $i = 2, \ldots, n-2$, where n is maximal index of spatial node. Values $J_{n,i}$ can be obtained from and $J_{0,i}$ by giving opposite sign. While the values of $J_{n-1,i}$ have the same sign as $J_{1,i}$.

References

[1] Wadhams P 1973 *Journal of Geophysical Research* **78** 3552–3563
[2] Weber J E 1987 *Journal of Physical Oceanography* **17** 2351–2361
[3] Liu A K and Mollo-Christensen E 1988 *Journal of Physical Oceanography* **18** 1702–1712
[4] Marchenko A 1988 *Journal of Applied Mathematics and Mechanics* **52** 180–183
[5] Kochanov M, Kudryashov N and Sinel'shchikov D 2013 *Journal of Applied Mathematics and Mechanics* **77** 25–32
[6] Taha T R and Ablowitz M J 1988 *Journal of Computational Physics* **77** 540–548

Developing new mathematical method for search of the time series periodicity with deletions and insertions

E V Korotkov[1,2], M A Korotkova[1]

[1] National Research Nuclear University MEPhI (Moscow Engineering Physics Institute), Kashirskoe highway, 31, 115409, Moscow, Russia
[2] Institute of Bioengineering, Research Center of Biotechnology of the Russian Academy of Sciences, Leninsky Ave. 33, bld. 2, 119071, Moscow, Russia

E-mail: genekorotkov@gmail.com

Abstract. The purpose of this study was to detect latent periodicity in the presence of deletions or insertions in the analyzed data, when the points of deletions or insertions are unknown. A mathematical method was developed to search for periodicity in the numerical series, using dynamic programming and random matrices. The developed method was applied to search for periodicity in the Euro/Dollar (Eu/$) exchange rate, since 2001. The presence of periodicity within the period length equal to 24 h in the analyzed financial series was shown. Periodicity can be detected only with insertions and deletions. The results of this study show that periodicity phase shifts, depend on the observation time. The reasons for the existence of the periodicity in the financial ranks are discussed.

1. Introduction.

Identification of the cyclic patterns in numeric time series and symbolical sequences may shed light on the processes occurring in systems of a different nature and give information about the structure of different time series. To identify periodicities in the time series and symbolical sequences, the methods mainly used are based on the Fourier transform, wavelet transform and dynamic programming, as well as some other method [1–10]. Previously, we proposed the method of informational decomposition, which allows the detection of periodicity in both symbolic and numeric sequences; however, the detection was severely impeded by the above approaches [11]. These difficulties emanate from the fact that methods based on Fourier and Wavelet transform, decompose the statistical significance of long periods (larger than the size of the analyzed sequence alphabet) for smaller periods with a multiple of the length [11]. Also, these methods are very sensitive to the insertion and deletion of symbols. This leads to the fact that spectral methods cannot detect periodicity at a statistically significant level, even in the presence of a few deletions or insertions. Dynamic programming, which allows the detection of deletions and insertions in the periods, cannot detect relatively "fuzzy" periodicity. This is due to the fact that this method is based on finding the similarity between pairs of periods in the studied sequence [4]. However, if a statistically significant similarity is absent between two separate periods, dynamic programming will fail to detect periodicity in the analyzed sequence. The lack of similarity between two separate periods can be observed for latent periodicity, where the periodicity occurs on the background of random noise [11]. To find such periodicity, the method of informational decomposition was used [11]. This method enabled the discovery of latent periodicity in the DNA sequences of many genes

[12,13] and revealed the latency of amino acid specific for the protein families [14,15]. These results suggest that latent periodicity with insertions and deletions can be detected in sequences of a different nature and in numerical sequences also.

Today, it is known that people, animals and plants have biological rhythms. The manifestation of these rhythms can be observed at all levels of biological organization. The interaction can be observed between the rhythms that affect the internal state of the person and on various social processes. A. Chizhevsky first drew attention to the influence of natural factors on social processes [16] The biological rhythms can also influence currency rates. If the periods in the exchange rates exist, then the various events in public life can affect them. Such events can lead to changes in the data and could be identified as a phase shift of the period. The phase shifts in the sequence could have resulted from deletion or insertion of values with respect to the existing period. Therefore, in order to detect this periodicity, it was necessary to develop a mathematical method for detecting the periodicity of the time series, taking into account the unknown location and unknown number of insertions or deletions, in the presence of large noise. It cannot be done with the help of all known mathematical methods. Previous studies have searched for insertions or deletions [17], but failed to find a periodicity with large noise [18]. It is right mainly for dynamic programming. Either of these methods works well in the presence of large noise, but fails to find periodicity in the presence of even small amounts of insertions and deletions [11].

This paper contributed to filling the gap in mathematical methods for periodicity search. A mathematical method was developed to find periodicity in the symbolical sequences, in the presence of insertions and deletions and the big noise. The method was developed using the random matrices of periodicity and the method of dynamic programming. The developed method was applied to search for periodicity in the exchange rate of the Euro to the US Dollar. To search for periodicity, the numerical sequence was converted to a symbolical sequence. The calculations show that there is a periodicity equal to 24 h in the Euro-Dollar exchange rate. This periodicity of the exchange rate contains a lot of insertions and deletions and could not be detected by previously developed mathematical methods.

2. Methods and Algorithms

For study, we took the numerical sequence of the Euro/US$ rate. The candle opening and closing was separated by an interval of 1 and 4 h. Let $x_1(i)$ - be the rate at the time of opening of the candle, and $x_2(i)$- the rate at the close of the candle. At the first numerical sequence A_1, we calculated the difference $a(i)=x_2(i)-x_1(i)$, where $x_1(i)$ and $x_2(i)$ are separated by a 1 h. The beginning of the candle falls on the start of each hour and the end of the candle at finish of each hour. The sequence A_1 was obtained from 14.08.2014 to 07.04.2015 and contained 3930 numbers ($a(i)$, i=1,...,3930). We used all days of the week except Saturday and Sunday. These days the trade is not made or the volume of trade is very low. Then, this sequence was translated into a symbolic form, as described below, and the sequence S_1 was obtained. The encoding is shown in Table 1.

At the sequence A_2, the difference $a(i)=x_2(i)-x_1(i)$, was calculated, where $x_1(i)$ and $x_2(i)$ are separated by 4 hours. This means that the beginning of the candle occurs at the beginning of each 4 hours, and the end of the candle is the end of each 4 hours. The sequence A_2 was obtained for data from 05.12.2014 to 07.04.2015 and it contained 4027 numbers. After translation of this sequence in the symbolic sequence, the sequence S_2 was obtained. We used all days of the week except Saturday and Sunday also. Numerical data were taken from the Meta Trader 4 Alfa-Forex client terminal, site http://www.alfa-forex.ru. These data can be received from http://www.finam.ru also.

The numeric sequences A_1 and A_2 of length N_1 and N_2 were transformed in the symbolic sequence S_1 and S_2 with alphabet of 20 letters (Table 1). To convert the numerical sequence, the minimum and maximum elements of the sequences A_1 and A_2 were determined. Then, this interval was divided into 20 intervals, the number of elements of the numeric sequence in each interval was approximately equal to $N_1/20$ and $N_2/20$, respectively. Each interval received the letter of the Latin alphabet. If a numeric sequence contained many of the equal values than the boundaries of the intervals were varied in such a way that the same numbers were encoded by the same symbol. The coding of the sequence A_1 is shown in Table 1.

Table 1. The coding of sequence A_1 to obtain the sequence S_1 is shown.

K	N	I	M	T	R	S	L	Y	F
-.01042 -0.00215	-0.00215 -0.00149	-0.00149 -0.00108	-0.00108 -0.00082	-0.00082 -0.00062	-0.00062 -0.00047	-0.00047 -0.00034	-0.00034 -0.00021	-0.00021 0.00011	-0.00011 0.00002
C	W	P	H	Q	V	A	D	E	G
0.00002 0.00008	0.00008 0.00017	0.00017 0.00029	0.00029 0.00040	0.00040 0.00052	0.00052 0.00071	0.00071 0.00097	0.00097 0.00130	0.00130 0.00202	0.00202 1.00000

To search the periodicity with insertions and deletions, the next algorithm was used. Firstly, a set of random matrices with size 20x*n* was generated, where *n* is the length of the period, and 20 is the alphabet size of the studied sequence. Then, a local alignment of the studied sequence (S_1 *or* S_2) was built relative to each of the generated random matrices. Dynamic programming was used to build the local alignment and in determining the similarity function *F*. Then, the matrices were transformed because the distribution of the maximum of the similarity function *F* for each of the matrices for random sequences (set *Q*, volume of the set is 1000 sequences) should be similar. The transformed matrix having the highest value of the similarity function *F*, with the studied sequence *S*, was chosen. Then, this matrix was optimized to achieve the highest value of the similarity function *F* (*maxF*) with the studied sequence *S* and the transformed matrix was called *T*. For this purpose we used the genetic algorithm [19,20]. Then, the value of *maxF* for each random sequence from the set *Q* and for matrix *T* was calculated. It allowed the mean value and variance for *maxF* to be determined. This algorithm was applied for periods of different lengths and for each length of the period *n*, the corresponding value of $Z = (maxF - \overline{maxF}) / D(maxF)^{0.5}$ was calculated. As a result of the algorithm, the dependence of *Z* on *n* was obtained and denoted as *Z(n)*. It should be noted that in this study, dynamic programming was used to find a local alignment. This means that the boundaries of the regions with *maxF*, may differ from the beginning and end of the studied sequence. It means also that the values of *Z(n)* for different *n* can be obtained for different fragments of the studied sequence. The boundaries of the fragments, obtained for relevant values of *Z(n)* are shown. More details of the applied mathematical method are shown in [19,20].

3. Results

Figure 1 shows the application of the developed approach to the random sequence and the sequence S_3 obtained from the sequence S_4=(EFKLWNMSTWRYLQKLWQSMETMQ)$_{16}$. Randomly, 75% substitutions were done in the sequence S_4, after which our developed approach was applied. The results of the analysis of random sequences show that *Z(n)* fluctuated around 4.0. Noticeable, a small trend toward increasing values of *Z* for the random sequence is only for lengths of periods of more than 70. From these results, it can be concluded that the periods with Z-values>7.0 were found to be interesting. The results of the study of sequence S_4 show that the developed approach was able to identify fuzzy frequency and determine the optimal weighting matrix [20] for the detection of periodicity with length equal to 24 symbols. It is possible to see periodicity equal to 48 and 72 symbols. However, large values of *Z* also received periods with lengths close to 48 and 72 symbols, because there is the possibility of insertions or deletions. It is always possible to do insertions, or deletions to align the sequence S_4 against the matrices and to receive periodicity near to 48 to 72 symbols with high *Z*. However, this leads to a slight drop in the values of *Z*, which is related to the penalties for insertions or deletions. So the graph around 48 and 72 symbols, looks like a gently sloping mountain.

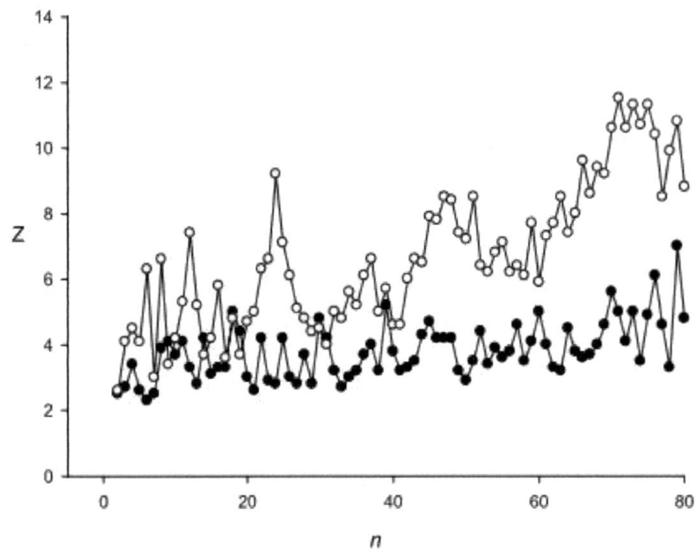

Figure 1. Black circles show the $Z(n)$ for the random sequence. White circles show the $Z(n)$ for the artificial sequence with a period of 24 symbols with total length of 384 symbols and 75% random substitutions (S_4). It is evident that despite the large number of random substitutions, the developed mathematical method could detect the periodicity with n=24.

Next, the developed approach was applied to the periodicity search for the symbolic sequence S_1. The spectrum of $Z(n)$ obtained for the sequence S_1 is shown in figure 2A. From this figure, it can be seen that the sequence S_1 has two periods. The first period equals 24 hours and has the largest value of Z from 341 to 3533, the hours of the sequence S_1. 39 insertions and deletions were made to find this period. The second period has a length equal to 23 hours and has the largest value of Z from 1044 to 3478 hours. The period equal to 24 h is the most statistically expressed ($Z \approx 10.3$) compared to the period having length of 23 h ($Z \approx 8.1$). It can be assumed that building a significant alignment for two periods is associated with the ability to create insertions or deletions of the symbols in the sequence S_1.

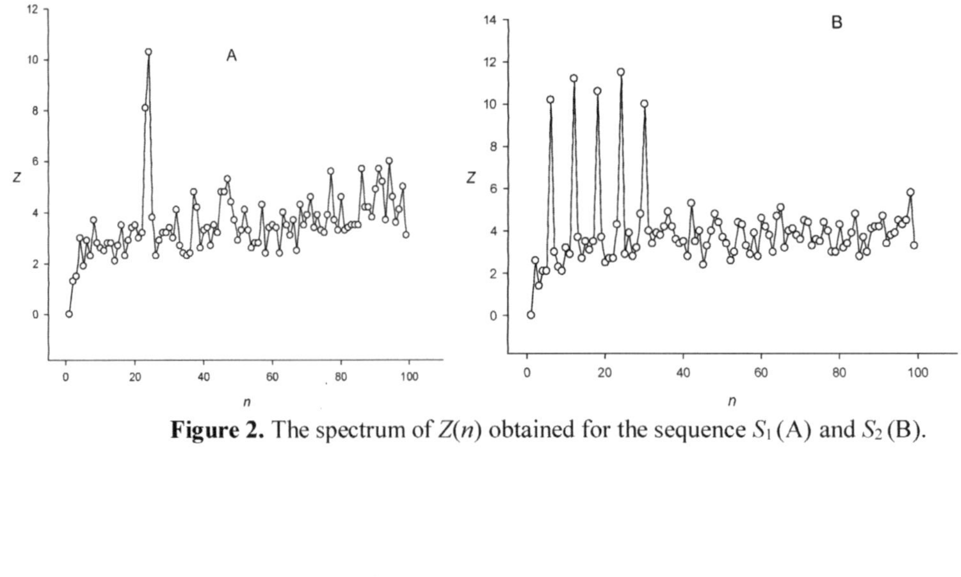

Figure 2. The spectrum of $Z(n)$ obtained for the sequence S_1 (A) and S_2 (B).

The figure 2B shows $Z(n)$ for the sequence S_2. For the sequence S_2, periodicity was observed for n=48 which corresponds to 24 h also. This period was observed for the 970 to 3893 numbers of the sequence S_2. 44 insertions and deletions were made to find this period. The periodicity of sequence S_3 is shown in figure 5. It is possible to see the periods for n=6, 12, 18, 24 and 30. It corresponds to 24, 48, 72, 96 and 120 hours. The period equal to 96 h is the most statistically expressed ($Z \approx 11.5$). These results demonstrate that the periodicity equal to 24 hours can be seen for different candles (half hour, one hour and four hours). It is also seen that there is a little period equal to 4 days apart from a period of 24 hours since Z(24)> Z(6).

The question arises about the nature of the periodicity discovered in this work. The different periodic processes influence currency exchange rates. Man himself has inherent rhythms of different frequencies; for example, Halberg [21] classified the biological rhythms of humans according to the period length. He identified several groups of rhythms: 1. Low frequency group has periods from 4 days to 12 months; 2. Mid-range group has period from 20 to 72 h; 3. High frequency area has period less than 20 h. It can therefore be assumed that the observed periodicity of the exchange rate at 24 h, reflects the influence of mid-frequency rhythms on the exchange rate.

An attempt was made to find the periodicity for the sequence $a(i)=x_2(i)-x_1(i)$,, where $x_1(i)$ и $x_2(i)$ are separated by minutes in the interval from 1 to 5 min. No statistically significant periodicity was found in a given interval, for a "minute" of the sequence. Sequences S_1 and S_2 were also built for other time intervals with the same length. In these cases, the $Z(n)$ and behavior of the phase shifts were similar.

It was also interesting to explain the presence of insertions or deletions of symbols (it was impossible to find frequency without it) and the behavior of the phase shift for the periodicity of 24 h. Probably, there is great instability in the behavior of large human masses and this instability may create phase shifts with insertions or deletions. Also, it can be assumed that some events of public life may be the cause of phase shifts, insertions, or deletions. For a more accurate consideration of this issue, a separate study is required. The correlations between the phase shifts, insertions and deletions in the sequences S_1 and S_2 should be searched and the events of public life and other factors, are of both social and physical nature.

References

[1] Stoica P, Moses R 2005 *Spectral Analysis of Signals* (New-York: Prentice Hall)
[2] Struzik Z R 2001 Phys. *A Stat. Mech. Its Appl.* **296** 307
[3] Hamilton J D 1994 *Time Series Analysis* (Princeton: Princeton University Press)
[4] Benson G 1999 *Nucleic Acids Res.* **27** 573
[5] Stankovic R S, Moraga C and Astola J 1987 *Fourier Analysis on Finite Groups with Applications in Signal Processing and System Design* (New-York: John Wiley & Sons)
[6] Marple S L 1987 *Digital Spectral Analysis: With Applications* (London:Prentice Hall)
[7] Dodin G, Vandergheynst P, Levoir P, Cordier C and Marcourt L 2000 *J. Theor. Biol.* **206** 323
[8] Jackson J H, George R and Herring P A 2000 *Biochem. Biophys. Res. Commun.* **268** 289
[9] Coward E and Drabløs F 1998 *Bioinformatics* **14** 498
[10] Chechetkin V R and Turygin AYu 1995 *J. Theor. Biol.* **175** 477
[11] Korotkov E V, Korotkova M A and Kudryashov N A 2003 *Phys. Lett. Sect. A* **312** 198
[12] Korotkov E V, Korotkova M A and Tulko J S 1997 *Comput. Appl. Biosci.* **13** 37
[13] Chaley M B, Korotkov E V and Skryabin K G 1999 *DNA Res.* **6** 153
[14] Turutina V P, Laskin A A, Kudryashov N A, Skryabin K G and Korotkov E V 2006 *J. Comput. Biol.* **13** 946
[15] Korotkov E V, Korotkova M A and Rudenko V M 1999 *J. Mol. Model.* **5** 103
[16] Chizhevsky A L 1976 *The Terrestrial Echo of Solar Storms* (Moscow: Misl')
[17] Rastogi S C,Mendiratta N and Rastogi P 2006 *Bioinformatics Methods and Applications: Genomics, Proteomics and Drug Discovery* (Delhi: PHI Learning)
[18] Suvorova Y M Korotkova M A and Korotkov E V 2014 *Comput. Biol. Chem.* **53** 43
[19] Pugacheva E, Korotkov E V and Korotkov A E 2016 in *Proc. 9th Int. Jt. Conf. Biomed. Eng. Syst. Technol. BIOSTEC* pp.117–127 (Setúbal: Scitepress)

[20] Pugacheva V M, Korotkov A E and Korotkov E V 2016 *Stat. Appl. Genet. Mol. Biol.* **15** (ahead of print).
[21] Halberg F 1969 *Annu. Rev. Physiol.* **31** 675

Symmetry analysis of the high-order equations for the description of the Fermi – Pasta – Ulam problem

N A Kudryashov, A K Volkov

National Research Nuclear University MEPhI (Moscow Engineering Physics Institute), 31 Kashirskoe Shosse, 115409 Moscow, Russian Federation

E-mail: nakudr@gmail.com

Abstract. Recently some new nonlinear equations for the description of the Fermi – Pasta – Ulam problem have been derived. The main aim of this work is to use the symmetry test to investigate these equations. We consider equations for the description of the α and $\alpha + \beta$ Fermi – Pasta – Ulam model. We find the infinitesimal operators and Lie groups, admitted by the equations. Using the groups we find the self-similar variables as well as the reductions to the ordinary differential equations. Some exact solutions are also constructed.

1. Introduction

The Fermi – Pasta – Ulam (FPU) model was introduced about 60 years ago in paper [5]. It describes perturbations in a one-dimensional lattice with a nonlinear potential of interaction. The authors supposed to observe a thermalization but the energy distribution was seen to be recurrent. This fact was called the Fermi – Pasta – Ulam paradox and have been studied until now.

Some results were achieved using the continuous limit approximation. In paper [1] it was shown that the α FPU model can be approximated by the Korteweg-de Vries equation. Then the recurrent dynamic can be explained. In [4] Kudryashov found out that the recurrence disappears if one uses more accurate approximation. Thus, it is important to study equations, for more accurate description of the Fermi – Pasta – Ulam problem. The fifth- and seventh-order equations for the description of the α FPU problem were derived in [3, 4]. The sixth-order equation was obtained to study two waves propagating in different directions in the α FPU model in [6]. And the fifth-order equation for the description of the $\alpha + \beta$ FPU problem was derived in [2]. It was shown that such equations does not pass the Painlevé test. Some exact solutions were found. The symmetry analysis of these equations has not been done so far. This is the main aim of the current work.

The rest of the work is organized as follows. In section 2 we apply the symmetry analysis for the fifth-order equation for the description of the α Fermi – Pasta – Ulam problem from the paper [4]. The seventh-order equation, derived in [3] is analyzed in section 3. Symmetries for the extended Boussinesq equation, derived in [6] are investigated in section 4. In section 5 we apply the symmetry analysis to investigate the equation for the description of the $\alpha + \beta$ FPU problem from the work [2]. In section 6 we briefly discuss our results.

2. Symmetry test of the fifth-order differential equation for the description of the α FPU model

If we consider the quadratic potential of interaction in the Fermi – Pasta – Ulam mass chain, then we can describe it using the fifth-order differential equation [4]. It has the form:

$$u_t + uu_x + \delta^2 u_{xxx} + 2\delta^2 u_x u_{xx} + \delta^2 u u_{xxx} + \delta^4 u_{xxxxx} = 0. \tag{1}$$

Let us search for the infinitesimal generator in the form:

$$X = \xi_x \frac{\partial}{\partial x} + \xi_t \frac{\partial}{\partial t} + \eta \frac{\partial}{\partial u}. \tag{2}$$

Its prolongation is:

$$X_5 = X + \xi_x \frac{\partial}{\partial u_x} + \xi_{2x} \frac{\partial}{\partial u_{2x}} + \xi_{3x} \frac{\partial}{\partial u_{3x}} + \xi_{4x} \frac{\partial}{\partial u_{4x}} + \xi_{5x} \frac{\partial}{\partial u_{5x}} + \xi_t \frac{\partial}{\partial u_t}. \tag{3}$$

We use generator (3) for equation (1) and the prolongation formulas for the components ξ_x, ξ_{2x}, ξ_{3x}, ξ_{4x}, ξ_{5x}, ξ_t, then we obtain the overdetermined linear system of partial differential equations:

$$\frac{\partial \xi_t}{\partial t} = \frac{\partial \xi_t}{\partial x} = \frac{\partial \xi_t}{\partial u} = \frac{\partial \xi_x}{\partial t} = \frac{\partial \xi_x}{\partial x} = \frac{\partial \xi_x}{\partial u} = \eta_u = 0. \tag{4}$$

As we solve (4) we derive components of the tangent vector field:

$$\xi_t = C_1, \quad \xi_x = C_2, \quad \eta_u = 0. \tag{5}$$

Thus the operator, admitted by (1) has the form:

$$X = C_1 \frac{\partial}{\partial t} + C_2 \frac{\partial}{\partial x}. \tag{6}$$

We can find from (6) that equation (1) admitts shift transformation $x' = x + a$ or $t' = t + a$. Thus we can use the traveling wave variables $u(x,t) = v(z), z = x - C_0 t$ to find the solution of the partial differential equation (1).

If we neglect term at δ^4, then equation (1) becomes the third-order partial differential equation:

$$u_t + uu_x + \delta^2 u_{xxx} + 2\delta^2 u_x u_{xx} + \delta^2 u u_{xxx} = 0. \tag{7}$$

In this case, the prolongation form for the infinitesimal generator has the form:

$$X_3 = X + \xi_x \frac{\partial}{\partial u_x} + \xi_{2x} \frac{\partial}{\partial u_{2x}} + \xi_{3x} \frac{\partial}{\partial u_{3x}} + \xi_t \frac{\partial}{\partial u_t}. \tag{8}$$

System of PDEs for the components of the tangent vector field is:

$$\frac{\partial^2 \xi_t}{\partial t^2} = \frac{\partial \xi_t}{\partial x} = \frac{\partial \xi_t}{\partial u} = \frac{\partial \xi_x}{\partial x} = \frac{\partial \xi_x}{\partial u} = 0, \quad \frac{\partial \xi_x}{\partial t} = -\frac{\partial \xi_t}{\partial t}, \quad \eta = -\frac{\partial \xi_t}{\partial t}(u+1). \tag{9}$$

From (9) we get:

$$\xi_t = C_1 t + C_2, \quad \xi_x = -C_1 t + C_3, \quad \eta_u = -C_1 u - C_1, \tag{10}$$

and the following infinitesimal generator:

$$X = (C_1 t + C_2) \frac{\partial}{\partial t} + (-C_1 t + C_3) \frac{\partial}{\partial x} - C_1 (u+1) \frac{\partial}{\partial u}. \tag{11}$$

Without loss of generality, we can take $C_1 = 1,\ C_2 = C_3 = 0$ in (11) and find the nontrivial transformations for (7):

$$x' = x + t\,(1 - e^a)\,, \quad t' = te^a, \quad u' = (u+1)e^{-a} - 1. \tag{12}$$

The following invariants correspond to this group:

$$I_1 = x + t, \quad I_2 = tu - x. \tag{13}$$

Thus we can search for the self-similar solution of equation (7) in the form:

$$u = \frac{f(\theta) + x}{t}, \quad \theta = t + x. \tag{14}$$

Substituting (14) into (7), we get ordinary differential equation for $f(\theta)$:

$$\theta f' + ff' + \delta^2\theta f''' + 2\delta^2 f''\left(1 + f'\right) + \delta^2 ff''' = 0 \tag{15}$$

Equation (15) is not an autonomous. Its infinitesimal generator has the form:

$$X = C_1\frac{\partial}{\partial\theta} - C_1\frac{\partial}{\partial f}. \tag{16}$$

Its corresponds to the shift transformation group:

$$\theta' = \theta + a, \quad f' = f - a. \tag{17}$$

However, we can not find any new solutions, using transformations (17).

Obvious solution for equation (15) is $f = c$, where c is a constant. We can use it to generate solution for equation (7) through transformation (14) in the form:

$$u = \frac{c + x}{t}. \tag{18}$$

It is a self-similar solution for equation (7).

3. Symmetry test of the seventh-order differential equation for the description of the α FPU model

If we take into account more terms in the Taylor series, then we can derive the seventh-order partial differential equation for the description of the Fermi – Pasta – Ulam model [3]. It has the form:

$$u_t + uu_x + \delta^2 u_{3x} + 2\delta^2 u_x u_{2x} + \delta^2 uu_{3x}\frac{2}{5}\delta^4 u_{5x} + 2\delta^4 u_{2x}u_{3x} + \frac{6}{5}\delta^4 u_x u_{4x} + \frac{3}{35}\delta^6 u_{7x} = 0. \tag{19}$$

As equation (1), equation (19) admits only trivial groups of transformations. If we take into account only members up to δ^4, then equation (19) takes the form:

$$u_t + uu_x + \delta^2 u_{3x} + 2\delta^2 u_x u_{2x} + \delta^2 uu_{3x}\frac{2}{5}\delta^4 u_{5x} + 2\delta^4 u_{2x}u_{3x} + \frac{6}{5}\delta^4 u_x u_{4x} = 0. \tag{20}$$

Equation (20) has nontrivial set of infinitesimals:

$$\xi_x = -t, \quad \xi_t = t, \quad \eta = -u - 1, \tag{21}$$

which corresponds to the generator:

$$X = t\frac{\partial}{\partial t} - t\frac{\partial}{\partial x} - (u+1)\frac{\partial}{\partial u}, \tag{22}$$

and the Lie group:

$$x' = x + t(1-e^a), \quad t' = te^a, \quad u' = (u+1)\,e^{-a} - 1. \tag{23}$$

Invariant transformation for equation (20) has the form:

$$u = \frac{f(\theta)+x}{t}, \quad \theta = x + t. \tag{24}$$

Transformation (24) is similar to (14). This fact can be explained as equation (19) is an extension of equation (1). Using (24) we can transform equation (20) to the ordinary differential equation:

$$\begin{aligned} &f'(\theta+f) + 2\delta^2(1+f')\,f'' + \delta^2\left(\theta + f + 2\delta^2 f''\right) f''' + \\ &+\frac{6}{5}\delta^4(1+f)\,f^{IV} + \frac{2}{5}\delta^4(\theta+f)\,f^{V} = 0. \end{aligned} \tag{25}$$

Equation (25) admits only trivial transformations. We can find a solution for equation (25) in the form $u = c$, where c is an arbitrary constant. Using (24) we get:

$$u = \frac{c+x}{t}. \tag{26}$$

We see that an self-similar solution for equation (20) is similar to the solution for equation (7).

4. Symmetry test of the sixth-order differential equation for the description of the α FPU model

If we study two waves, propagating to the different sides in the α or $\alpha+\beta$ Fermi – Pasta – Ulam model to the different sides, we can get the sixth-order differential equation [2, 6]:

$$\begin{aligned} \frac{d^2y}{dt^2} &= y_{xx} + y_x y_{xx} - \mu y_x^2 y_{xx} + \delta y_{xxxx} + 2\delta y_{xx} y_{xxx} + \delta y_x y_{xxxx} - \\ &-4\mu\delta y_x y_{xx} y_{xxx} - \mu\delta y_{xx}^3 - \mu\delta y_x^2 y_{xxxx} + \frac{2}{5}\delta^2 y_{xxxxxx}, \end{aligned} \tag{27}$$

We differentiate both parts of (27) with respect to x and substitute $y_x = u$:

$$\begin{aligned} \frac{d^2u}{dt^2} &= u_{xx} + (uu_x)_x - \mu\left(u^2u_x\right)_x + \delta u_{xxxx} + 2\delta\left(u_x u_{xx}\right)_x + \delta\left(uu_{xxx}\right)_x - \\ &-4\mu\delta\left(uu_x u_{xx}\right)_x - \mu\delta\left(u_x^3\right)_x - \mu\delta\left(u^2 u_{xxx}\right)_x + \frac{2}{5}\delta^2 u_{xxxxxx}. \end{aligned} \tag{28}$$

Equation (28) at any values of μ admits only trivial groups of transformations. We take $\mu = 0$ and take into account only members up to δ^2. As the result, equation (28) takes the form:

$$\frac{d^2u}{dt^2} = u_{xx} + (uu_x)_x + \delta u_{xxxx} + 2\delta\left(u_x u_{xx}\right)_x + \delta\left(uu_{xxx}\right)_x. \tag{29}$$

Equation (29) admits nontrivial set of infinitesimals:

$$\xi_x = 0, \quad \xi_t = t, \quad \eta_u = -2u - 2. \tag{30}$$

It corresponds to the infinitesimal generator:

$$X = t\frac{\partial}{\partial t} - 2\,(u+1)\,\frac{\partial}{\partial u}, \tag{31}$$

and transformation group:

$$x' = x, \quad t' = te^{a}, \quad u' = (u+1)\,e^{-2a} - 1. \tag{32}$$

We have the following invariants:

$$I_1 = x, \quad I_2 = t^2\,(u+1)\,. \tag{33}$$

Thus we can search for the self-similar solution of equation (29) in the form

$$u = \frac{v(x)}{t^2} - 1. \tag{34}$$

Using (34) we transform equation (29) to the ordinary differential equation for $v(x)$:

$$-6v + (vv')' + 2\delta(v'v'')' + \delta(vv''')' = 0. \tag{35}$$

We search for the particular solution of equation (35) in the form:

$$v = A + Bx + Cx^2, \tag{36}$$

where A, B, C are some constants. We substitute (36) into (35) and determine coefficients:

$$A = \frac{1}{4}B^2 + 2\delta, \quad C = 1. \tag{37}$$

Then we use (34) to get the solution for equation (29):

$$u = \frac{4x^2 + Bx + B^2 + 2\delta}{4t^2} - 1. \tag{38}$$

5. Symmetry test of the fifth-order differential equation for the description of the $\alpha + \beta$ FPU model

If we consider $\alpha + \beta$ Fermi – Pasta – Ulam problem, we can derive the following fifth-order partial differential equation [2]:

$$\begin{gathered} u_t + uu_x - \mu u^2 u_x + \delta^2 u_{xxx} + 2\delta^2 u_x u_{xx} + \delta^2 uu_{xxx} - \\ -4\delta^2 \mu uu_x u_{xx} - \delta^2 \mu u_x^3 - \delta^2 \mu u^2 u_{xxx} + \frac{2}{5}\delta^4 u_{xxxxx} = 0, \end{gathered} \tag{39}$$

The prolongation of the infinitesimal generator for equation (39) has the form:

$$X_5 = X + \xi_1\frac{\partial}{\partial u_x} + \xi_{11}\frac{\partial}{\partial u_{2x}} + \xi_{111}\frac{\partial}{\partial u_{3x}} + \xi_{1111} + \frac{\partial}{\partial u_{4x}} + \xi_{11111}\frac{\partial}{\partial u_{5x}} + \xi_2\frac{\partial}{\partial u_t}. \tag{40}$$

We get the system of partial differential equations:

$$\frac{\partial \xi_t}{\partial t} = \frac{\partial \xi_t}{\partial x} = \frac{\partial \xi_t}{\partial u} = \frac{\partial \xi_x}{\partial t} = \frac{\partial \xi_x}{\partial x} = \frac{\partial \xi_x}{\partial u} = \eta_u = 0. \tag{41}$$

Then we solve (41) and get components of the tangent vector field:

$$\xi_t = C_1, \quad \xi_x = C_2, \quad \eta_u = 0. \tag{42}$$

Infinitesimal generator, admitted by equation (39) is:

$$X = C_1\frac{\partial}{\partial t} + C_2\frac{\partial}{\partial x}. \tag{43}$$

We see from (43) that equation (39) admits only shift transformations and has invariant solution in the form of the traveling wave $u(x,t) = v(z), z = x - C_0 t$. These solutions have already been investigated in [2].

IOP Conf. Series: Journal of Physics: Conf. Series **788** (2017) 012020 doi:10.1088/1742-6596/788/1/012020

6. Conclusion

Nonlinear partial differential equations for the description of the Fermi – Pasta – Ulam model have been investigated using the symmetry analysis. We have found components of the tangent vector field, and groups, admitted by the equations. We have found the self-similar variables and the reductions to the ordinary differential equations. Using exact solutions for the ordinary differential equations, we have constructed new self-similar solutions for the investigated equations.

Acknowledgments

This research was supported by Russian Science Foundation Grant no. 14-11-00258.

References

[1] Zabusky N J and Kruskal M D 1965 *Phys. Rev. Lett.* **15** 240–243
[2] Kudryashov N A and Volkov A K 2017 *Communications in Nonlinear Sci. and Numerical Simul.* **42** 491–501
[3] Kudryashov N A 2016 *Rep. on Math. Phys.* **77** 57–67
[4] Kudryashov N A 2015 *Phys. Lett. A* **379** 2610–2614
[5] Fermi E, Pasta J and Ulam S 1955 *Los Alamos Report LA-1940* pp 978–988
[6] Kudryashov N A and Volkov A K 2016 *ICNAAM 2015* vol 1738 p 230005

Renormalization of the hydrogen sulfide properties due to the strong electron-phonon interaction

N A Kudryashov, A A Kutukov and E A Mazur

National Research Nuclear University MEPhI (Moscow Engineering Physics Institute), 31 Kashirskoye shosse, 115409, Moscow, Russia

E-mail: alexkutuk@gmail.com

Abstract. The normal state of a metal is described by generalized Eliashberg theory which takes into account the finite width of an electron band, strong electron-phonon coupling and electron-hole nonequivalence. Reconstructed parameters of the conduction band of the metallic hydrogen sulfide for both the real and imaginary parts of the mass renormalization of the electron Green's function and the real and imaginary parts of the renormalization of the chemical potential have been found.

1. Introduction

Superconducting state of the metallic phase of the hydrogen sulfide at a pressure of about 170 GPa and at the temperatures of $T_c = 190$ K and $T_c = 203$ K has been experimentally found [1, 2]. The study of the properties of the metallic hydrogen sulfide in a normal state allows us to determine the factors influencing the high-temperature superconductivity. The theory of the electron properties of the normal state of a crystal as well as the Eliashberg theory of the superconducting state [3, 4] do not take into account the reconstruction of the electron density of states. The reconstruction of the "bare" electron conduction band by the strong electron-phonon interaction with the account with the finiteness of the electron band width has been discussed in [5]. The reconstruction should be expected to be noticeable in the metallic hydrogen sulfide because of the high phonon energy associated with the collective oscillations of the quasi-two-dimensional metallic hydrogen stabilized with the sulfur sublattice simultaneously with the greater than one constant of the electron-phonon coupling.

2. Reconstruction of the conduction band of the metallic hydrogen sulfide

Describe the conduction band reconstruction in the metallic hydrogen sulfide with the use of the quantum field methods using the effect of the restructuring of the complex self-energy part $\Sigma(\omega)$ of the electron Green's function in the hydrogen sulfide together with the reconstruction of the electron density of states. For the imaginary part of the self-energy part of the electron Green's function the following expression was obtained in [6-8]:

$$\mathrm{Im}\Sigma(\omega) = -\pi \int_0^{+\infty} dz\,\alpha^2(z)F(z)\left\{\left[N(\omega-z)+N(\omega+z)\right]n_B(z)+N(\omega-z)f(z-\omega)+N(\omega+z)f(z+\omega)\right\} \quad (1)$$

where $n_B(z)$ is the Bose distribution function, $f(z)$ is the Fermi distribution function, $N(z)$ is the renormalized by the electron-phonon interaction electron density of states, $\alpha^2(z)F(z)$ is the spectral function of the electron-phonon coupling.

The expression for the real part of the electron Green's function self-energy part has the following form [6-8]

$$\mathrm{Re}\Sigma(\omega)=-P\int_0^{+\infty} dz\alpha^2(z)F(z)\int_0^{+\infty} dz'\left\{f(-z')\left(-\frac{N(-z')}{z'+z+\omega}+\frac{N(z')}{z'+z-\omega}\right)+f(z')\left(-\frac{N(-z')}{z'-z+\omega}+\frac{N(z')}{z'-z-\omega}\right)\right\} \quad (2)$$

Figure 1 shows the results of solving the system of equations (1) - (2) for both the real part $\mathrm{Re}\Sigma(\omega)$ and the imaginary part $\mathrm{Im}\Sigma(\omega)$ of the self-energy part of the electron Green's function of the hydrogen sulfide SH_3 phase in the range of the dimensionless energy variable from -40 to +40, which corresponds to the energy region about -9.36 eV to 9.36 eV.

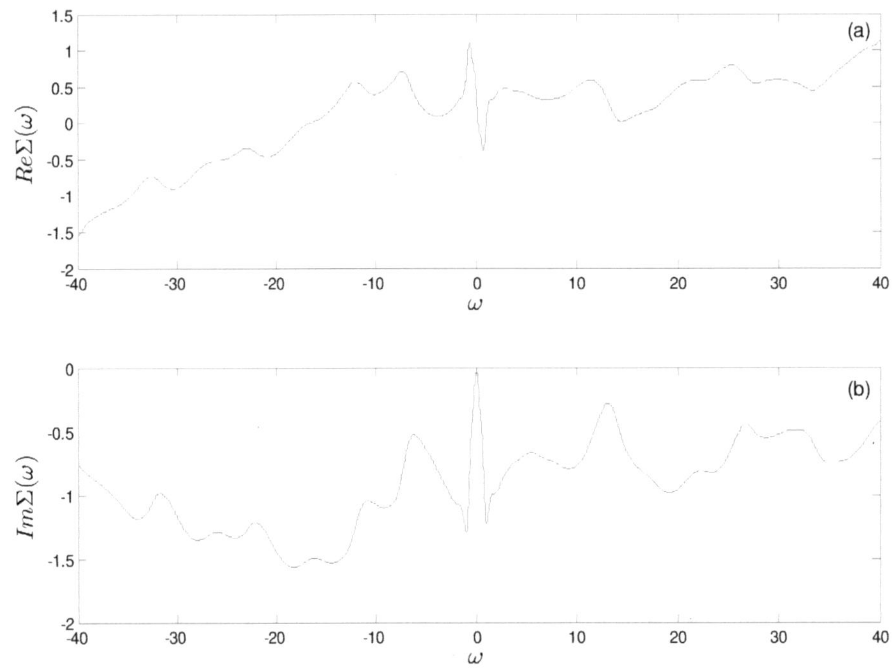

Figure 1. (a) The real part $\mathrm{Re}\Sigma(\omega)$ of the self-energy part of the electron Green's function in the SH_3 phase of the hydrogen sulphide. (b) The imaginary part $\mathrm{Im}\Sigma(\omega)$ of the self-energy part of the electron Green's function.

Figure 2 shows the results of calculation for the SH3 phase for the $\mathrm{Re}Z(\omega)$, $\mathrm{Im}Z(\omega)$ quantities of the hydrogen sulfide, which describe the real and imaginary parts of the renormalized mass of the electron, as well as the estimates for the renormalized by the strong interaction with phonons the real $\mathrm{Re}\chi(\omega)$ and imaginary $\mathrm{Im}\chi(\omega)$ parts of the renormalization of the chemical potential for the energy "distances" away from the Fermi level not exceeding 40 characteristic phonon energies, taken per unit.

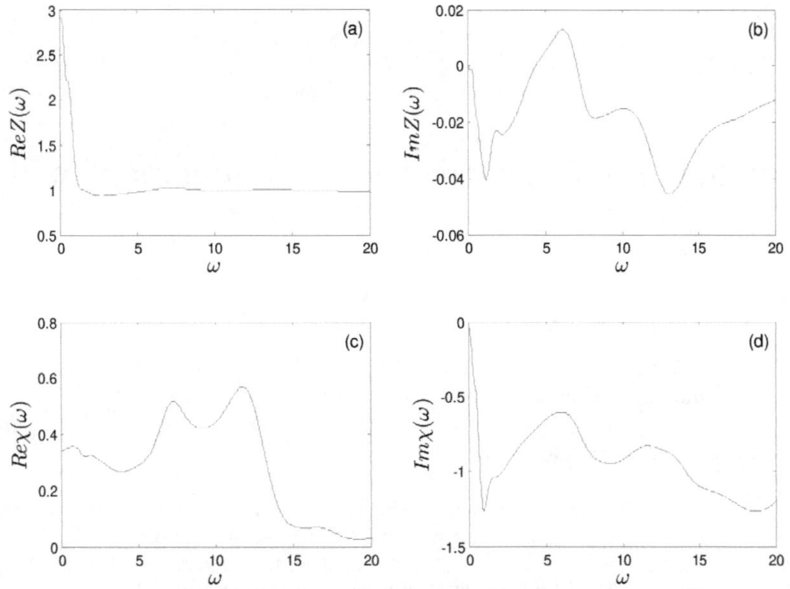

Figure 2. The real part $\mathrm{Re}Z(\omega)$ and imaginary part $\mathrm{Im}Z(\omega)$ of the electron mass renormalization in the self-energy part of the electron Green's function. (b) Renormalized by the electron-phonon interaction the real part $\mathrm{Re}\,\chi(\omega)$ and the imaginary part $\mathrm{Im}\,\chi(\omega)$ of the chemical potential renormalization.

All results (figure 1, 2) are obtained for a pressure of P = 225GPa and a temperature of T = 175K.

3. Conclusion

As a result of the self consistent consideration of the renormalization of the electron spectrum in the hydrogen sulphide SH_3 phase by the strong electron-phonon interaction we have found that the hydrogen sulfide in a SH_3 phase at a pressure of P = 225 GPa and a temperature of T = 175 K is a metal with a strong non-adiabatic effects.

Acknowledgments
The study was supported by a grant from the Russian Science Foundation (project №14-11-00258).

References

[1] Drozdov A P, Eremets M I and Troyan I A 2015 Conventional superconductivity at 190 K at high pressures *arXiv:1412.0460*
[2] Drozdov A P, Eremets M I, Troyan I A, Ksenofontov V and Shylin S I 2015 *Nature* **525** (7567) 73–6.
[3] Eliashberg G M 1960 *Sov. Phys. JETP* **11** 696-702.
[4] Eliashberg G M 1961 *Sov. Phys. JETP* **12** 1000-2.
[5] Marsiglio F and Carbotte J P 2008 *Superconductivity* vol 1, ed K H Bennemann and J B Ketterson (Springer, Berlin, Heidelberg).
[6] Mazur E A and Kagan Yu 2015 *J. Exp. Theor. Phys.* **121** (2) 237-45.
[7] Mazur E A and Kagan Yu 2013 *J. Supercond. Novel Magnet.* **26** (4) 1163-6.
[8] Mazur E A 2010 *Europhys. Lett.* **90** 47005-16.

IOP Conf. Series: Journal of Physics: Conf. Series **788** (2017) 012022 doi:10.1088/1742-6596/788/1/012022

Reconstruction of the electron spectrum in a metal hydrogen sulfide

N A Kudryashov, A A Kutukov and E A Mazur

National Research Nuclear University MEPhI (Moscow Engineering Physics Institute), 31 Kashirskoye shosse, 115409, Moscow, Russia

E-mail: alexkutuk@gmail.com

Abstract. Generalized Eliashberg theory of the normal properties of a metal electron-phonon system with a non constant electron density of states has been used to study the effect of the conduction band reconstruction. The electron density of states of the metallic phase of the hydrogen sulfide renormalized by the strong electron-phonon coupling at a pressure of P = 225 GPa has been calculated. It has been found that the reconstructed conduction band contains a series of narrow energy pockets.

1. Introduction

The presence of both the isotope effect and the Meissner effect in the hydrogen sulfide phase formed at a pressure of about 170 GPa has been experimentally demonstrated in [1]. In [2] the main candidate for the superconducting phase namely the compound with the stoichiometry of SH_3 and *IM-3M* symmetry is examined. In [3-4] it is shown that the main candidates for the role of the superconducting phase are the compounds with the stoichiometry of SH_3 and SH_2. It is shown in [5] that for the density of electron states for the SH3 phase at a pressure of P = 225 GPa the Fermi level is always located at the height of the electronic density of states.

2. Reconstruction of the electron spectrum of the metallic phase SH_3 of the hydrogen sulfide by phonons

The renormalized by the electron-phonon interaction electronic density of states $N(z')$ is expressed through the "bare" electron density of states $N_0(\xi)$ as follows

$$N(z') = -\frac{1}{\pi}\int_{-\mu}^{\infty} d\xi' N_0(\xi') \operatorname{Im} g_R(\xi', z') \tag{1}$$

where g_R is a retarded electron Green's function. The imaginary part $\operatorname{Im} g_R(\xi', z')$ of a retarded Green's function of the electron has the following form

$$\operatorname{Im} g_R(\xi, \varepsilon) = \frac{\operatorname{Im}\Sigma(\varepsilon)}{\left[\varepsilon - \xi - \operatorname{Re}\Sigma(\varepsilon)\right]^2 + \left[\operatorname{Im}\Sigma(\varepsilon)\right]^2} \tag{2}$$

where $\operatorname{Re}\Sigma(\varepsilon)$ is the real and $\operatorname{Im}\Sigma(\varepsilon)$ is the imaginary component of the electron Green's function self-energy part. $N_0(\xi)$ is a "bare" (not renormalized by the electron-phonon interaction) variable electron density of states defined by the following expression $\int_{S(\xi)} \frac{d^2 \mathbf{p}'}{v_{\xi \mathbf{p}'}} d\xi = \int_{S(\xi)} N_0(\xi, \omega') d\xi$ with the energy ξ of the "bare" electrons.

The spectral function of the electron-phonon interaction $\alpha^2(z)F(z)$ for the SH_3 phase of the hydrogen sulfide is presented in figure 1 [5]. The frequency ω is expressed in dimensionless units (as a fraction of the maximum frequency of the phonon spectrum equal to 0.234 eV).

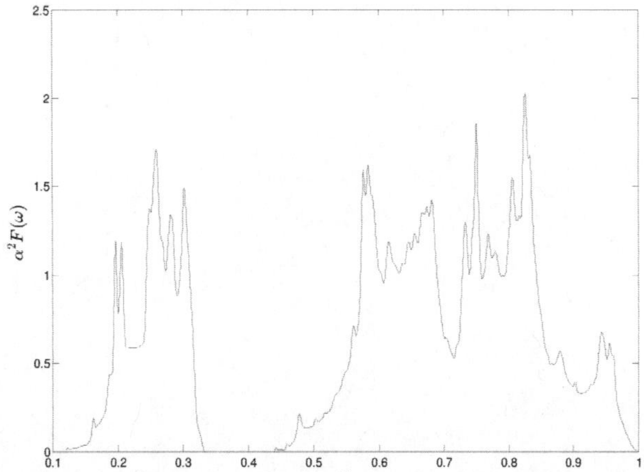

Figure 1. The spectral function of the electron-phonon interaction $\alpha^2(z)F(z)$ in the SH_3 hydrogen sulfide at a pressure of 225 GPa.

The "bare" electron density of states in the hydrogen sulfide which is not renormalized by the electron-phonon interaction is obtained in [5]. Such a density of states is presented in figure 2 (dimensionless argument is expressed in units of the maximum phonon frequency).

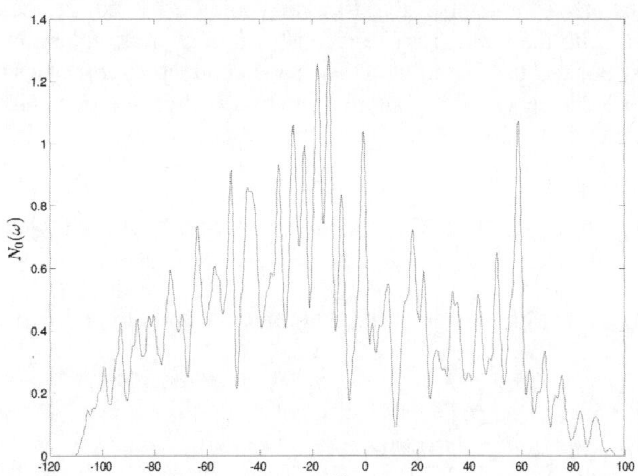

Figure 2. Dimensionless "bare" total electron density of states in SH_3 hydrogen sulfide at a pressure of 225 GPa

The program of the numerical simulation of the electronic density of states was presented in [6]. Figure 3 shows the comparison between the "bare" electron density of states (a) and the electron density of states reconstructed by the strong electron-phonon coupling (b) in the SH_3 hydrogen sulfide at a pressure of 225GPa.

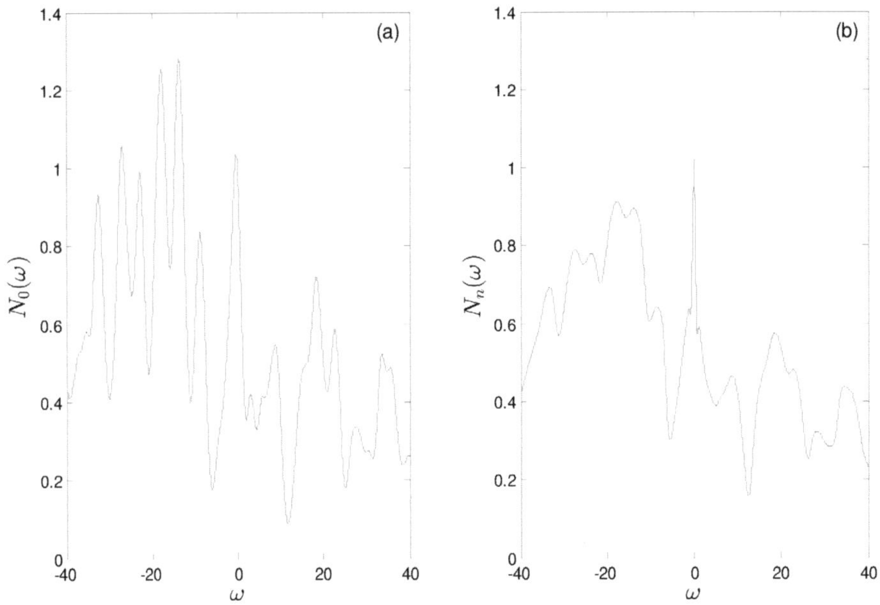

Figure 3. (a) "Bare" electron density of states in the conduction band and (b) the electron density of states reconstructed by the electron–phonon coupling.

3. Conclusion

The presented calculations completely reproduce the polaron effects at the multiple and strong interaction of electrons with the phonons. The reconstruction of the conduction band into a series of nonoverlapping energy pockets has been found. The superconducting properties of SH_3 phase with the *IM-3M* hydrogen sulfide lattice symmetry should be described by the generalized Eliashberg theory that takes into account the finiteness of the electron band gap.

Acknowledgments
The study was supported by a grant from the Russian Science Foundation (project №14-11-00258).

References

[1] Drozdov A P, Eremets M I, Troyan I A, Ksenofontov V and Shylin S I 2015 *Nature* **525** (7567) 73–6.
[2] Degtyarenko N N and Mazur E A 2015 *J. Exp. Theor. Phys.* **121** (2) 250–8.
[3] Migdal A B 1958 *Sov. Phys. JETP* **34** (7) 6 996-1001.
[4] Mazur E A 2010 *Europhysics Letters (EPL)* **90** 47005-16.
[5] Degtyarenko N N and Masur E A 2015 *J. Exp. Theor. Phys.* **121** (6) 1067–75.
[6] Kudryashov N A, Kutukov A A and Mazur E A 2016 *J. Exp. Theor. Phys* **123** (3) 481–8.

Multiscale numerical modeling of the spherically symmetric cryosurgery problem

N A Kudryashov, K E Shilnikov

Department of Applied Mathematics, National Research Nuclear University MEPhI 31 Kashirskoe Shosse, 115409 Moscow, Russian Federation

E-mail: nakudr@gmail.com

Abstract. The work is concerned with the numerical studying of the cryogenic biotissue destruction by a spherically symmetric tip. The multiscale bioheat transfer model is used for the describing of the biological solutions crystallization features. An explicit finite volume based approximation is applied for the numerical modeling of the processes taking place during the cryosurgery. The phase averaging method is applied as an computationally economic approach for the numerical modeling of the problem under study.

1. Introduction

During the last decades the cryosurgery methods has been considerably developed and has become a competitive approach for human cancer treatment. In the light of introduction of computer technologies into the wide range of science-based applications, nowadays, the prediction of the cryosurgery results is one of the actual problems of the computational medicine. The cryosurgical treatment is known to be based on the tissue necrosis effect. The last one is caused by the ice ball propagation from the tips of cryoneedles in the tissue to treat. However, the describing of the freezing interface evolution in terms of the classical Stephan problem for the phase change of the first kind meets a valuable difficulties because of the peculiarities of physical processes in the bio-tissues which take place under the cryogenic treatment on the cell scale.

2. Multiscale model of the cryogenic bioheat transfer

Let us consider the problem of the cryogenic bio-tissue destruction by a spherical cryotip with the radius R_0. The main peculiarity of the propagation of the freezing interface is the blurring of the last one in some temperature interval depending on the intensity of cryoigenic impact, i.e. local freezing speed. In particular, the cell dehydratation process caused by the occuring of the difference between the osmotically pressures from different sides of the cell membrane may become even negligible comparing with an intracellular ice nucleation. It is caused by the exponential reduction of the membrane permeability:

$$L_p = L_{pg} \exp\left[-\frac{E_{lp}}{R}\left(\frac{1}{T} - \frac{1}{T_r}\right)\right]. \quad (1)$$

In current work for the cell dehydratation process a model, obtained by Mazur [1] for the dynamics of the unfrozen intracellular water volume V_w, was used:

$$\frac{\partial V_w(T,t)}{\partial t} = -\frac{L_p ART}{\nu_w}\left(\ln\left[\frac{V_w - V_b}{(V_w - V_b) + \phi\nu_w n_s}\right] - \frac{L}{R}\left[\frac{1}{T_r} - \frac{1}{T}\right]\right) \quad (2)$$

The intracellular ice nucleation speed $\frac{dN}{dt}$ is assumed to be described in the following manner [2, 3]:

$$\frac{dN}{dt} = IN(V_w - V_b), \quad (3)$$

where I is the coefficient of the nucleation speed, satisfying the following equalities:

$$\begin{aligned} I &= \Omega(T)\exp\left(-\tfrac{\kappa}{\delta T^2 T^3}\right), \delta T = (T_r - T), \\ \Omega &= \Omega_0\left(\tfrac{T}{T_r}\right)^2\left(\tfrac{\eta_0}{\eta(T)}\right). \end{aligned} \quad (4)$$

Here η has sense of the viscosity and can be directly calculated:

$$\eta = 0.139 \cdot 10^{-3}\left[\frac{T}{225} - 1\right]^{-1.64}. \quad (5)$$

The extracellular ice nucleation can be described well by the temperature dependance of liquid phase fraction Λ_{ec} for NaCl solution:

$$\Lambda_{ec}(T) = \begin{cases} 1 & T \geq 272.62K \\ \max(\frac{0.53}{273.15-T}, b_e) & T < 272.62 \end{cases} \quad (6)$$

On this way the fraction of unfrozen water Λ can be estimated in the form:

$$\Lambda = f_{ic}\Lambda_{ic} + f_{ec}\Lambda_{ec}, \quad f_{ic,ec} = Const \quad (7)$$

Here f_{ic}, f_{ec} are the corresponding mass fractions and $\Lambda_{ic} = V_w/V_0$, where V_0 is the intracellular water volume under the normal conditions. Using the mass conservation law and introducing functions $X = V_w/V_0$ and $Y = V_{ice}/V_0$ for the intracellular ice one can obtain the governing system for the phase content of the cell, which in appropriate notation can be deduced to the form:

$$\frac{\partial X}{\partial t} = F(X,Y,T), \quad \frac{\partial Y}{\partial t} = G(X,Y,T) \quad (8)$$

Taking into account the mentioned above, it seems naturally to demand from macroscopic characteristics of heat capacity and thermal conductivity (which is known for fully unfrozen and frozen states) to be the functions of water content of the tissue. In this paper we assume that on the macroscopic scale the heat transfer can be described by the Pennes model [4]. In this way the problem under study can be written in following manner:

$$\begin{aligned} \overline{C}\frac{\partial T}{\partial t} &= \frac{1}{r^2}\frac{\partial}{\partial r}\left(r^2\overline{k}(T)\frac{\partial T}{\partial r}\right) + \rho_b\omega_b C_b(T_b - T) + q_{met} - \rho L\frac{\partial \Lambda}{\partial t}, \\ T(R_0,t) &= \Phi(t), \left.\frac{\partial T}{\partial r}\right|_{r=L} = 0, \ t > 0, \ L > r > R_0, \end{aligned} \quad (9)$$

where $\Phi(t)$ is the temperature of the cryotip surface neighborhood and the macroscopic characteristics are assumed as the weighted functions of the known values:

$$\begin{aligned}
\overline{C}(T) &= [f_{ic}\alpha_{fr} + f_{ec}\beta_{fr}]C_{fr}(T) + [f_{ic}\alpha_{unfr} + f_{ec}\beta_{unfr}]C_{unfr}(T),\\
\overline{k}(T) &= f_{ic}\alpha_{fr} + f_{ec}\beta_{fr}]k_{fr}(T) + [f_{ic}\alpha_{unfr} + f_{ec}\beta_{unfr}]k_{unfr}(T),\\
\frac{\partial X}{\partial t} &= F(X,Y,T), \quad \frac{\partial Y}{\partial t} = G(X,Y,T), \quad \Lambda = (f_{ic}X + f_{ec}\Lambda_{ec})\rho_w L,\\
\alpha_{fr} &= \frac{1-X}{1-V_b/V_0}, \ \alpha_{unfr} = \frac{X - V_b/V_0}{1-V_b/V_0}, \ \beta_{fr} = \frac{1-\Lambda_{ec}}{1-b_e}, \ \beta_{unfr} = \frac{\Lambda_{ec}-b_e}{1-b_e}.
\end{aligned} \tag{10}$$

The condition $\frac{\partial T}{\partial r}\big|_{r=L} = 0$ means the considered computational domain to be of spherical layer form with outer border enough distant from the cryotip surface.

For the numerical modeling of the thermal processes around the cryotip's working surface an explicit finite volume scheme was used. Introducing a computational mesh ω_{h_1,h_2}:

$$\begin{aligned}
\omega^{TM}_{h_1,h_2} &= [r_i : r_0 = R_0, r_N = L, r_i = r_{i-1} + h_i, i = \overline{1, N_r}\\
t^n &: t^0 = 0, t^N = T, t^j = t^{j-1} + \tau_M, j = \overline{1, N}],\\
h_1 &= \min_i h_i,\ h_2 = \max_i h_i
\end{aligned} \tag{11}$$

and defining a half step nodes $r_{i+1/2} = (r_{i+1} + r_i)/2$ one can integrate (9) upon the spherical layer enclosed by $r_{i\pm1/2}$ and obtain the following scheme:

$$\begin{aligned}
V_i\frac{T_i^{n+1} - T_i^n}{\tau_M} &= \frac{4\pi}{\overline{C}_i^n}\left[(r_{i+\frac{1}{2}})^2\overline{k}^n_{i+\frac{1}{2}}\frac{T^n_{i+1} - T^n_i}{r_{i+1} - r_i} - (r_{i-\frac{1}{2}})^2\overline{k}^n_{i-\frac{1}{2}}\frac{T^n_i - T^n_{i-1}}{r_i - r_{i-1}}\right] +\\
&+\rho_b\omega_b C_b(T_b - T_i^n)V_i + q_{met}V_i - \rho L\frac{\Lambda^n_i - \Lambda^{n-1}_i}{\tau_M},\\
V_i &= \frac{4\pi}{3}\left[(r_{i+\frac{1}{2}})^3 - (r_{i-\frac{1}{2}})^3\right],\\
\overline{k}^n_{i+\frac{1}{2}} &= \left(\overline{k}^n_i + \overline{k}^n_{i+1}\right)/2.
\end{aligned} \tag{12}$$

The semi-implicit Euler-Cauchy scheme was applied to evaluate the microscopic tissue parameters in each computational cell. During the numerical experiments the cryotip with working surface radius $R_0 = 0.0025$m and temperature $\Phi(t) = \max(300 - At, 100)K$ with $A = 80K/\text{min}$ was considered. The results of numerical computation for the evolution of the temperature profile in the tissue are shown on Fig.1 (left).

Unfortunately, the multiscale model is not applicable for practicing surgeons to obtain the prognosis of cryosurgery because of its computational economy even for the one dimensional reduction of the problem under study. Yet, it can be successfully used for the carrying out the scientific research of the cryogenic processes in bio-tissues.

In order to avoid the computational difficulties, arising when the multi-scale model is applied, and to save the consideration of the cryogenic processes on the cell scale the phase-averaged method can be used. This approach supposes the introduction of the biparametric family of an enthalpy models:

$$\begin{aligned}
&C(T; T_\alpha, T^*)\tfrac{\partial}{\partial t}T = \tfrac{1}{r^2}\tfrac{\partial}{\partial r}r^2\overline{k}(T; T_\alpha, T^*)\tfrac{\partial}{\partial r}T + \rho_b\omega_b C_b(T_b - T),\\
&t > 0, L > r > R_0.
\end{aligned} \tag{13}$$

IOP Conf. Series: Journal of Physics: Conf. Series **788** (2017) 012023 doi:10.1088/1742-6596/788/1/012023

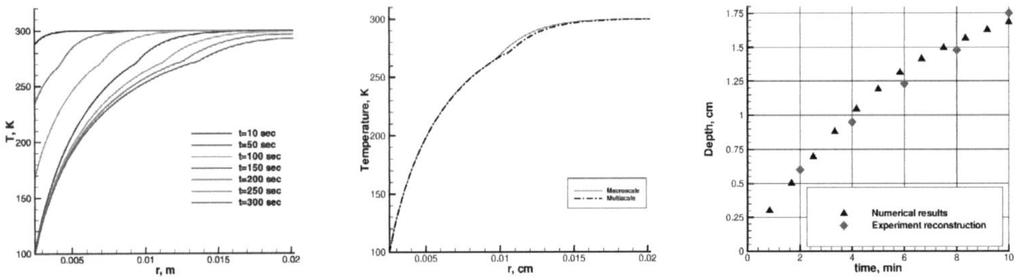

Figure 1. Left: the multiscale model based numerical results for the temperature profile evolution. Center: Comparison of the results obtained by multi-scale and phase-averaged model. Right: The dependence of the freezing zone propagation depth on the freezing time. Comparison of the numerical results with the experimental data reconstruction

Here the macroscopic parameters are determined by the interpolation equalities:

$$\overline{k}(T;T_\alpha,T^*) = \begin{cases} k_1(T), & T \le T_\alpha \\ k_1(T_\alpha) + \frac{k_2 - k_1(T_\alpha)}{T^* - T_\alpha}(T - T_\alpha) & T_\alpha < T < T^* \\ k_2, & T \ge T^* \end{cases}$$
$$\overline{C}(T;T_\alpha,T^*) = \begin{cases} C_1(T), & T \le T_\alpha \\ \frac{\rho L}{T_c - T_\alpha} + \frac{T^* - T_\alpha}{T_c - T_\alpha} C_1(T_\alpha) + \frac{T_c - T^*}{T_c - T_\alpha} C_2 & T_\alpha < T < T_c \\ C_2, & T \ge T_c \end{cases} \tag{14}$$

In this way the parameters of the phase averaging are the lower temperature of the phase change T_α and T^*, which has sense of the normalized phase change temperature in terms of the Stefan problem. The upper phase change temperature is assumed to be known value $T_c = 273K$. The numerical experiments have shown the possibility of choosing the averaging parameters which give the 98% agreement with the multiscale model based results. To validate the developed cryosurgery modeling tool the dependence of the freezing region propagation depth on the time was calculated and compared with the experimental data [5] for 3mm cryotip in the chicken breast (See Fig.1, Right). The discrepancy in the results was less than 5%.

3. Discussion

Although the multiscale model takes into account the features of the cryogenic processes on the cell scale its application in the medical practice meets the valuable difficulties caused by its computational cost. The phase-averaging approach has not got this disadvantage and can easily applied for the multidimensional problems [6], but it needs the knowing of the averaging parameters values.

It should be mentioned that the most of the cryosurgery protocols can be characterized by enough common thermal regimes. So, if the number of cryotips in the tissue to treat is not big (3-4) and their sizes are much lesser of the tumor tissue characteristic sizes ($\sim 2-4$mm versus $\sim 3-4$cm), the following values for the averaging parameters can be taken: $T_\alpha = 265K$, $T^* = 269K$.

4. Conclusion

The problem of the cryogenic biotissue destruction by a spherically symmetric tip was numerically studied. The multiscale bioheat transfer model was used for the describing of the biological solutions crystallization features. An explicit finite volume based approximation was applied for the numerical modeling of the processes taking place during the cryosurgery. The evolution of the temperature profile in the neighborhood of the cryotip surface was computed. The phase averaging method was applied as an computationally economic approach for the numerical modeling of the problem under study. The comparison between the multi-scale model and averaged model with the experimental data has shown the applicability of the proposed methods for the prediction of the cryosurgery results and carrying out the research of the cryogenic processes in living tissues.

5. Acknowledgments

This work was supported by Russian Foundation For Basic Research, grant no. 16-31-00348.

References

[1] Mazur P 1963 *J. Gen. Physiol.* **43** 347–69
[2] Toner M, Cravalho E G and Karel M 1990 *J. Appl. Phys.* **67** 1582–92
[3] Kudryashov N A and Shilnikov K E 2016 *Vestnik Natsionalnogo Issledovatelskogo Yadernogo Universiteta "MIFI"* **5** 88–96
[4] Pennes H H 1948 *Journal of Applied Physiology* **1** 93–122
[5] Jankun M, Kelly T J, Zaim A, Young K., Keck R W, Selman S H and Jankun J 1999 *Computer Aided Surgery* **4** 193
[6] Kudryashov N A and Shilnikov K E 2016 *Mathematical Models and Computer Simulations* **8** 680–8

IOP Conf. Series: Journal of Physics: Conf. Series **788** (2017) 012024 doi:10.1088/1742-6596/788/1/012024

About a mathematical model of market

D A Kulikov

Faculty of Mathematics, Yaroslavl State University, Sovetskaya str., 14, Yaroslavl, 150000, Russia

E-mail: kulikov_d_a@mail.ru

Abstract. In the paper a famous mathematical model of macroeconomics, which is called "market model" was considered. Traditional versions of this model have no periodic solutions and, therefore, they cannot describe a cyclic recurrence of the market economy. In the paper for the corresponding equation a delay was added. It allows obtaining sufficient conditions for existence of the stable cycles.

1. Introduction

The presence of cyclic recurrence in a market economy was mentioned in the works of economists of 19th century (see, for example, [1]). From the middle of the 20th century for the study of economic dynamics, the mathematical methods were widely used. Particularly, in macroeconomics a number of mathematical models: the aggregate demand-aggregate supply model (market model), the Solow model, the multiplier-accelerator model and so on were proposed (see [2-4]).

Let $p(t)$ is the price of the product at the moment t. Then the dynamics of the price are described by the following equation [2-4]

$$\dot{p} = D(p) - S(p), \tag{1.1}$$

where $D(p)$ - the demand function, $S(p)$ - the supply function. Usually, these functions are assumed with the following properties (see [4], p.146):

(1) non-negative and smooth functions $D(p)$, $S(p)$ are defined if $p \in (0, \infty)$;

(2) if $p \in (0, \infty)$ then the inequalities $D'(p) < 0$, $S'(p) > 0$ are hold;

(3) $\lim\limits_{p \to 0} D(p) = D_0$, where $D_0 >> 1$ or $D_0 = \infty$, $\lim\limits_{p \to 0} S(p) = S_0$, where $S_0 << 1$ or $S_0 = 0$;

(4) $\lim\limits_{p \to \infty} D(p) = D_\infty$, where $D_\infty << 1$, $\lim\limits_{p \to \infty} S(p) = S_\infty$, where $S_\infty \in R_+$ or $S_\infty = \infty$.

The inequality $D'(p) < 0$ means that demand for goods falls if the price of the product is rising. On the contrary, a supply is rising if the market price of the product is rising. Condition $\lim\limits_{p \to 0} S(p) = 0$ is natural since for a zero price the supply of the product is absent.

Equation (1.1) has only one positive equilibrium state ($p(t) = p_0 > 0$). Indeed, if we consider the function $F(p) = D(p) - S(p)$ for which $F'(p) = D'(p) - S'(p) < 0$, i.e. these functions are monotonically decreasing. Moreover, $\lim\limits_{p \to 0} F(p) = \infty$ or it equals the sufficiently large constant and $\lim\limits_{p \to \infty} F(p) < 0$. Therefore, the equation $F(p) = 0$ has only one root $p(t) = p_0 > 0$. The inequality $F'(p) < 0$ means that equilibrium state $p(t) = p_0$ is asymptotically stable. Hence, the equation (1.1) cannot have any cycles.

In the papers [5,6] the modification of the basic equation (1.1) was suggested. Instead of this equation we consider the delay-differential equation

$$\dot{p} = D(p) - S(p_h), \tag{1.2}$$

where $p = p(t)$, $p_h = p(t-h)$, i.e. we add a delay in the supply function $S(p)$. Now this function depends not on the price of trading day $p(t)$, but on the price in the previous period, i.e. the price of yesterdays trading day. This latest statement is only natural, because the manufacturer is guided by the price of the previous period in the supply of goods. Within this work we consider two variants of the choice of the functions $D(p)$, $S(p_h)$:

(1) $D_1(p) = a_\alpha p^{-\alpha}$, $S_1(p_h) = b_\alpha p_h$;

(2) $D_2(p) = \dfrac{a}{p+c}$, $S_2(p_h) = b\left(1 - \dfrac{c}{p_h + c}\right)$.

Above, $\alpha, a_\alpha, b_\alpha, a, b, c$ - positive constants. In the both variants the functions $D(p), S(p)$ belong to the corresponding class. There is one important difference between functions $S_1(p_h)$, $S_2(p_h)$: $\lim\limits_{p\to\infty} S_1(p_h) = \infty$, $\lim\limits_{p\to\infty} S_2(p_h) = b$. In the second variant for the equation (1.2) it used to say about the choice of the supply function with "saturation".

2. Analysis of the first model

Here, we consider the following equation

$$\dot{p} = a_\alpha p^{-\alpha} - b_\alpha p(t-h). \tag{2.1}$$

This equation after substitution $p = \mu p_1$, $\mu = (a_\alpha/b_\alpha)^{\alpha_1}$, $\alpha_1 = (1+\alpha)^{-1}$ has the form

$$\dot{p}_1(t) = b_\alpha \left[p_1^{-\alpha}(t) - p_1(t-h)\right].$$

Last differential equation has an equilibrium state $p_1(t) = p_0 = 1$.

Let $p_1(t) = 1 + x(t)$, $p_1(t-h) = 1 + y(t)$, $y(t) = x(t-h)$. As the result for the deviation $x(t)$ we obtain the equation

$$\dot{x}(t) = b_\alpha[-\alpha x - y] + a_2 x^2 + a_3 x^3 + ..., \; a_2 = b_\alpha \frac{\alpha(\alpha+1)}{2}, \; a_3 = -b_\alpha \frac{\alpha(\alpha+1)(\alpha+2)}{6}. \tag{2.2}$$

Here, as the points we denote terms having at zero an order of smallness greater than three.

2.1. Stability of an equilibrium state of the equation (2.2)

This equation has a zero equilibrium state. For the analysis of its stability we consider the linearized variant of the equation

$$\dot{x}(t) = -b_\alpha\left[\alpha x + y\right], \; x = x(t), \; y = x(t-h). \tag{2.3}$$

It is well known [7] that a question about stability of solutions of the equation (2.3) comes down to the analysis of location of the roots of the characteristic equation

$$\lambda + b_\alpha\Big(\alpha + \exp(-\lambda h)\Big) = 0. \tag{2.4}$$

At the same time the characteristic equation (2.4) has a countable number of roots $\lambda_j \in C$. If for them the inequalities $Re\lambda_j \le -\gamma_0 < 0$ are hold, then the solutions of the equation (2.3) are asymptotically stable and unstable if at least for one eigenvalue λ_k the inequality $Re\lambda_k > 0$ is hold. Note that the equation (2.4) has no zero root. Only critical case of a pair of pure imaginary roots of characteristic equation $\lambda_{1,2} = \pm i\sigma,\ \sigma > 0,\ Re\lambda_j \leq -\gamma_0 < 0\,(j = 3, 4, 5, \ldots)$ can take a place. For this σ from the following equation we have

$$cos\sigma h = -\alpha, \tag{2.5}$$

and $b_\alpha = b_{cr} = \min\limits_k\{b_{\alpha k}\} > 0$. In turn, $b_{\alpha k} = \sigma_k/(\sin\sigma_k h)$, where σ_k - one of the positive roots of the equation (2.5) for which $\sin\sigma_k h > 0$. It is elementary to verify that the appropriate pair (σ, b_{cr}) is the numbers $\sigma = \omega_\alpha/h,\ b_{cr} = \omega_\alpha/(h\sqrt{1-a^2})$, where $\omega_\alpha = \arccos(-\alpha) \in (\pi/2, \pi)$. There exist the corresponding solutions if $\alpha \in (0, 1)$. If $\alpha \geq 1$ for the differential equation (2.2), a critical case in the problem about stability cannot be realized.

2.2. Bifurcation problem

In the equation (2.2) we set $b_\alpha = b_{cr}(1 + \varepsilon),\ 0 < \varepsilon << 1$. As the result, we obtain the following equation which depends on ε

$$\dot{x} = A(\varepsilon)x + F_2(x, \varepsilon) + F_3(x, \varepsilon) + F_0(x, \varepsilon), \tag{2.6}$$

where

$$A(\varepsilon)x = -b_{cr}(1+\varepsilon)[\alpha x + y],\ F_2(x, \varepsilon) = b_{cr}(1+\varepsilon)\frac{\alpha(\alpha+1)}{2}x^2,$$

$$F_3(x, \varepsilon) = -b_{cr}(1+\varepsilon)\frac{\alpha(\alpha+1)(\alpha+2)}{6}x^3.$$

We denote $F_0(x, \varepsilon)$ as a function for which the inequality $|F_0(x, \varepsilon)| \leq M_0|x|^4,\ M_0 > 0$ is hold. For the equation (2.6) we add the initial condition

$$x(t, \varepsilon) = g(t),\ t \in [-h, 0], \tag{2.7}$$

where the given function is $g(t) \in C[-h, 0]$. The solution of the Cauchy problem (2.6), (2.7) with sufficiently small $g(t)(\parallel g(t) \parallel_{C[-h,0]} \leq r_0)$ generates a local semiflow. This semiflow has a two-dimensional local central manifold $M_2(\varepsilon)(dim M_2(\varepsilon) = 2)$. On this manifold the dynamics of the solutions of the equation (2.6) is reduced to the analysis of the system of two ordinary differential equations the Poincare-Dulac normal form (NF) (see [8,9]). We write this equation in a complex form

$$z' = \left[(\beta_1 + i\beta_2) + (l_1 + il_2)|z|^2\right] z + O(\varepsilon), \tag{2.8}$$

where $\beta_1, \beta_2, l_1, l_2, \in R$ and $z = z(s),\ s = \varepsilon t$ - "slow" time. All the rest solutions of the Cauchy problem (2.6), (2.7) with sufficiently small initial conditions tend to $M_2(\varepsilon)$ with a speed of exponent. Solutions on the $M_2(\varepsilon)$ may be reconstructed by using the solutions of the normal form (2.8). The link between solutions of the problem (2.6), (2.7) and NF (2.8) is established by using the algorithm which has the origin in a famous Krylov-Bogolyubov algorithm.

We will find the solutions of the equation (2.6) belonging to $M_2(\varepsilon)$ in the following form

$$x(t, \varepsilon) = \varepsilon^{1/2}x_1(t, z, \bar{z}) + \varepsilon x_2(t, z, \bar{z}) + \varepsilon^{3/2}x_3(t, z, \bar{z}) + o(\varepsilon^{3/2}), \tag{2.9}$$

where $z(s)$ - solutions of the equation (2.8), $x_j(t, z, \bar{z}),\ j = 1, 2, 3, ...$ which sufficiently smoothly depend on arguments. They have a period of $2\pi/\sigma$ of variable t. Also $x_1(t, \varepsilon) = z(s)\exp(i\sigma t) + \bar{z}(s)\exp(-i\sigma t)$ and for $x_2(t, z, \bar{z}),\ x_3(t, z, \bar{z})$ the identities $M_\pm(x_j) =$

$\sigma/(2\pi)\int\limits_0^{2\pi/\sigma} x_j \exp(\pm i\sigma t)dt = 0,\ j = 2,3,...$ are hold. Functions $x_2(t,z,\bar{z})$, $x_3(t,z,\bar{z})$ are determined as the solutions of the nonhomogeneous differential equations. For their formation we substitute the sum (2.9) in the equation (2.6) and equate the terms of $\varepsilon^{1/2}$, ε, $\varepsilon^{3/2}$ and so on. Equating the terms of ε we obtain the equation for $x_2(t)$

$$\dot{x}_2 + b_{cr}(\alpha x_2 + y_2) = b_{cr}\frac{\alpha(\alpha+1)}{2}x_1^2. \tag{2.10}$$

For $x_3(t)$ we obtain the following nonhomogeneous equation

$$\begin{gathered}\dot{x}_3 + b_{cr}(\alpha x_3 + y_3) = -b_{cr}\frac{\alpha(\alpha+1)(\alpha+2)}{6}x_1^3 + b_{cr}\alpha(\alpha+1)x_1x_2 + b_{cr}[-\alpha x_1 - y_1]+\\ +[b_{cr}h\exp(-i\sigma h) - 1]\psi(z,\bar{z})\exp(i\sigma t) + [b_{cr}h\exp(i\sigma h) - 1]\bar{\psi}(z,\bar{z})\exp(-i\sigma t),\\ \psi(z,\bar{z}) = \beta z + l|z|^2\bar{z},\ y_j = y_j(t,z,\bar{z}) = x_j(t-h,z,\bar{z}),\ z = z(s),\ j = 1,2,3,...\end{gathered} \tag{2.11}$$

Remark: nonhomogeneous differential equation $\dot{u} + b_{cr}[\alpha u + v] = \varphi(t),\ v = u(t-h)$, where $\varphi(t)$ - $2\pi/\sigma$ - periodic function has $2\pi/\sigma$ - periodic solution if

$$M_{\pm}(\varphi) = \frac{\sigma}{2\pi}\int\limits_0^{2\pi/\sigma}\varphi(t)\exp(\pm i\sigma t)dt = 0.$$

Equality $M_{\pm}(u) = 0$ determines one such solution.

Nonhomogeneous equation (2.10) has the following appropriate solution

$$x_2(t,z,\bar{z}) = \eta z^2\exp(2i\sigma t) + \xi + \bar{\eta}\bar{z}^2\exp(-2i\sigma t),$$

where $\xi = \alpha,\ \eta = \alpha/(2(2\alpha-1) + 4i\sqrt{1-\alpha^2})$. The solvability conditions of the nonhomogeneous differential equation (2.11) in the class of periodic functions let us to determine the coefficients of NF $\beta = \beta_1 + i\beta_2,\ l = l_1 + il_2$. After the corresponding calculations, we have

$$\beta_1 = b_{cr}\frac{(1-\alpha^2)^{3/2}w_\alpha}{L_0} > 0,\ \beta_2 = b_{cr}\frac{(1-\alpha^2)[\sqrt{1-\alpha^2} + \alpha w_\alpha]}{L_0},$$

$$L_0 = 1 - \alpha^2 + \omega_\alpha^2 + 2\alpha w_\alpha\sqrt{1-\alpha^2},\ l_1 = -b_{cr}\frac{\alpha(1-\alpha^2)^{3/2}}{(5-4\alpha)L_0}\Big[(5-\alpha)\sqrt{1-\alpha^2} + 6\alpha\omega_\alpha\Big] < 0,$$

$$l_2 = -b_{cr}\frac{\alpha(\alpha+1)(1-\alpha^2)}{(5-4\alpha)L_0}\Big[\alpha\sqrt{1-\alpha^2} + (6\alpha-5)\omega_\alpha\Big].$$

Consider a truncated form of the equation (2.8)

$$z' = (\beta_1 + i\beta_2)z + (l_1 + il_2)z|z|^2. \tag{2.12}$$

Lemma 1. Differential equation of the form (2.12) has a periodic solution

$$z(s) = \rho_0\exp(i\omega_0 s),\ \rho_0 = \sqrt{-\frac{\beta_1}{l_1}},\ \omega_0 = \beta_2 - \frac{l_2\beta_1}{l_1}$$

if $\beta_1 l_1 < 0$. This solution is stable if $\beta_1 > 0$ $(l_1 < 0)$.

A verifying of the statement is standard.

We note that in our case $\beta_1 > 0,\ l_1 < 0$, i.e. there exists a stable periodic solution.

Theorem 1. There exists $\varepsilon_0 > 0$ such that for all $\varepsilon \in (0,\varepsilon_0)$ the differential equation (2.1) has a stable cycle. This cycle generates the periodic solutions

$$\begin{gathered}p(t,\varepsilon) = \Big(\frac{a_\alpha}{b_\alpha}\Big)^{1/(1+\alpha)}\Big\{1 + \varepsilon^{1/2}\rho_0[\exp(i\sigma_\varepsilon t + i\varphi_0) + \exp(-i\sigma_\varepsilon t - i\varphi_0)]+\\ +\varepsilon\rho_0^2[\eta\exp(2i\sigma_\varepsilon t + 2i\varphi_0) + \xi + \bar{\eta}\exp(-2i\sigma_\varepsilon t - 2i\varphi_0)]\Big\} + o(\varepsilon),\ \sigma_\varepsilon = \sigma + \varepsilon\omega_0 + o(\varepsilon).\end{gathered}$$

The validity of the statement follows from the works [8,9].

3. Supply function with "saturation"

In this section we consider the second variant of model "aggregate demand-aggregate supply model" for which the function $S(p_h)$ is chosen with saturation ($\lim_{p\to\infty} S(p) = b < \infty$). The corresponding equation after renormalization $p = cp_2$ has the form

$$\dot{p}_2(t) = b_c\left[\frac{\alpha}{1+p_2(t)} - \left(1 - \frac{1}{1+p_2(t-h)}\right)\right], \ \alpha = \frac{a}{bc}, \ b_c = \frac{b}{c}. \qquad (3.1)$$

The equation (3.1) has only one positive equilibrium state $p_2(t) = \alpha$. Let

$$p_2(t) = \alpha + x_1(t),\ p_2(t-h) = \alpha + y_1(t),\ y_1(t) = x_1(t-h),\ x_1(t) = (1+\alpha)x(t)(y_1(t) = (1+\alpha)y(t)).$$

As the result of these two substitutions, we obtain the equation for $x(t)(y(t) = x(t-h))$

$$\dot{x} = b_\alpha\left[(-\alpha x - y) + (\alpha x^2 + y^2) - (\alpha x^3 + y^3)\right] + R_0(x,y), \qquad (3.2)$$

where $b_\alpha = b_c/(1+\alpha)^2$, $|R_0| \leq M_0[x^4 + y^4]$, $M_0 = const > 0$, in particular, $R_0(0,0) = 0$.

3.1. Stability of a zero equilibrium state of the equation (3.2)

For the investigation of a zero equilibrium state we consider a linearized at zero form of the equation (3.2)

$$\dot{x} = -b_\alpha[\alpha x + y], \qquad (3.3)$$

which is the same like for the equation (2.3) from the previous section. In particular, for $b_\alpha = b_{cr}$ a critical case is realized in the problem about stability of a zero solution. The spectrum of stability has a pair of pure imaginary eigenvalues $\pm i\sigma$, where $\sigma = \omega_\alpha/h$, $\omega_\alpha = \arccos(-\alpha)$, $b_{cr} = \omega_\alpha/(h\sqrt{1-\alpha^2})$

3.2. About bifurcation problem in the second variant of the supply function

In the equation (3.2) let $b_\alpha = b_{cr}(1+\varepsilon)$ and as the result we obtain the differential equation which is very close to the equation (2.6)

$$\dot{x} = A(\varepsilon)x + F_2(x,\varepsilon) + F_3(x,\varepsilon) + F_0(x,\varepsilon), \qquad (3.4)$$

$$A(\varepsilon)x = b_{cr}(1+\varepsilon)[-\alpha x - y],\ F_2(x,x) = b_{cr}(1+\varepsilon)(\alpha x^2 + y^2),$$
$$F_3(x,\varepsilon) = -b_{cr}(1+\varepsilon)(\alpha x^3 + y^3),\ y(t) = x(t-h).$$

In the equations (3.4) and (2.6) a quadratic and cubic terms differ. As in the second section, the question about existence of cycles, their stability is reduced to the analysis of NF

$$z' = \left[(\beta_1 + i\beta_2) + (l_1 + il_2)|z|^2\right]z + O(\varepsilon),\ z = z(s),\ s = \varepsilon t, \qquad (3.5)$$

β_1, β_2 are computed by using the formulas from the previous section. Here, in (3.5)

$$l_1 = -b_{cr}\frac{(1-\alpha)\sqrt{1-\alpha^2}}{L_0(5-4\alpha)}\left[(8\alpha-4)\sqrt{1-\alpha^2} + \omega_\alpha(7-\alpha+4\alpha^2)\right],$$

$$l_2 = -b_{cr}\frac{(1-\alpha^2)}{L_0(5-4\alpha)}\left[(7-4\alpha)\sqrt{1-\alpha^2} + \omega_\alpha(4-5\alpha+4\alpha^2)\right],$$

where $L_0 = (1-\alpha^2) + 2\alpha\omega_\alpha\sqrt{1-\alpha^2} + \omega_\alpha^2$, $\alpha \in (0,1)$. It is not difficult to see that $l_1 < 0$ for all values of the parameter $\alpha \in (0,1)$. Indeed, for the given α the expression has the form

$$(8\alpha - 4)\sqrt{1-\alpha^2} + \omega_\alpha(7 - \alpha + 4\alpha^2) \geq 6.89 > 0.$$

Therefore, the shortened NF $z' = (\beta_1 + i\beta_2)z + (l_1 + il_2)z|z|^2$ has the stable periodic solution $z(s) = \rho_0 \exp(i\omega s)$, where $\rho_0 = (-\beta_1/l_1)^{1/2}$ $(-\beta_1 l_1 > 0), \omega = \beta_2 - l_2\beta_1/l_1$.

Theorem 2. There exists $\varepsilon_0 > 0$ such that for all $\varepsilon \in (0, \varepsilon_0)$ the equation (3.1) has an attracting limit cycle which is generated by the periodic solutions

$$p(t,\varepsilon) = c\Big[\alpha + (1+\alpha)\varepsilon^{1/2}\rho_0[\exp(i\sigma_\varepsilon t + i\varphi_0) + \exp(-i\sigma_\varepsilon t - i\varphi_0)]\Big] + o(\varepsilon^{1/2}),\ \sigma_\varepsilon = \sigma + \varepsilon\omega + o(\varepsilon).$$

The validity of the statement follows from the works [7-9].

4. Conclusion

In the work we consider two variants of a mathematical market model which is more famous as "aggregate demand-aggregate supply model". As in the works [5,6] a classic form of the model was changed, and in the supply function the delay was added. It significantly changes a dynamics of the solutions. Without taking into account a delay, the corresponding equations have an equilibrium state and they cannot have the periodic solutions which describe the cyclic recurrence of a market economy. Taking into account a delay, the cycles can exist with a corresponding choice of parameters of the model (coefficients of the equation). The period of the cycles is close to $T(h) = 2\pi h/\omega_\alpha$, $\omega_\alpha = \arccos(-\alpha)$, where α - one of the main parameters for both of the models. The period is close to zero if a constant h is sufficiently small and, in opposite, sufficiently large with $h \to \infty$. In the second case when $h \to \infty$ the long cycles of the economic which were opened in the 1930s of the past century by a famous economist N.D. Kondratyev are realized. We will once again emphasize that the periodic solutions in the both variants of the choice of the demand function and the supply function exist, and they are stable for all realizable choices of the parameters of the problem. In the Solow-Swan model it was suggested to add a delay in the work [10], but in this work the analysis which is correct from a mathematical point of view is absent.

5. Acknowledgments

The author of the paper is grateful to Professor A.V. Kryanev for a discussing of the results of the article in the work of the section "Methods of Mathematical Physics" at the conference "Problems of Mathematical and Theoretical Physics and Mathematical Modeling" in National Research Nuclear University "MEPhI" (Moscow Engineering Physics Institute), Russia, Moscow, 5-7 april, 2016.

This work was supported by a grant of the President of the Russian Federation -5932.2015.1.

6. References

[1] Ricardo D 1971 *Principles of Political Economy and Taxation* (Harmondsworth: Penguin)
[2] Zhang W B 1991 *Synergetic Economics: Time and Change in Nonlinear Economics* (Berlin-Heidelberg: Springer-Verlag)
[3] Puu T 1990 *Economic Dynamics* (Berlin-Heidelberg: Springer-Verlag)
[4] Lebedev V V and Lebedev K V 2011 *Mathematicheskoe modelirovanie nestatsionarnych protsessov* in Russian (Moscow: eTest Publ.)
[5] Kulikov A N 2014 *Kolmogorovskie chtenia*, Yaroslavl **94**
[6] Kulikov A N and Kulikov D A 2015 *Taurida Journal of Computer Science Theory and Mathematics* **2(27)** 87
[7] Hale J 1977 *Theory of functional differential equations* (Berlin-Heidelberg: Springer-Verlag)
[8] Kolesov Yu S 1974 *Vestnik Yaroslav. Univ.* **3**
[9] Kulikov A N 1976 *Studies in Stability and the Theory of Oscillations* **67**
[10] Guerrini L 2012 *International Journal of Pure and Applied Mathematics* **78(7)** 993

Spinor representation of Maxwell's equations

D S Kulyabov[1,2], **A V Korolkova**[1], **L A Sevastianov**[1,3]
[1]Department of Applied Probability and Informatics,
RUDN University (Peoples' Friendship University of Russia),
6 Miklukho-Maklaya str., Moscow, 117198, Russia
[2]Laboratory of Information Technologies,
Joint Institute for Nuclear Research,
6 Joliot-Curie, Dubna, Moscow region, 141980, Russia
[3]Bogoliubov Laboratory of Theoretical Physics,
Joint Institute for Nuclear Research,
6 Joliot-Curie, Dubna, Moscow region, 141980, Russia

E-mail: yamadharma@gmail.com, akorolkova@sci.pfu.edu.ru, leonid.sevast@gmail.com

Abstract. Spinors are more special objects than tensor. Therefore possess more properties than the more generic objects such as tensors. Thus, the group of Lorentz two-spinors is the covering group of the Lorentz group. Since the Lorentz group is a symmetry group of Maxwell's equations, it is assumed to reasonable to use when writing the Maxwell equations Lorentz two-spinors and not tensors. We describe in detail the representation of the Maxwell's equations in the form of Lorentz two-spinors. This representation of Maxwell's equations can be of considerable theoretical interest.

1. Introduction

Maxwell's equations have a large number of representations [1–4]. The principle of the introduction of the following: every representation must simplify the concrete theoretical and practical study. In this paper, we consistently describe the Lorentz two-spinor [5] representation of Maxwell's equations. It is supposed that this form will be interested in theoretical studies [6–8].

The structure of the article is as follows. In the section 2 basic notations and conventions are introduced. Section 3 gives a brief description of the Maxwell equations. Section 4 gives the spinors of the electromagnetic field. Further, section 5 gives the Lorenz two-spinor representation of Maxwell's equations.

2. Notations and conventions

(i) The abstract indices notation [9] is used in this work. Under this notation a tensor as a whole object is denoted just as an index (e.g., x^i), components are denoted by underlined index (e.g., $x^{\underline{i}}$).

(ii) We will adhere to the following agreements . Greek indices (α, β) will refer to the four-dimensional space , in component form it looks like: $\underline{\alpha} = \overline{0,3}$. Latin indices from the middle of the alphabet (i, j, k) will refer to the three-dimensional space , in the component form it looks like: $\underline{i} = \overline{1,3}$.

IOP Conf. Series: Journal of Physics: Conf. Series **788** (2017) 012025 doi:10.1088/1742-6596/788/1/012025

(iii) The comma in the index denotes partial derivative with respect to corresponding coordinate ($f_{,i} := \partial_i f$); semicolon denotes covariant derivative ($f_{;i} := \nabla_i f$).
(iv) To write the equations of electrodynamics in the article is used CGS symmetrical system.

3. Maxwell's Equations

Maxwell's equations in 3-dimensional form are as follows:

$$\begin{aligned} \nabla_0 B^i &= -e^{ijk}\nabla_j E_k; \\ \nabla_i D^i &= 4\pi\rho; \\ \nabla_0 D^i &= e^{ijk}\nabla_j H_k - \frac{4\pi}{c} j^i; \\ \nabla_i B^i &= 0. \end{aligned} \tag{1}$$

where $e^{\underline{i}\,\underline{j}\,\underline{k}}$ is the alternating tensor expressed by Levi-Civita simbol $\varepsilon^{\underline{i}\,\underline{j}\,\underline{k}}$:

$$e_{\underline{i}\,\underline{j}\,\underline{k}} = \sqrt{{}^3g}\,\varepsilon_{\underline{i}\,\underline{j}\,\underline{k}}, \quad e^{\underline{i}\,\underline{j}\,\underline{k}} = \frac{1}{\sqrt{{}^3g}}\varepsilon^{\underline{i}\,\underline{j}\,\underline{k}}.$$

Let's rewrite (1) with the help of electromagnetic field tensors $F_{\alpha\beta}$ and $G_{\alpha\beta}$ [10]:

$$\begin{aligned} \nabla_\alpha F_{\beta\gamma} + \nabla_\beta F_{\gamma\alpha} + \nabla_\gamma F_{\alpha\beta} = F_{[\alpha\beta;\gamma]} &= 0, \\ \nabla_\alpha G^{\alpha\beta} &= \frac{4\pi}{c} j^\beta, \end{aligned} \tag{2}$$

where

$$F_{\underline{\alpha}\,\underline{\beta}} = \begin{pmatrix} 0 & E_1 & E_2 & E_3 \\ -E_1 & 0 & -B^3 & B^2 \\ -E_2 & B^3 & 0 & -B^1 \\ -E_3 & -B^2 & B^1 & 0 \end{pmatrix}, \quad F^{\underline{\alpha}\,\underline{\beta}} = \begin{pmatrix} 0 & -E^1 & -E^2 & -E^3 \\ E^1 & 0 & -B_3 & B_2 \\ E^2 & B_3 & 0 & -B_1 \\ E^3 & -B_2 & B_1 & 0 \end{pmatrix},$$

$$G^{\underline{\alpha}\,\underline{\beta}} = \begin{pmatrix} 0 & -D^1 & -D^2 & -D^3 \\ D^1 & 0 & -H_3 & H_2 \\ D^2 & H_3 & 0 & -H_1 \\ D^3 & -H_2 & H_1 & 0 \end{pmatrix}, \quad G_{\underline{\alpha}\,\underline{\beta}} = \begin{pmatrix} 0 & D_1 & D_2 & D_3 \\ -D_1 & 0 & -H^3 & H^2 \\ -D_2 & H^3 & 0 & -H^1 \\ -D_3 & -H^2 & H^1 & 0 \end{pmatrix},$$

$E_{\underline{i}}$, $H^{\underline{i}}$ are components of electric and magnetic fields intensity vectors; $D_{\underline{i}}$, $B^{\underline{i}}$ are components of vectors of electric and magnetic induction.

4. Spinors of electromagnetic field

Spinors are used in physics quite extensively. The following spinors are mainly used: Dirac four-spinors; Pauli three-spinors; quaternions. If Dirac four-spinors are used, the main difficulty is γ-matrices. The essence of these objects is that they serve to connect the spinor and tensor spaces and therefore have two types of indices: spinor and tensor ones. It would be logical to perform calculations in one of these spaces only. In this paper we use semispinors of Dirac spinors, Lorentz two-spinors.

The tensor of electromagnetic field $F_{\alpha\beta}$ and its components $F_{\underline{\alpha}\,\underline{\beta}}$, $\underline{\alpha}$ may be considered in spinor form (and similarly for $G_{\alpha\beta}$):

$$\begin{aligned} F_{\alpha\beta} &= F_{A\dot{A}B\dot{B}}; \\ F_{\underline{\alpha}\,\underline{\beta}} &= F_{\underline{A}\,\underline{\dot{A}}\,\underline{B}\,\underline{\dot{B}}}\, g_{\underline{\alpha}}{}^{\underline{A}\,\underline{\dot{A}}}\, g_{\underline{\beta}}{}^{\underline{B}\,\underline{\dot{B}}}, \end{aligned}$$

where $g_{\underline{\alpha}}{}^{\underline{A}\underline{\dot{A}}}$ are Infeld–van der Waerden symbols defined in real spinor basis $\varepsilon_{\underline{A}\underline{B}}$ in the following way [9]:

$$g_{\underline{\alpha}}{}^{\underline{A}\underline{\dot{A}}} := g_{\underline{\alpha}}{}^{\alpha}\varepsilon_{A}{}^{\underline{A}}\varepsilon_{\dot{A}}{}^{\underline{\dot{A}}}, \quad g_{\underline{A}\underline{\dot{A}}}{}^{\underline{\alpha}} := g^{\underline{\alpha}}{}_{\alpha}\varepsilon^{A}{}_{\underline{A}}\varepsilon^{\dot{A}}{}_{\underline{\dot{A}}}, \tag{3}$$

$$\varepsilon_{\underline{A}\underline{B}} = \varepsilon_{\underline{\dot{A}}\underline{\dot{B}}} = \begin{pmatrix} 0 & 1 \\ -1 & 0 \end{pmatrix}, \quad \varepsilon^{\underline{A}\underline{B}} = \varepsilon^{\underline{\dot{A}}\underline{\dot{B}}} = \begin{pmatrix} 0 & 1 \\ -1 & 0 \end{pmatrix}, \quad \varepsilon_{\underline{A}}{}^{A}\varepsilon_{A}{}^{\underline{B}} = \varepsilon_{\underline{A}}{}^{\underline{B}} = \begin{pmatrix} 1 & 0 \\ 0 & 1 \end{pmatrix}. \tag{4}$$

Let's $g_{\alpha\beta} = \mathrm{diag}(1,-1,-1,-1)$ is the Minkowski space metric. We use (4) as spinor space metric. Then the Infeld–van der Waerden symbols will have the following coordinate representation:

$$g_{\underline{0}}{}^{\underline{A}\underline{\dot{A}}} = g^{\underline{0}}{}_{\underline{A}\underline{\dot{A}}} = \frac{1}{\sqrt{2}}\begin{pmatrix} 1 & 0 \\ 0 & 1 \end{pmatrix}, \quad g_{\underline{1}}{}^{\underline{A}\underline{\dot{A}}} = g^{\underline{1}}{}_{\underline{A}\underline{\dot{A}}} = \frac{1}{\sqrt{2}}\begin{pmatrix} 0 & 1 \\ 1 & 0 \end{pmatrix},$$
$$g_{\underline{2}}{}^{\underline{A}\underline{\dot{A}}} = -g^{\underline{2}}{}_{\underline{A}\underline{\dot{A}}} = \frac{1}{\sqrt{2}}\begin{pmatrix} 0 & \mathrm{i} \\ -\mathrm{i} & 0 \end{pmatrix}, \quad g_{\underline{3}}{}^{\underline{A}\underline{\dot{A}}} = g^{\underline{3}}{}_{\underline{A}\underline{\dot{A}}} = \frac{1}{\sqrt{2}}\begin{pmatrix} 1 & 0 \\ 0 & -1 \end{pmatrix}.$$

The tensor $F_{\alpha\beta}$ is real and antisymmetric, it can be represented in the form

$$\begin{aligned} F_{\alpha\beta} &= \varphi_{AB}\varepsilon_{\dot{A}\dot{B}} + \varepsilon_{AB}\bar{\varphi}_{\dot{A}\dot{B}}, \\ {}^{*}F^{\alpha\beta} &= -\mathrm{i}\varphi^{AB}\varepsilon^{\dot{A}\dot{B}} + \mathrm{i}\varepsilon^{AB}\bar{\varphi}^{\dot{A}\dot{B}}. \end{aligned} \tag{5}$$

where φ_{AB} is a spinor of electromagnetic field:

$$\varphi_{AB} := \frac{1}{2}F_{ABC'}{}^{C'} = \frac{1}{2}F_{A\dot{A}B\dot{B}}\varepsilon^{\dot{A}\dot{B}} = \frac{1}{2}F_{\alpha\beta}\varepsilon^{\dot{A}\dot{B}}.$$

The components of electromagnetic field spinor:

$$\varphi_{\underline{A}\underline{B}} = \frac{1}{2}F_{\underline{\alpha}\underline{\beta}}\varepsilon^{\underline{\dot{A}}\underline{\dot{B}}}g^{\underline{\alpha}}{}_{\underline{A}\underline{\dot{A}}}g^{\underline{\beta}}{}_{\underline{B}\underline{\dot{B}}}.$$

Using the equations (3), (4) and notation $F_i = E_i - \mathrm{i}B^i$, we will get:

$$\begin{aligned} \varphi_{00} &= \frac{1}{2}(F_{31} + F_{01} - \mathrm{i}F_{32} - \mathrm{i}F_{02}) = \frac{1}{2}\left(F_1 - \mathrm{i}F_2\right), \\ \varphi_{01} = \varphi_{10} &= \frac{1}{2}(-F_{03} - \mathrm{i}F_{12}) = -\frac{1}{2}F_3, \\ \varphi_{11} &= \frac{1}{2}(F_{31} - F_{01} + \mathrm{i}F_{32} - \mathrm{i}F_{02}) = -\frac{1}{2}\left(F_1 + \mathrm{i}F_2\right). \end{aligned}$$

Similarly

$$\begin{aligned} G^{\alpha\beta} &= \eta^{AB}\varepsilon^{\dot{A}\dot{B}} + \varepsilon^{AB}\bar{\eta}^{\dot{A}\dot{B}}, \\ {}^{*}G_{\alpha\beta} &= -\mathrm{i}\eta_{AB}\varepsilon_{\dot{A}\dot{B}} + \mathrm{i}\varepsilon_{AB}\bar{\eta}_{\dot{A}\dot{B}}. \end{aligned} \tag{6}$$

where η^{AB} is a Minkowski spinor:

$$\eta^{AB} := \frac{1}{2}G^{ABC'}{}_{C'} = \frac{1}{2}G^{A\dot{A}B\dot{B}}\varepsilon_{\dot{A}\dot{B}} = \frac{1}{2}G^{\alpha\beta}\varepsilon_{\dot{A}\dot{B}}.$$

The components of the spinor η^{AB}:

$$\eta^{\underline{A}\underline{B}} = \frac{1}{2}G^{\underline{\alpha}\underline{\beta}}\varepsilon_{\underline{\dot{A}}\underline{\dot{B}}}g_{\underline{\alpha}}{}^{\underline{A}\underline{\dot{A}}}g_{\underline{\beta}}{}^{\underline{B}\underline{\dot{B}}}.$$

Using the equations (3), (4) and notation $G^i = D^i - \mathrm{i}H_i$, we will get:

$$\eta^{00} = \frac{1}{2}(G^{31} + G^{01} + \mathrm{i}G^{32} + \mathrm{i}G^{02}) = \frac{1}{2}\left(G^1 - \mathrm{i}G^2\right),$$
$$\eta^{01} = \eta^{10} = \frac{1}{2}(-G^{03} + \mathrm{i}G^{12}) = -\frac{1}{2}G^3,$$
$$\eta^{11} = \frac{1}{2}(G^{31} - G^{01} - \mathrm{i}G^{32} + \mathrm{i}G^{02}) = -\frac{1}{2}\left(G^1 + \mathrm{i}G^2\right).$$

5. Spinor Form of Maxwell's Equations

Let's write Maxwell's equations using the spinors.

Replacing in (2) abstract indices α by $A\dot{A}$ and β by $B\dot{B}$, we can write:

$$\nabla_{A\dot{A}} G^{A\dot{A}B\dot{B}} = \frac{4\pi}{c} j^{B\dot{B}}.$$

Using (6) we will get

$$\nabla^{A\dot{B}} \eta^B_A + \nabla^{B\dot{A}} \bar{\eta}^{\dot{B}}_{\dot{A}} = \frac{4\pi}{c} j^{B\dot{B}}.$$

Similarly, from (5) it follows

$$\nabla^{\dot{A}B} \varphi^A_B - \nabla^{A\dot{B}} \bar{\varphi}^{\dot{A}}_{\dot{B}} = 0.$$

In so doing the system of Maxwells equations can be written as

$$\begin{aligned} \nabla^{\dot{A}B} \varphi^A_B - \nabla^{A\dot{B}} \bar{\varphi}^{\dot{A}}_{\dot{B}} &= 0, \\ \nabla^{A\dot{B}} \eta^B_A + \nabla^{B\dot{A}} \bar{\eta}^{\dot{B}}_{\dot{A}} &= \frac{4\pi}{c} j^{B\dot{B}}. \end{aligned} \tag{7}$$

In the vacuum case (no medium), we can put ${\eta^A}_B = {\varphi^A}_B$. Then we can write the equations (7) as follows:

$$\nabla^{\dot{A}B} \varphi^A_B = \nabla^{A\dot{B}} \bar{\varphi}^{\dot{A}}_{\dot{B}},$$
$$\nabla^{A\dot{B}} \varphi^B_A + \nabla^{B\dot{A}} \bar{\varphi}^{\dot{B}}_{\dot{A}} = \frac{4\pi}{c} j^{B\dot{B}}.$$

Thus, the spinor form of Maxwell's equations system in vacuum can be written in the form of one equation:

$$\nabla^{A\dot{B}} \varphi^B_A = \frac{2\pi}{c} j^{B\dot{B}}.$$

6. Conclusions

Thus, in the article, we have proposed a representation of Maxwell's equations in the form of Lorentz 2-spinors. We consider that the given representation might be interested in in theoretical studies.

Acknowledgments

The work is partially supported by RFBR grants No's 14-01-00628, 15-07-08795, and 16-07-00556. Also the publication was supported by the Ministry of Education and Science of the Russian Federation (the Agreement No 02.a03.21.0008).

IOP Conf. Series: Journal of Physics: Conf. Series **788** (2017) 012025 doi:10.1088/1742-6596/788/1/012025

References

[1] Kulyabov D S, Korolkova A V and Korolkov V I 2012 *Bulletin of Peoples' Friendship University of Russia. Series "Mathematics. Information Sciences. Physics"* 96–106 (*Preprint* 1211.6590)
[2] Laporte O and Uhlenbeck G E 1931 *Physical Review* **37** 1380–1397
[3] Aste A 2012 *Journal of Geometry and Symmetry in Physics* 47–58 (*Preprint* arXiv:1211.1218v3)
[4] Bialynicki-Birula I 1996 *Progress in Optics* vol 36 (Elsevier) chap 5, pp 245–294 (*Preprint* 0508202)
[5] Kulyabov D S and Ul'yanova A G 2009 *Physics of Particles and Nuclei Letters* **6** 546–549 (*Preprint* 1312.6655)
[6] Moses H E 1959 *Physical Review* **113** 1670–1679
[7] Luneburg R K 1964 *Mathematical Theory of Optics* (Berkeley & Los Angeles: University of California Press)
[8] Khan S A 2005 *Physica Scripta* **71** 440–442 (*Preprint* 0205083v1)
[9] Penrose R and Rindler W 1987 *Spinors and Space-Time: Volume 1, Two-Spinor Calculus and Relativistic Fields* vol 1 (Cambridge University Press)
[10] Minkowski H 1910 *Math. Ann.* 472–525

On integration of the first order differential equations in a finite terms

M D Malykh

Department of Applied Probability and Informatics, Peoples' Friendship University of Russia, 6, Miklukho-Maklaya str., Moscow, 117198, Russia.
Moscow State University Materials Science Department, Leninskie Gory, Moscow, 119991, Russia.

E-mail: malykhmd@yandex.ru

Abstract. There are several approaches to the description of the concept called briefly as integration of the first order differential equations in a finite terms or symbolical integration. In the report three of them are considered: 1.) finding of a rational integral (Beaune or Poincaré problem), 2.) integration by quadratures and 3.) integration when the general solution of given differential equation is an algebraical function of a constant (Painlevé problem). Their realizations in Sage are presented.

1. Introduction

Algorithms realizing in popular computer algebra systems, for example, Maple, can solve many differential equations. However if system can't solve the equation user doesn't obtain any of information about this equation. Here we consider three approaches when the negative answer is interesting too. In each of them we fix some algebraic properties of solution.

2. Beaune problem

Problem 1. For an ordinary differential equation

$$pdx + qdy = 0, \quad p, q \in \mathbb{Q}[x, y], \tag{1}$$

the problem, attributed below to Beaune, consists of finding a rational integral w in all cases when such integral exist.

Note. In modern terms Florimond de Beaune asked Descartes how to find the algebraical equation of integral curves for the differential equation $y' = y$ and Descartes wanted to find the solution by selection of coefficients in the equation

$$y^m = a_0 + a_1 x + \cdots + a_n x^n,$$

but stopped on $n, m \simeq 100$ [1]. Authors of the 19th century associated this problem with Poincaré, written about this question several big articles [2].

Modern computer algebra systems can't solve the problem 1. Let u, v be polynomials from $\mathbb{Z}[x, y]$. Then quotient $w = u/v$ is an integral of eq.

$$\left(v\frac{\partial u}{\partial x} - u\frac{\partial v}{\partial x}\right) dx + \left(v\frac{\partial u}{\partial y} - u\frac{\partial v}{\partial y}\right) dy = 0 \tag{2}$$

Put for example

$$u = (x^2 + y)^5(x - y^6 + 1) + 1, \quad v = (13xy^8 + y^5 + 3xy + 2)(x^2 + y)^4.$$

Standard solver `dsolve` in Maple can solve the eq. (2) in quadratures

$$\int rdx + sdy = C, \quad r, s \in \mathbb{Q}(x, y),$$

but Maple can't calculate those quadratures. Furthermore `symgen` from DETools package [3] finds only one integrating divisor.

If eq. (1) has a rational integral then integral curves constitute a irreducible linear pencil

$$u(x, y) + c \cdot v(x, y) = 0, \quad u, v \in \mathbb{C}[x, y],\ c \in \mathbb{C},$$

but we can't estimate pencil order. So in all known algorithms for solution of problem 1 the user must give a boundary for degree of required integral [4, 5]. This boundary has simple geometric sense so we can state the following variant of Beaune problem.

Problem 2. For an ordinary differential equation (1) the problem consists of finding an integral curves pencil of given degree N or less in all cases when such pencil exist.

We present here solution for the problem 2 in Sage [6] founded on Lagutinski method [7, 8, 9, 10, 11, 12], the scientific biography and bibliography of M.N. Lagutinski can be found in [13]. Required package Lagutinski can be downloaded from the author home page [14].

Let R be a polynomial ring with differentiation D and constants field $\mathbb{Q}$. Countable set

$$B = \{\varphi_1, \varphi_2, \ldots, \}, \quad \varphi_n \in R,$$

is called a basis of the ring R iff

(i) any element ψ from the ring R can be described as a linear combination of elements from B under constants field, that is

$$\psi = c_{n_1}\varphi_{n_1} + \cdots + c_{n_m}\varphi_{n_m}, \quad c_{n_1}, \ldots, c_{n_m} \in k, \tag{3}$$

(ii) for any integers i, j there is an integer $n > \max(i, j)$ such that

$$\varphi_i\varphi_j = \varphi_n.$$

Maximum from numbers $n_1, \ldots, n_m$ in (3) will be called an order of ψ with respect to the basis B. Determinant of the matrix

$$\begin{pmatrix} \varphi_1 & \varphi_2 & \cdots & \varphi_n \\ D\varphi_1 & D\varphi_2 & \cdots & D\varphi_n \\ \vdots & \vdots & \ddots & \vdots \\ D^{n-1}\varphi_1 & D^{n-1}\varphi_2 & \cdots & D^{n-1}\varphi_n \end{pmatrix} \tag{4}$$

is called Lagutinski determinant of degree n and will be denoted bellow as Δ_n.

Theorem 1 (M.N. Lagutinski, 1911). A rational integral of degree N exists iff $\Delta_N = 0$. In this case some rational integral is constructed as ratio of two minors in Δ_N.

In presented package we can calculate Lagutinski determinants and minors mentioned in the theorem 1. For solving of problem 2 we put $R = \mathbb{Q}[x, y]$ and

$$D = q\frac{\partial}{\partial x} - p\frac{\partial}{\partial y},$$

also choose standard glex-basis $\{1, y, x, y^2, \dots\}$ as a basis B. Then problem 2 has a solution iff

$$\Delta_{\frac{(n+1)(n+2)}{2}} = 0.$$

Example. Are the integrated curve of equation

$$xy' - (2x+1)y + y^2 = -x^2$$

conic sections? — Yes, because $\Delta_6 = 0$. Furthermore, in Sage we can calculate equation for the pencil, exactly

$$-x^2 + xy - x = C(y - x).$$

Computation of Lagutinski determinant of a big order is a difficult task [4] but for the solution of the problem 2 it is enough to calculate determinant in one point with random integer coordinates. If

$$\Delta_{\frac{(n+1)(n+2)}{2}}$$

at that point is not equal to zero then there isn't integral curves pencil of degree n or less, otherwise there is *almost certainly* integral curves pencil of degree n or less.

Example. Are the integral curve of equation

$$(x - y^2)y' - (2x+1)y + y^2 = -x^2$$

curves of 9 degree or less? — No, because $\Delta_{55} \neq 0$ at random point.

This idea allows to distinguish quickly in a text-book all equations with algebraic integral curves pencil. For example, all that equations in the standard Russian textbook on differential equations by A.F. Filippov [15] can be recognize by testing Δ_{55}. So Lagutinski method give a practical solution of the problem 1.

3. Integration in quadratures

If differential 1-form $udx + vdy$ is exact then expressions

$$\mathrm{S}(udx + vdy) = \lim \sum (u_n \Delta x + v_n \Delta y) = \int udx + vdy$$

and

$$\mathrm{P}(1 + udx + vdy) = \lim \prod (1 + u_n \Delta x + v_n \Delta y) = e^{\int udx + vdy}$$

are functions of variables x, y.

Note. First construction was introduced by Leibniz and second by Volterra and Schlesinger many years later [16]. P-integrals allow to state the further theory without any notion of elementary functions.

Definition 1. Let integral curves pencil can be represented by an equation

$$F(\alpha_n, \alpha_{n-1}, \dots, x, y, C) = 0,$$

here F is an algebraic function of a constant C, variables x, y and quadratures $\alpha_1, \dots, \alpha_n$; this quantities are connected by equations of the form

$$\alpha_i = \mathrm{S}\left(f_i(x, \alpha_1, \dots, \alpha_{i-1})dx + g_i(x, \dots)dy\right)$$

or

$$\alpha_i = \mathrm{P}\left[1 + f_i(x, \alpha_1, \dots, \alpha_{i-1})dx + g_i(x, \dots)dy\right],$$

where f_i, g_i are algebraic functions of the arguments. In this case we say for brevity that differential equation (1) can be integrated by n quadratures.

Problem 3. For an ordinary differential equation (1) the problem consists of integrating by quadratures in all cases when such integral exist.

Theorem 2 (M. Singer, 1992). Differential equation (1) is integrated by several quadratures iff there is an integrating factor among P-quadratures, that is

$$\mu = \mathrm{P}(1 + udx + vdy) = e^{\int udx + vdy}, \tag{5}$$

where $udx + vdy$ is exact 1-form and $u, v \in \mathbb{Q}(x, y)$.

Note. This theorem was proved by M. Singer, but in [17] the concept of integration by quadratures is entered by different way; this theorem can be proved as a corollary of Liouville principle by classical methods [18].

Let the rational function v be a root of an equation

$$F(x, y, v) = 0.$$

Expression (5) is integrating factor for the form $pdx + qdy$ iff F is Darboux polynomial for differentiation

$$D_v = q^2 \frac{\partial}{\partial x} - pq \frac{\partial}{\partial y} + \left(vq^2 \frac{\partial}{\partial y} \frac{p}{q} + q^2 \frac{\partial}{\partial y} \frac{1}{q} \left(\frac{\partial p}{\partial y} - \frac{\partial q}{\partial x} \right) \right) \frac{\partial}{\partial v} \tag{6}$$

of ring $\mathbb{Q}[x, y, v]$. This polynomial can be found again by Lagutinski method.

Theorem 3 (M.N. Lagutinski, 1911). A Darboux polynomial F of degree N exist iff $\Delta_N = 0$ or F is factor of Δ_N.

For finding v in (5) with given degree N with respect to standard glex-basis we can calculate linear under v factor of Δ_{2N} with respect to appropriate basis of the ring $R = \mathbb{Q}[x, y, v]$. This idea is realised in Lagutinski package [14] as the routine.

Example. We can integrate equation

$$x^2 y' = y(x + y)$$

at $N = 5$. In doing so, the equation has the integrating factor

$$\mu = \exp - \int \frac{dx}{x} + \frac{2dy}{y} = e^{-\ln(xy^2)} = \frac{1}{xy^2}$$

and the general solution

$$\int \frac{-y(x + y)dx + x^2 dy}{xy^2} = C.$$

We tested the routine on some tasks from the textbook by A.F. Filippov [15], for integration of vast majority we can take $N = 4 \div 5$. However this method don't give full solution of the problem 3 because user has to set the boundary N for degree of v.

Note. Of course, these tasks are solved in elementary functions and Prelle-Singer algorithm [19, 20] is more natural way of their solution.

IOP Conf. Series: Journal of Physics: Conf. Series **788** (2017) 012026 doi:10.1088/1742-6596/788/1/012026

4. Painlevé problem

In the Stockholm lectures [21] Painlevé gave on the example of the equations of the 1st and 2nd order property which is common for all equations, solvable in elementary, special and abelian functions: the general solutions of these equations depend on integration constants algebraically [22, 23, 24].

Problem 4. For a ordinary differential equation (1) the problem consists of finding an integral in all cases when its general solution depends on an integration constant algebraically.

Theorem 4. Let K be the field of analytical functions of the variable x. The general solution of (1) depends on integration constants algebraically iff there is a *unique* substitution

$$z = u(x, y), \quad r \in K(y)$$

mapping the initial equation (1) in Riccati equation

$$z' = p(x) + q(x)z + r(x)z^2$$

and preserving three points $y = 0, 1, \infty$.

For solving problem 4 with the help of this theorem we have again to put the boundary N for degree of u. For $N = 2$ we can state the problem of finding an substitution

$$z = \frac{y(y + a_1)}{a_2 y + (1 + a_1 - a_2)} \tag{7}$$

mapping given equation (1) in Riccati equation.

In generally we can write this condition as some system (S) of algebraical equations between a_1, a_2 and its derivatives, but by theorem 4 there is an unique solution of (S) or no solutions. So by eliminating other variables we have a linear equation for a_1 iff such substitution exist.

Example. Problem: to find a substitution (7) mapping the equation

$$(x^4 - 2x^3y + 2x^2y^2 + 2xy^3 + y^4 - y^2)dx + (x^2 + 2xy - y^2)dy = 0 \tag{8}$$

in Riccati equation. — In Sage we have an excellent set of instruments for elimination of variables in algebraic systems. So we have equation

$$(x - a_1)(a_1 + 1) = 0$$

for a_1 and

$$a_2(a_2 x - a_2 + x + 1) = 0$$

for a_2. Direct check shows that substitution

$$z = \frac{y(y + x)}{\frac{x+1}{1-x}y + 1 + x - \frac{x+1}{1-x}} = \frac{1 - x}{1 + x}\frac{y(y + x)}{y - x}$$

is the required one. Existence of parasitic roots is unexpected, but non fatale for the solution of the problem. It should be noted that Maple can't solve (8).

5. Conclusion

All the three discussed problems 1, 3, and 4 can be solved in computer algebra systems if we put some restrictions on degree required polynomials. With this reservation all three approach to symbolic integration can be successfully applied to solving ordinary differential equations in practice.

The research is executed within the agreement no. 02.a03.21.0008 from 4/24/2016 between the Ministry of Education and Science of the Russian Federation and the RUDN University. The reported study was funded by RFBR according to the research project no. 16-07-00556.

References

[1] Descartes R 1938 *Geometry with the appendix of some works of P. Fermat and Descartes's correspondence* (Moscow-Leningrad, in Russian)
[2] Poincaré H 1934 *Œuvres* vol 3 (Paris: Gautier)
[3] Cheb-Terrab E S 1997 Computer algebra solving of first order ODEs *Computer physics communications* **101** 254
[4] Chèze G 2011 Computation of Darboux polynomials and rational first integrals with bounded degree in polynomial time *Journal of Complexity* **27** 246–62
[5] Bostan A, Chèze G, Cluzeau T and Weil J-A 2016 Efficient algorithms for computing rational first integrals and Darboux polynomials of planar polynomial vector fields *Math. Comp.* **85** 1393–425
[6] Stein W 2016 *SageMath, the Sage Mathematics Software System Version 7.3* (The Sage Developers, `http://www.sagemath.org`)
[7] Lagutinski M N 1911 The application of polar operations to integration of the ordinary differential equations in finite terms *Soobshh. Har'kov. matem. obshh.* 2 ser. **12** 111–243 (in Russian)
[8] Lagutinski M N 1912 On some polynoms and their aplication for algebraic integration of ordinary differential algebraic equations *Soobshh. Har'kov. matem. obshh.* 2 ser. **13** 200–24 (in Russian)
[9] Maciejewski A J and Strelcyn J-M 2008 On the algebraic non-integrability of the Halphen system (*Preprint* arXiv:950.5002)
[10] Christopher C, Llibre J and Vitòrio Pereira J 2007 Multiplicity of invariant algebraic curves in polynomial vector fields *Pacific J. Math.* **229** 63–117
[11] Malykh M D 2016 On M.N. Lagutinsky's method for computation of rational integrals of ordinary differential equations systems *Vestnik natsional'nogo issledovatel'skogo yadernogo universiteta MIFI* **5** 35–44 (in Russian.)
[12] Malykh M D 2016 On M.N. Lagutinski method for integration of ordinary differential equations *International conference Polynomial Computer Algebra'2016* (VVM Publ.) pp 57–8
[13] Dobrovol'skij V A, Lokot' N V, and Strelcyn J-M 1998 Mikhail Nikolaevich Lagutinskii (1871–1915): Un Mathématicien Méconnu. *Historia Mathematica* **25** 245–64
[14] `http://malykhmd.neocities.org`
[15] Filippov A F 2000 *Collection of tasks of the differential equations* (Moscow-Izhevsk: R&C)
[16] Schlesinger L 1922 *Einführung in die Theorie der Gewöhnlichen Differentialgleichungen auf Funktionentheoretischer Grundlage* (De Gruyter, Berlin-Leipzig)
[17] Singer M F 1992 Liouvillian first integrals of differential equations *Trans. Amer. Math. Soc.* **333** 673–88
[18] Junzhi Lei 2011 Nonlinear differential Galois theory (*Preprint* Arxiv:0608492v2)
[19] Gorieli A 2001 *Integrability and Nonintegrability of Dynamical Systems.* (Singapore).
[20] Avellar J, Duarte L G S, Duarte S E S and da Mota L A C P 2007 Determining Liouvillian first integrals for dynamical systems in the plane *Computer Physics Communications* **177** 584–96
[21] Painlevé P 1895 Leçons sur la théorie analytique des équations différentielles, professées à Stockholm (septembre, octobre, novembre 1895) sur l'invitation de S. M. le roi de Suède et de Norwège *Œuvres de Painlevé* vol **1** (Paris: CNRS, 1971)
[22] Umemura H 1990 Birational automorphism groups and differential equations *Nagoya Math. J.* **119** 1–80
[23] Malykh M D 2014 On Integration of the Differential Equations in Abelian and other Transcendental Functions *International conference Polynomial Computer Algebra'2014* (VVM Publ.) pp 52–3
[24] Malykh M D 2015 On transcendental functions arising from integrating differential equations in finite terms *Journal of Mathematical Sciences* **209** 935–52

Multicriteria adaptation of robotic groups to dynamically changing conditions

S Yu Misyurin[1], A P Nelyubin[2] and V I Ivlev[2]

[1] National Research Nuclear University MEPhI, 31 Kashirskoe shosse, Moscow 115409, Russia
[2] Mechanical Engineering Research Institute RAS, 4 Malyi Kharitonievski pereulok, Moscow 101990, Russia

nelubin@gmail.com

Abstract. A new approach is proposed to design complex robotic systems composed of many robots that can operate under different conditions and perform various tasks. Bio-inspired ideas of adaptation of robotic groups are discussed.

1. Introduction

In this paper we consider the problem of designing a wide class of robotic systems composed of several or many robots that interact with each other and perform common tasks. In addition, it is assumed that the designed robotic system should operate in various environmental conditions and perform various tasks.

The examples of such systems are groups of mobile robots that are designed to explore difficult to reach areas (under water, under earth, in space, under high radiation, etc.) and to perform certain operations with objects in these areas [1]. Such technologies of the future as "smart dust" [2], self-organizing robot swarm [3], or surgery with microscopic robots can also be regarded. Another example is military robots such as drones.

The design (configuration) of a particular robot includes a specified mechanical structure, the values of the mechanical parameters and the control system of the robot. The configuration of a complex robotic system consisting of many robots includes the configuration of all robots, which in general case can differ, and the control system of the whole group of robots.

The quality of the whole robotic system is evaluated according to several criteria (indicators, performance indices). The set of these criteria of quality and the relative preference between them depend on the conditions of functioning of the robotic system and the task being performed at the moment. In some conditions, one criteria are important, in other conditions the focus is put on other criteria.

2. Multicriteria adaptation of robotic systems

One approach to designing a universal robotic system is to find compromise values of quality criteria to ensure satisfactory operation of the system in all possible conditions. However, this approach leads to the fact that the resulting system in each mode is not functioning as efficiently as possible. Universality is achieved by reducing efficiency.

More promising is the design of adaptive reconfigurable robotic systems [4-6]. Ideally, such a system reconfigures itself and changes its parameters depending on the operation conditions, so that in each mode it operates most efficiently. Reconfiguration means changing the structure and values of parameters of the mechanisms, changing the rules and parameters of control systems, changing the interaction of the robots among themselves.

In the case when all the modes of operation of the robotic system are known and there is a finite number of such modes, the reconfiguration of the system between these modes can be programmed during the design phase. To do this, one should define the best configuration of the system in each mode of operation and introduce a mechanism of changeover (toggle) between these modes. In addition to the changeover mechanism, the control system of the robotic system should have a mechanism for determining that the conditions of operation changed, and the mode to be configured. In some problems these changeover mechanisms can be organized with the involvement of the operator.

In the case when the operation conditions of the robotic system change continuously, it is hard or even not possible to provide a mechanism for a smooth changeover of a system configuration. This problem can be solved by discretization of the range of variation conditions.

The situation is more complex when not all the possible operation conditions of the robotic system are known in advance. In this case, the real-time adaptation of the system is necessary.

In this paper we propose multicriteria approach to adaptation of robotic systems. Figure 1 shows an example of evaluation of a system designs by two criteria (objective functions to be minimized). These designs were obtained during parameters optimization procedure by uniformly sampling of parameter space. Each circle corresponds to the variant of design. And solid circles correspond to Pareto optimal designs.

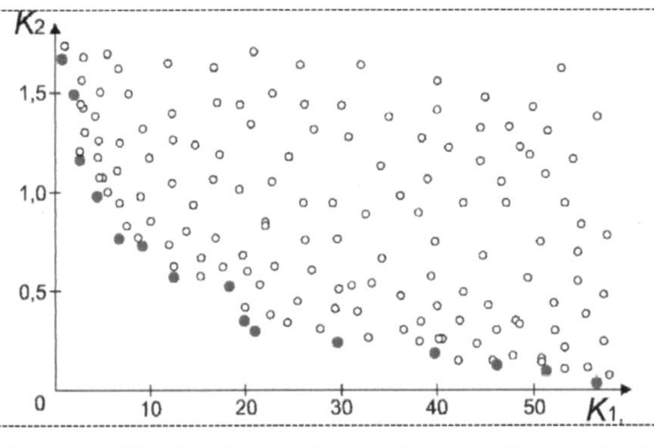

Figure 1. Typical Pareto frontier in case of two criteria.

The best variant of design should be chosen from Pareto optimal designs. This choice depends on the relative preferences between the criteria. Usually the preferences are somehow determined and fixed on the design phase and a compromise decision is made. But these preferences depend on environmental conditions and tasks of the system which can change dynamically during the system operation. So it will be better to save an option to change system configuration in accordance with changing preferences.

3. Bio-inspired approach to adaptation of robotic groups

In this paper we propose to use a bio-inspired approach to the adaptation of group of robots to the changing conditions. Adaptation of biological species to changing environmental conditions is achieved through the maintenance of biological diversity within the population. Similarly, we can use a group of robots, differing in their configuration. Using multicriteria approach we can take Pareto optimal designs as basic configurations of robots in group. Then some subgroup of robots will be better in one conditions, and other subgroups will be more efficient in other conditions.

Robots most adapted to the current conditions are put in the forefront and do most of the work. Other robots can do the same work, or can perform helper functions. Partial reconfiguration of other robots in accordance with the parameters of the fittest robots is also possible. This will increase the number of robots in the main leading group and increase the efficiency of the entire system. However, there must be diversity of configurations of robots in the population.

Changing of conditions enables the mechanism of reconfiguration of the entire robotic system. This moment can be detected by the fact that a new group of robots starts to work more efficient than the previously leading group. When this occurs, the leading group is changed. The advantage of this reconfiguration mechanism is that it can be carried out automatically without operator intervention. In addition, this process of adaptation can happen fairly smoothly if there is sufficient diversity in the population of robots. This diversity also can be maintained by regular random changes of design parameters of the individual robots, similar to the mutation of living organisms. This means that partial optimization of robotic systems can be made on-line in new and unknown conditions. We can apply here ideas form various global optimization techniques such as Genetic Algorithms [7] or Particle Swarm Optimization [8].

In addition to the change of environmental conditions, the tasks performed by a robotic system can also be changed. Moreover, multiple tasks can be performed simultaneously. This is achieved through various specializations of the robots in the system. It also can be given a counterpart of living systems – specialization in populations of ants or bees. In the case of the specialization, the approach described above with the change of leading subgroups of robots can also be used. However, the change of specialization of individual robot may require more radical rebuild of its configuration than adaptation to the changing conditions.

4. Conclusion

The proposed ideas of multicriteria adaptation of robotic systems and groups can enhance flexibility and robustness of such systems in dynamically changing conditions. For these purposes reconfigurable robots and their groups are better than fixed. This allows to transfer part of optimization routine from design phase to the real-time phase of system operation. So we can change configurations in accordance with changing conditions and multicriteria preferences, or even optimize some parameters on-line.

The study was financially supported by the RFBR project No. 16-29-04401.

References

[1] Yu C, Nagpal R 2010 A Self-Adaptive Framework for Modular Robots in Dynamic Environments: Theory and Applications *Int Journal of Robotics Research **30** 1015–36*

[2] Warneke B, Last M Liebowitz B and Pister K S J 2001 Smart Dust: Communicating with a Cubic-Millimeter *Computer* **34** 44–51

[3] Rubenstein M, Cornejo A and Nagpal R 2014 Programmable self-assembly in a thousand-robot swarm *Science* **345** 795–99

[4] Yim M, Shen W, Salemi B, Rus D, Moll M, Lipson H, Klavins E and Chirikjian G S 2007 Modular Self-Reconfigurable Robot Systems IEEE Robotics & Automation Magazine pp 43–52

[5] Christoforou E G, Muller A, Phocas M C, Matheou M and Arnos S 2015 Design and Control Concept for Reconfigurable Architecture *J. of Mechanical Design* **137** 042302

[6] Suzuki Y, Inou N, Kimura H and Koseki M 2006 Reconfigurable group robots adaptively transforming a mechanical structure *Proc. of the 2006 IEEE/RSJ Int. Conf. on Intelligent Robots and Systems (Beijing, China)* pp 2200–2205

[7] Mitchell M 1996 An Introduction to Genetic Algorithms (Cambridge, MA: MIT Press)

[8] Poli R, Kennedy J and Blackwell T 2007 Particle Swarm Optimization: An Overview *Swarm Intelligence* **1** 33–57

Parametrically defined differential equations

A D Polyanin[1,2,3], **A I Zhurov**[1,4]

[1] Institute for Problems in Mechanics, Russian Academy of Sciences, 101 Vernadsky Avenue, bldg 1, 119526 Moscow, Russia

[2] Bauman Moscow State Technical University, 5 Second Baumanskaya Street, 105005 Moscow, Russia

[3] National Research Nuclear University MEPhI, 31 Kashirskoe Shosse, 115409 Moscow, Russia

[4] Cardiff University, Heath Park, Cardiff CF14 4XY, UK

E-mail: zhurovai@cardiff.ac.uk

Abstract. The paper deals with nonlinear ordinary differential equations defined parametrically by two relations. It proposes techniques to reduce such equations, of the first or second order, to standard systems of ordinary differential equations. It obtains the general solution to some classes of nonlinear parametrically defined ODEs dependent on arbitrary functions. It outlines procedures for the numerical solution of the Cauchy problem for parametrically defined differential equations.

Keywords: parametrically defined differential equations, nonlinear differential equations, numerical methods, Cauchy problem, exact solutions

1. Introduction

1.1. Preliminary remarks

Von Mises and Crocco type transformations are used to reduce the order of boundary layer equations and some other nonlinear PDEs [1–6]. Such transformations suggest choosing suitable first- or second-order partial derivatives to be new independent variables. Sometimes, the reduced equations admit exact solutions in parametric form [7]. As a result, finding exact solutions to the original PDEs reduces to integrating parametric ODEs. Moreover, parametrically defined nonlinear differential equations also arise directly in the theory of ODEs [7].

The study [7] describes a few classes of nonlinear parametric ordinary differential equations of the first and second order for which the authors managed to find general solutions. Note that parametric ODEs are non-classical and are not treated in the literature dealing with standard ODEs (e.g., see [8–12]). The general solutions obtained in [7] were further used to construct exact solutions to unsteady axisymmetric boundary-layer equations.

1.2. Ordinary differential equations defined parametrically

In general, a parametrically defined nth-order ordinary differential equation is given by two equations of the form [7]

$$y_x^{(m)} = F(x, y, y'_x, \dots, y_x^{(m-1)}, t), \quad y_x^{(n)} = G(x, y, y'_x, \dots, y_x^{(n-1)}, t), \tag{1}$$

where $y = y(x)$ is the unknown function with $y_x^{(n)}$ denoting the nth derivative $d^n y/dx^n$, $t = t(x)$ is a functional parameters, and $n \geq m$. In what follows, we assume that the parameter t cannot generally be eliminated from Eqs. (1).

2. First-order parametrically defined ODEs

2.1. Reduction to a standard system of ODEs. An example of an exact solution

Consider parametric first-order ODEs defined by

$$F(x, y, t) = 0, \quad y'_x = G(x, y, t). \tag{2}$$

In the theory of differential-algebraic equations (DAEs), equations of the form (2) are referred to a system of semi-explicit DAEs or ODEs with constraints [13, 14]. Normally, such equations are reduced to a standard system of ODEs for $y = y(x)$ and $t = t(x)$ by differentiating the first equation in (2) with respect to x [13]. However, there is a more convenient approach that suggests using an alternative system of ODEs, for $y = y(t)$ and $x = x(t)$, which is derived below.

Eqs. (2) can be rewritten in terms of differentials as

$$F_x\, dx + F_y\, dy + F_t\, dt = 0, \quad dy = G(x, y, t)\, dx, \tag{3}$$

where F_x, F_y, and F_t are respective partial derivatives of $F = F(x, y, t)$. By eliminating dy and then dx from (3), we arrive at

$$(F_x + GF_y)x'_t + F_t = 0, \tag{4}$$

$$(F_x + GF_y)y'_t + GF_t = 0. \tag{5}$$

Eqs. (4) and (5) represent a standard system of first-order equations for $x = x(t)$ and $y = y(t)$. If we manage to solve this system, we automatically obtain a solution to the original equations (2) in parametric form. In some cases, it suffices to use either Eq. (4) or (5) and the first equation in (2).

Example 1. Let us look at the nonlinear first-order ODE defined parametrically

$$x = f(t)y, \quad y'_x = g(t), \tag{6}$$

where $f = f(t)$ and $g = g(t)$ are arbitrary functions ($fg - 1 \neq 0$).

Eq. (6) is a special case of Eq. (2) with $F = x - fy$ and $G = g$. Then Eq. (5) becomes $(1 - fg)y'_t - gf'_t y = 0$. Integrating this equation in view of the first relation in (6) yields the general solution to Eq. (6) in parametric form

$$x = Cf \exp\biggl(\int \frac{gf'_t\, dt}{1 - fg}\biggr), \quad y = C \exp\biggl(\int \frac{gf'_t\, dt}{1 - fg}\biggr), \tag{7}$$

where C is an arbitrary constant.

Remark 1. If equations (4) and (5) are used, isolated solutions satisfying $F_x + GF_y = 0$ may be lost; this issue calls for further analysis.

2.2. Cauchy problem. Numerical solution procedure

Consider the Cauchy problem for the parametric first-order ODE (2) with the initial condition

$$y(x_0) = y_0. \tag{8}$$

The initial value of the parameter, $t = t_0$, can be found from the algebraic (generally transcendental) equation

$$F(x_0, y_0, t_0) = 0, \tag{9}$$

where x_0 and y_0 are the values appearing in the initial condition (8).

Using the technique described in Section 2.1, one first reduces the parametric ODE (2) to the standard system of first-order ODEs (4) and (5) for $x = x(t)$ and $y = y(t)$. Then this system, subject to the initial conditions

$$x(t_0) = x_0, \quad y(t_0) = y_0, \tag{10}$$

can be solved numerical by the Runge–Kutta or another suitable method [10, 14–16].

Remark 2. The algebraic (or transcendental) equation (9) can generally have more than one root. In this case, the original problem (2), (8) will have the same number of solutions.

3. Second-order parametrically defined ODEs

3.1. Reduction to a standard system of first-order ODEs

Consider parametric second-order ODEs of the form

$$y'_x = F(x, y, t), \quad y''_{xx} = G(x, y, t). \tag{11}$$

On differentiating the first equation with respect to t, we get $(y'_x)'_t = F_x x'_t + F_y y'_t + F_t$. Taking into account that $y'_t = F x'_t$ and $(y'_x)'_t = x'_t y''_{xx}$, we further obtain

$$x'_t y''_{xx} = F_x x'_t + F F_y x'_t + F_t. \tag{12}$$

On eliminating the second derivative y''_{xx} with the aid of the second equation in (11), we arrive at the first-order equation

$$(G - F_x - F F_y) x'_t = F_t. \tag{13}$$

In view of $y'_t = F x'_t$, Eq. (13) becomes

$$(G - F_x - F F_y) y'_t = F F_t. \tag{14}$$

Equations (13) and (14) represent a standard system of first-order equations for $x = x(t)$ and $y = y(t)$. If we manage to solve this system, we automatically get a solution to the original equation (11) in parametric form. In some cases, it suffices to use either Eq. (13) or (14) and the first equation in (11).

Remark 3. If equations (13) and (14) are used, isolated solutions satisfying $G - F_x - F F_y = 0$ may be lost; this issue calls for further analysis (see Example 3, Item 2 below).

Example 2. Consider the nonlinear second-order ODE defined parametrically

$$y'_x = f(t) y, \quad y''_{xx} = g(t) y, \tag{15}$$

where $f = f(t)$ and $g = g(t)$ are arbitrary functions ($g - f^2 \neq 0$).

Eq. (15) is a special case of Eq. (11) with $F = fy$ and $G = gt$. Then Eqs. (13) and (14) are independent first order linear equations

$$(g - f^2) x'_t = f'_t, \quad (g - f^2) y'_t = f f'_t y.$$

General solution in parametric form:

$$x = \int \frac{f'_t(t)\, dt}{g(t) - f^2(t)} + C_1, \quad y = C_2 \exp\left[\int \frac{f(t) f'_t(t)\, dt}{g(t) - f^2(t)}\right],$$

where C_1 and C_2 are arbitrary constants.

Example 3. Consider the equation

$$y'_x = f(t)x + g(t), \qquad y''_{xx} = h(t). \tag{16}$$

where $f = f(t)$, $g = g(t)$, and $h = h(t)$ are arbitrary functions.

Eq. (16) is a special case of Eq. (11) with $F = fx + g$ and $G = h$. Then Eqs. (13) and (14) take the form

$$\begin{aligned} (h-f)x'_t &= f'_t x + g'_t, \\ (h-f)y'_t &= (fx+g)(f'_t x + g'_t). \end{aligned} \tag{17}$$

1. Suppose $h - f \not\equiv 0$. Then the general solution to the first (linear) equation of system (17) has the form

$$x = C_1 E + E \int \frac{g'_t\,dt}{E(h-f)}, \qquad E = \exp\biggl(\int \frac{f'_t\,dt}{h-f}\biggr),$$

where C_1 is an arbitrary constant. Substituting this expression of x into the second equation of the system, we arrive at a separable equation for $y = y(t)$ (its solution is omitted).

2. Suppose $h - f \equiv 0$. On differentiating the first equation in (16) with respect to x and comparing with the second equation, we see that $t'_x = 0$. Hence, $t = C_1$, where C_1 is an arbitrary constant. By replacing t in the first equation in (16) t with C_1 followed by integrating with respect to x, we obtain the general solution to the original equation

$$y = \frac{1}{2} f(C_1)x^2 + g(C_1)x + C_2,$$

where C_1 and C_2 are arbitrary constants.

3.2. Cauchy problem. Numerical solution procedure

Consider the Cauchy problem for the parametric second-order ODE (11) with the initial conditions

$$y(x_0) = y_0, \quad y'_x(x_0) = y_1. \tag{18}$$

The initial value of the parameter, $t = t_0$, is determined from the algebraic (generally transcendental) equation

$$y_1 = F(x_0, y_0, t_0), \tag{19}$$

where x_0, y_0, and y_1 are the values appearing in the initial conditions (18); Eq. (19) follows from the first equation in (11).

Using the technique outlined in Section 3.1, we first reduce the parametric differential equation (11) to the standard system of first-order ODEs (13), (14) for $x = x(t)$ and $y = y(t)$. Then this system, subject to the initial conditions (10), can be solved numerically using, for example, the Runge–Kutta or another suitable method [10, 14–16].

Remark 4. The algebraic (transcendental) equation (19) can generally have more than one root. In this case, the original Cauchy problem (11), (18) will have the same number of solutions.

3.3. First boundary value problem. Numerical solution procedure

Let us look at the first boundary value problem for the parametric second-order ODE (11) in the range $x_1 \le x \le x_2$ with the boundary conditions

$$y(x_1) = y_1, \quad y(x_2) = y_2. \tag{20}$$

Below we present the main idea of a numerical procedure to solve this kind of problem.

Consider two auxiliary Cauchy problems for the equation (see the second equation in Eq. (11))

$$y''_{xx} = G(x, y, t) \tag{21}$$

subject to the initial conditions

$$y(x_1) = y_1, \quad y'_x(x_1) = F(x_1, y_1, t) \qquad \text{(problem 1);} \tag{22}$$

$$y(x_2) = y_2, \quad y'_x(x_2) = F(x_2, y_2, t) \qquad \text{(problem 2).} \tag{23}$$

By choosing a specific value of the parameter, $t = t_k$, we solve the auxiliary Cauchy problems numerically (e.g., by the Runge–Kutta method [10, 14–16]) to obtain $y^1 = y^1(x, t_k)$ and $y^2 = y^2(x, t_k)$, respectively (the superscripts indicate the problem number). To any t_k there corresponds a point (x_k, y_k) in the (x, y) plane at which the curves corresponding to the solutions $y^1 = y^1(x, t_k)$ and $y^2 = y^2(x, t_k)$ intersect. By choosing a different value, t_{k+1}, we find another point, (x_{k+1}, y_{k+1}). The discrete set of points (x_k, y_k) with $k = 0, 1, 2, \ldots$ defines an approximation to the solution $y = y(x)$ of the original boundary value problem (11), (20).

4. Higher-order parametrically defined ODEs. Cauchy problem. Numerical solution procedure

Below we outline a procedure for the numerical solution of the Cauchy problem for a parametric nth-order equation defined by two relations of the form (1), with $n > m$, subject to the initial conditions

$$y(x_0) = y_0, \quad y'_x(x_0) = y_1, \quad \ldots, \quad y_x^{(n-1)}(x_0) = y_{n-1}. \tag{24}$$

By relying on Eqs. (1) directly, let us consider the following two auxiliary Cauchy problems:

$$y_x^{(m)} = F(x, y, y'_x, \ldots, y_x^{(m-1)}, t),$$

$$y(x_0) = y_0, \quad y'_x(x_0) = y_1, \quad \ldots, \quad y_x^{(m-1)}(x_0) = y_{m-1} \quad \text{(problem 1);} \tag{25}$$

$$y_x^{(n)} = G(x, y, y'_x, \ldots, y_x^{(n-1)}, t),$$

$$y(x_0) = y_0, \quad y'_x(x_0) = y_1, \quad \ldots, \quad y_x^{(n-1)}(x_0) = y_{n-1} \quad \text{(problem 2).} \tag{26}$$

Let $y_F = y_F(x, t)$ and $y_G = y_G(x, t)$ denote their respective solutions. We look at the solution difference

$$\Delta(x, t) = y_G(x, t) - y_F(x, t). \tag{27}$$

By choosing a specific value of the parameter, $t = t_k$, we find solutions $y_F(x, t_k)$ and $y_G(x, t_k)$ numerically using, for example, the Runge–Kutta method [10, 14–16]. Further, by varying x, we find an x_k such that $\Delta(x_k, t_k) = 0$. To this value of x_k there corresponds a value of the desired function: $y_k = y_F(x_k, t_k) = y_G(x_k, t_k)$. Hence, to each t_k there corresponds a point (x_k, y_k) in the (x, y) plane at which the curves $y_F = y_F(x, t_k)$ and $y_G = y_G(x, t_k)$ intersect.

By choosing a different value t_{k+1}, we find, in the same manner as above, the corresponding point (x_{k+1}, y_{k+1}). The discrete set of points (x_k, y_k), $k = 0, 1, 2, \ldots$, defines an approximation to the solution $y = y(x)$ of the original problem (1), (24).

The initial value of the parameter, $t = t_0$, is determined from the algebraic (transcendental) equation

$$y_m = F(x_0, y_0, y_1, \ldots, y_{m-1}, t_0), \tag{28}$$

where x_0, y_0, y_1, $\ldots$, y_m are the values appering in the initial conditions (24).

Remark 5. The algebraic (transcendental) equation (28) can generally have more than one root. In this case, the original Cauchy problem (1), (24) will have the same number of solutions.

5. Brief conclusions

For nonlinear parametrically defined ordinary differential equations of the form (1), we have proposed techniques to reduce such equations, of the first and second order, to standard systems of ordinary differential equations, in which x and y are the unknown functions and t is the independent variable. We have constructed the general solution to some classes of equations in question and suggested procedures for the numerical solution of the Cauchy problem for parametrically defined differential equations of the first, second, and higher order.

References

[1] Schlichting H 1981 *Boundary Layer Theory* (New York: McGraw-Hill)

[2] Loitsyanskiy L G 1995 *Mechanics of Liquids and Gases* (New York: Begell House)

[3] Saccomandi G 2004 *J. Phys.* A *Math. Gen.* **37** 7005–17

[4] Polyanin A D and Zhurov A I 2012 *Commun. Nonlinear Sci. Numer. Simulat.* **17** 536–44

[5] Polyanin A D and Zhurov A I 2012 *Int. J. Non-Linear Mech.* **47** 413–7

[6] Polyanin A D and Zaitsev V F 2012 *Handbook of Nonlinear Partial Differential Equations, 2nd ed* (Boca Raton: CRC Press)

[7] Polyanin A D and Zhurov A I 2016 *Appl. Math. Lett.* **55** 72–80

[8] Murphy G M 1960 *Ordinary Differential Equations and Their Solutions* (New York: D Van Nostrand)

[9] Kamke E 1977 *Differentialgleichungen: Lösungsmethoden und Lösungen* I *Gewöhnliche Differentialgleichungen* (Leipzig: B G Teubner)

[10] Korn G A and Korn T M 2000 *Mathematical Handbook for Scientists and Engineers, 2nd ed* (New York: Dover Publ.)

[11] Polyanin A D and Zaitsev V F 2003 *Handbook of Exact Solutions for Ordinary Differential Equations, 2nd ed* (Boca Raton–London: Chapman & Hall/CRC Press)

[12] Kudryashov N A 2004 *Analytical Theory of Nonlinear Differential Equations* (Moscow–Izhevsk: Institute of Computer Science) in Russian

[13] Brenan K E, Campbell S L and Petzold L R 1996 *Numerical Solution of Initial-Value Problems in Differential-Algebraic Equations* (Philadelphia: SIAM)

[14] Ascher U M and Petzold L R 1998 *Computer Methods for Ordinary Differential Equations and Differential-Algebraic Equations* (Philadelphia: SIAM)

[15] Schiesser W E 1994 *Computational Mathematics in Engineering and Applied Science: ODEs, DAEs, and PDEs* (Boca Raton: CRC Press)

[16] Butcher J C 1987 *The Numerical Analysis of Ordinary Differential Equations: Runge-Kutta and General Linear Methods* (New York: Wiley-Interscience)

Control of coupled localized nonlinear wave solutions

A V Porubov[1,2,3], **I D Antonov**[3]

[1]Institute of Problems in Mechanical Engineering, Bolshoy 61, V.O., Saint Petersburg 199178, Russia

[2]St. Petersburg State University, 7-9, Universitetskaya nab., V.O., Saint Petersburg 199034, Russia

[3]St. Petersburg State Polytechnical University, Polytechnicheskaya st., 29, Saint Petersburg 195251, Russia

E-mail: alexey.porubov@gmail.com

Abstract. A method of forced localization of non-linear wave by a feedback control is developed for coupled equations accounting for non-linear dynamic processes in complex lattices. It is shown, that the control of the shape and velocity of the wave function of macro-strain allows to achieve localization of the shape of the function describing variations of defects in the lattice. Moreover, change of the sign of the amplitude of the last wave may be achieved by variation of the parameters of the control function but independent of the initial conditions.

Introduction

Exact localized traveling wave solutions to non-linear wave equations may be found by various methods, see, e.g., [1, 2]. Their existence results from a balance between contributions of the terms of the equations which in turn reflect a an influence of physical factors such as nonlinearity, dispersion, dissipation, diffraction. However, existence and stable propagation of localized waves also depend on the choice of initial and boundary conditions. The consistent initial conditions should coincide with the exact solutions and their corresponding derivatives at $t = 0$. For coupled equations, an extra factor concerns relative positions of the inputs for the coupled variables [3, 4, 5]. Variations in the shape of the initial conditions and their initial velocities result in failing of localization, arising of oscillations, serious variations in the shape of the waves.

Among the methods allowing support of stable propagation of localized waves one can note the control methods [6, 7, 8]. Recently a distributive feedback control algorithm has been developed for single non-linear equations [9, 10, 11]. It was found that both the kink-shaped and envelope wave solutions may be achieved for the sine- Gordon equation if both the tendencies to a desired shape and velocity of the wave are included into the control function. The control function is physically reasonable for the sine- Gordon equation describing a crystalline lattice continuum limit if it is related to an external loading. When the wave achieves the shape and the velocity of the desired or target function, the control function becomes very small. Thus its addition to the equation does not seriously change it.

In this paper, a feedback control algorithm is developed for non-linear coupled equations arising for continuum description of highly non-linear dynamic processes in complex lattices [12, 13, 14, 15]. The only one function may be physically reasonable controlled - the function of macro-strain. Another function in the equations accounts for internal variations in the crystalline

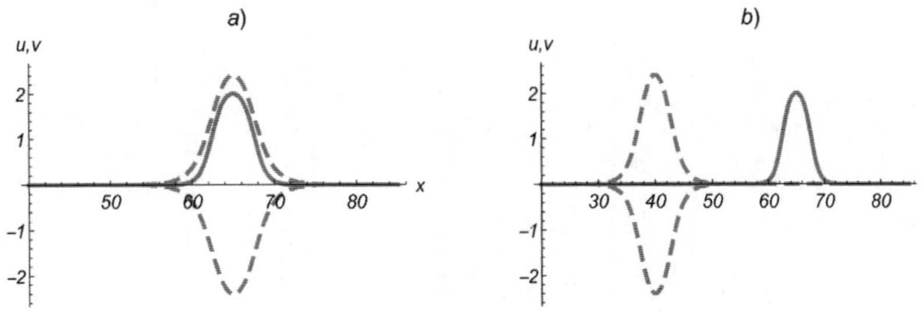

Figure 1. Initial profiles, dashed line for u, solid line for v: a) $x_{11} = x_{12} = 65$; b) $x_{11} = 65$, $x_{12} = 40$.

structure. In Sec. 1 the equations and their known exact traveling wave solutions are discussed. Special attention is paid on variations in the solutions when the positions of the inputs for the functions in the equation do not coincide. Next section is devoted to application of the feedback distributive control algorithm for the macro-strain function and numerical solutions demonstrating recovering of the localization of directly non-controlled function of internal variation in the crystalline structure. Also change of the sign of the amplitude of the localized wave out of the control is achieved by modification of the control of another function in the equations.

1. Non-linear coupled equations for complex lattices

Recently nonlinear coupled equations were obtained in Refs. [12, 13, 14, 15] to account for highly nonlinear dynamic processes in crystals in the continuum approach,

$$v_{tt} - c_L^2 v_{xx} = \frac{S}{\rho}(\cos(u))_{xx}, \tag{1}$$

$$u_{tt} - c_l^2 u_{xx} = \frac{1}{\mu}(Sv - P)\sin(u). \tag{2}$$

Here v is longitudinal macro-strain, u is micro-displacement characterizing movement of defects in a di- atomic crystals [13, 14] or an influence of an additional rotational degree of freedom in an atomic chain [12].

Exact localized traveling wave solution represents coupled functions u, v [15],

$$v = \frac{A}{Q\cosh(k\,(x - Vt - x_{11})) + 1}, \tag{3}$$

$$u = \pm\arccos\left(\frac{\rho(V^2 - c_L^2)v^*}{S} + 1\right), \tag{4}$$

where $v = v^*$ and the parameters are

$$A = \frac{4S}{\rho(c_0^2 + c_L^2 - V^2)},\ Q = -\frac{c_L^2 - c_0^2 - V^2}{c_L^2 + c_0^2 - V^2}, k = 2\sqrt{\frac{P}{c_l^2 - V^2}}.$$

One has to note that initial positions of both waves are defined by one and the same parameter x_{11}, see Fig. 1(a), and their phase velocity V is the same. Also the function u may have an amplitude of either sign. The initial conditions may be chosen in the form of exact solution for

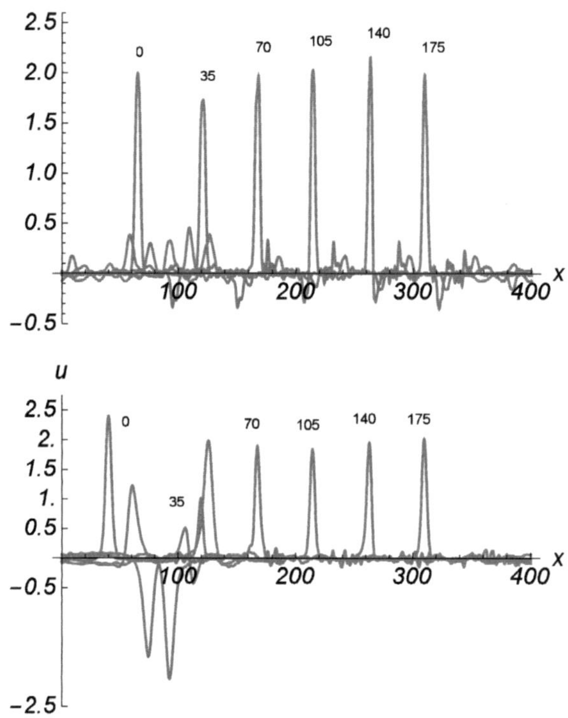

Figure 2. Evaluation of the waves at $x_{11} = 65$, $x_{12} = 40$ without control. Times are noted at corresponding maxima/minima of the wave profiles.

v (3), but with initial function for u with

$$v^* = \frac{A}{Q \cosh(k\ (x - Vt - x_{ij})) + 1}, \tag{5}$$

where x_{ij} may be chosen not equal to x_{11}, say, equal to x_{12}. This is shown in Fig. 1(b). Numerical simulations in [3, 4, 5] revealed failing of stable propagation of the exact traveling wave solution (3), (4) at initial condition with v^* defined by Eq. (5). In particular change of the sign of the amplitude of the wave u happens depending on the difference between x_{11} and x_{12}.

Consider the case with the values of the coefficients $c_L = 1.6, c_l = 2, c_0 = 1, S = 1, P = 1, \rho = 1$ and with the values of the parameters of the initial conditions $V = 1.3, x_{11} = 65, x_{12} = 40$. Numerical simulations are performed using the NDSolve tool of the Wolfram Mathematica 11. Shown in Fig. 2 is the evolution of the waves that demonstrates an influence of the difference between initial positions on stable propagation of localized waves. Left profiles correspond to the initially mutually spaced localized inputs. One can see variation in the amplitudes for the wave v and appearance of oscillations. The profile for the wave u suffers even temporal change of the sign of the amplitude. Finally the maxima of the wave come to the same position, and the waves propagate together with one and the same velocity. Similar behavior is observed for negative amplitude input for u.

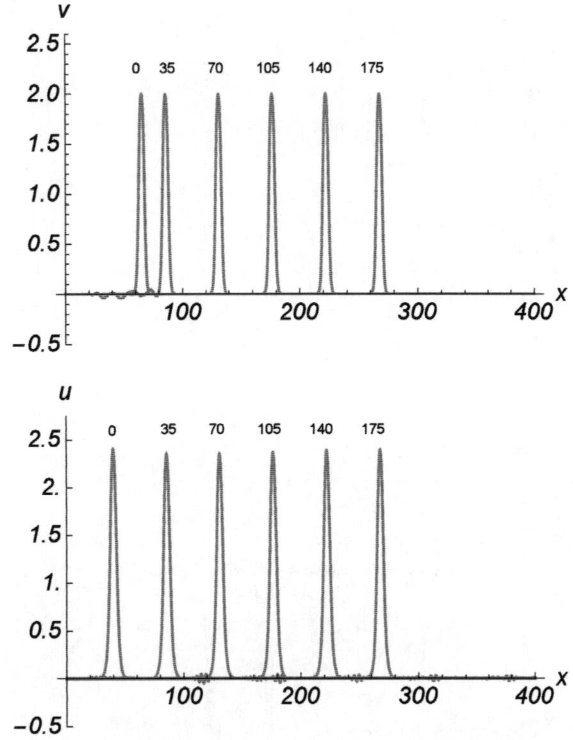

Figure 3. Support of the waves localization by the control. Times are noted at corresponding maxima/minima of the wave profiles.

2. Control of localization of the coupled waves

Physically reasonable control concerns only the macro-strain function v that may be explained as an external loading. The control of the internal structure variable u is unlikely. Then a control function $w(x, t)$ is introduced only in one equation (1),

$$v_{tt} - c_L^2 v_{xx} = \frac{S}{\rho}(\cos(u))_{xx} + w(x, t), \tag{6}$$

while Eq. (2) does not contain any control term. The function w controls only the behavior of the function v.

Following Refs. [9, 10, 11] let us choose the goal function in the form of exact solution for v (3), but with initial phase defined as x_{ij} in Eq. (5). The function $v_t^*(x, t)$ is defined as the first temporal derivative of Eq.(5). Then one defines

$$e(x, t) = v(x, t) - v^*(x, t), \tag{7}$$

$$e_t(x, t) = v_t(x, t) - v_t^*(x, t). \tag{8}$$

and

$$\epsilon(x, t) = \alpha e(x, t) + e_t(x, t), \tag{9}$$

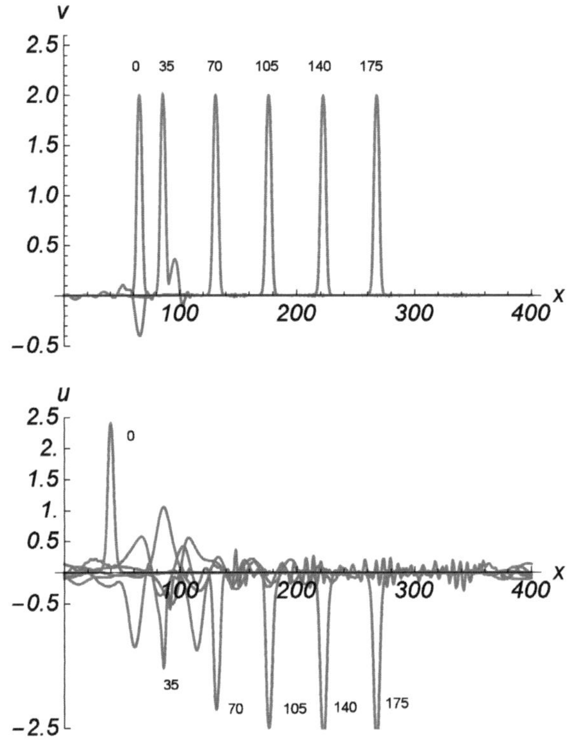

Figure 4. Change of the sign of the amplitude of localized wave due to the control. Times are noted at corresponding maxima/minima of the wave profiles.

Like in [9, 10, 11], the function w is defined as

$$w(x,t) = -\gamma\epsilon(x,t), \tag{10}$$

where $\gamma > 0$.

Numerical simulations are shown in Fig.3 for the choice of the control parameters $\gamma = 0.16, \alpha = 160$. Comparing with Fig.2, one can see recovering of stable propagation of localized wave u with no more variations in the sign of the amplitude, and no more variations in the amplitude of v is seen. The oscillations almost disappear on the profiles of both functions.

One can see an interval for evolution of u in Fig. 2 when the amplitude changes its sign. However, then the original sign of the amplitude is recovered. One can switch on the control at time corresponding to the interval of the negative sign by modifying the control function,

$$w(x,t) = -\gamma\epsilon(x,t)H(t-t_b), \tag{11}$$

where H is the unit-step function, t_b is time for switch on. Shown in Fig. 4 is numerical simulations with modified control function (11) and with $t_b = 19$. One can see that the wave u changes its sign and continue to propagate like in the case without control when the initial condition is chosen with negative sign, see Fig. 1. The choice of t_b is sensitive to small variations. In particular, choice of $t_b = 18$ results only in a more longer negative amplitude wave propagation than in Fig. 2 with forthcoming recovering of the sign of the amplitude. An increase in the value of t_b, say, $t_b = 23$, gives rise to the scenario similar to that shown in Fig. 3 without change of the sign of the amplitude of wave u.

3. Conclusions

A distributive feedback control may support localization of nonlinear coupled equations even if only one of the functions is controlled. The sign of the localized wave may be changed by the control independent of the initial conditions. For the equations considered, it means that an external loading plays the role of the the control of the shape and velocity of the macro-strain wave in the lattice, while another coupled function accounting for the movement of defects of the crystalline structure also suffers serious variations.

Acknowledgments

The work was performed in IPME RAS, supported by the Russian Science Foundation (grant 14-29-00142).

[1] Kudryashov N A 2012 *Commun. Nonlinear Sci Numer Simulat.* **17** 2248
[2] Kudryashov N A and Demina M V 2009 *Applied Mathematics and Computation* **210** 551
[3] Porubov A V and Andrievsky B R 2011 *Commun. Nonlinear Sci Numer Simulat.* **16** 3964
[4] Porubov A V, Maugin G A and Andrievsky B R 2011 *Wave Motion* **48** 773
[5] Porubov A V and Andrievsky B R 2012 *Phys. Rev.* E **85** 046604
[6] Lober J and Engel H 2014 *Phys. Rev. Lett.* **112** 148305
[7] Puebla H, Martin R, Alvarez-Ramirez J and Aguilar-Lopez R 2009 *Chaos, Solitons and Fractals* **39** 971
[8] Kobayashi T 2004 *Math. Meth. Appl. Sci.* **27** 957
[9] Porubov A V, Fradkov A L and Andrievsky B R 2015 *Applied Mathematics and Computation* **269** 17
[10] Porubov A V, Fradkov A L, Bondarenkov R S and Andrievsky B R 2016 *Commun. Nonlinear Sci Numer Simulat.* **39** 29
[11] Porubov A V, Fradkov A L, Andrievsky B R and Bondarenkov R S 2016 *Wave Motion* **65** 147
[12] Sayadi M K and Pouget J 1991 *J. Phys.* A: *Math. Gen.* **24** 2151
[13] Aero E L 2002 *J. Eng. Math.* **55** 81
[14] Aero E L and Bulygin A N 2007 *Mechanics of Solids* **42** 807
[15] Porubov A V, Aero E L and Maugin G A 2009 *Phys. Rev.* E **79** 046608

IOP Conf. Series: Journal of Physics: Conf. Series **788** (2017) 012030 doi:10.1088/1742-6596/788/1/012030

Feedback control of monotonic shocks

A V Porubov[1,2,3], **R S Bondarenkov**[2], **D Bouche**[4], **A L Fradkov**[1,2]
[1]Institute of Problems in Mechanical Engineering, Bolshoy 61, V.O., Saint Petersburg 199178, Russia
[2]St. Petersburg State University, 7-9, Universitetskaya nab., V.O., Saint Petersburg 199034, Russia
[3]St. Petersburg State Polytechnical University, Polytechnicheskaya st., 29, Saint Petersburg 195251, Russia
[4] CMLA, ENS de Cachan, Cachan, France
E-mail: alexey.porubov@gmail.com

Abstract. The feedback control algorithm is developed to suppress oscillations caused by high resolution schemes. The stable propagation of the shock of the nonlinear advection equation is studied using the Lax-Wendroff and Warming- Beam second- order schemes. It is shown that the addition of one and the same control artificial term in both schemes results in efficient suppression of oscillations. The switch on/off the control is studied to demonstrate the role of the control in the stable monotonic shock propagation.

Introduction

An important problem in utilization of higher-order shock capturing schemes is their ability to maintain the total Variation Diminishing (TVD) condition, see. e.g., [1]. In other words, the shock profile should not contain oscillations caused by the features of the scheme. Second-order schemes, e.g., the Lax-Wendroff and Warming-Beam schemes, are not TVD schemes, and an improvement of the numerical algorithm is needed.

As noted in [1] natural way to suppress the oscillations is to add artificial viscosity term in the scheme keeping approximate consistence with the original equation. It turns out that artificial viscosity may help to suppress oscillations caused by the dispersion of the scheme, however, it also produces unnecessary smearing of the front of the shock. Reducing of smearing requires further development of artificial viscosity by so-called adaptive viscosity [2, 3] which is introduced only in the areas where it is necessary to suppress dispersive oscillations. Despite some progress in this direction problem of adding optimal amount of viscosity is still difficult [1] while improvements are rather complicated [2].

Another modification is utilization of the so-called limiters whose role is to act as a nonlinear switching between numerical methods applied for equation under study. A variety of limiters may be found in the literature, see, e.g., [1, 4, 5, 6, 7] and references therein. There is no universal limiter, and their use is computationally expensive [2]. A TVD flux-limiter may be developed for nonlinear hyperbolic equations [7], it improves the similarity with moving monotonic discontinuity in comparison with, e.g., the Minmod limiter, however, there still remains small differences between required and numerically obtained shocks.

Method of equivalent equation (or differential approximation) [8, 9] may be used to analyze both artificial viscosity [9] and limiters [6]. Also asymptotic and exact solutions of the equivalent equation may help for finding suitable artificial additional terms [10, 11, 12]. However, equivalent equations are usually very complicated for simulations of nonlinear equations and coupled equations that makes their analysis impossible.

Recently [13, 14], it was shown that the feedback control may provide stable propagation of the waves with monotonic shape for the sine- Gordon equation. It was found that oscillations caused by imperfect initial conditions may be suppressed by inclusion of additional control terms in the equation. An additional control term may be added to the discrete scheme of equation following the methods of control [15, 16], however, the addition of the term should be justified. It may be done using the speed-gradient control approach [15].

In this paper, a distributed feedback control algorithm will be developed to achieve suitable monotonic moving shock wave solution to the nonlinear advection equation solved by the second-order numerical methods. The paper is organized as follows. In Sec.1 the known Lax-Wendroff and Warming -Beam schemes for the advection equation are presented. Their utilization leads to oscillatory shock wave solution. Next Sec.2 is devoted to the development of the control algorithm, and artificial additions to the schemes are found. In the next Section, numerical simulations are performed for the control that may be switched on/off at any time. Conclusions summarize the paper.

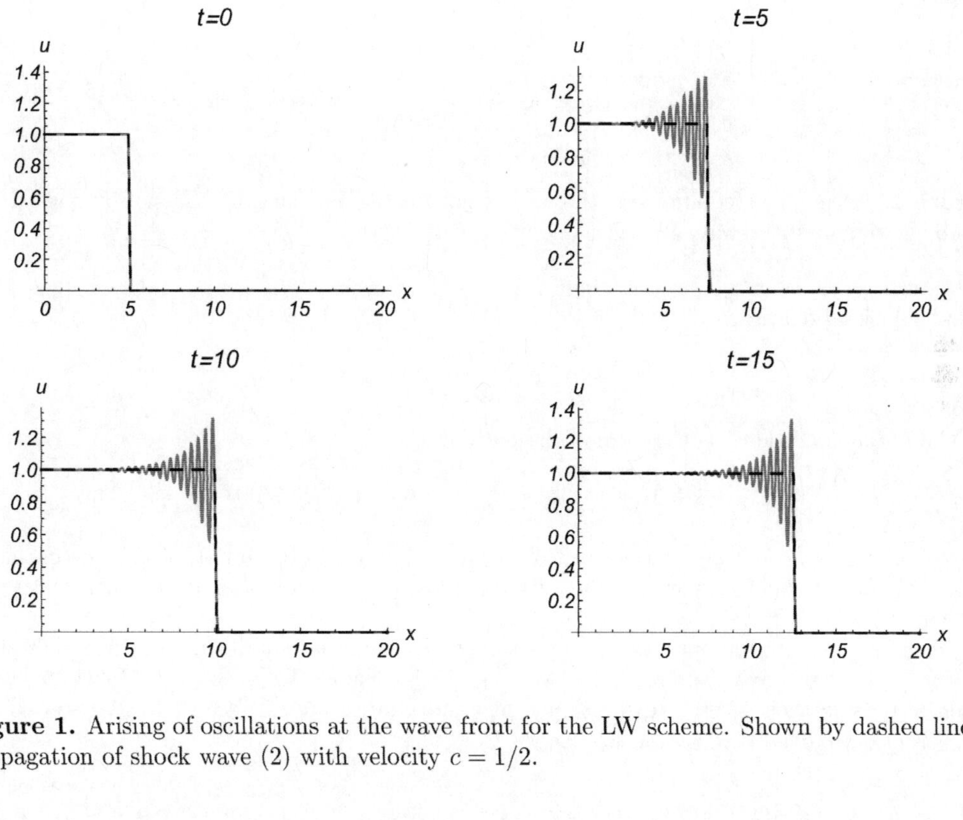

Figure 1. Arising of oscillations at the wave front for the LW scheme. Shown by dashed line is propagation of shock wave (2) with velocity $c = 1/2$.

1. Statement of the problem

Consider the nonlinear advection equation,

$$u_t + u\, u_x \;=\; 0. \tag{1}$$

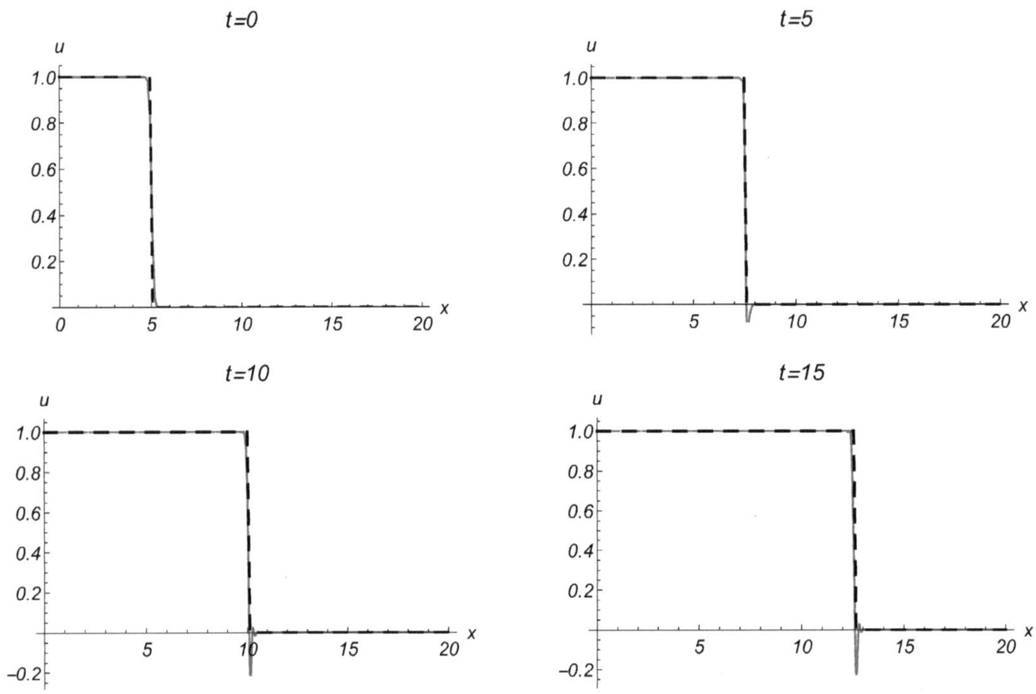

Figure 2. Arising of oscillations at the wave front for the WB scheme. Shown by dashed line is propagation of shock wave (2) with velocity $c = 1/2$.

whose solution behaves as

$$u \to 1 \text{ at } x \to -\infty,$$

$$u \to 0 \text{ at } x \to \infty.$$

The initial condition has the form of discontinuity,

$$u_0 = 1 \text{ at } x - x_0 \leq 0, \;\; u_0 = 0 \text{ at } x - x_0 > 0, \tag{2}$$

x_0 is a constant accounting for the position of discontinuity. Propagation of this discontinuity is carried out with velocity equal to the average of the values before and after discontinuity [17] and equal to $c = 1/2$ in our case.

Correct numerical description of the shock wave propagation requires use of the higher-order schemes, e.g., the Lax- Wendroff (LW) or Warming- Beam (WB) schemes. However, these schemes possess their internal dispersion that results in an appearance of oscillations on the wave front. In particular, the LW scheme for Eq. (1) is

$$\frac{u_i^{n+1} - u_i^n}{\Delta t} + \frac{1}{2\Delta x}\left(f_{i+1}^n - f_{i-1}^n\right) - \frac{\Delta t}{2\Delta x^2}\left(u_{i+1/2}\left(f_{i+1}^n - f_i^n\right) - u_{i-1/2}\left(f_i^n - f_{i-1}^n\right)\right) = 0, \tag{3}$$

where $f_i^n = (u_i^n)^2/2$, $u_{i\pm1/2}^2 = (u_{i\pm1}^n + u_i^n)/2$, Δt and Δtx are the temporal and spatial steps respectively. Simulations of Eq. (3) were realized using the Wolfram Mathematica programm. One can see in Fig. 1 that oscillations appear at the upper side of the wave front as time goes

IOP Conf. Series: Journal of Physics: Conf. Series **788** (2017) 012030 doi:10.1088/1742-6596/788/1/012030

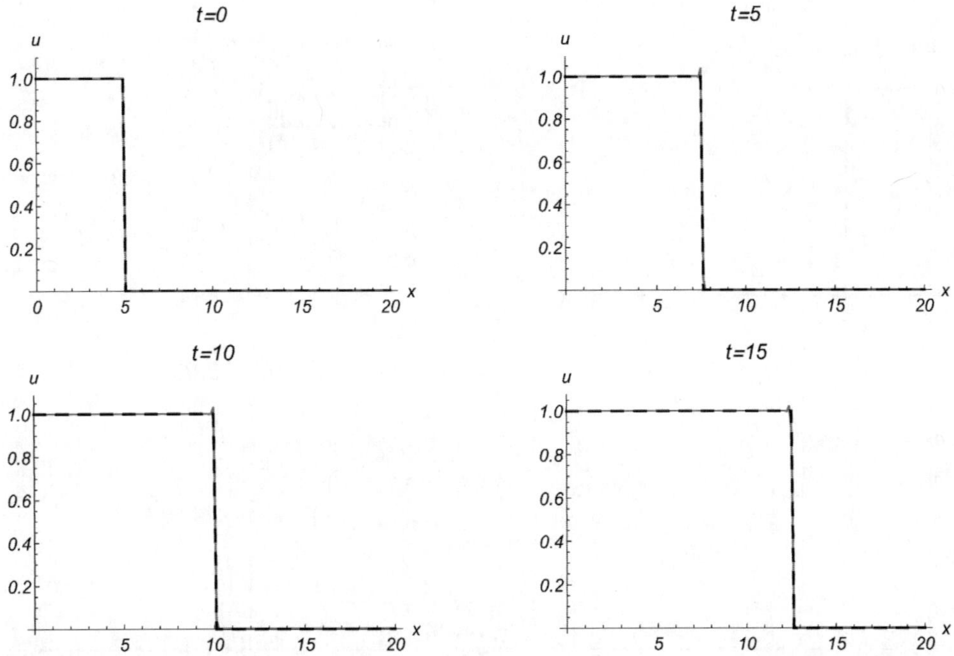

Figure 3. Control action (8) for both the LW and WB schemes. Shown by dashed line is propagation of shock wave (2) with velocity $c = 1/2$.

on while the slope and the velocity of the wave are similar to those of the desired wave shown by dashed line.

For the WB scheme one obtains

$$\frac{u_i^{n+1} - u_i^n}{\Delta t} + \frac{1}{2\Delta x}\Big(3(f_i^n - f_{i-1}^n) - (f_{i-1}^n - f_{i-2}^n)\Big) +$$
$$\frac{\Delta t}{2\Delta x^2}\Big(u_{i-3/2}(f_{i-1}^n - f_{i-2}^n) - u_{i-1/2}(f_i^n - f_{i-1}^n)\Big) \;=\; 0. \tag{4}$$

Simulations of Eq. (4) reveal oscillations at the lower side of the front of the shock as shown in Fig. 2. Again the slope and the velocity of the wave are similar to those of the desired wave shown by dashed line.

There is a need in a modification of the schemes that keeps the steepness of the profile of the shock and its velocity but suppresses parasitic oscillations caused by the scheme (shock capturing).

2. Feedback speed- gradient control method

To suppress scheme oscillations caused by the LW and WB schemes, a control algorithm is developed. Let us add an artificial control function, $w(x,t)$ in Eq. (1),

$$u_t + \; u \; u_x + w(x,t) \;=\; 0, \tag{5}$$

The *distributed* error of the shape of the wave is

$$e(x,t) = u(x,t) - u_0(x,t). \tag{6}$$

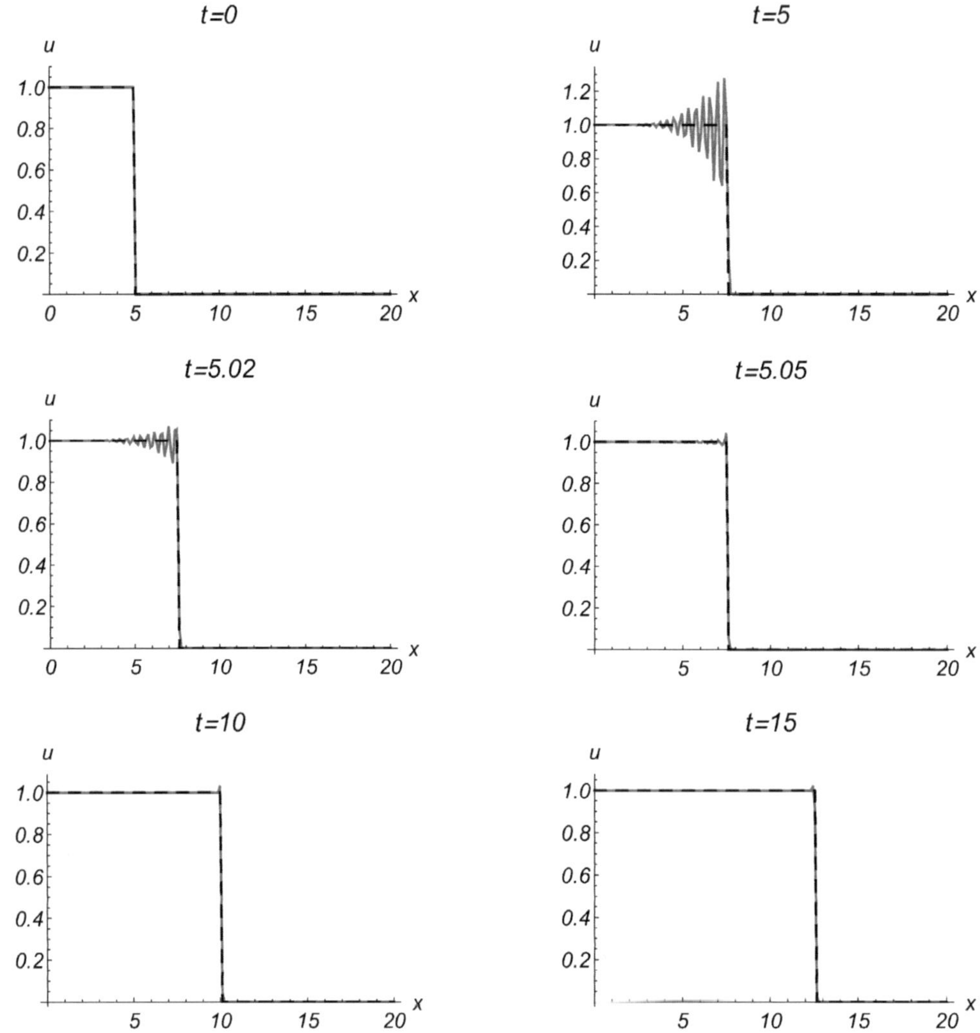

Figure 4. Control action (9) for the LW scheme switched on at $t_b = 5$. Shown by dashed line is propagation of shock wave (2) with velocity $c = 1/2$.

where u_0 is the desired wave profile, e.g., moving discontinuity like Eq. (2). Then the objective functional Q is

$$Q(u) = \frac{1}{2}e(x,t)^2. \tag{7}$$

Let us introduce an auxiliary control goal: to diminish the functional (7). However, it does not depend *explicitly* on the control function w. To involve the dependence, consider the first derivative of Q with the use of Eq.(5):

$$Q_t(u) = e(x,t)e_t(x,t) \;\; = \;\; -\, e(x,t)(uu_x + w)$$

Then $\partial Q_t(u)/\partial w$ is evaluated to characterize decrease in $Q_t(u)$, and the distributed control function w is assumed to be

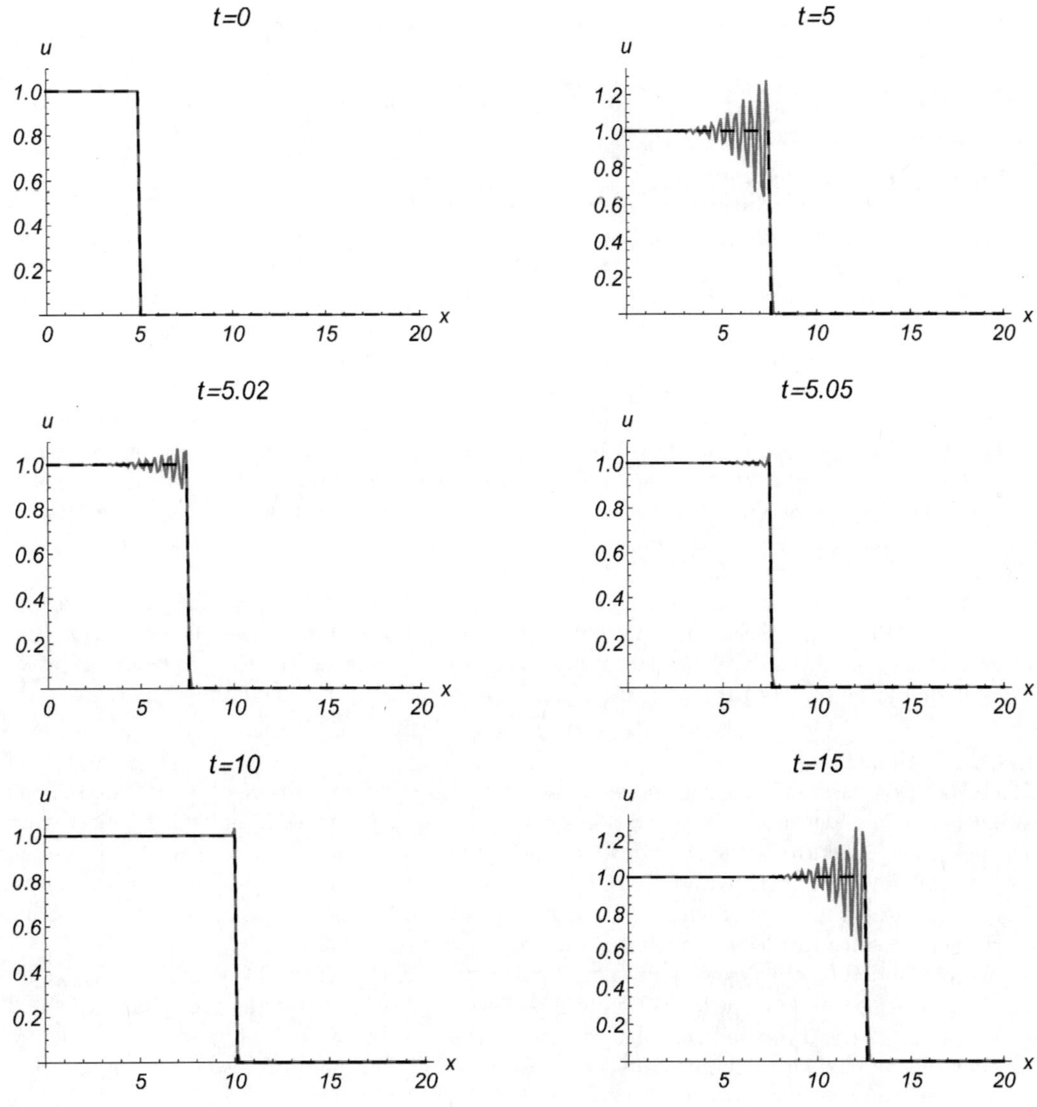

Figure 5. Control action (10) for the LW scheme switched on at $t_b = 5$ and switched off at $t_f = 10$. Shown by dashed line is propagation of shock wave (2) with velocity $c = 1/2$.

$$w(x,t) = -\gamma(u(x,t) - u_0(x,t)), \tag{8}$$

$\gamma > 0$ is the parameter of the algorithm. The algorithm does not contain derivatives of the function u and may be easily incorporated as artificial addition in both Eqs. (3), (4) in one and the same form.

The control may be switched on at some time $t = t_b$. In this case the distributed control function w is

$$w(x,t) = -\gamma(u(x,t) - u_0(x,t))H(t - t_b), \tag{9}$$

where H is the unit-step function. Similarly, the control may be switched off at some time $t_f > t_b$, in this case we get

$$w(x,t) = -\gamma(u(x,t) - u_0(x,t))H(t - t_b)H(t_f - t). \quad (10)$$

3. Shock capturing by control

Let us choose the target function u_0 in the form of shock wave (2) propagating with velocity $c = 1/2$. Then LW scheme (3) modified by addition of control function (8) at $\gamma = 0.5$ results in description of stable propagation of monotonic shock shown in Fig. 3. Similar results are obtained using simulations of the WB scheme (4) modified by addition of control function (8). Due to definition of the control function, it is small when u is almost equal to u_0 and tends to zero as soon as these functions coincide. Therefore, the addition of the control function gives rise to a small deviation from the original discrete equations. However, even very small control function term does not mean switching off the control.

To see it, let us consider first the control which is realized according to Eq. (9). One can see in Fig. 4 for the LW scheme that oscillations are developing by $t = t_b$ when the control is switched on. Then oscillations are suppressed by the control very fast, see the second row in Fig. 4, and again stable propagation of the shock carries out, see last two sketches in Fig. 4. Similar behavior is observed for the WB scheme.

Further, the control may be both switched on and switched off using control function in the form of Eq. (10). One can see in Fig. 5 that oscillations recover after switching off the control. Therefore, despite the control term is very small after coincidence of the wave profiles after $t = 5$, it should be kept to support further stable propagation of the monotonic shock.

4. Conclusions

The developed feedback algorithm provides fast and efficient suppression of the scheme oscillations. The algorithm does not contain derivatives that makes it universal for different schemes. The structure of the control function results in very small addition to the scheme in the areas where the numerical and desired wave profiles coincide, then the equation with control slightly deviates from that of without control. The method does not require knowledge of the exact solution since the desired function is chosen according our choice.

The LW and WB schemes were used to demonstrate the efficiency of the method. Certainly it may be extended to more advanced finite difference approaches using the procedure explained in Sec. 2. Also the algorithm may be extended by modification of the higher- order schemes used for the coupled gas dynamic equations and to two-dimensional problems although, in the last case, it may be difficult to get the reference solution for the control function. It will be the subject of future work.

5. Acknowledgment

The work of AVP, RSB and ALF was performed in IPME RAS, supported by the Russian Science Foundation (grant 14-29-00142).

[1] Randall J and LeVeque 1992 *Numerical methods for conservation laws. Lectures in Mathematics* (Zurich: ETH)
[2] Kurganov A and Liu Yu 2012 *J. Comp. Phys.* **231** 8114
[3] Popov I V and Fryazinov I V 2009 *Mathematical Models and Computer Simulations* **1** 493
[4] Sweby P K 1984 *SIAM J. Num. Anal.* **21** 995
[5] Rider W J and Drikakis D 2002 High resolution methods for computing turbulent flows *Turbulent flow computation* ed D Drikakis and BJ Geurts (New York: Kluwer) pp 43–74

[6] Margolin L G and Rider W J 2007 Numerical Regularization: the numerical analysis of implicit subgrid models *Implicit Large Eddy Simulation. Computing turbulent fluid dynamics* eds F Grinstein, LG Margolin and WJ Rider WJ (Cambridge:Cambridge Univ. Press) pp 195–221
[7] Galliano S J and Zapata M U 2010 *J. Comp. Appl. Math.* **234** 1395
[8] Lerat A and Peyret R 1975 *La Recherche Aérospatiale* **No2** 61
[9] Shokin Yu 1983 *The Method of Differential Approximation* (Berlin: Springer)
[10] Porubov A V, Bouche D and Bonnaud G 2008 *Intern. J. Finite Volumes* **5** 1
[11] Porubov A V, Bouche D and Bonnaud G 2010 *Trans. on Comput. Sci. VII, LNCS* **5890** 122
[12] Porubov A V, Bouche D and Bonnaud G 2013 *Commun. Nonlinear Sci. Numer. Simulat.* **18** 2679
[13] Porubov A V, Fradkov A L, Andrievsky B R and Bondarenkov R S 2016 *Wave Motion* **65** 147
[14] Porubov A V, Fradkov A L, Bondarenkov R S and Andrievsky B R 2016 *Commun. Nonlinear Sci Numer Simulat.* **39** 29
[15] Fradkov A L 2007 *Cybernetical Physics: From Control of Chaos to Quantum Control* (Berlin- Heidelberg: Springer)
[16] Chernousko F L, Ananievski I M and Reshmin S A 2008 *Control of Nonlinear Dynamical Systems Methods and Applications* (Berlin -Heidelberg: Springer-Verlag)
[17] Whitham GB 1974 *Linear and Nonlinear Waves* (New York: Wiley)

Adiabatic shear bands localization in materials undergoing deformations

P N Ryabov, N A Kudryashov, R V Muratov

National Research Nuclear University MEPhI (Moscow Engineering Physics Institute), Kashirskoe shosse 31, Moscow, Russia

E-mail: nakudr@gmail.com, pnryabov31@gmail.com

Abstract. We consider the adiabatic shear banding phenomenon in composite materials undergoing the high speed shear deformations. The mathematical model of adiabatic shear banding in thermo-visco-plastic material is given. New two step numerical algorithm which is based on the Courant-Isaacson-Rees scheme that allows one to simulate fully localized plastic flow from initial stage of localization is proposed. To test this numerical algorithm we use three benchmark problems. The testing results show the accuracy and efficiency of proposed algorithm. The features of adiabatic shear bands formation in composites are studied. The existence of characteristic depth of localization in composites is shown. Influence of initial temperature distribution on the processes of adiabatic shear bands formation in composites is considered.

1. Introduction

The study of the strength properties of different technologically important materials subjected to shear deformations is one of the most important sections of mechanics. One of the methods used for this purpose is the torsional Kolsky-bar method that loads the sample in high rate of simple shear [1,2]. In work [1] authors have shown that geometrical defects in the sample of material undergoing deformations leads to the plastic flow localization with subsequent formation of adiabatic shear bands (ASB) in the area of defect. The studying of the adiabatic shear bands formation processes is an extremely important task, since these processes lead to the failure of materials. The following phenomenon was observed in shock loading, metal forming, ballistic impact and etc [3-5].

The adiabatic shear bands are narrow regions where high temperature and deformations are exceeds in a short period of time. The reason of the ASB formation is a loss of stability of plastic flow occurs due to the thermal softening effect. In conditions close to adiabatic the work of plastic deformation converted into heat which causes thermal softening of material and leads to adiabatic shear bands farmation.

Besides the geometrical defects which leads to the ASB formation, it turned out that the following processes may take place if we have thermal heterogeneity in heating of material. However, most of the works devoted to study this effect, concentrated on the formation of ASB in simple material. Thus the studying the processes of ASB formation in composite material is of interest.

2. Mathematical model of ASB formation in composites

We consider the simple shear deformation of an infinite sample which consist of two thermo-visco-plastic materials. The height of the composite sample is H, where y_1 and y_2 are a heights of materials used in it. To convinience we define the following notations: y_1 mm – y_2 mm.

The governing equations that describe this physical process have the following form [6-9]

$$v_t = \frac{1}{\rho}\tau_y, \tag{1}$$

$$\tau_t - \mu v_y = -\mu \dot{\varepsilon}^p, \tag{2}$$

$$\tau = \Phi(T, \psi, \dot{\varepsilon}^p), \tag{3}$$

$$\psi_t = \frac{\tau \dot{\varepsilon}^p}{\kappa(\psi)}, \tag{4}$$

$$C\rho T_t = (kT_y)_y + \beta\tau\dot{\varepsilon}^p, \tag{5}$$

where y, t is the coordinate and time, τ is a stress, v is a velocity, T is a temperature, ψ is a strain hardening factor, ρ is the material density, μ is the shear module, C, k are the specific heat and thermal conductivity respectively and β is the Taylor-Quinney parameter.

The composite is free of strain at $t = 0$. Initial strain rate $\dot{\varepsilon}^p = \dot{\varepsilon}_0$. We use the following initial conditions

$$v(y,0) = v_0(y), \quad T(y,0) = T_0(y), \quad \tau(y,0) = \tau_0(y), \quad \psi(y,0) = \psi_0, \quad \varepsilon(y,0) = 0. \tag{6}$$

Velocity on the boundaries is used in the form

$$v(0,t) = 0, \quad v(H,t) = v_{up}. \tag{7}$$

We use two types of conditions on temperature on boundaries. The first one has the form

$$T(y=0,t) = 0, \quad T(y=H,t) = 0. \tag{8}$$

The second one is

$$T_y(y=0,t) = 0, \quad T_y(y=H,t) = 0. \tag{9}$$

The plastic flow (3) low is used in the following form

$$\tau = \kappa_0 g(T)\left(1 + \left(\frac{\psi}{\psi_0}\right)^n\right)\left(1 + \frac{\dot{\varepsilon}^p}{\dot{\varepsilon}_y}\right)^m. \tag{10}$$

The thermal softening factor $g(T)$ is used in the form

$$g(T) = \exp(-aT) \quad \text{or} \quad g(T) = (1 - aT)^3, \tag{11}$$

depending on what material we use.

The boundary value problem (1)–(9) is solved numerically. The numerical algorithm consists of two stages. The first stage devoted to solution of mechanical part of the problem. For this purpose we use the Courant-Isaacson-Rees scheme with the Newton iterations to system (1)–(3), (6)–(8). The strain hardening is defining from the solution of Cauchy problem (4), (6) using the fourth order Runge-Kutta method. On the second stage, we solve thermal part of the problem, i.e. we define the temperature of composite solving mixed boundary value problem (5), (8), (9). The proposed numerical algorithm was tested on three benchmark problems. The results of testing prove its accuracy and efficiency.

3. Results of numerical simulation

We consider the composite which is made from steel and copper parts. The height of the steel is y_1 and the height on copper is y_2. Numerical experiments show that the evolution of stress, velocity and temperature on initial stage of localization does not depend on values of y_1, y_2 and $\dot{\varepsilon}_0$.

Since initial stress in steel more than in the copper approximately on ten times, there is a big gradient of stress in composite. This causes wave of initial disturbance, which moves to point $y = 0$. Than it reflects from the boundary $y = 0$ goes to the boundary between steel and copper. The velocity in steel oscillates. We can estimate the period and frequency of its oscillations using the following formulaes

$$T = 4y_1\sqrt{\frac{\rho_{st}}{\mu_{st}}}, \quad \omega = \frac{\pi}{2y_1}\sqrt{\frac{\mu_{st}}{\rho_{st}}}, \tag{12}$$

where μ_{st}, ρ_{st} – shear module and density of steel. After a while these oscillations are damped. The disturbance goes to copper part of composite and localized in it. The number of localization areas depends on the value of initial strain rate $\dot{\varepsilon}_0$ and y_1, y_2. We found that there is a characteristic depth ξ on which the plastic flow localized (see Fig. 1). The value of ξ inversely proportional $\dot{\varepsilon}_0$, that can be seen in Fig. 2.

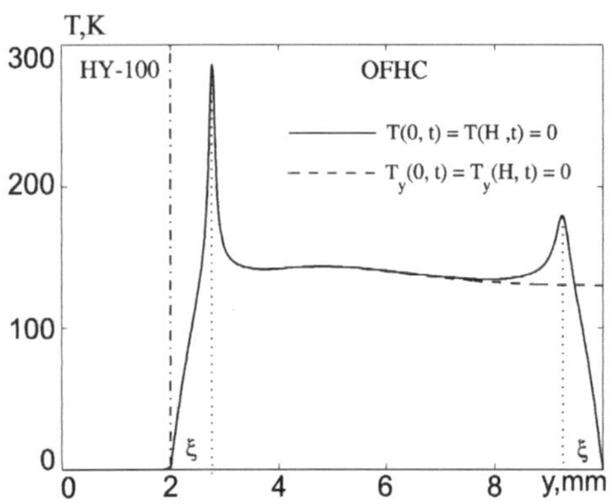

Figure 1. Distribution of temperature T through the sample. Solid line corresponds to (8) and doted line to (9).

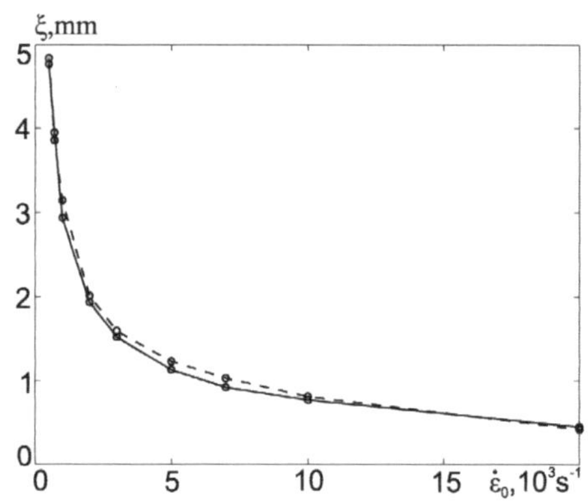

Figure 2. Dependence of depth ξ from $\dot{\varepsilon}_0$ in slab 2 mm - 8 mm. Solid line corresponds to (8) and doted line to (9).

In works [6-9] authors have shown that nonuniform distribution of temperature initiates the process of adiabatic shear bands formation. Here we consider the influence of initial temperature distribution on the ASB formation process in composite material. The initial temperature distribution was used in the form

$$T(y, 0) = T_{max} \exp\left[-500\left(\frac{y}{H} - \delta\right)^2\right], \tag{13}$$

where T_{max} is a maximum of initial distribution, $y = \delta H$ is a point where this maximum is reached.

We consider the composite 3 mm - 3 mm with thermal isolated boundaries. The initial strain rate is equal to $\dot{\varepsilon}_0 = 10^4\ \mathrm{s}^{-1}$. If there is no initial disturbance ASB localized on the upper moving boundary [9]. If we place the initial disturbance in the point $\delta = 0.75$, at $T_{max} = 0.5 - 500$ K, the shear band formed in the point $y = 0.75H$. Increasing the value of T_{max} decries the

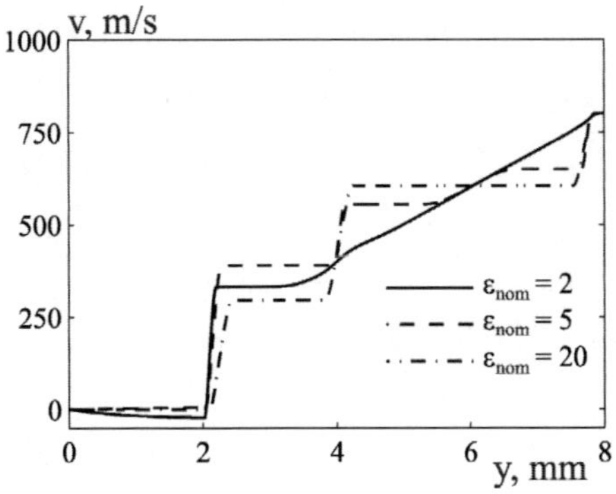

Figure 3. Velocity distribution in slab 2 mm - 6 mm at different time moments ($\dot{\varepsilon}_0 = 10^5$ s^{-1}).

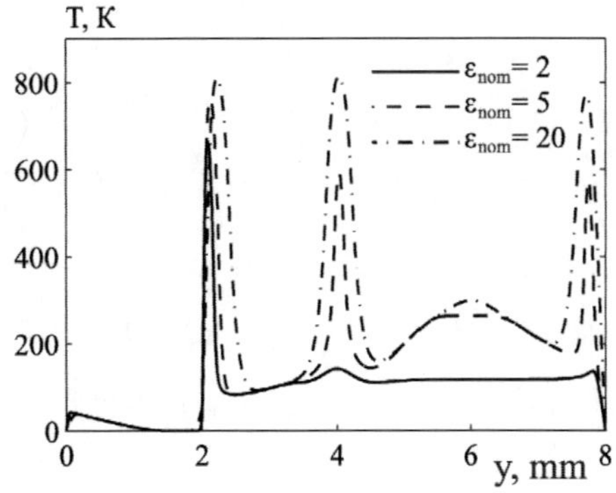

Figure 4. Temperature distribution in slab 2 mm - 6 mm at different time moments ($\dot{\varepsilon}_0 = 10^5$ s^{-1}.

localization time. If we move the maximum of initial disturbance to the boundary between materials ($\delta = 0.5$) the ASB formed near boundary between materials. At a high value of T_{max}, deformations localized on a boundary and then moves to the copper part of the composite.

If we change the height of the composite with zero temperature on the boundaries and consider 2 mm - 6 mm sample, the formation of two ASB is observed in the left and right boundary of copper [9]. The velocity distribution becomes stepwise (Fig. 3). If we place initial disturbance in the middle of the copper part of the sample, the third ASB formes in that area. Whatever disturbance we use, the shear bands on the boundaries of copper are observed (see Fig. 4).

4. Acknowledgments

This work was supported by the Grant for support of young Russian scientists 7207.2016.1.

References

[1] Marchand A and Duffy J 1988 *J. Mech. Phys. Solids* **36** 251
[2] Duffy J, Campbell J D and Hawley R H 1971 *J. Appl. Mech.* **38** 83
[3] Schneider J and Nunes J A 2004 *Metall. Mater. Trans.* B **35** 777
[4] Seidel T and Reynolds A 2001 *Metall. Mater. Trans.* A **32** 2879
[5] Rogers H C 1979 *Annu. Rev. Mater. Sci.* **9** 283
[6] Walter J W 1992 *Int. J. Plast.* **8** 657
[7] Wright T W and Batra R C 1985 *Int. J. Plast.* **1** 205
[8] Kudryashov N A, Ryabov P N and Zakharchenko A S 2015 *J. Mech. Phys. Sol.* **76** 180
[9] Kudryashov N A, Muratov R V and Ryabov P N 2016 *Model. Anal. Inf. Sys.* **23** 298 (in Russian)

Nonlinear waves described by the generalized Swift-Hohenberg equation

P N Ryabov and N A Kudryashov

National Research Nuclear University MEPhI (Moscow Engineering Physics Institute), Kashirskoe shosse 31, Moscow, Russia

E-mail: nakudr@gmail.com, pnryabov31@gmail.com

Abstract. We study the wave processes described by the generalized Swift–Hohenberg equation. We show that the traveling wave reduction of this equation does not pass the Kovalevskaya test. Some solitary wave solutions and kink solutions of the generalized Swift–Hohenberg equation are found. We use the pseudo–spectral algorithm to perform the numerical simulation of the wave processes described by the mixed boundary value problem for the generalized Swift–Hohenberg equation. This algorithm was tested on the obtained solutions. Some features of the nonlinear waves evolution described by the generalized Swift–Hohenberg equation are studied.

1. Introduction

The Swift-Hohenberg equation is a fundamental partial differential equation that describes pattern formation in different physical systems. For the first time this equation was found by J. Swift and P. Hohenberg for describing the hydrodynamic fluctuations at the convective instability [1] and has the form

$$u_t + 2u_{xx} + u_{xxxx} = \alpha u + \beta u^2 - \gamma u^3, \tag{1}$$

where α, β and γ are constant parameters of equation. This equation arises in a variety fields of science. For example it describes the mechanism of shear microbands formation in nanocrystalline materials, the amplitude of optical electric field inside the cavity, the patterns inside thin vibrated granular layers [2–4] and etc. Thus, it is very important to study the wave processes described by this equation and its generalizations.

One of the interesting generalizations of equation (1), arises in literature, have the following form

$$u_t + 2u_{xx} - \sigma u_{xxx} + u_{xxxx} = \alpha u - \gamma u^{m+1}, \tag{2}$$

where σ, α and γ are the parameters. In work [5] authors define the existence condition of non stationary meromorphic solutions of equation (2), that corresponds to $\sigma \neq 0$. Using this fact they were succeed to find elliptic and simple periodic solutions. Another work, devoted to studding the equation (2) is a work [6], where the so-called snakes-and-ladders structure of equation is studied.

Here we review our results of studding the wave processes described by equation (2).

2. Exact solutions of generalized Swift–Hohenberg equation

Since equation (2) admits spatiotemporal translational symmetry, we try to find the solutions of (2) using the traveling waves variables $u(x,y) = u(z), z = kx - \omega t$

$$-\omega y_z + 2k^2 y_{zz} - \sigma k^3 y_{zzz} + k^4 y_{zzzz} = \alpha y - \gamma y^{m+1}. \tag{3}$$

We show that equation (3) does not have the Painlevé property [4]. Thus the finding partial solutions of equation (3) becomes an actual task.

Equation (3) contains the term γy^{m+1} with an arbitrary parameter m. It prevents to use known algorithms for exact solutions searching. To overcome this difficulty we use the method introduced in works [7, 8]. Then after the variable transformation we apply an approach presented in works [9–12] that allows us to find two types of exact solution of equation (3). First of them is a kink solution in the form

$$y(z) = \left(-\frac{32(m+2)(m+4)(3m+4)}{\gamma(11m^2+72m+96)^2}\right)^{1/m}\left(\frac{1}{1+\mathrm{e}^{-z}}\right)^{4/m}, \tag{4}$$

where $m \neq 0, -2, -4, -4/3, -8/3$. This solution exists when

$$\begin{gathered} k^{(1,2)} = \pm\frac{2m}{\sqrt{22m^2+144m+192}}, \quad \omega^{(1,2)} = \pm\frac{8m(3m+8)(m^2+12m+16)}{(11m^2+72m+96)^2}, \\ \sigma = \frac{2\sqrt{2}(3m+8)}{\sqrt{11m^2+72m+96}}, \quad \alpha = -\frac{32(m+4)(3m+4)(m+2)}{(11m^2+72m+96)^2}. \end{gathered} \tag{5}$$

Another exact solution is solitary wave solution which reads as

$$y(z) = \left(\frac{32(3m^3+22m^2+48m+32)}{\gamma(4m^4+60m^3+297m^2+540m+324)}\right)^{1/m}\left(\frac{\mathrm{e}^{-\frac{1}{4}z}}{1+\mathrm{e}^{-z}}\right)^{4/m}. \tag{6}$$

This solution exists when

$$\begin{gathered} k^{(3,4)} = \pm\frac{2m}{\sqrt{4m^2+30m+36}}, \quad \omega^{(3,4)} = \pm\frac{4m^2(4m+9)}{(m+6)^2(2m+3)^2}, \\ \sigma = \frac{\sqrt{2}(3m+8)}{\sqrt{2m^2+15m+18}}, \quad \alpha = \frac{12(m+3)}{(m+6)(2m^2+15m+18)}. \end{gathered} \tag{7}$$

The graphical representation of obtained solutions is given on Fig. 1, 2.

It should be noted that equation (3) admits the transformation $k \to -k, \omega \to -\omega$ and $z \to -z$. Thus the solutions (4) and (6) are the solutions if we make the above mentioned transformation.

3. Main results of numerical simulation

Here we consider the periodic boundary value problem for equation (2) in the following form

$$\begin{gathered} u_t + 2u_{xx} - \sigma u_{xxx} + u_{xxxx} = \alpha u - \gamma u^{m+1}, \\ u(x,0) = u_0(x), \quad u(x,t) = u(x+H,t), \end{gathered} \tag{8}$$

where $u_0(x)$ is an initial disturbance.

Numerical solution of problem (8) is constructed using the pseudo-spectral method [13–15]. The applied numerical approach was tested using the obtained solutions [4].

First of all, it should be noted that the wave behavior, described by the mixed boundary value problem (8), strongly depends on the parity of parameter m. If we take in computations

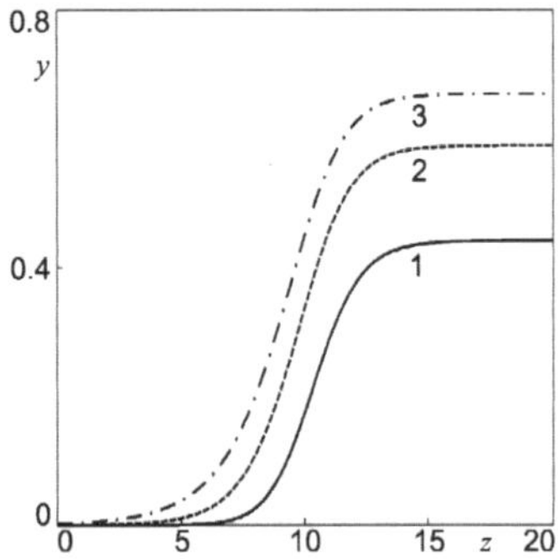

Figure 1. Exact solution (4) at $z_0 = 10, \gamma = -1, m = 3, 5, 7$ (curves 1–3).

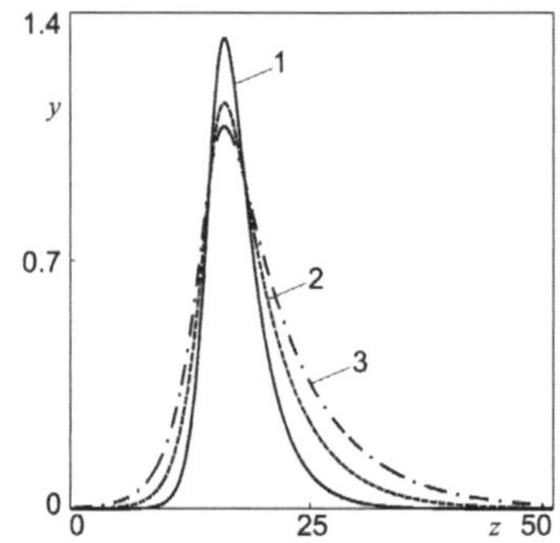

Figure 2. Exact solution (6) at $z_0 = 10, \gamma = 0.1, m = 3, 5, 7$ (curves 1–3).

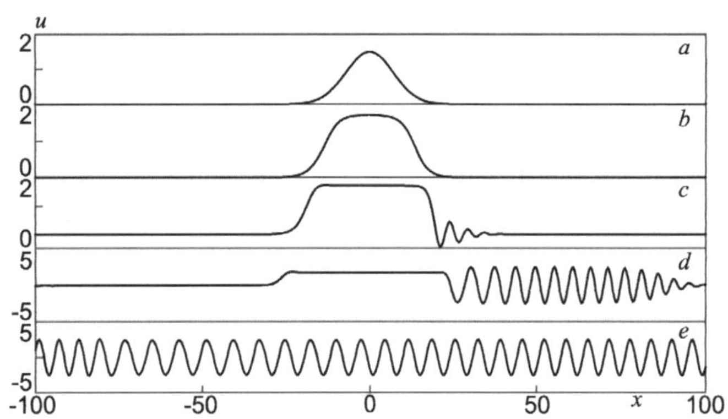

Figure 3. Evolution of initial "hat" disturbance at $m = 2$. (a)–(e) – t=0, 1, 2.5, 6, 45.

an even value of parameter m, initial disturbance transforms in a regular structure after a cascade of transformations. For example, if we take the disturbance of trivial solution in the form $\sim a\exp(-\beta x^2)$, we observe three stages of transformations. The first stages corresponds to the formation of horizontal plateau on the top of disturbance. The formation of a "snake" structure is observed on the next stage. This structure travels thought the whole computational domain. Finally, this structure becomes regular (see Fig. 3).

However, if the value of parameter m is odd, we observe the blow-up phenomenon.

4. Acknowledgments

This work was supported by the Grant for support of young Russian scientists 7207.2016.1, the Grant for scientific schools 6748.2016.1 and RFBR grant 14-01-00493.

References

[1] Swift J and Hohenberg P C 1977 *Physical Review A* **15** 319
[2] Malygin G A 2009 *Physics of the Solid State* **51** 1814
[3] Blair D, Aranson I S, Crabtree G W, Vinokur V, Tsimring L S and Josserand C 2000 *Phys. Rev.* E **61** 5600
[4] Kudryashov N A, Ryabov P N 2016 *Appl. Math. Comp.* **286** 171
[5] Kudryashov N A, Sinelshchikov D I 2012 *Commun. Nonlinear. Sci. Numer. Simulat.* **17** 26

[6] Burke J, Houghton S M and Knobloch E 2009 *Phys. Rev.* E **80** 036202
[7] Kudryashov N A and Demina M V 2009 *Appl. Math. Comput.* **210** 551
[8] Kudryashov N A and Ryabov P N 2011 *Fluid Dynamics* **46** 425
[9] Kudryashov N A 1988 *Prikl. Math. Mech.* **52** 361
[10] Kudryashov N A 2012 *Commun. Nonlinear. Sci. Numer. Simulat.* **17** 2248
[11] Ryabov P N, Sinelshchikov D I and Kochanov M B 2011 *Appl. Math. Comput.* **218** 3965
[12] Ryabov P N 2010 *Appl. Math. Comput.* **217** 3585
[13] Cox S M and Matthews P C 2002 *J. Comp. Phys.* **176** 430
[14] Kassam A, Trefethen L N 2005 *J. Sci. Comput.* **26** 1214
[15] Kudryashov N A, Ryabov P N, Fedyanin T E and Kutukov A A 2013 *Phys. Lett.* A **377** 753

On the general traveling wave solutions of some nonlinear diffusion equations

D I Sinelshchikov and N A Kudryashov

National Research Nuclear University MEPhI (Moscow Engineering Physics Institute), Moscow, Russia

E-mail: nakudr@gmail.com, disine@gmail.com

Abstract. We consider a family of nonlinear diffusion equations with nonlinear sources. We assume that all nonlinearities are polynomials with respect to a dependent variable. The traveling wave reduction of this family of equations is an equation of the Lienard–type. Applying recently obtained criteria for integrability of Lienard–type equations we find some new integrable families of traveling wave reductions of nonlinear diffusion equations as well as their general analytical solutions.

1. Introduction

In this work we consider the following family of equations

$$u_t = (u^m u_x)_x + a_1 u^k + a_2 u^l + a_3 u^p, \qquad (1)$$

where x, t and $u(x,t)$ are independent and dependent variables correspondingly, a_i, $i = 1,2,3$, $a_1^2 + a_2^2 + a_3^2 \neq 0$ and m, k, l, p, $m^2 + k^2 + l^2 + p^2 \neq 0$ are arbitrary real parameters.

Equation (1) can be considered as a nonlinear diffusion or nonlinear heat equation with polynomial sources, which are both frequent in many applications (see, e.g. [1]). For example, the generalized Fisher equation, the Newell–Whitehead equation and Zeldovich equation belong to family (1) [1–3]. It is known (see, e.g. [1,2,4] and references therein) that traveling wave solutions of equations like (1) are very important for understanding of both underlying phenomenon and solutions' behaviour. Therefore, a problem of constructing general analytical traveling wave solutions of equations from family (1) is worth studying.

Assuming that $u(x,t) = y(z)$, $z = x - C_0 t$ in (1) we obtain its traveling wave reduction

$$y_{zz} + \frac{m}{y} y_z^2 + C_0 y^{-m} y_z + a_1 y^{k-m} + a_2 y^{l-m} + a_3 y^{p-m} = 0, \qquad (2)$$

where $C_0 \neq 0$ is traveling waves speed. Equation (2) is a nonlinear ordinary differential equation of the Lienard type. Some approaches for finding integrable cases of such equations have recently been proposed (see, e.g. [5–14] and references therein). In work [14] several new integrability conditions for the Lienard–type equations were found. In this work we apply these conditions for finding integrable cases of (2). As a result, we find four subcases of (2) for which the closed–from general analytical solution can be constructed. We believe that these families of integrable traveling wave reductions of (1) have not been reported previously.

2. Integrable traveling wave reductions of nonlinear diffusion equation (1)

In this section we apply integrability criteria obtained in [14] for finding integrable cases of (2). The approach of work [14] is based on using of the following nonlocal transformations

$$w = F(y), \quad d\zeta = G(y)dz, \quad F_y G \neq 0, \tag{3}$$

where w and ζ are new dependent and independent variables. Connections between the Lienard–type equations and Painleve–Gambier equations (see, e.g. [15])of type III via (3) were studied in [14] and four integrability criteria were found. Below we apply these integrability criteria for finding general analytical solutions for some members of equations family (2). Let us remark that throughout this work we assume that the parameter κ is equal to 0 in all criteria from work [14]. Note also that below we use Ince's notation for the Painleve–Gambier equations [15].

We start with the criterion for integrability of (2) that connects Lienard–type equations with Painleve–Gambier equation XXIII (see Theorem 1, [14]). It can be shown that equation (2) satisfies this criterion when $k = 5 - m$, $l = 1 - m$, $a_2 = 3C_0^2/16$ and $a_3 = 0$. As a result, we obtain that the following integrable equation from family (2)

$$y_{zz} + \frac{m}{y} y_z^2 + C_0 y^{-m} y_z + a_1 y^{5-2m} + \frac{3C_0^2}{16} y^{1-2m} = 0, \tag{4}$$

can be transformed into Painleve–Gambier equation XXIII

$$w_{\zeta\zeta} - \frac{3}{4w} w_\zeta^2 + 2\alpha w_\zeta + w^2 + 3\alpha^2 w = 0, \tag{5}$$

via transformations (3) with $F = 16\alpha^2 a_1 C_0^{-2} y^4$ and $G = (C_0/2\alpha) y^{-m}$. Here $\alpha \neq 0$ is an arbitrary parameter.

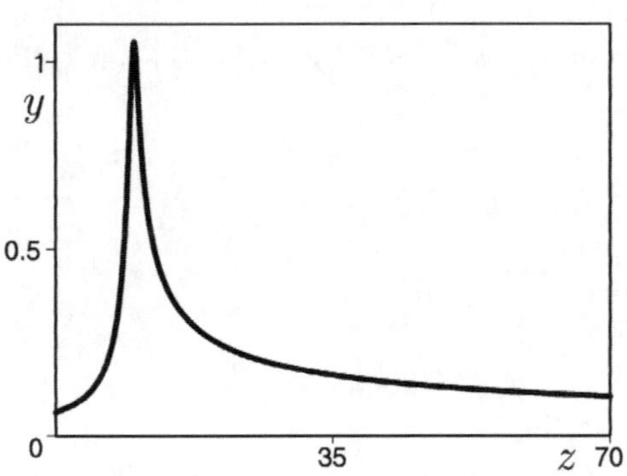

Figure 1. Solution (6) at $m = -2$, $\zeta_0 = 1.1$ and $\alpha = C_0 = g_3 = a_1 = 1$.

Taking into account the general solution of equation (5) we find the general solution of (4)

$$y = \left[\frac{3C_0 g_3}{16 a_1} e^{-2\alpha(\zeta - \zeta_0)} \wp^{-2}\{e^{-\alpha(\zeta-\zeta_0)}, 0, g_3\} \right]^{1/4}, \quad z = \frac{2\alpha}{C_0} \int y^m d\zeta, \tag{6}$$

where $\wp$ is the Weierstrass elliptic function, $g_3 \neq 0$ and ζ_0 are arbitrary constants. We demonstrate a plot of solution (6) for certain values of the parameters in Fig.1. One can see that in this case solution (6) describes a pulse–type structure.

Now we proceed to the next criterion form [14] that connects Lienard–type equations with equation XXV of the Painleve–Gambier type (see Theorem 2, [14]). This criterion holds for equation (2) if $k = -m - 3/5$, $l = -m + 1$, $p = -m - 7/5$ and $a_2 = 5C_0^2/36$. Consequently, the following equation from family (2) is integrable:

$$y_{zz} + \frac{m}{y} y_z^2 + C_0 y^{-m} y_z + a_1 y^{-2m-3/5} + \frac{5C_0^2}{36} y^{1-2m} + a_3 y^{-2m-7/5} = 0. \tag{7}$$

This equation can be transformed into Painleve–Gambier equation XXV

$$w_{\zeta\zeta} - \frac{3}{4w} w_\zeta^2 + \frac{3}{2} w w_\zeta + \frac{1}{4} w^3 + \frac{9a_1}{5} w + \frac{18C_0 a_3}{10} = 0, \tag{8}$$

with the help of transformations (3) with $F = C_0 y^{4/5}$ and $G = (2/3) y^{-4/5-m}$,

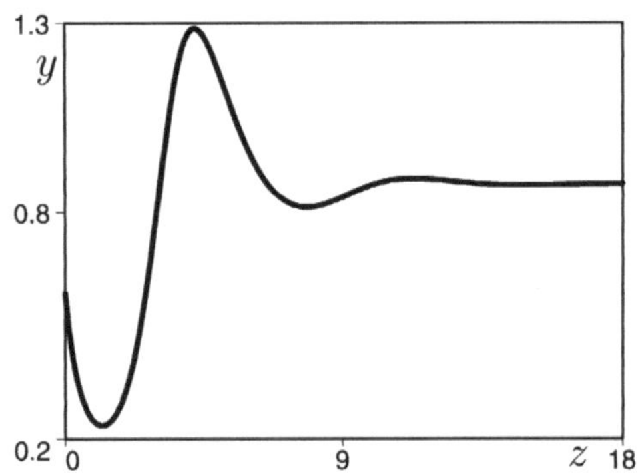

Figure 2. Solution (9) at $m = a_3 = -1$ and $C_0 = 1$.

One can obtain the general solution of (7) by virtue of (3) and the general solution of (8) as follows

$$y = \left(\frac{w}{C_0}\right)^{5/4}, \quad z = \frac{3}{2} \int y^{m+4/5} d\zeta, \quad w = -\frac{9C_0 a_3}{10 v_\zeta + 5v^2 + 9a_1}, \quad v = \frac{\Psi_\zeta}{\psi}, \tag{9}$$

where Ψ is the general solution of the equation

$$\Psi_{\zeta\zeta\zeta} + \frac{9}{5} a_1 \Psi_\zeta + \frac{9C_0}{10} a_3 \Psi = 0. \tag{10}$$

A plot of solution (9) for certain values of the parameters is shown in Fig.2. We see that solution (9) can describe pulse–type structures with oscillations.

Let us consider criterion of equivalence between Lienard–type equations and Painleve–Gambier equation XXVIII (see Theorem 3 work [14]). Equation (2) satisfies this criterion at $a_1 = (n-1)\epsilon n^2/(n-2)^2$, $a_3 = 0$, $a_2 = -(n-1)C_0^2/(n-2)^2$, $k = -m - (n+1)/(n-1)$ and $l = 1 - m$ and in this case it has the form

$$y_{zz} + \frac{m}{y} y_z^2 + C_0 y^{-m} + \frac{(n-1)\epsilon n^2}{(n-2)^2} y^{-2m-\frac{n+1}{n-1}} - \frac{(n-1)C_0^2}{(n-2)^2} y^{1-2m} = 0. \tag{11}$$

Equation (11) can be transformed into

$$w_{\zeta\zeta} - \frac{n-1}{n} \frac{w_\zeta^2}{w} + \frac{n-2}{n} \frac{w_\zeta}{w} + \frac{1}{nw} - n\epsilon w = 0, \tag{12}$$

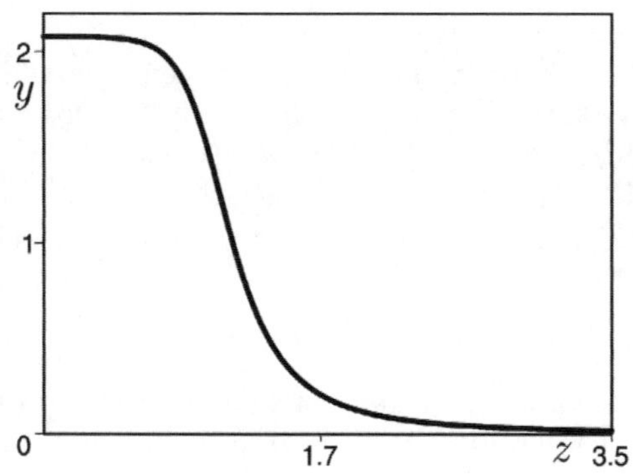

Figure 3. Solution (13) at $n = 3$, $m = -2$, $C_0 = 1$, $C_1 = 0$, $C_2 = 1$ and $C_3 = 10000$.

where $\epsilon \neq 0$ is an arbitrary parameter, $n \neq 0, \pm1, \pm2$ is an integer number, with the help of transformations (3) with $F = y^{-\frac{n}{n-1}}/C_0$, $G = ny^{-m-\frac{n}{n-1}}/(n-2)$.

Therefore, we find the general solution of equation (12)

$$y = (C_0 w)^{-\frac{n-1}{n}}, \quad z = \frac{n-2}{n}\int y^{m+\frac{n}{n-1}} d\zeta, \tag{13}$$

where w is defined by the relation

$$w = \psi^n \left(C_1 + \int \psi^{-n} d\zeta \right), \quad \psi = C_2 e^{\sqrt{\epsilon}\zeta} + C_3 e^{-\sqrt{\epsilon}\zeta}, \tag{14}$$

where C_1, C_2 and C_3 are arbitrary constants. We demonstrate a plot of solution (13) for particular values of the parameters in Fig. 3. We see that this solution can describe kink–type structures.

Finally, we consider connections via (3) between equation (2) and Painleve–Gambier XVIII equation. Using Theorem 4 from [14] we find that equation (2) satisfies this criterion in the case of $a_3 = 0$, $a_1 = 108\beta/C_0^2$, $a_2 = -3C_0^2/4$, $k = -m - 5/3$ and $l = -m - 1$. Consequently, we get the corresponding equation from family (2)

$$y_{zz} + \frac{m}{y} y_z^2 + C_0 y^{-m} y_z + \frac{108\beta}{C_0^2} y^{-2m-5/3} - \frac{3C_0^2}{4} y^{-2m+1} = 0, \tag{15}$$

that can be transformed into

$$w_{\zeta\zeta} - \frac{1}{2w} w_\zeta^2 + w w_\zeta - \frac{1}{2} w^3 + \frac{72\beta}{w} = 0 \tag{16}$$

by virtue of with (3) $F = C_0 y^{2/3}$ and $G = y^{-2/3-m}$. Here $\beta \neq 0$ is an arbitrary parameter. The general solution of (15) can be found via the general solution of (16) and corresponding transformations (3)

$$y = \left[\frac{6(\wp^2\{\zeta - \zeta_0, 12\beta, g_3\} - \beta)}{C_0 \wp_z(\zeta - \zeta_0, 12\beta, g_3)} \right]^{3/2}, \quad z = \int y^{m+2/3} d\zeta, \tag{17}$$

where g_3 and ζ_0 are arbitrary constants. Solution (17) as long as $\beta \neq 0$ and $g_3 \neq 0$ is periodic, although it may be singular on the real line.

3. Conclusion

In this work we have considered a family of nonlinear diffusion equations with polynomial sources. Traveling wave reductions of these equations are nonlinear ordinary differential equations of Lienard type. Applying recently obtained integrability conditions for the Lienard–type equations, we have found four new integrable families of traveling wave reductions of nonlinear diffusion equations with polynomial sources. We have also constructed the corresponding general solutions in the explicit form.

4. Acknowledgments

This research was partially supported by grant for the state support of young Russian scientists 6624.2016.1 and by grant for Scientific Schools 6748.2016.1.

References

[1] Polyanin A D and Zaitsev V F 2011 Handbook of Nonlinear Partial Differential Equations (Boca Raton-London-New York: Chapman and Hall/CRC)
[2] Harko T and Mak M K 2015 Exact travelling wave solutions of non-linear reaction-convection-diffusion equationsAn Abel equation based approach J. Math. Phys. 56 111501
[3] Kudryashov N A and Zakharchenko A S 2014 A note on solutions of the generalized Fisher equation Appl. Math. Lett. 32 536
[4] Gilding B H and Kersner R 2012 Travelling waves in nonlinear diffusion-convection reaction (Basel: Birkhuser)
[5] Nakpim W and Meleshko S V 2010 Linearization of Second-Order Ordinary Differential Equations by Generalized Sundman Transformations Symmetry, Integr. Geom. Methods Appl. 6 111
[6] Nucci M C and Tamizhmani K M 2010 Lagrangians for dissipative nonlinear oscillators: the method of Jacobi last multiplier J. Nonlinear Math. Phys. 17 16778
[7] Mancas S C and Rosu H C 2013 Integrable dissipative nonlinear second order differential equations via factorizations and Abel equations Phys. Lett. A 377 14348
[8] Harko T, Lobo F S N and Mak M K 2014 A class of exact solutions of the Linard-type ordinary nonlinear differential equation J. Eng. Math. 89 193205
[9] Gubbiotti G and Nucci M C 2014 Noether symmetries and the quantization of a Linard-type nonlinear oscillator J. Nonlinear Math. Phys. 21 24864
[10] Gubbiotti G and Nucci M C 2015 Quantization of quadratic Linard-type equations by preserving Noether symmetries J. Math. Anal. Appl. 422 123546
[11] Tiwari A K, Pandey S N, Senthilvelan M and Lakshmanan M 2015 Lie point symmetries classification of the mixed Linard-type equation Nonlinear Dyn. 82 195368
[12] Kudryashov N A and Sinelshchikov D I 2015 On the connection of the quadratic Lienard equation with an equation for the elliptic functions Regul. Chaotic Dyn. 20 48696
[13] Kudryashov N A and Sinelshchikov D I 2016 On the criteria for integrability of the Linard equation Appl. Math. Lett. 57 11420
[14] Kudryashov N A and Sinelshchikov D I 2016 On the integrability conditions for a family of the Lienard-type equations Regul. Chaotic Dyn. 21 54957
[15] Ince E L 1956 Ordinary differential equations (New York: Dover)

Effect of the condensed particles on a flow of combustion products in a Laval nozzle

A V Teterev[1,2], P A Mahdrik[1], L V Rudak[1] and N I Misuchenko[1]

[1] Belarusian State University, Nezavisimosti av., 4, 220030 Minsk, Belarus
[2] National Research Nuclear University «MEPhI», Kashirskoe av., 31, 115409 Moscow, Russian

E-mail: teterev@bsu.by

Abstract. The specialized multifunctional software package is developed for the numerical simulation of the flow of products of combustion in a Laval nozzle. The complex has a friendly graphical user interface of input-output and wide possibilities for visualization of simulation results directly in the process of calculating. Particles moving in the gas stream are exposed to the action of hydro-mechanical forces due to its interaction with the carrier gas environment, as well as mass forces caused by the gravitational field. Performed parametric calculations have shown that the interaction of particles with the combustion products, even with a relatively small volume fraction may lead to a qualitative change in the internal flow in the Laval nozzle, and thereby influence the characteristics of the nozzle.

1. Introduction

Multiphase flow is significantly non-equilibrium process. The various phases are not necessarily in the equilibrium with each other [1] even if each phase is in local thermodynamic equilibrium. Therefore, that the system of equations describing such a flow were closing, it is necessary to determine the transfer velocities of mass, energy and momentum between the phases. The forces of various natures act on a particle moving in a gas flow. These forces are hydromechanics and mass forces [2, 3]. Hydromechanics forces are caused by the interaction of the particles with the carrier gas phase. The forces of resistance due to friction and pressure are forces of this type. Also forces arising from parameter gradients existing in the carrier-phase flow are such forces. This type includes rising forces caused by acceleration and deceleration of the particles in the gas flow. Mass forces are determined external physical fields such as gravitational, electrostatic and other fields. Forces acting on the particle as it move along a curved path and also forces generated during rotary motion of the carrier or discrete phases, for example, Coriolis forces are the same type of forces. In general, all of the listed forces are no stationary. The degree of influence on the particle depends on its relative size, structure of the gas flow in the channel with variable cross section and its mode parameters [3-6]. In most cases, it is possible to take into account the aerodynamic drag force only. Two-speed model are typically used to account for the interaction of particles with a flow of combustion products. In this model the particle field is represented continuous medium with the appropriate density [7, 8]. This approach works well; however it has several disadvantages. At first it is not possible to describe the situation where the particles move in different directions in single point of space, which can occur when the particles rebound from the obstacle. At second the description of the particle size distribution leads to the necessity of considering multiple continuums.

2. The physical model of the motion of particles in the gas flow

In the construction of mathematical and computational models that describe the interaction of condensed particles with combustion products in channels with variable cross section, we shall proceed from the following basic assumptions:

- the movement of the carrier medium is described by the system of gas dynamics equations;
- the particles of condensed phase have a spherical shape;
- the particles undergo aerodynamic drag force and the force of gravity only;
- the processes of collision, fragmentation and coagulation can be neglected;
- an exchange of energy, impulse and mass in general case is a result of the interaction of individual particles with the gas flow;
- mutual particles affect at each other is through the interaction of the gas flow.

As a basic model, which describes the interaction of the particles with the gas flow of the combustion products the Sand Bag model [9] was taken. This model has been successfully used in the description of the dynamics of entry into the dense layers of the Earth's atmosphere the originally destroyed body [10] which do not experience further fragmentation, as well as cratering model for the description of the dynamics of soil release from crater [11]. Individual pieces of the object or release interact with the atmosphere, resulting in accelerated or decelerated, thus exchanging with impulse, energy, and, in general, mass. In addition to the force interaction and heat and mass transfer the change of the density of the gas component is taken into account in the model by reducing its volume by the amount of volume of the condensed phase of matter. The same situation may occur when the condensed particles are moving in a flow of combustion products in a Laval nozzle.

All the variety of particles can be divided into separate conglomerates according to their size, speed and location. In this case, any interaction of conglomerate with the gas stream can be described by a single representative which bearing of all the average parameters of conglomerate particle, and also the number of particles in the form of the total mass of the conglomerate. Since the processes of collision, crushing and particle coagulation are neglected in this model, the only possibility of their interaction with each other is mutual influence through the changed by them gas-dynamic parameters of the carrier medium.

3. Mathematical model of the interaction of particles with the flow

Let's take into account only the mechanical interaction of particles with the gas flow. A mathematical model describing the gas flow perturbation due to the interaction of particles in a Laval nozzle in the two-dimensional axially symmetric approximation is a gas-dynamic system of equations with the additional terms describing the exchange of energy and impulse. This system in a conservative form is:

$$\frac{\partial \rho}{\partial t}+\frac{1}{r}\frac{\partial(r\rho v)}{\partial r}+\frac{\partial(\rho u)}{\partial z}=0$$

$$\frac{\partial(\rho u)}{\partial t}+\frac{1}{r}\frac{\partial(r\rho v u)}{\partial r}+\frac{\partial(\rho u^2)}{\partial z}=-\frac{\partial p}{\partial z}+\delta I_z \quad (1)$$

$$\frac{\partial(\rho v)}{\partial t}+\frac{1}{r}\frac{\partial(r\rho v^2)}{\partial r}+\frac{\partial(\rho u v)}{\partial z}=-\frac{\partial p}{\partial r}+\delta I_r \qquad (1)$$

$$\frac{\partial(\rho E)}{\partial t}+\frac{1}{r}\frac{\partial(r v(\rho E+p))}{\partial r}+\frac{\partial(u(\rho E+p))}{\partial z}=\delta E$$

where u and v – axial and radial components of velocity $\vec{w}$ respectively; δI_z, δI_r, δE - changing the impulse component and the gas flow energy due to the interaction with a condensed phase material; ρ - the total density of the carrier gas flow, which generally in the case when the flow is a mixture of N component, is given by

$$\rho = \sum_{i=1}^{N} \rho_i \tag{2}$$

To close the system (1) the equation for the pressure in a medium are used

$$p = \sum_i p_i , \quad p_i = p_i(\rho_i, \varepsilon_i) \tag{3}$$

where pressure of each components of the medium p_i is determined by the equation of state given in analytical form or in tabular form. In addition, if in the task the parameters of air at a given height h are required, then they are determined by the standard tables of the Earth's atmosphere

$$\varepsilon_a = \varepsilon_a(h), \qquad \rho_a = \rho_a(h) \tag{4}$$

To solve the system of equations (1) - (4) the large particles method was used [12]. The main idea of the method of large particles is splitting into the physical processes of the original non-stationary system of equations. At the first stage the system is solved under the assumption instantaneous retardation substances i.e. without convective terms and at the second stage the mass transfer and corresponding impulse and energy transfer are calculated.

Consider now a collection of particles that occupy at the moment a certain volume. The total number of particles that need to be considered in the problem, can be very large, therefore, it is advantageous to combine the individual particles which are placed in a small spatial region and which have close on the value parameters such as size, speed, etc. into a single conglomerate. Interaction of any conglomerate with the gas flow can be described by its single representative. It's necessary to calculate the interaction of the representative with the gas flow, and then get the change in the total energy, mass and momentum of the conglomerate by multiplying the corresponding changes in the values obtained for a single representative on the actual number of particles in the conglomerate.

Dynamics of a particle-representative having a spherical shape according to our assumption is described by a system of ordinary differential equations, including the equation of motion, momentum change equation due to hydrodynamic and exchange forces and energy change equation:

$$\begin{aligned} \frac{dr_i}{dt} &= \vec{w}_i \\ \frac{d(m_i \vec{w}_i)}{dt} &= m_i(\vec{F}_i + \vec{g}) \\ \frac{d(m_i E_i)}{dt} &= m_i \vec{F}_i \vec{w}_i \end{aligned} \tag{5}$$

where m_i – mass, $\vec{w}_i$ – velocity, E_i – full energy, $\vec{r}_i$ – the radius vector of the particle-representative, respectively.

The strength of the aerodynamic braking per unit mass of the i-th particle-representative is given by

$$\vec{F}_i = \frac{3\rho}{4\rho_i} \frac{C_{Di}}{r_i} \frac{|\vec{a}|\vec{a}}{2} \tag{6}$$

where $\vec{a} = \vec{w} - \vec{w}_i$, ρ_i – density of condensed substance of particle.

To determine the resistance coefficient of the particle moving in a gas stream, it is necessary to know the Reynolds number, which defined by its size

$$\mathrm{Re} = \rho_1 |\vec{a}| d_i \eta^{-1} \tag{7}$$

For a spherical solid particle in the region of small Reynolds numbers (Re = 10^{-4} – 1.0) the Stokes formula could be used

$$C_{Di} = 24\,\mathrm{Re}^{-1} \tag{8}$$

In a wide range of Reynolds numbers (Re_∞ = 10 – 10^3) the empirical formula could be used [3]

$$C_{Di} = 12/\sqrt{\mathrm{Re}} \tag{9}$$

This system of equations is solved locally at the point in space (computational cell) inside of which the considered particle-representative is placed. If at the same time at this cell there are several particle-representatives, there shall be a joint decision of several systems. Obtained change of mass, momentum and energy of representative particles which are calculated in the same cell are added, whereby they become sources of the corresponding values for the carrier gas environment. It is necessary to satisfy a number of conditions that limit of the integration time step for stability of calculations in accordingly with described algorithm.

At the diffuser part of the nozzle the particles intensely collide with the walls. Therefore, the Sand Bag model was modified to take into account of the adhesion processes and reflection particles to/from the nozzle walls. The sticking coefficient can vary from zero (no adhesion) to one (complete adhesion). When the reflection factor equal one the specular reflection occurs, while a decrease reflection occurs at a smaller angle and partial adhesion.

4. Testing of model

When the particles are so small that their size is comparable to the mean free path of the particles, it is necessary to take into account an Cunningham amendment or the Millikan resistance factor. The simplest expression of this amendment [13] is:

$$C_c = 1 + 1,275\frac{2\lambda}{d} \tag{10}$$

where λ - free path, d - particle diameter.

As a test of the model the calculations were conducted to determine the steady-state rate of subsidence in the air for a particle under the influence of gravity. Two variants of calculations were performed. At the first case the initially resting particles with diameter 0,1- 10 m are accelerated in a gravitational field till stationary velocity, while at the second case the initially moving particles slowed down to it. In both cases the times of getting settling rate were the same and coincide with the data of [13].

5. Results of simulations

Parametric calculations were performed for the same size of the particles, which ranged from 1 μm to 100 μm. The initial volume fraction of the particles coming to the nozzle input changed from 10^{-7} to 10^{-4}. Only the force interaction was studied. In some calculations the number of particles-representatives reached about 10^5. With the some volume fraction smaller particles have a significant influence on the flow of products of combustion in connection with considerably intensive exchange. Characteristic changes in the flow of combustion products are following: appearance of subsonic zone inside the diffuser of nozzle in close proximity its axis of symmetry; appearance of the field of reverse flow inside the convergent part of the nozzle in close proximity the place where the nozzle wall begins to narrow; total or partial removal of the sound line in the critical section of the nozzle at high volume fractions of small particles; increased volume fraction of particles in the axial region of the expanding part of the nozzle; oscillation of output nozzle parameters, including traction. The presence of particles leads to loss of traction to 5% in some calculations. Another feature of the flow in the nozzle is a sharp expansion of the gas flow in the expanding supersonic part of the nozzle. Particles-representatives disperse in this region to the extent that not every calculation cell there is at least one representative. This leads to spatial fluctuations of environmental parameters, caused by nonuniformity of particle-representatives distribution in the nozzle channel. Analysis of the simulation results revealed the need of algorithm for separation peripheral conglomerates into several parts when they move towards the axis of the nozzle.

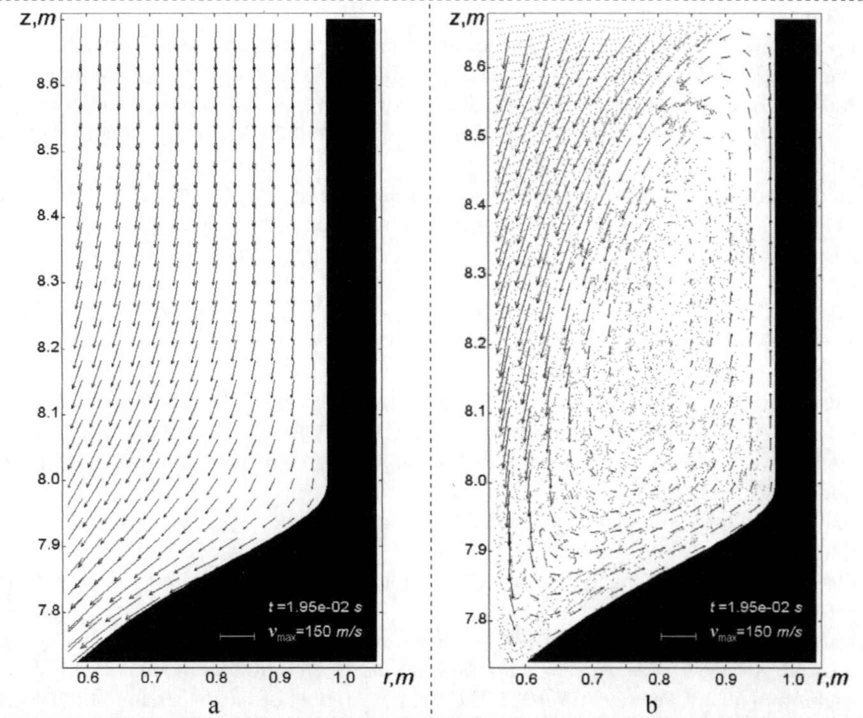

Figure 1. The velocity field in the cylindrical part of the nozzle: a – in the absence of the particles, b - particle diameter is 1 μm, the initial volume fraction is 10^{-5}.

The figure 1 illustrates the results of the effect of particle influence at the velocity field into the vicinity of the cylindrical part of the nozzle immediately prior to constriction. The dots indicate the position of the particle-representatives. The presence of particles increases the effective viscosity of the gas + particles substances whereby there is a swirling motion with the strong upflow near the nozzle wall. Inside of this vortex some of the particles move on loop-like path.

The distribution of representative particles is shown in Figure 2 at the moment of their release on the steady-state solution. Here clearly visible the region of formed vortex motion, in which the volume fraction of the particles is significantly lower than the input value. Also note that on the nozzle axis in its output the volume part of particles fractions is strongly increased, forming spatial accumulations of the condensed phase.

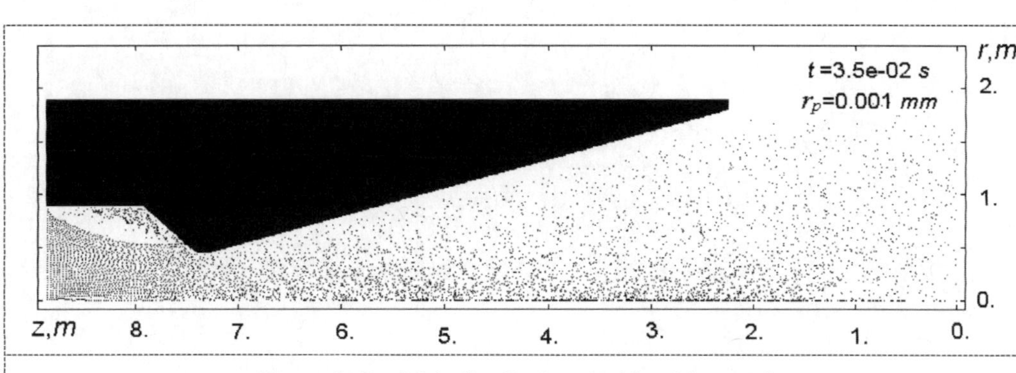

Figure 2. Particle's distributions inside of the nozzle.

6. Conclusion

The developed software complex allows performing calculations internal flows in channels of variable section in the 1D-3D approximations. There is the possibility of simulating multiphase flows and multi-medium. Physics-chemical properties of the medium may be given in tabular form directly or calculated during the simulation. The complex is equipped with a convenient GUI for easy input of the parameters of simulated tasks, and dynamic graphical visualization of the calculated gas-dynamic parameters.

Conducted parametric calculations have shown that the interaction of particles with the combustion products, even with a relatively small volume fraction may lead to a qualitative change in the internal flow in the Laval nozzle, and thereby could influence at it characteristics.

References

[1] Oran E S, Boris J P 2001 *Numerical simulation of reacting flow. Second Eition* (Cambridge: University Press) p 529
[2] Boothroyd R G 1971 *Flowing Gas-Solids Suspensions* (London: Chapman and Hall Ltd.)
[3] Saltanov G A 1972 *Supersonic two-phase flow* (Minsk: Vysheishaya shkola) p 479 (in Russian)
[4] Pirumov U G, Roslyakov G S 1986 *Gas flow in nozzles* (Berlin-Heidelberg-New York-Tokyo: Springer-Verlag) p 368
[5] Deych M Ye and Filippov G A 1981 *The gas dynamics of two-phase media* (Moscow: Energoizdat) p 472 (In Russian)
[6] Tsiklauri G V, Danilin V S and Selezlev L I 1973 *Adiabatic two-phase flow* (Moscow: Atomizdat) p 447 (In Russian)
[7] Stewart H B, Wedroff B 1984 Two-Phase Flows: Models and Methods *J. Comp. Phys.* **56** 363
[8] Nigmatulin R I 1990 *Dynamics of Multiphase Media* (New York: Hemisphere) p 532
[9] Nemtchinov I V and Teterev A V 1993 The sand bag model of the dispersion of the cosmic body in the atmosphere *Twenty-fourth Lunar and Planetary Science Conf.* (Houston, Texas) pp 1415-6
[10] Nemtchinov I V, Svetcov V V and Teterev A V 1995 Disintegration of large meteoroids in earth's atmosphere: theoretical models *Icarus* **116** 131-53
[11] Teterev A V 1999 Cratering model of Asteroid and Comet impact on a planetary surface *Int. J. Impact Eng.* **23** 921-2
[12] Belotserkovskii O M and Davydov Yu M 1973 Numerical Approach for Investigating Some Transonic Flow *Lect. Notes Phys.* (Springer-Verlag) **19** 25-32
[13] Reist P C 1984 *Introduction to aerosol science* (New York: Macmillan) p 299

The Painlevé test for nonlinear system of differential equations with complex chaotic behavior

V Tsegel'nik

Belarusian State University of Informatics and Radioelectronics, P. Brovki Str., 6, Minsk, 220013, Belarus

E-mail: tsegvv@bsuir.by

Abstract. The Painlevé-analysis was performed for solutions of nonlinear third-order autonomous system of differential equations with quadratic nonlinearities on their right-hand sides. At certain values of two constant parameters incorporated into the system, the latter exhibits complex chaotic behavior. When the parameters attain the values corresponding to complex chaotic behavior, the system was found not to possess the Painlevé property.

The system of differential equations

$$\dot{x} = z, \quad \dot{y} = -ay - xz, \quad \dot{z} = z - bz^2 + xy \tag{1}$$

with unknown functions x, y, z of the independent variable t and constant parameters a, b belongs [1, 2] to the class of chaotic systems

$$\dot{x} = y, \ \ \dot{y} = -x + yz, \ \ \dot{z} = -x - axy - bxz, \tag{LE_1}$$

$$\dot{x} = y, \ \ \dot{y} = -x + yz, \ \ \dot{z} = -y - axy - bxz, \tag{LE_2}$$

$$\dot{x} = y, \ \ \dot{y} = -x + yz, \ \ \dot{z} = x^2 - axy - bxz, \tag{LE_3}$$

$$\dot{x} = y, \ \ \dot{y} = -x + yz, \ \ \dot{z} = -axy - bxz - yz, \tag{LE_4}$$

$$\dot{x} = y, \ \ \dot{y} = -ax + yz, \ \ \dot{z} = -x^2 + y^2 - bxy, \tag{LE_5}$$

$$\dot{x} = y, \ \ \dot{y} = -x + yz, \ \ \dot{z} = ay^2 - xy - bxz, \tag{$LE6$}$$

$$\dot{x} = z, \ \ \dot{y} = x + yz, \ \ \dot{z} = ax^2 - xy - byz, \tag{LE_7}$$

$$\dot{x} = z, \ \ \dot{y} = x - yz, \ \ \dot{z} = -ax^2 + xy + bxz, \tag{LE_8}$$

with hidden attractors and line equilibrium. Such systems are important and potentially problematic in engineering applications, because they allow unexpected and potentially disastrous responses to perturbations in a structure like a bridge or an airplane wing. In case $a = 1.62$, $b = 0.2$ the system (1) possesses a hidden attractor. Considering, that t is a complex variable, find out, at what values a, b the general solution of the system (1) has no movable critical singular points, i.e., whether for (1) the so-called Painlevé property is carried out.

To solve this problem let us apply to the system (1) formal Painlevé test [3].

The following statements are true

Theorem 1 The system (1) passes the Painlevé test under parameters values $a = -1, b = 0$.

Theorem 2 The system (1) when $a = -1, b = 0$ is a system of Painlevé-type and does not possess chaotic behavior.

The validity of **Theorem 2** follows from the facts, that

1. The system (1) is equivalent to the equation

$$x\dddot{x} + (a-1)x\ddot{x} + 2bx\dot{x}\ddot{x} - \dot{x}\ddot{x} - b\dot{x}^3 + abx\dot{x}^2 + \dot{x}^2 + x^3\dot{x} - ax\dot{x} = 0 \tag{2}$$

2. The equation (2) when $a = -1, b = 0$ has the first integral

$$(\ddot{x} - \dot{x})^2 = C^2x^2e^{2t} - x^2\dot{x}^2, \tag{3}$$

where C is the arbitrary constant. The equation (3) possesses the Painlevé property, since its solutions are expressed through solutions of the third Painlevé equation under the private values of the parameters.

3. The unknown functions y, z, incorporated into the system (1), in a rational way are expressed through x, $\dot{x}$.

The statement of **Theorem 2** is consistent with the well-known hypothesis [4], according to which the execution for the system Painlevé property with a high degree of confidence is considered to be incompatible with the randomness of its behavior.

Theorem 3 The equation (2) under parameters values $a = 1.62$, $b = 0.2$ possesses chaotic behavior and does not pass the Painlevé test.

Theorem 4 The system (1) in case $a = 0$ does not possess chaotic behavior.

Indeed, when $a = 0$ the system (1) has autonomous first integral $\frac{x^2}{2} + y = C$, where C is the constant of integration. The presence of this integral allows us to reduce (1) to two-dimensional autonomous system. According to [5] solutions of two-dimensional autonomous systems cannot be chaotic.

Theorem 5 The equation

$$x\dddot{x} - 2x\ddot{x} - Ax^3\dot{x} - Bx\dot{x} + H(\dot{x}\ddot{x} - \dot{x}^2) = 0 \tag{4}$$

under parameters values

a) either $H = 0$;

b) or $H = -1$, $A = B = -1$

is the equation of Painlevé-type.

Indeed, when $H = 0$ the equation (4) has first integral

$$\ddot{x} = Ax^3 + Kx^2 - Bx, \tag{5}$$

where K is the constant of integration. Integration of the equation (5) is carried out in elliptic functions.

In case b) the equation (4) coincides with (2) when $a = -1$, $b = 0$.

References

[1] Jafari S and Sprott J C 2013 Simple chaotic flows with a line equilibrium *Chaos, solitons and fractals* **57** 79–84

[2] Jafari S and Sprott J C 2015 Erratum to: Simple chaotic flows with a line equilibrium *Chaos, solitons and fractals* **77** 341–42

[3] Ablowitz M J, Ramani A and Segur H 1980 A connection between nonlinear evolution equations and ordinary differential equations of P-type *J. Math. Phys.* **21** 715–21

[4] Goriely A 2001 *Integrability and nonintegrability of dynamical systems* (*Advanced series in nonlinear dynamics* vol 19) (Singapore: World Scientific Publishing Company)

[5] Moon F C 1987 *Chaotic vibrations* (New York: John Wiley and Sons)

Modelica-based TCP simulation

T R Velieva[1], E G Eferina[1], A V Korolkova[1], D S Kulyabov[1,2], L A Sevastianov[1,3]

[1]Department of Applied Probability and Informatics,
RUDN University (Peoples' Friendship University of Russia),
6 Miklukho-Maklaya str., Moscow, 117198, Russia
[2]Laboratory of Information Technologies,
Joint Institute for Nuclear Research,
6 Joliot-Curie, Dubna, Moscow region, 141980, Russia
[3]Bogoliubov Laboratory of Theoretical Physics,
Joint Institute for Nuclear Research,
6 Joliot-Curie, Dubna, Moscow region, 141980, Russia

E-mail: trvelieva@gmail.com, eg.eferina@gmail.com, akorolkova@sci.pfu.edu.ru, yamadharma@gmail.com, leonid.sevast@gmail.com

Abstract. For the study and verification of our mathematical model of telecommunication systems a discrete simulation model and a continuous analytical model were developed. However, for various reasons, these implementations are not entirely satisfactory. It is necessary to develop a more adequate simulation model, possibly using a different modeling paradigm. In order to modeling of the TCP source it is proposed to use a hybrid (continuous-discrete) approach. For computer implementation of the model the physical modeling language Modelica is used. The hybrid approach allows us to take into account the transitions between different states in the continuous model of the TCP protocol. The considered approach allowed to obtain a simple simulation model of TCP source. This model has great potential for expansion. It is possible to implement different types of TCP.

1. Introduction

While complex systems modeling there is the problem of choosing a model approach.When using a discrete-continuous dichotomy, we always have some inappropriate elements. These elements of the model do not well correspond to the selected approach. During simulation of TCP protocol the serious problems arised. As it turned out adequate TCP models are simply missing. Not even a common method for its modeling (see [1–3]) exists.

For modeling we used the continuous (fluid) model for TCP (see [4, 5]). However, this approach allowed us to model the TCP protocol only partially. In addition, we received differential inclusions instead of differential equations.

For further development of our model, it was decided to use a hybrid approach (see [6–10]). This article discusses a general approach to a hybrid modeling of TCP. The implementation is demonstrated on the basis of Modelica language (see [11, 12]).

The structure of the article is as follows. The section 2 describes the hybrid paradigm of mathematical modeling. Also the general information about the language of physical modeling named Modelica is given in the same section. Modelica implements continuous and hybrid paradigms. The congestion control mechanism in the TCP Reno protocol are presented in the section 3. In the section 4 we construct the UML-diagram for TCP discrete transitions. In the section 5 we record the discrete equations for TCP window in a continuous form. All sections are illustrated by the program fragments in the Modelica language.

2. The hybrid approach to modeling

The hybrid (see [6–10]) system has both continuous and discrete aspects of behavior. The hybrid behavior may be due to different reasons.

- Hybrid behavior is due to the joint operation of the continuous and discrete objects. For example, the automatic control system with continuous control object and discrete control device.
- Hybrid behavior is caused by changes in the structure of the system. A system with variable number of components may be considered as an example.
- Hybrid behavior may be caused by instant qualitative changes in a continuous object. In this case, qualitative changes during the simulation of continuous systems are presented as discrete events. Asesult, the hybridism is not an inherent characteristic of the system, but the modeling technique.

Hybrid systems may be considered as discrete-continuous or continuous-discrete systems.

- The waiting time for the next input and the duration of the output action can be taken into account in discrete systems.
- The coexistence of instant and long-term processes in a continuous-time model.

We will add discrete elements to the initially developed continuous dynamic model.

Modelica language (see [11,12]) is developed by a nonprofit organization Modelica, which also develops a free library for this language. Modelica is positioned as an object-oriented physical modeling language. The equations are a special entity in the language. The number of equations and variables in the program must be the same.

Modelica supports continuous and hybrid (continuous-discrete) paradigms. However, the discrete elements are also present in the language.

3. TCP congestion control mechanism

The TCP protocol uses a sliding window mechanism to avoid a congestion. The implementation of this mechanism depends on the particular type of TCP protocol.

Since the original model (see [13–15]) is based on the TCP Reno protocol [16,17], then this particular protocol will be simulated.

In TCP Reno protocol the congestion control mechanism consists of the following phases: slow start, congestion avoidance, fast recovery, and timeout. Dynamics of changes in congestion window size (CWND) depends on the specific phase (see Fig. 1).

If the total network arrival rate exceeds the maximum network rate, there is a network overload. If the overload condition lasts long enough for the buffer overflows, the data loss occurs in the network. Lost segments are retransmitted by TCP and promote continued growth of congestion. The transport protocol control algorithm should prevent the occurrence of congestion and facilitate the end of the network overload conditions in the case of its occurrence.

TCP uses a greedy algorithm for the use of bandwidth. At first it attempts to use all the available bandwidth. To do this, it increases the window exponentially (multiplicative increase). In the case of packet loss, the TCP first reduces the window size. Then TCP increases the window size linearly. TCP determines the occurrence of an overload condition expecting that the cause of packet loss is the network congestion.

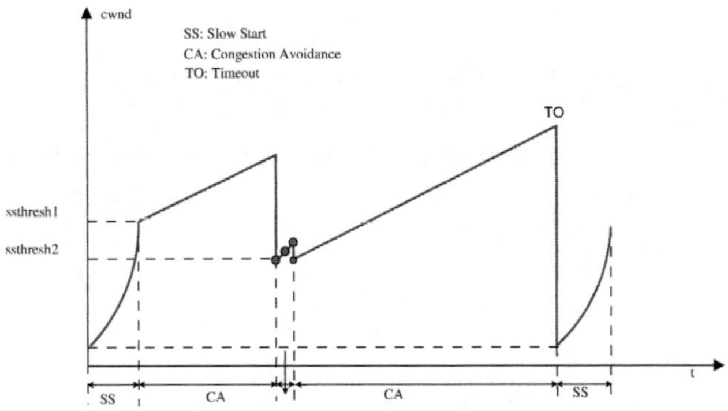

Figure 1. TCP phases

Overall congestion control algorithm belongs to AIMD algorithms type (Additive Increase, Multiplicative Decrease) — an additive increase of the window size and multiplicative decrease of it.

The congestion avoidance algorithm is described in RFC 5681 [17].

Each time the source receives a delivery notification (Acknowledge, ACK), in slow start phase congestion window is increased. The source increases the congestion window size depending on the number of confirmed segments (Segment Size, SS): $cwnd = cwnd + 1$ for each transmitted ACK. Initial congestion window size (Maximum Segment Size, MSS) can take a value of 1, 2 or 10 segments. The receiver sends an ACK for each packet, but in reality it can be assumed that confirmation come together at the end of the double turnaround time (Round-Trip Time, RTT). Thus, the congestion window is doubled after the round-trip time.

When TCP Reno window size takes the certain value the protocol mechanism enters the congestion avoidance phase. In this phase, the congestion window is increased by the amount of $1/cwnd$ for each acknowledgment ACK, which is equivalent to increasing of the window by one packet for the double-turn.

Protocol TCP Reno monitors two options of packet loss:

- Triple Duplicate ACK (TD). Let n-th package is not delivered, and subsequent packets ($n+1$, $n+2$, etc.) are delivered. For each packet delivered in violation of prioritization(for $n+1$, $n+2$, and so on), the recipient sends ACK message for the last undelivered (n-th) package. With receiving three such packets the source resends the n-th package. In addition, the window size is decreased by 2 times $cwnd \to cwnd/2$.
- Timeout (TO). While sending a package the timeout timer is started. After receiving the confirmation the timer is restarted. Wherein the window size is set to the initial value of the congestion window. The first lost package is resent. The protocol passes into a slow start phase.

The TCP protocol uses multiple timers. The main one is *Retransmission timer.* It restarts each time when a segment is sended. If an acknowledgment does't received before the timer triggering, then the segment is retransmitted.

The retransmission timer doesn't change in the slow start phase and in the congestion avoidance phase. It linearly decreases over time in the fast recovery phase and in the timeout phase.

Thus we can write the equations for the window evolution in TCP protocol in Modelica language (Listing 1).

Listing 1. Window evolution in TCP protocol

```
// Fast Recovery
when (pre(state) == TCPState.slowStart or pre(state) ==
   TCPState.congestAvoid) and state == TCPState.fastRecov then
  reinit(retr_timer, o.RTT);
  reinit(ssth, w / 2);
  reinit(w, w / 2);
end when;
// Timeout
when (pre(state) == TCPState.slowStart or pre(state) ==
   TCPState.congestAvoid) and state == TCPState.timeOut then
  reinit(retr_timer, RTO);
  reinit(ssth, w / 2);
  reinit(w, 1);
end when;
```

4. Modeling of discrete state transitions

Let's describe the TCP state transitions. Note that since we have already described the internal states of each phase as the equations in Modelica, we can't use the equations to describe the state transitions. Otherwise, the resulting system will be an overdetermined. For such cases, Modelica language provides the *algorithm* statement.

The slow start is the initial state. We use the `ssth` variable for transition from a slow start phase to a congestion avoidance phase. when a new connection is opened, this variable is initialized to the maximum possible size of the window. The slow start phase continues until the value of the `cwnd` reaches the `ssth`, then a transition occurs to the congestion avoidance phase.

When packet loss occurs the TCP goes either into a fast recovery phase, or into a timeout phase. We model this transition empirically, depending on the window threshold (`timeout_th`) [18].

Similarly, the TCP protocol can either go from the congestion avoidance phase into a fast recovery phase, or into a timeout phase.

After retransmission time the TCP protocol goes from a fast recover phase into a congestion avoidance phase, and from a timeout phase into a slow start phase.

Based on the description of the state transitions we can construct the UML-diagram (Fig. 2). The resulting chart can be converted to a Modelica program. We give a fragment of the listing (Listing 2). Here we demonstrate only the state transition algorithm for TCP. As can be seen, it is made by almost verbatim copying of UML-diagrams (it is theoretically possible to carry out the code generation based on the corresponding chart).

5. The transition to the continuous model

Because we want to construct the hybrid continuous-discrete model, we need pass to the continuous-time model in order to describe the operation of each TCP phase. The transition between the phases will be described by discrete states.

Using the results of section 3, the behavior of our model may be formalized. A congestion window change is described by an elementary event, which corresponds to a single acknowledgment or confirmation of all. Let us assume that the elementary event is the arrival of all acknowledgments that occurs during the round-trip time (RTT).

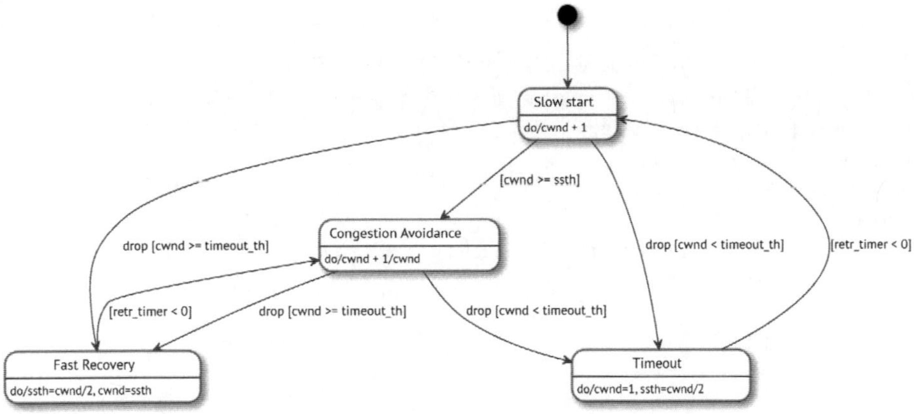

Figure 2. TCP state diagram

Listing 2. State transition algorithm for TCP protocol

```
algorithm
  state := TCPState.slowStart;
  when edge(drop_delay) and w >= timeout_th and (state ==
     TCPState.slowStart or state == TCPState.congestAvoid) then
    state := TCPState.fastRecov;
  elsewhen w >= ssth and state == TCPState.slowStart then
    state := TCPState.congestAvoid;
  elsewhen edge(drop_delay) and w < timeout_th and (state ==
     TCPState.slowStart or state == TCPState.congestAvoid) then
    state := TCPState.timeOut;
  elsewhen retr_timer < 0 and state == TCPState.fastRecov then
    state := TCPState.congestAvoid;
  elsewhen retr_timer < 0 and state == TCPState.timeOut then
    state := TCPState.slowStart;
  end when;
```

In the slow start phase a congestion window size increases with each occurrence of confirmation (ACK):

$$W(t_n^{ACK} + \Delta t^{ACK}) = W(t_n^{ACK}) + 1. \tag{1}$$

We now rewrite (1) relative to the round-trip time T:

$$W(t_n + \Delta t) = W(t_n) + 1 \cdot W(t_n),$$
$$\frac{W(t_n + \Delta t) - W(t_n)}{\Delta t} = \frac{W(t_n)}{\Delta t}.$$

Assuming that $\Delta t = T$, we obtain

$$\frac{\mathrm{d}W}{\mathrm{d}t} = \frac{W}{T},$$
$$\mathrm{d}\ln W = \frac{\mathrm{d}t}{T}, \quad \ln W = \frac{1}{T}; \quad W = \exp\left\{\frac{1}{T}\right\}.$$

Listing 3. Equations for TCP protocol

```
if state == TCPState.congestAvoid then
  // additive increase
  der(w) = 1 / o.RTT;
  o.rate = w * L / o.RTT;
  der(retr_timer) = 0;
elseif state == TCPState.fastRecov then
  der(w) = 0;
  o.rate = w * L / o.RTT;
  der(retr_timer) = -1;
elseif state == TCPState.timeOut then
  der(w) = 0;
  o.rate = 0.001;
  der(retr_timer) = -1;
else
  // state == slowStart
  // exponential increase
  der(w) = w / o.RTT;
  o.rate = w * L / o.RTT;
  der(retr_timer) = 0;
end if;
```

Thus, the window grows exponentially, as it should be in a slow start phase in accordance with the TCP description.

Similarly, we examine congestion avoidance phase. For each occurrence of an ACK the window size is increased:

$$W(t_n^{ACK} + \Delta t^{ACK}) = W(t_n^{ACK}) + \frac{1}{W(t_n^{ACK})}.$$

We rewrite this for a round-trip time:

$$W(t_n + \Delta t) = W(t_n) + \frac{1}{W(t_n)} W(t_n);$$

$$\frac{W(t_n + \Delta t) - W(t_n)}{\Delta t} = \frac{1}{\Delta t}.$$

Assuming that $\Delta t = T$, we have

$$\frac{\mathrm{d}W}{\mathrm{d}t} = \frac{1}{T},$$

$$\mathrm{d}W = \frac{\mathrm{d}t}{T}, \quad W = \frac{1}{T}.$$

The result is a linear increase of the window, as described in the specification of the TCP.

Using these equations, we can write the equations for the TCP protocol in Modelica language (Listing 3).

6. Conclusions

We investigated the TCP protocol as part of a larger system. Some mathematical models (both analytical and simulation) of this mechanism using different paradigms and techniques were presented. On closer inspection, the presented modeling techniques have shown their shortcomings.

The considered in the article hybrid (continuous-discrete) approach seems to be the most appropriate for network protocol modeling.

Acknowledgments

The work is partially supported by RFBR grants No's 14-01-00628, 15-07-08795, and 16-07-00556. Also the publication was supported by the Ministry of Education and Science of the Russian Federation (the Agreement No 02.a03.21.0008).

References

[1] Paxson V and Floyd S 1997 *Proc. of the 29th conference on Winter simulation - WSC '97* (New York, New York, USA: ACM Press) pp 1037–1044
[2] Paxson V and Floyd S 1995 *IEEE/ACM Transactions on Networking* **3** 226–244
[3] Leland W E, Taqqu M S, Willinger W and Wilson D V 1994 *IEEE/ACM Transactions on Networking* **2** 1–15
[4] Demidova A V, Korolkova A V, Kulyabov D S and Sevastyanov L A 2015 *6th Int. Congress on Ultra Modern Telecommunications and Control Systems and Workshops (ICUMT)* (IEEE Computer Society) pp 557–562 (*Preprint* 1504.00576)
[5] Eferina E G, Korolkova A V, Gevorkyan M N, Kulyabov D S and Sevastyanov L A 2014 *Bulletin of Peoples' Friendship University of Russia. Series "Mathematics. Information Sciences. Physics"* 46–59 (*Preprint* 1503.07342)
[6] Maler O 1992 *Workshop on Theory of Hybrid Systems* (Lyndby, Denmark: Springer-Verlag)
[7] Maler O 2002 *Annual Reviews in Control* **26** 175–187
[8] Färnqvist D, Strandemar K, Johansson K H and Hespanha J P 2002 *The 2nd Int. Modelica Conf.* pp 209–213
[9] Hespanha J P, Bohacek S, Obraczka K and Lee J 2001 *Lncs* 2034 pp 291–304
[10] Bohacek S and Lee J 2001 *Proc. of the 39th Annual Allerton Conf. on Communication, Control, and Computing* pp 1–10
[11] Fritzson P 2003 *Principles of Object-Oriented Modeling and Simulation with Modelica 2.1* (Wiley-IEEE Press)
[12] Fritzson P 2011 *Introduction to Modeling and Simulation of Technical and Physical Systems with Modelica* (Hoboken, NJ, USA: John Wiley & Sons, Inc.)
[13] Misra V, Gong W B and Towsley D 1999 *Proc. of PERFORMANCE* **99**
[14] Misra V, Gong W B and Towsley D 2000 *ACM SIGCOMM Computer Communication Review* **30** 151–160
[15] Velieva T R, Korolkova A V and Kulyabov D S 2015 *6th Int. Congress on Ultra Modern Telecommunications and Control Systems and Workshops (ICUMT)* (IEEE Computer Society) pp 570–577 (*Preprint* 1504.02324)
[16] Fall K and Floyd S 1996 *ACM SIGCOMM Computer Communication Review* **26** 5–21
[17] Allman M, Paxson V and Blanton E 2009 TCP Congestion Control Tech. rep.
[18] Padhye J, Firoiu V, Towsley D and Kurose J 1998 *ACM SIGCOMM Computer Communication Review* **28** 303–314

Exact solutions to nonlinear delay differential equations of hyperbolic type

Andrei V Vyazmin[1] **and Vsevolod G Sorokin**[2,3]

[1] Moscow State University of Mechanical Engineering, 38 Bolshaya Semenovskaya Street, 107023 Moscow, Russia

[2] Institute for Problems in Mechanics, Russian Academy of Sciences, 101 Vernadsky Avenue, bldg 1, 119526 Moscow, Russia

[3] Bauman Moscow State Technical University, 5 Second Baumanskaya Street, 105005 Moscow, Russia

E-mail: av1958@list.ru, vsesor@gmail.com

Abstract. We consider nonlinear delay differential equations of hyperbolic type, including equations with varying transfer coefficients and varying delays. The equations contain one or two arbitrary functions of a single argument. We present new classes of exact generalized and functional separable solutions. All the solutions involve free parameters and can be suitable for solving certain model problems as well as testing numerical and approximate analytical methods for similar and more complex nonlinear differential-difference equations.

Keywords: delay reaction-diffusion equations, hyperbolic delay equations, differential-difference equations, exact solutions

1. Introduction

The classical parabolic heat-conduction and diffusion equation has a phisically paradoxical property, namely, an infinite disturbance propagation rate, which is not observed in nature. This problem does not appear in the heat-conduction and diffusion equation of hyperbolic type [1, 2]

$$\tau T_{tt} + T_t = a\Delta T,$$

which have a finite disturbance propagation rate at $\tau > 0$. Here T is temperature, t is time, a is thermal diffusivity, Δ is the Laplace operator, and τ is a relaxation (delay) time. The thermal and diffusion relaxation times can vary in extremely wide limits from milliseconds (or less) to several tens of seconds [3–5] and should be taken into account while solving numerous heat- and mass-transfer problems.

The second important feature of evolutionary processes is that the rate of change of the unknowns in chemical, biological, biomedical, ecological and other systems generally depends not only on the current state of the system but also on its entire previous evolution [3, 6]. These systems are called *hereditary systems*. In particular, when the state is determined by a specific moment in the past, rather than the entire evolution, the system is referred to as *a delayed feedback system.*

Delayed feedback systems are frequently modeled by delay reaction-diffusion equations with the kinetic function F (the rate of a chemical or biochemical reaction) that depends on both $u = u(x,t)$ and its delayed counterpart $w = u(x, t - \tau)$. The special case when $F(u,w) = f(w)$ implies a simple physical interpretation: a heat- and mass-transfer process in local non-equilibrium media posesses inertial properties, i.e., the system responds to an impact not immediately at the time t, as opposed to the classical local-equilibrium case, but the delay time τ later.

The paper deals with the nonlinear delay reaction-diffusion equations of hyperbolic type

$$\varepsilon u_{tt} + \sigma u_t = a u_{xx} + F(u, w), \qquad w = u(x, t - \tau), \tag{1}$$

where $a > 0$, $\varepsilon \geq 0$, $\sigma \geq 0$ $(\varepsilon + \sigma \neq 0)$. We will present several exact solutions to the equations of the form (1), which have been obtained by applying the methods of generalized and functional separation of variables [7, 8] as well as the functional constraints method [9, 10]. All the solutions involve free parameters and can be suitable for solving certain model problems as well as testing numerical and approximate analytical methods for similar and more complex nonlinear differential-difference equations.

2. The concept of 'exact solution'

The term 'exact solution' (see [9]) with regard to nonlinear delay partial differential equations is used in cases when the solution is expressible in terms of:

(i) elementary functions or definite/indefinite integrals;

(ii) solutions of ordinary differential equations with or without delay (or systems of these equations);

(iii) solutions of linear partial differential equations.

3. The functional constraints method. Some examples

Now we are going to briefly describe the functional constraints method from the article [9]. For definiteness, we consider the nonlinear delay reaction-diffusion equations of hyperbolic type (1) with the kinetic function that depends on one argument:

$$\varepsilon u_{tt} + \sigma u_t = a u_{xx} + F(z), \quad z = z(u, w), \quad w = u(x, t - \tau), \tag{2}$$

where $z = z(u, w)$ is a function to be determined.

We search for generalized separable solutions of the form

$$u = \sum_{n=1}^{N} \varphi_n(x) \psi_n(t). \tag{3}$$

The simplest cases are a multiplicative, $u = \varphi(x)\psi(t)$, and an additive, $u = \varphi(x) + \psi(t)$, separable solution.

The form of the argument $z = z(u, w)$ of the arbitrary function $F(z)$ is determined by the method of functional constraints. It is based on searching for exact solutions of the form (3) that satisfy one of two functional constraints:

$$z(u, w) = p(x), \qquad w = u(x, t - \tau); \tag{4}$$

$$z(u, w) = q(t), \qquad w = u(x, t - \tau). \tag{5}$$

The constraints represent difference equations in t with x playing the role of a free parameter.

Below we give some examples of constructing exact solutions to the equations of the form (2) by use of the functional constraints method.

We consider the equation

$$\varepsilon u_{tt} + \sigma u_t = a u_{xx} + bu + f(u - w), \quad w = u(x, t - \tau), \tag{6}$$

where b is an arbitrary constant, f is an arbitrary function.

The functional constraint of the first kind (4) has the form

$$u - w = p(x), \quad w = u(x, t - \tau). \tag{7}$$

The difference equation (7) can be satisfied with the generalized separable solution

$$u = t\varphi(x) + \psi(x), \tag{8}$$

which leads to $p(x) = \tau\varphi(x)$. Substituting (8) into (6) and separating the variables, we get ordinary differential equations for $\varphi(x)$ and $\psi(x)$:

$$\begin{aligned} &a\varphi''(x) + b\varphi(x) = 0, \\ &a\psi''_{xx}(x) + b\psi(x) + f(\tau\varphi(x)) - \sigma\varphi(x) = 0. \end{aligned} \tag{9}$$

The general solution of equation (9) has the form

$$\varphi(x) = \begin{cases} C_1\cos(\lambda x) + C_2\sin(\lambda x), & \lambda = \sqrt{b/a} \quad b > 0; \\ C_1\exp(-\lambda x) + C_2\exp(\lambda x), & \lambda = \sqrt{-b/a} \quad b < 0, \end{cases}$$

where C_1 and C_2 are arbitrary constants.

The functional constraint of the second kind (5) has the form

$$u - w = q(t), \quad w = u(x, t - \tau). \tag{10}$$

The difference equation (10) can be satisfied with the additive separable solution

$$u = \varphi(x) + \psi(t), \tag{11}$$

which leads to $q(t) = \psi(t) - \psi(t - \tau)$. Substituting (11) into (6), we get an equation for $\varphi(x)$ that coincides with equation (9) and an equation for $\psi(t)$:

$$\varepsilon\psi''(t) + \sigma\psi'(t) = b\psi(t) + f(\psi(t) - \psi(t - \tau)).$$

4. Exact solutions

4.1. Equations of the form (1)

In Table 1 we present some exact solutions obtained by use of the method described above. Details about these solutions are given below. Other solutions to the equations of the form (1) and more complex equations can be found in [11, 12]. Functions f, g, h are arbitrary.

1–2. Here C_1, C_2, λ are arbitrary constants and the function $\psi(t)$ is described by the ordinary delay differential equation

$$\varepsilon\psi''(t) + \sigma\psi'(t) = -a\lambda^2\psi(t) + \psi(t)f\big(\psi(t - \tau)/\psi(t)\big).$$

3. Here $\alpha, \beta, \lambda, \gamma$ are arbitrary constants and the function $\theta(z)$ satisfies the ordinary delay differential equation

$$\begin{aligned} &(a\lambda^2 - \varepsilon\gamma^2)\theta''(z) + (2a\alpha\lambda - 2\varepsilon\beta\gamma - \sigma\gamma)\theta'(z) + \\ &\qquad + (a\alpha^2 - \varepsilon\beta^2 - \sigma\beta)\theta(z) + \theta(z)f\big(e^{-\beta\tau}\theta(z - \delta)/\theta(z)\big) = 0, \qquad \delta = \gamma\tau. \end{aligned}$$

4–5. Here b, β, γ are arbitrary constants, $\lambda = \sqrt{|b|/a}$ and the function $\theta(z)$ is determined by the ordinary delay differential equation

$$(\varepsilon\gamma^2 - a\beta^2)\theta''(z) + \sigma\gamma\theta'(z) = b\theta(z) + f(\theta(z) - \theta(z - \delta)), \qquad \delta = \gamma\tau.$$

Table 1. Exact solutions of the equations $\varepsilon u_{tt} + \sigma u_t = a u_{xx} + F(u, w)$

N	The kinetic function $F(u,w)$	The form of the solution $u(x,t)$
1	$uf(w/u)$	$(C_1\cos(\lambda x) + C_2\sin(\lambda x))\psi(t)$
2	$uf(w/u)$	$(C_1 e^{-\lambda x} + C_2 e^{\lambda x})\psi(t)$
3	$uf(w/u)$	$e^{\alpha x+\beta t}\theta(z),\ z = \lambda x + \gamma t$
4	$bu + f(u-w),\ b>0$	$C_1\cos(\lambda x) + C_2\sin(\lambda x) + \theta(z),\quad z = \beta x + \gamma t$
5	$bu + f(u-w),\ b<0$	$C_1 e^{-\lambda x} + C_2 e^{\lambda x} + \theta(z),\quad z = \beta x + \gamma t$
6	$f(u-w)$	$C_1 x^2 + C_2 x + \theta(z),\quad z = \beta x + \gamma t$
7	$uf(u-kw)+$ $+wg(u-kw) + h(u-kw)$	$e^{ct}\sum_{n=1}^{N}\big(\varphi_n(x)\cos(\beta_n t) + \psi_n(x)\sin(\beta_n t)\big) + e^{ct}\theta(x) + \xi(x)$
8	$uf(u+kw)+$ $+wg(u+kw) + h(u+kw)$	$e^{ct}\sum_{n=1}^{N}\big(\varphi_n(x)\cos(\beta_n t) + \psi_n(x)\sin(\beta_n t)\big) + \xi(x)$

6. Here C_1, C_2, β, γ are arbitrary constants and the function $\theta(z)$ satisfies the ordinary delay differential equation

$$(\varepsilon\gamma^2 - a\beta^2)\theta''(z) + \sigma\gamma\theta'(z) = 2C_1 a + f(\theta(z) - \theta(z-\delta)), \qquad \delta = \gamma\tau.$$

7. Here $k > 0$ is an arbitrary constant, N is a positive integer, $c = (\ln k)/\tau,\ \beta_n = 2\pi n/\tau$, the functions $\varphi_n(x)$ and $\psi_n(x)$ are described by the system

$$a\varphi_n'' + \Big[f(\eta) + \frac{1}{k}g(\eta) + \varepsilon(\beta_n^2 - c^2) - \sigma c\Big]\varphi_n - (2\varepsilon c + \sigma)\beta_n\psi_n = 0,$$
$$a\psi_n'' + \Big[f(\eta) + \frac{1}{k}g(\eta) + \varepsilon(\beta_n^2 - c^2) - \sigma c\Big]\psi_n + (2\varepsilon c + \sigma)\beta_n\varphi_n = 0, \quad \eta = (1-k)\xi,$$

the function $\theta(x)$ is determined by the ordinary differential equation

$$a\theta'' + \Big[f(\eta) + \frac{1}{k}g(\eta) - \varepsilon c^2 - \sigma c\Big]\theta = 0,$$

and the function $\xi(x)$ satisfies the independent equation

$$a\xi'' + [f(\eta) + g(\eta)]\xi + h(\eta) = 0, \quad \eta = (1-k)\xi.$$

8. Here $k > 0$ is an arbitrary constant, N is a positive integer, $c = (\ln k)/\tau,\ \beta_n = (2n-1)\pi/\tau$, the functions $\varphi_n(x)$ and $\psi_n(x)$ are described by the system

$$a\varphi_n'' + \Big[f(\eta) - \frac{1}{k}g(\eta) + \varepsilon(\beta_n^2 - c^2) - \sigma c\Big]\varphi_n - (2\varepsilon c + \sigma)\beta_n\psi_n = 0,$$
$$a\psi_n'' + \Big[f(\eta) - \frac{1}{k}g(\eta) + \varepsilon(\beta_n^2 - c^2) - \sigma c\Big]\psi_n + (2\varepsilon c + \sigma)\beta_n\varphi_n = 0, \quad \eta = (1+k)\xi,$$

and the function $\xi(x)$ is determined by the independent equation

$$a\xi'' + [f(\eta) + g(\eta)]\xi + h(\eta) = 0, \quad \eta = (1-k)\xi.$$

Remark 1. Equations of the form (1) with varying delay times $\tau = \tau(t)$ admit the solutions from lines 1, 2. A few other solutions for these equations are presented in [11].

Table 2. Exact solutions of the equations $\varepsilon u_{tt} + \sigma u_t = a\big(G(u)u_x\big)_x + F(u, w)$

N	The transfer coefficient $G(u)$	The kinetic function $F(u, w)$	The form of the solution $u(x, t)$
1	u^n	$bu^{n+1} + uf(w/u)$, $b(n+1) > 0$	$\big(C_1 \cos(\lambda x) + C_2 \sin(\lambda x)\big)^{1/(n+1)}\psi(t)$
2	u^n	$bu^{n+1} + uf(w/u)$, $b(n+1) < 0$	$(C_1 e^{-\lambda x} + C_2 e^{\lambda x})^{1/(n+1)}\psi(t)$
3	u^n	$uf(w/u) + u^{n+1}h(w/u)$	$e^{\lambda t}\varphi(x)$
4	u^{-1}	$b + uf(w/u)$	$C_1 \exp\Big(-\frac{b}{2a}x^2 + C_2 x\Big)\psi(t)$
5	$u^{-1/2}$	$f(u^{1/2} - w^{1/2}) +$ $+ u^{1/2}h(u^{1/2} - w^{1/2})$	$\big(t\varphi(x) + \psi(x)\big)^2$
6	$e^{\beta u}$	$f(u - w)$	$\frac{1}{\beta}\ln(C_1 x^2 + C_2 x + C_3) + \psi(t)$
7	$e^{\beta u}$	$be^{\beta u} + f(u - w)$, $b\beta > 0$	$\frac{1}{\beta}\ln\big(C_1 + C_2\cos(\lambda x) + C_3 \sin(\lambda x)\big) + \psi(t)$
8	$e^{\beta u}$	$be^{\beta u} + f(u - w)$, $b\beta < 0$	$\frac{1}{\beta}\ln\big(C_1 + C_2\exp(-\lambda x) + C_3 \exp(\lambda x)\big) + \psi(t)$

4.2. Equations with varying transfer coefficients

In Table 2 we present some exact solutions to nonlinear delay hyperbolic reaction-diffusion equations with varying transfer coefficients of the form

$$\varepsilon u_{tt} + \sigma u_t = a\big(G(u)u_x\big)_x + F(u, w). \tag{12}$$

The solutions have been obtained by use of the method briefly described above (for more explanation see [10]). Details about these solutions are given below. Other solutions to the equations of the form (12) can be found in [11, 12]. Functions f and h are arbitrary.

1–2. Here C_1, C_2, b, n are arbitrary constants, $\lambda = \sqrt{|b(n+1)|/a}$, and the function $\psi(t)$ is determined by the ordinary delay differential equation

$$\varepsilon\psi''(t) + \sigma\psi'(t) = \psi(t)f(\psi(t-\tau)/\psi(t)). \tag{13}$$

Equation (13) yields a particular solution $\psi(t) = Ae^{\beta t}$, where A is an arbitrary constant and β is a solution of the algebraic (transcendental) equation

$$\varepsilon\beta^2 + \sigma\beta - f(e^{-\beta\tau}) = 0.$$

3. Here n is an arbitrary constant, λ is a solution of the algebraic (transcendental) equation

$$\varepsilon\lambda^2 + \sigma\lambda = f(e^{-\lambda\tau}),$$

and the function $\varphi(x)$ is described by the ordinary differential equation

$$a(\varphi^n\varphi'_x)'_x + \varphi^{n+1}h(e^{-\lambda\tau}) = 0.$$

At $n \neq 1$, a substitution $\theta = \varphi^{n+1}$ leads to a linear ordinary differential equation of the second order; at $n = -1$, one should make a substitution $\theta = \ln\varphi$.

4. Here C_1, C_2, b are arbitrary constants and the function $\psi(t)$ is determined by equation (13).

5. Here the functions $\varphi(x)$ and $\psi(t)$ satisfy the ordinary differential equations

$$2a\varphi''_{xx} + \varphi h(\tau\varphi) - 2\sigma\varphi^2 = 0,$$
$$2a\psi''_{xx} + \psi h(\tau\varphi) - 2\sigma\varphi\psi - 2\varepsilon\varphi^2 + f(\tau\varphi) = 0.$$

A particular solution of this system is

$$\varphi = A, \quad \psi = \frac{2\varepsilon A^2 - f(A\tau)}{4a}x^2 + C_1 x + C_2,$$

where C_1, C_2 are arbitrary constants and the constant A is describes by the algebraic (transcendental) equation

$$h(A\tau) - 2\sigma A = 0.$$

6. Here C_1, C_2, C_3, β are arbitrary constants and the function $\psi(t)$ satisfies by the ordinary delay differential equation

$$\varepsilon\psi''(t) + \sigma\psi'(t) = \frac{2aC_1}{\beta}e^{\beta\psi(t)} + f(\psi(t) - \psi(t-\tau)).$$

7–8. Here b, β, C_1, C_2, C_3 are arbitrary constants, $\lambda = \sqrt{|b\beta|/a}$, and the function $\psi(t)$ is determined by the ordinary delay differential equation

$$\varepsilon\psi''(t) + \sigma\psi'(t) = bC_1\beta e^{\beta\psi(t)} + f(\psi(t) - \psi(t-\tau)).$$

Acknowledgments
The authors thank A. D. Polyanin for his attention to the work and for very helpful comments. The work is partically supported by the Russian Foundation for Basic Research (project No. 16-08-01252).

References

[1] Baumeister K J and Hamill T D 1969 *J. Heat Transfer* **91** 542-8
[2] Taitel Y 1972 *Int. J. Heat Mass Transfer* **15** 369-71
[3] Demirel Y 2007 *Nonequilibrium Thermodynamics: Transport and Rate Processes in Physical, Chemical and Biological Systems* 2nd ed (Amsterdam: Elsevier)
[4] Roetzel W, Putra N and Das S K 2003 *Int. J. Therm. Sci* **42** 541-52
[5] Polyanin A D and Vyazmin A V 2013 *Theor. Found. Chem. Eng.* **47** 217-24
[6] Jou D, Casas-Vázquez J and Lebon G 2010 *Extended Irreversible Thermodynamics* 4th ed (New York: Springer)
[7] Galaktionov V A and Svirshchevskii S R 2006 *Exact Solutions and Invariant Subspaces of Nonlinear Partial Differential Equations in Mechanics and Physics* (Boca Raton: Chapman & Hall/CRC Press)
[8] Polyanin A D and Zaitsev V F 2012 *Handbook of Nonlinear Partial Differential Equations* 2nd ed (Boca Raton: Chapman & Hall/CRC Press)
[9] Polyanin A D and Zhurov A I 2014 *Commun. Nonlinear. Sci. Numer. Simul.* **19** 417-30
[10] Polyanin A D and Zhurov A I 2014 *Int. J. Non-Linear Mech.* **67** 267-77
[11] Polyanin A D and Sorokin V G 2014 *Vestnik Natsional'nogo Issledovatel'skogo Yadernogo Universiteta MIFI* 141-48
[12] Polyanin A D, Sorokin V G and Vyazmin A V 2015 *Theor. Found. Chem. Eng.* **49** 622-35

Analytical solution of the second Stokes problem with variable amplitude on behavior of gas over an oscillation surface

E A Bedrikova and A V Latyshev

Moscow State Regional University, 105005, Moscow, Radio str., 10-A

E-mail: bedrikova@mail.ru, avlatyshev@mail.ru

Abstract. The analytical solution of the second Stokes problem is found. The case of the variable amplitude of fluctuation of a surface is considered. The linear kinetic equation with boundary conditions is derived. Eigen solutions of the equation are found. Properties of dispersion function are investigated. The general solution of the kinetic equation with boundary conditions in terms of the eigen solutions decomposition is composed.

In the last time the second Stokes problem has a great interest of scientists. It is connected to development of technologies and in particular nanotechnologies. There are many articles devoted to the second Stokes problem [1-5], where the numerical solution is given [3-5]. The problem has the analytical solution is shown in the present work. In this article we continue works by V A Akimova, A V Latyshev and A A Yushkanov, in which the second Stokes problem has solution in case of the constant amplitude of fluctuation of a surface [1]. In this paper we consider the variable amplitude of fluctuation of a surface.

Let one-atomic rarefied gas fills half-space $x>0$. Surface makes harmonical oscillations lengthwise an axis Oy under the law $u_s(t)=u_0e^{-i\omega t}$, where $\omega=\omega_1+i\omega_2$. Here ω_1 is the oscillation frequency of plates and $A(t)=u_0e^{\omega_2 t}$ is the variable amplitude of fluctuation of a surface.

We will consider a model kinetic BGK equation (Bhatnagar, Gross, Krook)

$$\frac{\partial f}{\partial t}+\mathrm{v}_x\frac{\partial f}{\partial x}=\frac{f_{eq}-f}{\tau},$$

where $\nu = 1/\tau$ is the collision frequency of gaseous molecules, τ is the time between two consecutive collisions of molecules, f_{eq} is the equilibrium distribution function,

$$f_{eq}=n\left(\frac{m}{2\pi kT}\right)^{\frac{3}{2}}\exp\left[-\frac{m(\mathbf{v}-\mathbf{u})^2}{2kT}\right].$$

Here $u_y(t,x)$ is the mass velocity of gas, k is the Boltzmann constant, m is the mass of molecule of gas, n is the concentration of gas, T is the temperature of gas.

Considering, that the velocity of molecule of gas much less than the thermal velocity of molecule $|u_y(t_1,x_1)|<<\upsilon_T$, where $\upsilon_T=1/\sqrt{\beta}$ is the thermal velocity of molecule of gas, then problem can be linearized. Concentration of gas and temperature are considered as constants. We will search the distribution function in the form

$$f = f_M(\mathrm{v})\bigl(1+\varphi(t,x,\mathbf{v})\bigr).$$

$f_M(\mathrm{v})$ is the absolute Maxwellian

$$f_M(\mathrm{v}) = n\left(\frac{m}{2\pi kT}\right)^{\frac{3}{2}}\exp\left[-\frac{m\mathrm{v}^2}{2kT}\right],$$

We introduce dimensionless parametres: dimensionless velocity of molecules $\mathbf{C}=\sqrt{\beta}\mathbf{v}=\dfrac{\mathbf{v}}{\upsilon_T}$, $\beta=\dfrac{m}{2kT}$, dimensionless mass velocity $U_y(t,x)=\sqrt{\beta}u_y(t,x)$, dimensionless coordinate $x_1 = x\nu\sqrt{\beta}$, dimensionless time $t_1 = t\nu$.

Making necessary calculations, we obtain the kinetic equation

$$\frac{\partial\varphi}{\partial t_1}+C_x\frac{\partial\varphi}{\partial x_1}+\varphi(t,x_1,C_x)=\frac{2C_y}{\pi^{3/2}}\int\exp(-\mathbf{C}'^2)C'_y\,\varphi(t_1,x_1,\mathbf{C}')d^3\mathbf{C}', \qquad (1)$$

with the boundary conditions

$$\varphi(t_1,0,C_x)=2qC_yU_s(t_1)+(1-q)\varphi(t_1,0,-C_x,C_y,C_z),\quad C_x>0, \qquad (2)$$

and $\qquad \varphi(t_1,x_1\to\infty,\mathbf{C})=0.$

Considering, that plate oscillations are considered along an axis y, we will search function $\varphi(t,x_1,\mathbf{C})$ in the form

$$\varphi(t,x_1,\mathbf{C})=C_ye^{-i\Omega t_1}h(x_1,C_x).$$

Thus, the equation (1), can be rewritten as

$$C_x\frac{\partial h}{\partial x_1}+(1-i\Omega)h(x_1,C_x)=\frac{1}{\sqrt{\pi}}\int\limits_{-\infty}^{\infty}\exp(-C'^2_x)h(x_1,C'_x)dC'_x, \qquad (3)$$

and boundary conditions (2) can be written as

$$h(0,C_x)=2qU_0+(1-q)h(0,-C_x),\quad C_x>0,$$

and $\qquad h(x_1\to\infty,C_x)=0.$

We will seek the solution to equation (3) in the form

$$h_\eta(x_1,\mu)=\exp\left(-\frac{x_1z_0}{\eta}\right)\Phi(\eta,\mu),\qquad \frac{1}{z_0}\int\limits_{-\infty}^{\infty}\exp\left(-\mu'^2\right)\Phi(\eta,\mu')d\mu'=1. \qquad (4)$$

Here η is the spectral parameter or the separation parameter, it is complex one. Using these equalities (4), we obtain from equation (3) the characteristic equation

$$(\eta-\mu)\Phi(\eta,\mu)=\frac{\eta}{\sqrt{\pi}},\quad \eta\in C.$$

Then, we find the eigen functions of the characteristic equation

$$\Phi(\eta,\mu)=\frac{1}{\sqrt{\pi}}\eta P\frac{1}{\eta-\mu}+\exp(\eta^2)\lambda(\eta)\delta(\eta-\mu),$$

where $-\infty<\eta,\mu<+\infty$, *δ(x)* is the Dirac delta function, $\lambda(z)$ is the dispersion function, which present in the form

$$\lambda(z)=1-i\Omega+\frac{z}{\sqrt{\pi}}\int\limits_{-\infty}^{\infty}\frac{\exp(-\tau^2)d\tau}{\tau-z}.$$

or $\qquad \lambda(z)=-i\Omega+\lambda_0(z),$

where $$\lambda_0(z) = \frac{1}{\sqrt{\pi}} \int_{-\infty}^{\infty} \frac{e^{-\tau^2}\tau\, d\tau}{\tau - z}.$$

Using the Sokhotsky formulas [6], we will find boundary value of dispersion function

$$\lambda^{\pm}(\mu) = \pm i\sqrt{\pi}\,\mu\, e^{-\mu^2} - i\Omega + \frac{1}{\sqrt{\pi}} \int_{-\infty}^{\infty} \frac{e^{-\tau^2}\tau\, d\tau}{\tau - \mu}.$$

Let's separate at function $\lambda^{\pm}(\mu)$ the real and imaginary parts (figure 1 and 2)

$$\operatorname{Re}\lambda^{+}(\mu) = \Omega_2 + \lambda_0(\mu), \quad \operatorname{Im}\lambda^{+}(\mu) = -\Omega_1 \pm \sqrt{\pi}\,\mu\, e^{-\mu^2},$$

$$s(\mu) = \sqrt{\pi}\,\mu e^{-\mu^2}, \quad \lambda_0(\mu) = \frac{1}{\sqrt{\pi}} \int_{-\infty}^{\infty} \frac{e^{-\tau^2}\tau\, d\tau}{\tau - \mu}.$$

It's clear that the function $\lambda_0(\mu)$ has two complex zero $\pm\mu_0$, $\mu_0 = 0.924$ on the real axis, which differing only signs.

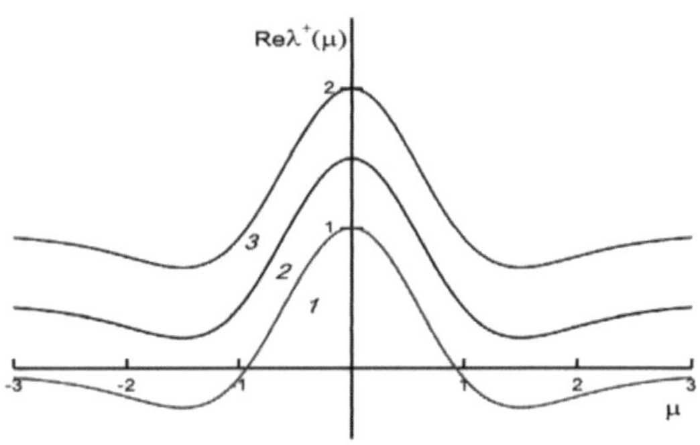

Figure 1. Real part of dispersion function $\lambda^{+}(\mu) = \Omega_2 + \lambda_0(\mu)$ on the real axis. Curves 1, 2, 3 correspond to values of parameter $\Omega_2 = 0, 0.5, 1$.

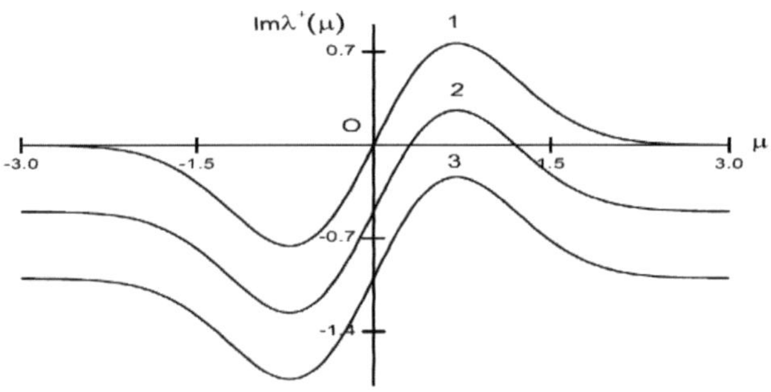

Figure 2. Imaginary part of dispersion function $\lambda^{+}(\mu) = -\Omega_1 + s(\mu)$ on the real axis. Curves 1, 2, 3 correspond to values of parameter $\Omega_1 = 0, 0.5, 1$.

Let's expand dispersion function in asymptotic series on negative degrees variable z at the vicinity of infinitely remote z

$$\lambda(z) = -i\Omega - \frac{1}{2z^2} - \frac{3}{4z^4} - \frac{15}{8z^6} - ..., \quad z \to \infty. \tag{5}$$

From expansion (5) it is visible, that at small values Ω dispersion function has two complex zero differing only signs

$$\pm \eta_0^{(0)}(\Omega) = \frac{1+i}{2\sqrt{\Omega}}.$$

Considering a family of curves on the complex plane

$$\Gamma = \Gamma(\Omega): \; G(\mu) = \frac{\lambda^+(\mu)}{\lambda^-(\mu)}, \; 0 \le \mu \le +\infty,$$

we note that $G(0) = 1$, $\lim_{\mu \to +\infty} G(\mu) = 1$. It means that curves $\Gamma(\Omega)$ are closed.

Making necessary calculations and considering principle of argument, we obtain that

$$N = \frac{1}{\pi}[\arg G(\tau)]_0^{+\infty} = 2\chi(G) \text{ или } N = 2\chi(G),$$

where $\chi = \chi(G)$ is the index of function $G(\mu)$, i.e. is the number of revolution, which are made by curve in the positive direction from the beginning of coordinates.

We will construct the frequency plane (Ω_1, Ω_2) (figure 3).

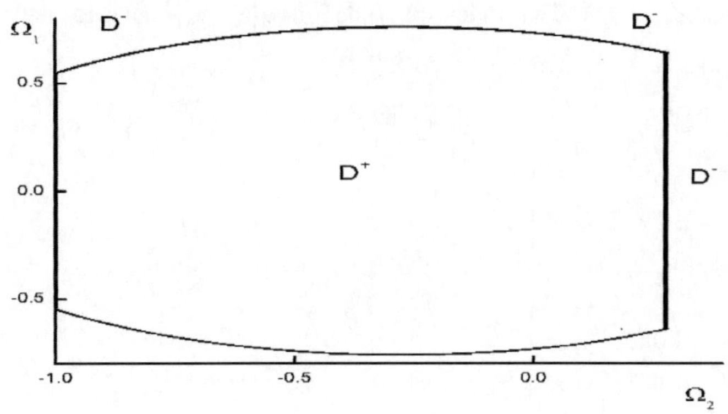

Figure 3. Regions D^+ and D^-. If $(\Omega_1, \Omega_2) \in D^+$ then $\chi = 1$, $(\Omega_1, \Omega_2) \in D^-$ then $\chi = 0$.

On the frequency plane (Ω_1, Ω_2) there is such region D^+, that for the points (Ω_1, Ω_2), lying in this region, the index of the problem is equal to unit: $\chi(\Omega_1, \Omega_2) = 1$, $(\Omega_1, \Omega_2) \in D^+$, and the points (Ω_1, Ω_2), belonging region D^-, the index of the problem is equal to zero, i.e. $\chi(\Omega_1, \Omega_2) = 0$, $(\Omega_1, \Omega_2) \in D^-$. The border of this region ∂D^+ is called the line of critical frequency.

The region D^+ is built following in the way

$$D^+ = \{(\Omega_1, \Omega_2) : |\Omega_1| < \Omega_1^*(\Omega_2), \; -1 < \Omega_2 < \max_{\mu}\{-\lambda_0(\mu)\}\}.$$

Here $|\Omega_1| = \Omega_1^*(\Omega_2) = \max_{\mu} \sqrt{\{s^2(\mu) - [\Omega_2 + \lambda_0(\mu)]^2\}}$ is the line of critical frequency.

If $(\Omega_1,\Omega_2)\in D^+$, then the dispersion function has zero equal two, and ifи $(\Omega_1,\Omega_2)\in D^-$, then the dispersion function has no zero.

Let's separate at function $G(\mu)$ the real and imaginary parts and build the covers $\Gamma(\Omega)$ on the frequency plane, which are defined by the following equations (figure 4 и 5)

$$\Gamma(\Omega):\ x=\operatorname{Re}G(\mu),\ \ y=\operatorname{Im}G(\mu),\ \ 0\le\mu\le+\infty.$$

In the case when $\Omega_1=\Omega_2=0$, then curve $\Gamma(0)$ once cover the beginning of coordinates. The function $\lambda_0(\mu)$ has single zero $\mu_0\approx 0.924$ on the real axis.

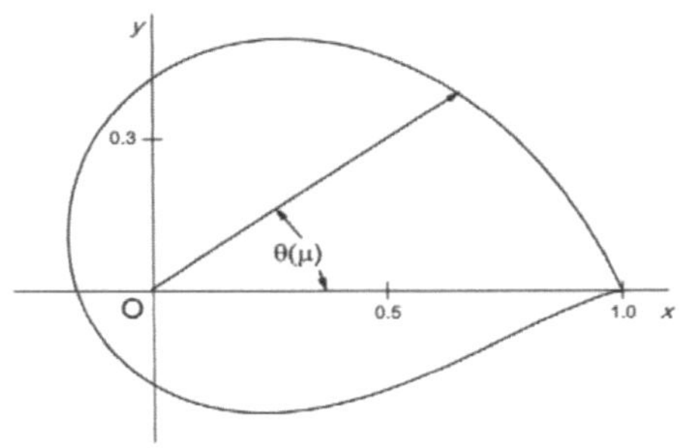

Figure 4. Curve $\Gamma(\Omega)$ cover the beginning of coordinates, when $(\Omega_1,\Omega_2)\in D^+$ The index of function $G(\mu)$ is equal to unit and dispersion function has zero equal two.

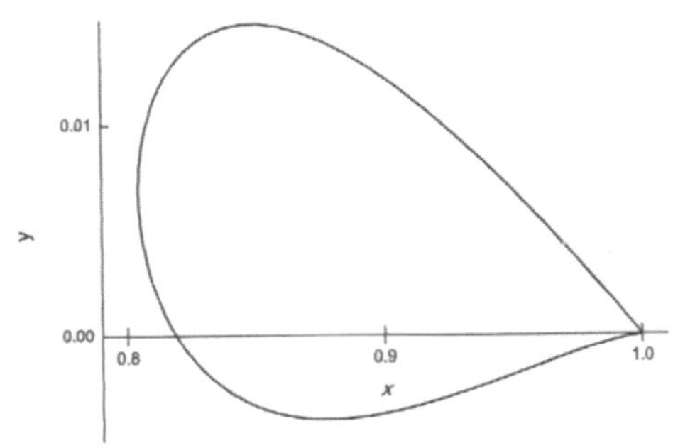

Figure 5. Curve $\Gamma(\Omega)$ doesn't cover the beginning of coordinates, when $(\Omega_1,\Omega_2)\in D^-$. The index of function $G(\mu)$ is equal to zero and dispersion function has no zero.

At $(\Omega_1,\Omega_2)\in D^-$ the discrete spectrum is the empty set. At $(\Omega_1,\Omega_2)\in D^+$ the characteristic equation has two solutions, provided the eigen functions of the characteristic equation

$$\Phi(\pm\eta_0(\Omega),\mu)=\frac{1}{\sqrt{\pi}}\frac{\pm\eta_0(\Omega)}{\pm\eta_0(\Omega)-\mu}$$

and has two eigen solutions of the kinetic equation

$$h_{\pm\eta_0(\Omega)}(x_1,\mu)=\exp\left(-\frac{x_1 z_0}{\pm\eta_0(\Omega)}\right)\frac{1}{\sqrt{\pi}}\frac{\pm\eta_0(\Omega)}{\pm\eta_0(\Omega)-\mu}.$$

At the heart of the analytical solution of boundary problems the kinetic theory lays the solution of the homogeneous boundary value Riemann problem [6] with coefficient $G(\mu)=\lambda^+(\mu)/\lambda^-(\mu)$

$$\frac{X^+(\mu)}{X^-(\mu)}=G(\mu),\quad \mu>0 \quad \text{or} \quad \frac{X^+(\mu)}{X^-(\mu)}=\frac{\lambda^+(\mu)}{\lambda^-(\mu)},\quad \mu>0. \tag{6}$$

Homogeneous boundary value Riemann problem (6) is called also the factorization problem of coefficient $G(\mu)$. Its solution [5], has the form

$$X(z)=\frac{1}{z}\exp V(z)$$

in the case, when the index of problem is equal to unit, where

$$V(z)=\frac{1}{2\pi i}\int_0^\infty\frac{\ln|G(\mu)|+i\varsigma(u)}{u-z}du,$$

In the case, when the index of problem is equal to unit. The solution of Riemann problem (6) is given by a formula

$$X(z)=\exp V(z),\quad V(z)=\frac{1}{2\pi i}\int_0^\infty\frac{\ln G(\tau)d\tau}{\tau-z}.$$

Here $\ln G(\tau)$ is the principal branch of logarithm, fixed at zero by condition $\ln G(0)=0$, angle is the principal value of argument. We will mark that $\ln G(\infty)=0$.

We can seek the general solution to problem with the boundary condition in the form

$$h(x_1,\mu)=\frac{a_0}{\eta_0-\mu}\exp\left(-\frac{x_1 z_0}{\eta_0}\right)+\int_0^\infty\exp\left(-\frac{x_1 z_0}{\eta}\right)\Phi(\eta,\mu)a(\eta)d\eta, \tag{7}$$

$a(\eta)$ is the coefficient of the continuous spectrum, a_0 is the coefficient of the discrete spectrum.

Adding eigen functions of the characteristic equation in the formula (7), we can write it in a classical form

$$h(x_1,\mu)=\frac{\eta_0 a_0}{\sqrt{\pi}(\eta_0-\mu)}\exp\left(-\frac{x_1 z_0}{\eta_0}\right)+\frac{1}{\sqrt{\pi}}\int_0^\infty\exp\left(-\frac{x_1 z_0}{\eta}\right)\frac{\eta a(\eta)d\eta}{\eta-\mu}+$$
$$+\exp\left(-\frac{x_1 z_0}{\mu}+\mu^2\right)\lambda(\mu)a(\mu)\theta_+(\mu),$$

where $\theta_+(\mu)$ is the Heaviside function: $\theta_+(\mu)=1,\ \mu>0,\ \theta_+(\mu)=0,\ \mu<0$.

This expression is proved with the help analytical methods [6], as shown in article [1, 7].

Conclusions. In this work the second Stokes problem in the case of the variable amplitude of fluctuation of a surface is formulated. Zero of the dispersion function are calculated and investigated. The region of critical frequency is built. Eigen functions of the characteristic equation and eigen solution of kinetic equation ate found. The general solution to problem with the boundary condition can be present the form of the sum of the discrete solution and the integral over the continuous spectrum of eigen solutions corresponding to the continuous spectrum is shown.

References

[1] Akimova V A, Latyshev A V and Yushkanov A A 2013 *Fluid Dynamics* **48** 1 109-122
[2] Asghar S, Nadeem S, Hanif K and Hayat T 2006 *Math. Prob. Eng.* V Article ID 72468 8
[3] Siewert C E and Sharipov F 2002 *Physics of Fluids* **14** 12 4123-4129

[4] Dudko V V, Yushkanov A A and Yalamov Yu I 2005 *J. Tech. Phys.* **75** 4 134-135
[5] Dudko V V, Yushkanov A A and Yalamov Yu I 2009 *High Temperature* **47** 2 262-268
[6] Latyshev A V and Yushkanov A A 2008 Analytical methods in kinetic theory (Monograph) *Moscow State Regional University* 280
[7] Latyshev A V and Yushkanov A A 2013 *Comp. Math. and Math. Phys.* **53** 3 336-349

Depolarization of circularly polarized light in the Mie resonance region

E E Gorodnichev, A I Kuzovlev and D B Rogozkin

National Research Nuclear University MEPhI, Kashirskoe Shosse 31, 115409 Moscow, Russia

E-mail: gorodn@theor.mephi.ru

Abstract. We show that a disordered ensemble of dielectric particles near the Mie resonances has anomalous depolarizing properties. Under the first Kerker condition the depolarization length of circularly polarized light reaches its peak value, and can be ten times greater than the transport mean free path. The second Kerker condition is shown to be satisfied as the refractive index of particles increases. In this case, the depolarization length is minimum and almost coincides with the mean free path.

1. Introduction

The effect of slow decay of circular polarization is inherent in wave propagation through media with large scattering inhomogeneities [1] (see also [2] and references therein) and, as has been shown recently [3], through media with resonant Mie particles. The latter case is considered below.

Between the first two Mie resonances a single scattered wave can retain its circular polarization [3]. This effect arises provided that the electric dipolar contribution to the scattering amplitude becomes equal to the magnetic dipolar one (e.g., for silicon spheres, refractive index $n = 3.5$, the circular polarization memory can be observed at size parameter $k_0 a = 0.784$, k_0 and a are the wavenumber and the particle radius). The equality between the electric and magnetic dipolar contributions corresponds to the so-called first Kerker condition [3] under which the differential scattering cross-section is equal to zero in the backward direction. Numerical simulations [3] have indicated that multiply scattered light remains also completely polarized. However, as shown below, the results [3] can be considered as a first approximation. The inclusion of the quadrupolar and higher order contributions violates the exact fulfillment of the Kerker condition. The depolarization cross section proves to be small, but nonzero, and the circular polarization decays at scales that are more than ten times greater than the transport mean free path.

2. General relations

Consider a beam of polarized light incident on the surface of a scattering medium. The medium is assumed to be a statistically isotropic disordered ensemble of particles. The polarization state of scattered light can be described by the Stokes vector $\hat{S} = (I,\ Q,\ U,\ V)$ which obeys the vector radiative transfer equation [4],

$$\left\{\mathbf{n}\frac{\partial}{\partial\mathbf{r}} + n_0\sigma_{tot}\right\}\hat{S}(z,\mathbf{n}) = n_0 \int d\mathbf{n}'\hat{Z}(\mathbf{n},\mathbf{n}')\hat{S}(z,\mathbf{n}') \tag{1}$$

where $\sigma_{tot} = \sigma + \sigma_a$ is the cross section of total extinction, σ and σ_a are the the cross sections of scattering and absorption, respectively; vectors $\mathbf{n}$ and $\mathbf{n}'$ denote the directions of wave propagation, n_0 is the number of scattering particles per unit volume. The phase matrix $\hat{Z}(\mathbf{n}, \mathbf{n}')$ entering into Eq.(1) can be expressed in terms of the scattering matrix (see [4]),

$$\hat{Z}(\mathbf{n}, \mathbf{n}') = \hat{L}(\pi - \beta)\hat{F}(\mathbf{nn}')\hat{L}(-\beta') \tag{2}$$

The scattering matrix $\hat{F}(\mathbf{nn}')$ describes the intrinsic properties of the medium. The matrices $\hat{L}(-\beta')$ and $\hat{L}(\pi - \beta)$ describe the transformation of the Stokes parameters under rotations in the space of directions $\mathbf{n}$ (regarding the definition of the matrix $\hat{L}$ and angles β, β', see [4]).

For a macroscopically isotropic and symmetric medium, the scattering matrix $\hat{F}(\mathbf{nn}')$ appearing in Eq.(2) has the block-diagonal structure [4]:

$$\hat{F}(\mathbf{nn}') = \begin{pmatrix} a_1(\mathbf{nn}') & b_1(\mathbf{nn}') & 0 & 0 \\ b_1(\mathbf{nn}') & a_1(\mathbf{nn}') & 0 & 0 \\ 0 & 0 & a_2(\mathbf{nn}') & b_2(\mathbf{nn}') \\ 0 & 0 & -b_2(\mathbf{nn}') & a_2(\mathbf{nn}') \end{pmatrix} \tag{3}$$

The element a_1 in matrix (3) is the differential cross section of elastic scattering. It is subject to the condition

$$\int d\mathbf{n}' a_1(\mathbf{nn}') = \sigma \tag{4}$$

For spherical particles of given radius and refractive index, the matrix elements a_1, a_2 and b_1, b_2 are equal to

$$\begin{aligned} a_1(\cos\gamma) &= \frac{1}{2}\left(|A_{\|}(\cos\gamma)|^2 + |A_{\perp}(\cos\gamma)|^2\right), \; a_2(\cos\gamma) = \mathrm{Re}A_{\|}(\cos\gamma)A_{\perp}^{*}(\cos\gamma) \\ b_1(\cos\gamma) &= \frac{1}{2}\left(|A_{\|}(\cos\gamma)|^2 - |A_{\perp}(\cos\gamma)|^2\right), \; b_2(\cos\gamma) = \mathrm{Im}A_{\|}(\cos\gamma)A_{\perp}^{*}(\cos\gamma) \end{aligned} \tag{5}$$

where $\cos\gamma = \mathbf{nn}'$, $A_{\|}$ and $A_{\perp}$ are the amplitudes of the components polarized parallel and perpendicularly to the scattering plane. The values of $A_{\|}$ and $A_{\perp}$ are calculated with the Mie theory [5].

In the case of the normal incidence of the circularly polarized light the Stokes parameters I, Q, U and V obey two independent systems of transfer equations

$$\left\{\mu\frac{\partial}{\partial z} + n_0\sigma_{tot}\right\}\begin{pmatrix} I(z,\mu) \\ Q(z,\mu) \end{pmatrix} = n_0 \int d\mathbf{n}' \begin{pmatrix} a_1 & b_1\cos\beta \\ b_1\cos\beta' & a_1\cos\beta\cos\beta' - a_2\sin\beta\sin\beta' \end{pmatrix}\begin{pmatrix} I(z,\mu') \\ Q(z,\mu') \end{pmatrix}, \tag{6}$$

$$\left\{\mu\frac{\partial}{\partial z} + n_0\sigma_{tot}\right\}\begin{pmatrix} V(z,\mu) \\ U(z,\mu) \end{pmatrix} = n_0 \int d\mathbf{n}' \begin{pmatrix} a_2 & -b_2\cos\beta' \\ b_2\cos\beta & a_2\cos\beta\cos\beta' - a_1\sin\beta\sin\beta' \end{pmatrix}\begin{pmatrix} V(z,\mu') \\ U(z,\mu') \end{pmatrix}, \tag{7}$$

where $\mu = \mathbf{nn}_{in}$, $\mathbf{n}_{in}$ is the inward normal to the surface, $\psi = \varphi - \varphi'$ is the difference between the azimuthal angles of vectors $\mathbf{n}$ and $\mathbf{n}'$. The degree of circular polarization of light is defined as $P_C = V/I$.

For a unit incident flux, the boundary conditions to Eqs.(6) and (7) have the form

$$I(z=0,\mu) = V(z=0,\mu) = \frac{1}{2\pi}\delta(1-\mu) \tag{8}$$

The other quantities, Q and U, should be put equal to zero at $z = 0$.

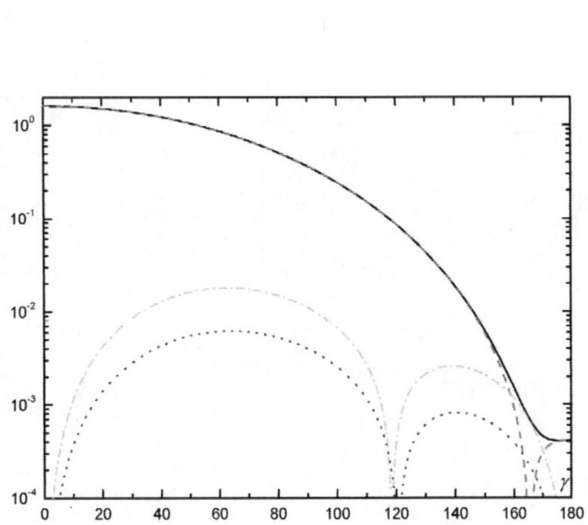

Figure 1. Angular dependence of the elements a_1 (solid black line), $|a_2|$ (dashed red line), $|b_1|$ (dash-dotted green line) and $|b_2|$ (dotted blue line) at the first Kerker point. The numerical calculations were carried out with the Mie theory for a silicon sphere ($n = 3.5$, $k_0a = 0.784$).

3. First Kerker point. Basic mode approximation

Between the first two Mie resonances the differential scattering cross section can tend to zero in the backward direction [6]. This corresponds to the so-called first Kerker condition [6, 7] and occurs when the electric and magnetic dipolar contributions to the scattering amplitude coincide to each other.

Under the first Kerker condition the elements a_1 and a_2 appearing in Eqs.(3) and (5) prove to be very nearly equal to each other and vastly greater than the elements b_1 and b_2 (see Fig. 1). Therefore, in the vicinity of the first Kerker point we can neglect the off-diagonal elements of the scattering matrix, and the systems (6) and (7) reduce to two independent equations [2]. Within such an approximation, the specific intensity I and the fourth Stokes parameter V are subject to the transfer equations [2]

$$\left\{\mu\frac{\partial}{\partial z} + n_0\sigma_{tot}\right\} I(z,\mu) = n_0 \int d\mathbf{n}' a_1(\mathbf{n}\mathbf{n}')I(z,\mu') \tag{9}$$

$$\left\{\mu\frac{\partial}{\partial z} + n_0\sigma_{tot}\right\} V(z,\mu) = n_0 \int d\mathbf{n}' a_2(\mathbf{n}\mathbf{n}')V(z,\mu') \tag{10}$$

Equations (9) and (10) correspond to the basic mode approximation [2] in the vector radiative transfer equation. The difference between the elements a_1 and a_2 is responsible for attenuation of ratio V/I due to depolarization of circularly polarized light in the medium. The cross-section of depolarization [2, 9]

$$\sigma_{dep} = \int d\mathbf{n}' \left(a_1(\mathbf{n}\mathbf{n}') - a_2(\mathbf{n}\mathbf{n}')\right) = \frac{1}{2}\int d\mathbf{n}' \left|A_{\parallel}(\mathbf{n}\mathbf{n}') - A_{\perp}(\mathbf{n}\mathbf{n}')\right|^2 \tag{11}$$

acts as the cross section of an "additional absorption" in Eq.(10). The elements a_1 and a_2 differ from each other only in the vicinity of the backward direction (see Fig. 1).

If only the electric and magnetic dipolar contributions to the scattering amplitudes are taken into account, the amplitudes $A_{\parallel}$ and $A_{\perp}$ and, consequently, the elements a_1 and a_2 can be expressed in terms of the electric α_e and magnetic α_m particle polarizabilities and written in the form

$$A_{\parallel} = k_0^2\left(\alpha_e\cos\gamma + \alpha_m\right), \; A_{\perp} = k_0^2\left(\alpha_e + \alpha_m\cos\gamma\right) \tag{12}$$

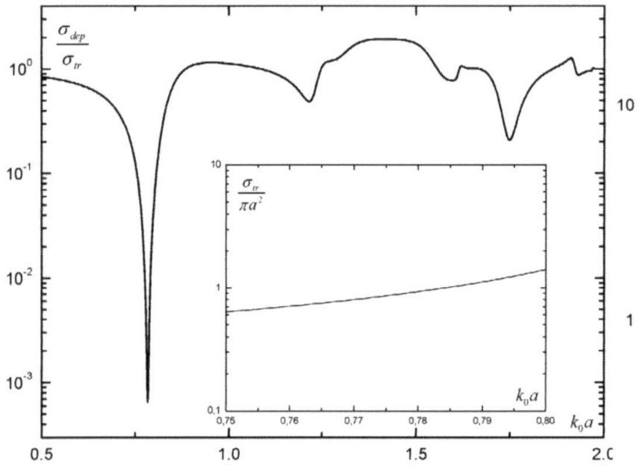

Figure 2. Depolarization cross section σ_{dep} for silicon spheres as a function of size parameter k_0a. The inset illustrates the smooth behavior of σ_{tr} near of the first Kerker point.

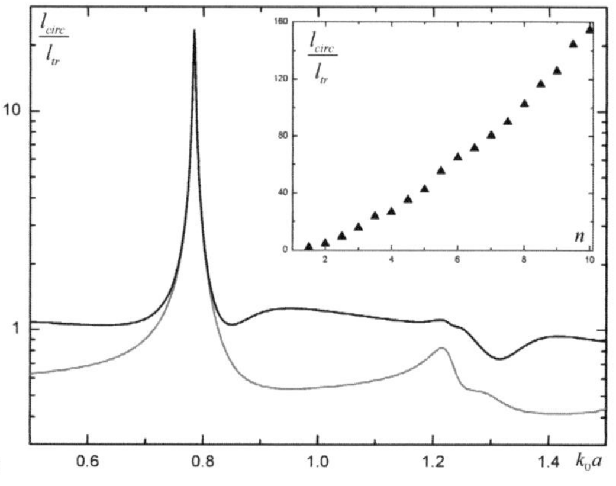

Figure 3. Depolarization length l_{circ} for an ensemble of silicon spheres as a function of size parameter k_0a. The upper curve is the result of numerical calculations with the characteristic transfer equation. The lower curve is the diffusion result (15). The inset shows the peak value of l_{circ}/l_{tr} as a function of the refractive index of scattering particles.

$$a_{1,2} = \frac{k_0^4}{2}\left(|\alpha_e + \alpha_m|^2 (1+\cos\gamma)^2 \pm |\alpha_e - \alpha_m|^2 (1-\cos\gamma)^2\right) \tag{13}$$

From Eqs.(12) and (13) it follows that under the first Kerker condition, $\alpha_e = \alpha_m$ [6], the equalities $A_\parallel = A_\perp$ and $a_1 = a_2$ are valid. In this approximation the cross section of depolarization $\sigma_{dep} = 0$, and the circular polarization is retained at an arbitrary depth z [3].

The difference between a_1 and a_2 shown in Fig. 1 is due to the higher-order multipolar contributions to the scattering amplitudes. The cross section σ_{dep} is small but nonzero. The wavelength dependence of σ_{dep} is illustrated in Fig. 2. The numerical calculations were carried out with the Mie theory [5]. A sharp dip in σ_{dep} corresponds to the first Kerker point. In the vicinity of this point, the value of σ_{dep} is much less than the transport cross-section σ_{tr}, resulting in the effect of circular polarization memory.

As z increases, the fourth Stokes parameter decays as $V(z,\mu) \sim \exp(-z/l_{circ})$, where the depolarization length l_{circ} can be found from the characteristic equation corresponding to Eq.(10) (for details see [2]). For a medium with no absorption ($\sigma_a = 0$) this equation has the form

$$\det\left(n_0(\sigma - a_2(l))\,\delta_{l,m} - \frac{1}{(2l+1)l_{circ}}\left(l\delta_{l-1,m} + (l+1)\delta_{l+1,m}\right)\right) = 0 \tag{14}$$

where $a_2(l)$ $(l = 0, 1, \ldots)$ are the expansion coefficients of the element $a_2(\cos\gamma)$ in the Legendre polynomials.

For silicon spheres, the results of numerical calculations of l_{circ} in the vicinity of the first Kerker point are shown in Fig. 3. For a number of materials with high refractive index [8], the values of l_{circ} are also presented in Table 1. From the obtained results it follows that ratio l_{circ}/l_{tr} at the first Kerker point increases with the refractive index of particles. For different materials, the position of the first Kerker point is well approximated by the relation $k_0an = 2.744$.

Table 1. Depolarization length at the first Kerker point for high refractive index materials

Material	Refractive index n ($\lambda = 1.6\ \mu m$) [8]	k_0an	σ_{dep}/σ_{tr}	l_{circ}/l_{tr}
AlAs	2.9	2.743	0.0016	14.6
GaP	3.05	2.745	0.0012	16.5
InP	3.15	2.744	0.0010	17.9
AlSb	3.28	2.745	0.0009	19.3
GaAs	3.37	2.743	0.0008	20.8
Si	3.47	2.745	0.0007	22.0
Ge	4.24	2.743	0.0003	32.6

Under conditions of the circular polarization memory, $\sigma_{dep} \ll \sigma_{tr}$, the fourth Stokes parameters V falls off at scales that are much greater than the transport mean free path l_{tr}, and the diffusion approximation can be applied to calculations of $V(z, \mu)$ [9]. Within such an approximation, the depolarization length l_{circ} of circularly polarized light is equal to [2, 9]

$$l_{circ} = \sqrt{l_{tr} l_{dep}/3} \tag{15}$$

where $l_{dep} = (n_0\sigma_{dep})^{-1}$ is the mean free path with respect to depolarizing collisions. Comparison of Eq.(15) with the results of numerical calculations is illustrated in Fig. 3. As follows from Fig. 3, the diffusion formula (15) is valid wherever the inequality $l_{circ} > l_{tr}$ is fulfilled.

In practice, spread of particles in size reduces the peak value of l_{circ}/l_{tr}. However, for relatively small deviations from the Kerker condition, the circular polarization memory effect remains observable.

4. Second Kerker point

The second Kerker condition implies that the scattering cross section tends to zero in the forward direction [6]. Owing to the optical theorem, this condition can be fulfilled only approximately.

With allowance for the electric and magnetic dipolar contributions, the optical response of the Mie particles was studied in [6,10]. Within such an approximation, the scattering amplitudes $A_\parallel$ and $A_\perp$ at the second Kerker point, $\alpha_e = -\alpha_m$ [6], satisfy the equality $A_\parallel = -A_\perp$. In this case $a_1 = -a_2$ and the cross section a_1 is proportional to $(1 - \cos\gamma)^2$ (see Eq.(13)). As a result, the scattering to the backward hemisphere dominates, and the mean cosine of single-scattering angle reaches its minimum value $\langle\cos\gamma\rangle = -0.5$. The cross section of depolarization σ_{dep} peaks at the second Kerker point. The maximum value of σ_{dep} is equal to 2σ.

If, in addition to the electric and magnetic dipolar contributions, the higher-order multipolar contributions are taken into account, the second Kerker condition can not be satisfied exactly. As follows from our calculations, the suppression of the forward scattering turns out to be slightly pronounced for relatively moderate values of the refractive index. However, the fulfillment of the second Kerker condition is achieved asymptotically as the refractive index increases, $n \gg 1$ (see Fig. 4). The values of σ_{dep} and $\langle\cos\gamma\rangle$ tend to their limiting values at $n \geq 5$ (see Fig. 5).

Owing to rather great values of σ_{dep}, multiple scattering near the second Kerker point is accompanied by fast depolarization of light. The results of numerical calculations of l_{circ} with the characteristic equation (14) are presented in Fig. 5. As the refractive index increases, the depolarization length tends to the value close to the mean free path, $l_{circ} = 1.038l$. This value corresponds to the limit $n \gg 1$ where $a_1(\cos\gamma) = a_1(180^0)(1 - \cos\gamma)^2/4$.

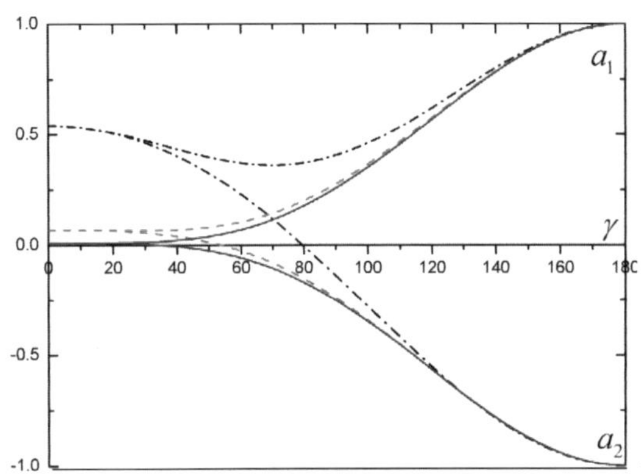

Figure 4. Angular dependence of the elements a_1 (upper lines) and a_2 (lower lines) at the second Kerker point for particles with the refractive index $n = 3.5$ (dash-dotted black lines), $n = 5$ (dashed red lines) and $n = 7$ (solid blue lines). The curves are normalized to the value of $a_1(180^0)$.

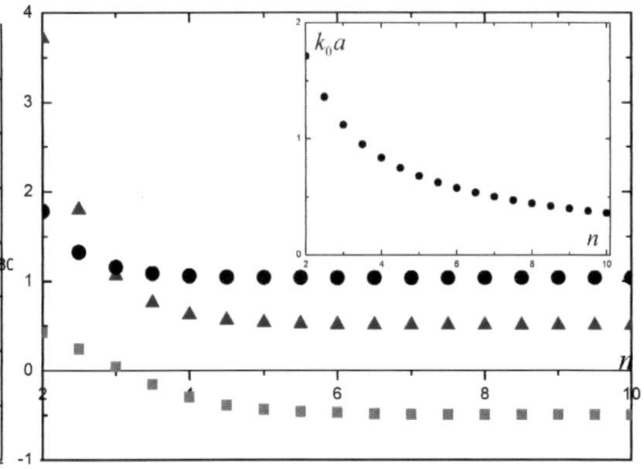

Figure 5. Mean cosine $\langle\cos\gamma\rangle$ (■), ratios σ/σ_{dep} (▲) and l_{circ}/l (●) at the second Kerker point as functions of the refractive index. The inset shows the refractive index dependence of the size parameter k_0a corresponding to the second Kerker point.

From the obtained results it follows that an ensemble of Mie particles near the second Kerker point can be considered as the medium that maximally scatters and depolarizes the electromagnetic radiation.

5. Conclusions

In conclusion, we have shown that a disordered ensemble of dielectric spheres with a relatively high refractive index exhibits anomalous depolarizing properties in the Mie resonance region. For the first and second Kerker points, the depolarization lengths of circularly polarized light have been first calculated. The decay of polarization in the vicinity of the first Kerker point has been shown to occur deep in the diffusion regime. It has also been found that the second Kerker condition can be satisfied as the refractive index of particles increases. In this case, the depolarizing ability of the medium is maximum. The depolarization length of circularly polarized light is very close to the mean free path.

References

[1] Bicout D, Brosseau C, Martinez A S and Schmitt J M 1994 *Phys. Rev. E* **49** 1767
[2] Gorodnichev E E, Kuzovlev A I and Rogozkin D B 2014 *Phys.Rev E* **90** 043205
[3] Schmidt M K, Aizpurua J, Zambrana-Puyalto X, Vidal X, Molina-Terriza G and Saenz J J 2015 *Phys.Rev.Lett.* **114** 113902
[4] Mishchenko M I, Travis L D and Lacis A A 2002 *Scattering, Absorption and Emission of Light by Small Particles* (Cambridge: Cambridge University Press).
[5] Bohren C F and Huffman D R 1998 *Absorption and Scattering of Light by Small Particles* (Toronto: Wiley Professional Paperbacks).
[6] Nieto-Vesperinas M, Gomez-Medina R and Saenz J J 2011 *J. Opt. Soc. Am. A* **28** 54
[7] Kerker M., Wang D-S and C. L. Giles 1983 *J. Opt. Soc. Am.* **73** 765
[8] Garcia-Camara B, Gomez-Medina R, Saenz J J and Sepulveda B 2013 *Opt. Express* **21** 23007 (2013).
[9] Gorodnichev E E, Kuzovlev A I and Rogozkin D B 1998 *JETP Lett.* **68** 22
[10] Gomez-Medina R, Froufe-Perez L S, Yepez M, Scheffold F, Nieto-Vesperinas M and Saenz J J 2012 *Phys. Rev. A* **85** 035802

Mesoscopic fluctuations in diffusive transport of circularly polarized light

E E Gorodnichev, A I Kuzovlev and D B Rogozkin
National Research Nuclear University MEPhI, Kashirskoe Shosse 31, 115409 Moscow, Russia
E-mail: gorodn@theor.mephi.ru

Abstract. We study how the effect of circular polarization memory in a disordered ensemble of resonant Mie particles reveals itself in mesoscopic intensity fluctuations. It is shown that polarization of light enhances the fluctuations. In the vicinity of the first Kerker point, sharp changes in the depolarization rate result in a quasiresonant dependence of the variance of transmission coefficient fluctuations on the wavelength of light.

1. Introduction

Mesoscopic transport in disordered systems is accompanied by long-range correlations in the transmitted flux [1]. For quantum electronic transport, these correlations are responsible for the universal conductance fluctuations in small metal samples. In the optical case (light or microwaves) the long-range correlations are observed in sample specific fluctuations of the speckles formed by multiply scattered waves.

Most of the theoretical results concerned with the problem of correlations between fluctuations of intensity in the speckles were derived within the scalar wave approximation [1] (see, also [2–10]). Such an approach, as applied to diffusive transport of electromagnetic waves, is believed to be valid provided that correlations between the cross-polarized fields are neglected and the polarization of waves is accounted for by doubling the total number of propagating channels (see, e.g., [11,12]).

However, as was shown in Refs. [13,14] for media with large inhomogeneities, the correlations between the cross-polarized waves result in a more complex, as compared with [1], picture of the mesoscopic fluctuations. These correlations are responsible for the polarization state of light and, therefore, until the scattered light is depolarized completely, the contribution from the cross-polarized field correlations to the variance of intensity fluctuations proves to be of the same order as the contribution obtained within the scalar wave theory.

Since the imaginary part of the correlation function of the cross-polarized fields is directly related to the helicity of light, the effect of circularly polarization memory found recently [15] in scattering by resonant Mie particles gives the opportunity to take a new look at mesoscopic phenomena in transport of electromagnetic waves through disordered media. Virtually complete retention of the circular polarization in single scattering results in the depolarization of multiply scattered light to occur deep in the diffusion regime [16]. In such a situation the cross-polarized field correlations should be taken into account in studying the diffusive wave transport of light and the related mesoscopic effects.

Below we analyze the effect of polarization memory on the mesoscopic fluctuations. Our calculations are based on the single-Hikami-box diagram for the intensity correlations. The

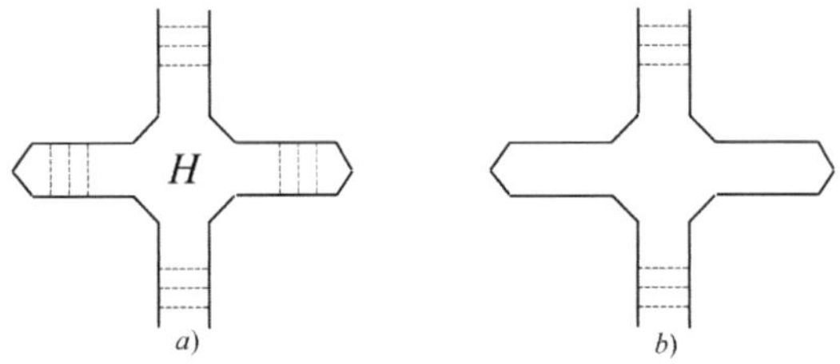

Figure 1. Feynman diagrams for the moment $\langle I(z_f,\mathbf{r}_\parallel)I(z_f,\mathbf{r}_{1\parallel})\rangle$ (a) and the short-range correlation function (b). Converging solid lines to a point denotes integration over directions of propagation and summation over polarization, H is the Hikami vertex [19].

spatial spectrum of correlations between the local values of the flux density are calculated beyond the scalar wave approximation and expressed in terms of the Stokes parameters of multiply scattered waves. The inclusion of the cross-polarized field correlations results in the enhancement of intensity fluctuations. Under the first Kerker condition [15,16] the variance of fluctuations can be twice greater compared to that obtained within the scalar theory. As the cross-polarized field correlations decay and the transmitted light is depolarized, the crossover between two regimes of diffusion occurs. In the vicinity of the first Kerker condition the crossover reveals itself as a sharp resonant dependence of the variance of transmission coefficient fluctuations on the wavelength of light.

2. Correlation function and spectrum of intensity fluctuations

Consider a plane electromagnetic wave incident on a sample composed of randomly positioned scattering particles. Interference of multiply scattered waves results in a random distribution of irradiance at the sample boundaries which is known as speckle. The speckle pattern is related directly to the spatial arrangement of scattering particles in the sample and fluctuates from one sample to another.

Sample specific fluctuations in the speckles are strongly correlated. The correlation function between the local values of the flux density (or irradiance) is defined as

$$C\left(z_f,\boldsymbol{\rho}=\mathbf{r}_\parallel-\mathbf{r}_{1\parallel}\right)=\langle I(z_f,\mathbf{r}_\parallel)I(z_f,\mathbf{r}_{1\parallel})\rangle-\langle I(z_f,\mathbf{r}_\parallel)\rangle\langle I(z_f,\mathbf{r}_{1\parallel})\rangle \tag{1}$$

where $I(z_f,\mathbf{r}_\parallel)$ is the spatial distribution of irradiance over the output boundary of the sample ($z_f=L$ for transmission through the slab of thickness L). Brackets $\langle\ldots\rangle$ appearing in Eq.(1) denote averaging over disorder. For brevity, the function $C(z_f,\boldsymbol{\rho})$ is conventionally referred to as the intensity correlation function.

In addition to the correlation function, it is convenient to introduce the spatial spectrum of intensity fluctuations [7, 17]

$$M(z_f,\mathbf{q})=\int d\boldsymbol{\rho}\exp(-i\mathbf{q}\boldsymbol{\rho})C(z_f,\boldsymbol{\rho}) \tag{2}$$

The q-dependence of the spectrum is related directly to the spatial behavior of the correlation function and is sensitive to the regime of wave propagation in disordered media [7, 17]. The value of the spectrum at $q=0$ determines the variance of total transmission coefficient T.

The intensity correlations are described by the diagram shown in Fig.1(a) (see, e.g., [1]). Based on the results of diagrammatic calculations [13,14] we can express the spectrum of intensity fluctuations in terms of the Stocks parameters of scattered light,

$$M(z_f, \mathbf{q}) = \left(\frac{2\pi}{k_0}\right)^2 n_0 \int_0^L dz \int d\mathbf{n}d\mathbf{n}' |I_{\mathbf{q}}(z_f|z, \mathbf{n}) - I_{\mathbf{q}}(z_f|z, \mathbf{n}')|^2 S_m(z, \mathbf{n}) Z_{mn}(\mathbf{n}, \mathbf{n}') S_n(z, \mathbf{n}') \quad (3)$$

where n_0 is the number of scattering particles per unit volume, $k_0 = 2\pi/\lambda$, λ is the wavelength of light, $Z_{mn}(\mathbf{n}, \mathbf{n}')$ is the phase matrix [16,18], $\mathbf{n}$ and $\mathbf{n}'$ are the directions of wave propagation. Quantities S_m $(m = 1 \div 4)$ entering into Eq.(3) are the components of the Stokes vector $\hat{S} = (I,\ Q,\ U,\ V)$ which is subject to the vector radiative transfer equation [18]. Equation (3) differs from the corresponding result [7,17] obtained within the scalar wave approximation by the factor $S_m(z, \mathbf{n}) Z_{mn}(\mathbf{n}, \mathbf{n}') S_n(z, \mathbf{n}')$ which takes into account the polarization state of scattered light.

Formula (3) establishes the interrelation of the mesoscopic intensity fluctuations with the polarization state of the incident light and optical characteristics of the scattering medium, including those that are responsible for depolarization of light.

The spectrum (3) involves all spatial scales of intensity correlations. The correlations at scales of the order of the wavelength λ are described by the diagram with no scattering event in the "outgoing" propagators and the Hikami-vertex (see Fig. 1(b)) [17]. The short-range contribution to the correlation function can be presented in the form

$$C^{(sh)}(z_f, \boldsymbol{\rho}) = \frac{1}{4} \int d\mathbf{n} \int d\mathbf{n}' e^{ik_0 \boldsymbol{\rho}(\mathbf{n}-\mathbf{n}')} \left[\mathbf{n}_{in} \left(\mathbf{n} + \mathbf{n}'\right)\right]^2 \left(S_m(z_f, \mathbf{n}) D_{mn}(\mathbf{n}, \mathbf{n}') S_n(z_f, \mathbf{n}')\right) \quad (4)$$

where integration is carried out over the forward hemisphere, $\mathbf{n}_{in}$ is the inward normal to the output boundary, the matrix $\hat{D}$ is obtained from the matrix $\hat{Z}$ by substitution of the matrix elements a_1, a_2, b_1 and b_2 in the form $a_1(\mathbf{nn}') = (1/2)\left(1 + (\mathbf{nn}')^2\right), a_2(\mathbf{nn}') = (\mathbf{nn}'), b_1(\mathbf{nn}') = (1/2)\left((\mathbf{nn}')^2 - 1\right), b_2 = 0$.

Further simplifications in Eqs.(3) and (4) are based on the assumptions concerned with the polarization state of light and optical characteristics of the medium. In scattering of circularly polarized light by resonant Mie particles only two components of the Stokes vector, I and V, survive near the first Kerker point [16], and

$$S_m(z, \mathbf{n}) Z_{mn}(\mathbf{n}, \mathbf{n}') S_n(z, \mathbf{n}') = \frac{1}{2} \left(a_1 \left(\mathbf{nn}'\right) I(z, \mathbf{n}) I(z, \mathbf{n}') + a_2 \left(\mathbf{nn}'\right) V(z, \mathbf{n}) V(z, \mathbf{n}')\right) \quad (5)$$

An analogous formula is also valid for the factor $S_m D_{mn} S_n$ entering into Eq.(4).

3. Short-range correlations in diffusive transport

In propagation of light through thick samples of a weakly absorbing medium ($L \gg l_{tr}$ and $l_{tr} \ll l_a$, where l_{tr} is the transport mean free path and l_a is the absorption mean free path), the angular distribution of the transmitted radiation turns out to be virtually isotropic. In this case the intensity I can be calculated with the well-known diffusion approximation [20]. Such an approximation is also applicable for the fourth Stokes parameter V near the first Kerker point where the mean free path with respect to depolarizing collisions l_{dep} is much greater than the transport mean free path l_{tr} [16].

Within the diffusion approximation, the short-range correlation function (see Eq.(3)) can be presented in the form

$$C^{(sh)}(L, \rho) = \frac{1}{2} \langle T \rangle^2 \left(1 + P_C^2(L)\right) \frac{J_1^2(k_0\rho)}{(k_0\rho)^2} \quad (6)$$

where $J_1(x)$ is the Bessel function, $\langle T \rangle$ is the average total transmission coefficient and $P_C(L)$ is the degree of polarization of transmitted light. In a medium with no absorption

$$\langle T \rangle = \sqrt{3}\,\frac{l_{tr}}{L},\ P_C(L) = \frac{(L/l_{circ})}{\sinh(L/l_{circ})}. \tag{7}$$

Within the diffusion approximation, the depolarization length l_{circ} is determined by the relation $l_{circ} = \sqrt{l_{tr} l_{dep}/3}$ [21]. From Eq.(6) it follows that intensity fluctuations at short ranges are positively correlated. The ρ-dependence of the correlation function is insensitive to the polarization state of light. The spatial scale of correlations proves to be of the order of the wavelength λ.

The correlation function (6) is proportional to the square of the transmission coefficient and includes two contributions, from the intensity and the fourth Stokes parameter. The polarization contribution to the function $C^{(sh)}(\rho)$ depends explicitly on the degree of polarization $P_C(L)$ and, therefore, vanishes simultaneously with the depolarization of light. Near the first Kerker point, the wavelength dependence of the depolarization length l_{circ} has a resonance shape [16]. This leads to a similar quasiresonant behavior of the correlation function (6).

Equation (6) differs from the corresponding result of Ref. [22] because of Eq.(6) describes correlations of the flux-flux type (not the density-density correlations as in Ref. [22]).

4. Long-range correlations of intensity fluctuations

The long-range correlations are governed by the q-dependence of the spectrum (3) at relatively low spatial frequencies $q < l_{tr}^{-1}$. Near the first Kerker point where the depolarization length l_{circ} can exceed noticeably the transport mean free path l_{tr} [16] the contributions from the Stokes parameters I and V to the spectrum can be calculated within the diffusion approximation, to give

$$M(z_f, \mathbf{q}) = \frac{l_{tr}}{48\pi k_0^2} \int_0^L dz \left[\left| \frac{\partial}{\partial z} I_{\mathbf{q}}(z_f|z) \right|^2 + \mathbf{q}^2 \left| I_{\mathbf{q}}(z_f|z) \right|^2 \right] \left(I^2(z) + V^2(z) \right) \tag{8}$$

where quantities $I_{\mathbf{q}}(z_f|z)$, $I(z)$ and $V(z)$ are expressed in terms of the integrals of $I_{\mathbf{q}}(z_f|z, \mathbf{n})$, $I(z, \mathbf{n})$ and $V(z, \mathbf{n})$ over directions and subject to to the diffusion equation (see, e.g., [20]). For purely elastic scattering the spectrum of transmitted intensity fluctuations is written as [13, 14]

$$M(L, q) = \frac{9\pi}{k_0^2}\frac{l_{tr}}{L} \left(F(qL, 0) + F\left(qL, \frac{L}{l_{circ}}\right) \right) \tag{9}$$

where

$$F(x, y) = \frac{xy\ (x \sinh 2y - y \sinh 2x)}{4(x^2 - y^2) \sinh^2 x \sinh^2 y}, \tag{10}$$

The Fourier transform of Eqs.(9) and (10) leads to the correlation function of the form

$$C(\rho) = \frac{9\pi l_{tr}}{k_0^2 L^3} \left[\nu\left(\frac{\rho}{L}, 0\right) + \nu\left(\frac{\rho}{L}, \frac{L}{l_{circ}}\right) \right] \tag{11}$$

In transmission of light through a thick depolarizing slab, $L > l_{circ}$, the polarization effect reveals itself at all ρ. For relatively small ρ, $\rho < l_{circ}$, the scalar and polarization contributions to the correlation function decrease in inverse proportion to ρ, but with the different amplitudes. In this case $\nu(\rho/L, L/l_{circ}) / \nu(\rho/L, 0) = (L/l_{circ})^2 / \sinh^2(L/l_{circ})$, and the polarization term proves to be suppressed as compared to the scalar one (see Fig. 2). For $l_{circ} < \rho < L$ the polarization contribution $\nu(\rho/L, L/l_{circ})$ can be estimated as l_{circ}/L times smaller than the scalar one $\nu(\rho/L, 0)$. Just these values of ρ contribute to the polarization term of the variance of transmission coefficient fluctuations at $L > l_{circ}$.

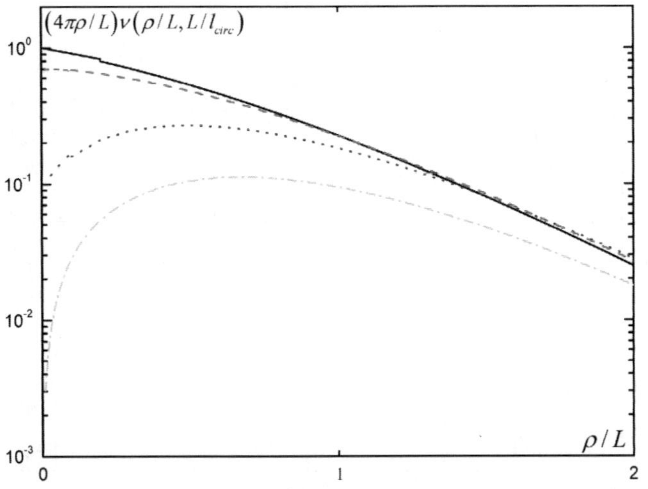

Figure 2. Scalar $\nu(\rho/L, 0)$ (solid black line) and polarization $\nu(\rho/L, L/l_{circ})$ (other color lines) contributions to the correlation function of transmitted intensity fluctuations. From upper to lower curves $L/l_{circ} = 0,\ 1,\ 3,\ 7$.

5. Transmission coefficient fluctuations

As follows from definition (2) the value of the spectrum at $q = 0$ determines the variance of the total transmission coefficient [7, 17]

$$\langle(\delta T)^2\rangle = \frac{1}{A} M(L, \mathbf{q} = 0), \tag{12}$$

where A is the area of the slab surface, T is the ratio of the total transmitted flux to the incident one. From Eqs.(9), (10) and (12) it follows that the variance of the transmission coefficient is determined by the expression

$$\frac{\langle(\delta T)^2\rangle}{\langle T\rangle^2} == \frac{3}{2N}\frac{L}{l_{tr}}\left(\frac{1}{3} + \frac{\sinh 2y - 2y}{4\ y\ \sinh^2 y}\right) \tag{13}$$

where $N = Ak_0^2/(2\pi)$ is the number of propagating modes [12], $y = L/l_{circ}$. Equation (13) describes the crossover in transmission fluctuations which is caused by decay of correlations between the cross-polarized fields and, correspondingly, by depolarization of the incident light. For $L < l_{circ}$ the depolarization of circularly polarized light is of little significance and the variance tends to the value that is twice greater compared to the result $\langle(\delta T)^2\rangle/\langle T\rangle^2 = L/(2Nl_{tr})$ of the scalar theory [1] which neglects correlations between the cross-polarized fields. As the sample thickness increases, $L > l_{circ}$, the polarization contribution to the relative value of the variance tends to an L-independent value (i.e., the polarization contribution is a non-self-averaging quantity):

$$\frac{\langle(\delta T)^2\rangle}{\langle T\rangle^2} = \frac{1}{2N}\left(\frac{L}{l_{tr}} + \frac{3}{2}\frac{l_{circ}}{l_{tr}}\right) \tag{14}$$

According to Eq.(14), the polarization contribution to the variance of the transmission coefficient is noticeable even for large thickness L where the degree of polarization (see Eq.(7))

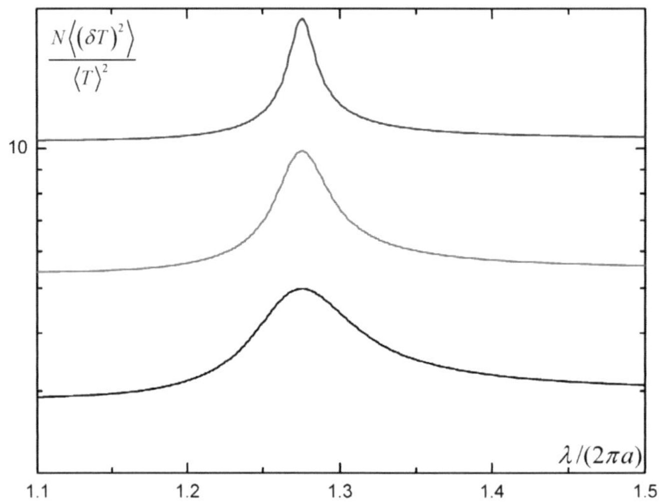

Figure 3. Wavelength dependence of the variance of relative transmission fluctuations for an ensemble of silicon spheres near the first Kerker point. From lower to upper curves $L/l_{tr} = 5,\ 10,\ 20$.

and, correspondingly, the polarized component of the radiation transmitted through the sample are exponentially small.

Near the first Kerker point, the depolarization length l_{circ} varies sharply with the wavelength [16]. Hence, depending on the wavelength, the sample with given parameters (thickness L, radius of particles, their refractive index, etc.) can either retain the initial polarization of the incident light ($L < l_{circ}$) or completely depolarize the light ($L > l_{circ}$). The crossover between these two regimes of propagation manifests itself in the variance of transmission coefficient fluctuations. The wavelength dependence of $\langle(\delta T)^2\rangle/\langle T\rangle^2$ is illustrated in Fig. 3. As follows from the results of calculations, the fluctuations are enhanced near the Kerker point. If the thickness L is smaller than the length l_{circ}, the maximum value of the variance is twice greater than the result of the scalar theory [1]. The peak in the wavelength dependence of the variance is narrowed with increasing L, and for thick samples repeats the dependence of the depolarization length on the wavelength λ (see Eq.(14)).

6. Effect of absorption

A number of materials with a high refractive index is characterized by noticeable absorption [23] which should be taken into account in studying the depolarization of light and, correspondingly, the intensity fluctuations.

In an absorbing medium, one more spatial scale appears [20]. The length of wave trajectories turns out to be restricted by the absorption mean free path l_a. As a result, the intensity of light falls off at depths z of the order of the diffusion length $l_d = \sqrt{l_{tr}l_a/3}$ [20]. The presence of absorption affects also the depolarization length l_{circ}. Decay of the degree of circular polarization $P_C = V/I$ is governed by the competition between attenuation of I and V, and occurs at scales of the order

$$l_{circ} = l_d\left(\frac{l_{dep}}{l_a}\right)\left(1+\sqrt{1+\frac{l_a}{l_{dep}}}\right) = \begin{cases}\sqrt{(l_{tr}l_{dep}/3)}, & l_{dep} < l_a,\\ 2l_d\,(l_{dep}/l_a), & l_{dep} > l_a\end{cases} \tag{15}$$

IOP Conf. Series: Journal of Physics: Conf. Series **788** (2017) 012040 doi:10.1088/1742-6596/788/1/012040

From Eq.(15) it follows that in the case of relatively high absorption, $l_a < l_{dep}$, the depolarization length l_{circ} increases as $l_{circ} \sim l_{dep}\sqrt{l_{tr}/l_a}$, and exceeds the diffusion length l_d.

As L increases, the variance of the transmission coefficient tends to

$$\frac{\langle(\delta T)^2\rangle}{\langle T\rangle^2} = \frac{1}{2N}\left(\frac{3L}{4l_{tr}} + \frac{3l_d}{4l_{tr}}\frac{2\alpha^2 - 1}{\alpha(\alpha^2-1)}\right) \tag{16}$$

where $\alpha = 1 + (l_d/l_{circ})$. A relative role of the polarization contribution to the variance of transmission fluctuations increases with absorption. If the radiation is depolarized faster than absorbed, $l_a > l_{dep}$, the polarization contribution to the variance of the transmission coefficient coincides with that entering into Eq.(14). In the opposite case of comparatively high absorption, $l_a < l_{dep}$, the polarization contribution to the variance increases noticeably. This contribution gains additional factor $\sqrt{l_{dep}/l_a}$.

7. Conclusions

As has been shown above, the anomalously slow depolarization of circularly polarized light near the first Kerker point [15,16] opens new possibilities for studying optical mesoscopic effects which were previously considered [1] in the scalar approximation. Accounting for the vector nature of light we have obtained the additional contribution to the intensity fluctuations which results from correlations between the cross-polarized fields. Under conditions of circular polarization memory, this contribution survives at great distances because the correlation function of the cross-polarized field involves, as part, the fourth Stokes parameter. We have shown that the main observable effect consists in the enhancement of fluctuations. In the vicinity of the first Kerker point the variance of the transmission coefficient varies resonantly with the wavelength. The peak value of the variance is two times as great as the result of the scalar wave theory [1].

References

[1] Akkermans E and Montambaux G 2011 *Mesoscopic Physics of Electrons and Photons* (Cambrige: Cambrige University Press)
[2] Shapiro B 1986 *Phys. Rev. Lett.* **57** 2168
[3] Zyuzin A Yu and Spivak B Z 1987 *Sov. Phys. JETP* **66** 560
[4] Stephen M J and Cwilich G 1987 *Phys. Rev. Lett.* **59** 285
[5] Feng S, Kane C, Lee P A and Stone A D 1988 *Phys. Rev. Lett.* **61** 834
[6] Pnini R and Shapiro B 1991 *Phys. Lett.* A **157** 265
[7] Rogozkin D B and Cherkasov M Yu 1993 *JETP Lett.* **58** 585
[8] van Rossum M C W and Nieuwenhuizen Th M 1999 *Rev. Mod. Phys* **71** 313
[9] Garcia-Martin A, Scheffold F, Nieto-Vesperinas M and Saenz J J 2002 *Phys. Rev. Lett.* **88** 143901
[10] Froufe-Perez L S, Garcia-Martin A, Cwilich G and Saenz J J 2007 *Physica* A **386** 625
[11] de Boer J F, Albada M P and Lagendijk A 1992 *Phys. Rev.* B **45** 658
[12] Chabanov A A, Tregoures N P, van Tiggelen B A and Genack A Z 2004 *Phys. Rev. Lett.* **92** 173901
[13] Gorodnichev E E, Kuzovlev A I and Rogozkin D B 2009 *JETP Lett.* **89** 547
[14] Gorodnichev E E, Kuzovlev A I and Rogozkin D B *JOSA* A **33** 95
[15] Schmidt M K, Aizpurua J, Zambrana-Puyalto X, Vidal X, Molina-Terriza G and Saenz J J 2015 *Phys.Rev.Lett.* **114** 113902
[16] Gorodnichev E E, Kuzovlev A I and Rogozkin D B 2016 *J. Physics: Conf. Series* (published in this issue)
[17] Rogozkin D B and Cherkasov M Yu 1995 *Phys. Rev.* B **51**, 12256
[18] Mishchenko M I, Travis L D and Lacis A A 2002 *Scattering, Absorption and Emission of Light by Small Particles* (Cambridge: Cambridge University Press)
[19] Hikami S 1981 *Phys. Rev.* B **24** 2671
[20] Ishimaru A 1997 *Wave Propagation and Scattering in Random Media* (New York: Institute of Electrical and Electronics Engineers)
[21] Gorodnichev E E, Kuzovlev A I and Rogozkin D B 1998 *JETP Lett.* **68** 22
[22] Dogariu A and Carminati R 2015 *Phys. Rep.* **559** 1
[23] Garcia-Camara B, Gomez-Medina R, Saenz J J and Sepulveda B 2013 *Opt. Expr.* **21** 23007

Nonlinear Interaction of a Shock Wave with an Anisotropic Entropy Perturbation Field

K E Gorodnichev[1], S E Kuratov[1] and E E Gorodnichev[2]
[1]Dukhov Research Institute of Automatics, Sushchevskaya st. 22, Moscow, 127055, Russia
[2]National Research Nuclear University MEPhI, Kashirskoe Shosse 31, Moscow, 115409, Russia
E-mail: cyrgo85@gmail.com

Abstract. The problem of the interaction of a shock wave with an anisotropic entropy perturbation field has been solved including second-order corrections to hydrodynamic quantities. It has been shown that nonlinear interactions between acoustic waves result in the localization of acoustic perturbations behind the shock front. This effect is observed when sound attenuation is absent in the linear approximation. The problem of the propagation of the shock wave in an incident sample, where the spatially anisotropic density perturbation field initially exists, has been numerically solved in application to the collision of two plates. Numerical calculations confirm the results of the theoretical analysis.

1. Introduction

The interaction of planar shock waves with various perturbation fields in a medium has been studied in the last half century in numerous theoretical and experimental works (see, e.g., [1–12]). In particular, in the linear approximation, the interaction of a shock wave with acoustic waves was considered in [2, 11], with an isotropic turbulent vorticity field in [8], and with the perturbation of the density in [9, 12]. In all these cases, acoustic and entropy–vortex perturbations appear behind the shock front in addition to the initial fields. The interaction of the shock wave with the anisotropic density perturbation field was considered in the linear approximation in [13]. It was shown that the regime of propagation of sound behind the shock front depends on the angle between the axis of anisotropy of the perturbation field and the velocity of the shock wave. In the case of relatively small angles θ_0 between the axis of anisotropy and the inward normal to the shock front $\mathbf{n}_0$, the projection of the velocity of an acoustic wave c_x on the $Ox \parallel \mathbf{n}_0$ axis is positive. With an increase in the angle θ_0, the projection c_x decreases and, at $\theta_0 = \theta_0^{**}$ (θ_0^{**} is determined by the parameters of the medium behind and ahead of the shock front [13]), becomes equal to the velocity of the medium. In the angular interval $\theta_0^{**} < \theta_0$, the acoustic wave moves behind the shock wave. When the angle θ_0 reaches the threshold value θ_0^*, the velocity projection c_x becomes equal to the velocity of the shock wave. With a further increase in the angle θ_0 ($\theta_0 > \theta_0^*$), acoustic perturbations behind the shock wave are described by surface acoustic waves [2]. As follows from the described results [13], when $\theta_0 = \theta_0^*$, the velocity projection c_x is equal to the velocity of the shock wave. A spatial decay of acoustic perturbations at $\theta_0 = \theta_0^*$ is absent also. Consequently, the amplitude of these perturbations behind the shock front should be zero. The spatial structure of the acoustic field under these conditions can be correctly described only beyond the linear approximation.

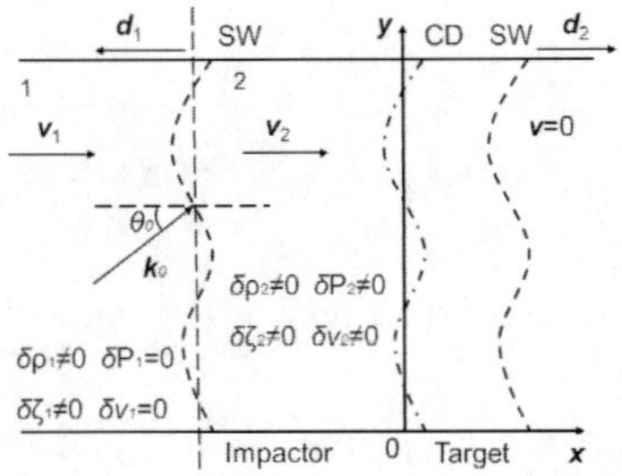

Figure 1. Illustration of the problem.

Below, the problem of the interaction of the shock wave with the anisotropic density perturbation field is solved including second-order corrections to hydrodynamic quantities. In this approximation, hydrodynamic quantities satisfy inhomogeneous wave equations [14]. The application of standard perturbation theory to the solution of the equation for the pressure results in the appearance of secular (growing in time) terms. In this work, we propose a new analytical method for solving such equations. This method generalizes the approach used in the theory of nonlinear ordinary differential equations [15] and involves the introduction of an additional phase into a linear solution. Using this method, it is shown that damping acoustic perturbations exist behind the shock front at $\theta_0 = \theta_0^*$. The problem of the propagation of the shock wave in an incident sample, where the spatially anisotropic density perturbation field initially exists, is numerically solved in application to the collision of two plates. The results of the calculations confirm qualitative conclusions made in the quadratic approximation concerning the character of propagation of sound in the immediate vicinity of the angle θ_0^* ($\theta_0^* - \theta_0 \ll \theta_0^*$). It is found that the width of the region where acoustic perturbations are observed increases with the time. It is also shown that the damping acoustic field behind the shock front is observed not only in the immediate vicinity of the angle θ_0^*, but also in the entire range $\theta_0^{**} > \theta_0 \geq \theta_0^*$.

2. Interaction of the shock wave with the density perturbation field. Linear approximation

We consider the collision of two semi-infinite plates one of which is at rest (see Fig. 1). A spatially anisotropic density perturbation field initially exists in the incident plate (impactor). Collision results in the appearance of shock waves diverging in both plates from the contact discontinuity (see Fig. 1). It is assumed that the velocity of the impactor is quite high (about 5 km/s). For this reason, the elastoplastic properties of the colliding plates can be neglected [16] and the interaction of the shock wave with the density perturbation field in the impactor can be described by the equations of hydrodynamics

$$\rho \frac{d\mathbf{v}}{dt} + \nabla P = 0, \tag{1}$$

$$\frac{d\rho}{dt} + \rho \nabla \mathbf{v} = 0, \tag{2}$$

$$\frac{d\varepsilon}{dt} - \frac{P}{\rho^2}\frac{d\rho}{dt} = 0. \tag{3}$$

Equation (3) implies that the motion of the medium is isentropic.

The system of Eqs. (1)–(3) should be supplemented by the Rankine–Hugoniot conditions at the shock front [17]:

$$v_{1\tau} = v_{2\tau}, \qquad v_{1n} - v_{2n} = \sqrt{(P_2 - P_1)\left(\frac{1}{\rho_1} - \frac{1}{\rho_2}\right)}, \tag{4}$$

$$j^2 = \rho_1\rho_2\frac{P_2 - P_1}{\rho_2 - \rho_1}, \qquad \varepsilon_1 - \varepsilon_2 = \frac{1}{2}\left(\frac{1}{\rho_2} - \frac{1}{\rho_1}\right)(P_1 + P_2), \tag{5}$$

where $v_{1\tau}$ (v_{1n}) and $v_{2\tau}$ (v_{2n}) are the tangential (normal) components of the velocity ahead of and behind the front, respectively; $j = v_1\rho_1 = v_2\rho_2$ is the mass flux; and ε_1 and ε_2 are the specific internal energies on both sides of the shock wave.

Following [18], we represent the specific internal energy ε in the form of the sum $\varepsilon = \varepsilon_{\text{el}} + \zeta$, where ε_{el} is the elastic component of the internal energy and ζ is the specific thermal energy. In this case, Eq. (3) can be written in the form

$$\frac{d\zeta}{dt} - \frac{\Gamma\zeta}{\rho}\frac{d\rho}{dt} = 0. \tag{6}$$

The equation of state of the medium is used in the form of the Mie–Grüneisen equation [18]:

$$P = \frac{\rho_0 c_0^2}{n}\left(\left(\frac{\rho}{\rho_0}\right)^n - 1\right) + \Gamma\rho\zeta, \tag{7}$$

where ρ_0 is the density of the uncompressed ("cold") medium, c_0 is the speed of sound in this medium, ρ is the current density, Γ is the Grüneisen coefficient, and n is a constant (e.g., for Fe – $\rho_0 = 7.885\ g/cm^3$, $c_0 = 3.887\ km/s$, $n = 4.3$, $\Gamma = 1.6$ and for Al – $\rho_0 = 2.710\ g/cm^3$, $c_0 = 5.333\ km/s$, $n = 3.5$, $\Gamma = 2.13$).

Let the initial density perturbation field in the impactor be described by the expression

$$\delta\rho_1 = A_{\delta\rho_1}\exp\left(ik_{0x}x + ik_{0y}y\right). \tag{8}$$

The axis of anisotropy for such perturbation is directed along the vector $\mathbf{k_0}$ (see Fig. 1). The shock wave moves from the right to the perturbation at the velocity $(\mathbf{d_1} - \mathbf{v_1})$ in the negative direction of the x axis ($d_1 - v_1 < 0$). The angle between the vector $\mathbf{k_0}$ and the Ox axis is θ_0 ($\tan\theta_0 = k_{0y}/k_{0x}$). At the interaction of the density perturbation field with the shock wave, acoustic and entropy–vortex waves are formed behind the shock front [1]. In the linear approximation, the system of Eqs. (1)–(3) behind the shock front has the form (see, e.g., [13])

$$\frac{\partial\delta v_{2x}}{\partial t} + v_2\frac{\partial\delta v_{2x}}{\partial x} = -\frac{1}{\rho_2}\frac{\partial\delta P_2}{\partial x}, \tag{9}$$

$$\frac{\partial\delta v_{2y}}{\partial t} + v_2\frac{\partial\delta v_{2y}}{\partial x} = -\frac{1}{\rho_2}\frac{\partial\delta P_2}{\partial y}, \tag{10}$$

$$\frac{\partial\delta\rho_2}{\partial t} + v_2\frac{\partial\delta\rho_2}{\partial x} + \rho_2\left(\frac{\partial\delta v_{2x}}{\partial x} + \frac{\partial\delta v_{2y}}{\partial y}\right) = 0, \tag{11}$$

$$\frac{\partial\delta\zeta_2}{\partial t} + v_2\frac{\partial\delta\zeta_2}{\partial x} - \frac{\Gamma_1\zeta_2}{\rho_2}\left(\frac{\partial\delta\rho_2}{\partial t} + v_2\frac{\partial\delta\rho_2}{\partial x}\right) = 0, \tag{12}$$

where $\delta\mathbf{v}_2$, δP_2, and $\delta\zeta_2$ are the first-order corrections to the initial values of the velocity, pressure, and internal energy, respectively. The system of Eqs. (9)-(12) should be supplemented by the Rankine–Hugoniot conditions written in the linear approximation (4), (5) (details see, e.g., in [13]).

It follows from Eqs. (9)–(12) that the x projection of the wave vector of acoustic waves is given by the expression [13]

$$(k_x)_\pm = \frac{k_{0x}(d_1 - v_1)(v_2 - d_1) \pm c_2\sqrt{k_{0x}^2(v_1 - d_1)^2 - \left(c_2^2 - (v_2 - d_1)^2\right)k_{0y}^2}}{c_2^2 - (v_2 - d_1)^2}. \tag{13}$$

According to Eq. (13), at angles of incidence in the interval

$$0 < \theta_0 < \theta_0^{**},\ \theta_0^{**} = \arctan\left(\frac{v_1 - d_1}{c_2}\right), \tag{14}$$

the solution $(k_x)_-$ corresponds to the incidence of acoustic waves from the right on the shock front. Sound sources behind the shock front that can be responsible for the appearance of such waves are absent. For this reason, we below consider the acoustic field with the projection of the wave vector $(k_x)_+ \equiv k_x$ and the frequency [13, 17]

$$\omega = k_x v_2 + c_2\sqrt{(k_x)^2 + k_{0y}^2}. \tag{15}$$

If $\theta_0 = \theta_0^{**}$, the projection of the wave vector k_x is equal to zero ($k_x = 0$) and acoustic waves propagate parallel to the shock front.

At large angles of incidence in the interval

$$\theta_0^* < \theta_0 < \frac{\pi}{2}, \tag{16}$$

the projection of the wave vector k_x has a positive imaginary part and acoustic perturbations are described by surface waves [2].

The most interesting angular range is

$$\theta_0^{**} < \theta_0 < \theta_0^*,\ \theta_0^* = \arctan\left(\frac{v_1 - d_1}{\sqrt{c_2^2 - (v_2 - d_1)^2}}\right). \tag{17}$$

In this range, the projection k_x has a negative real part and zero imaginary part and sound, in contrast to angles of incidence in interval (14), should propagate at the velocity

$$c_{2x} = v_2 + k_x c_2/\sqrt{k_x^2 + k_{0y}^2} < 0 \tag{18}$$

towards the shock front [13]. This is inconsistent with the condition of the absence of sound sources behind the shock front. This contradictory can be resolved by introducing a virtual source. The amplitude of the virtual wave at $x = c_{2x}t$ should be equal in absolute value to the amplitude of the actual wave and should damp as approaching the shock front. For the correct introduction of the virtual source into Eqs. (9)–(12), it is necessary to go beyond the linear approximation and to take into account nonlinear interactions between "behind-front" perturbations of hydrodynamic quantities.

In the next section, a solution of the system of Eqs. (1), (2), and (4)-(6) is found in the quadratic approximation when only the interaction between acoustic waves is taken into account.

3. Nonlinear interaction between acoustic waves

3.1. Additional-phase method

Inhomogeneous equations for second-order corrections to hydrodynamic quantities can be obtained from the system of Eqs. (1), (2), and (4)-(6) by standard perturbation theory [14]. In the general case, these equations include three types of sources. The first type corresponds to the interaction between acoustic waves. The second and third types describe the interaction of entropy–vortex waves with each other and with acoustic fields. The analysis shows that the main second-order contribution to the acoustic perturbation field comes from the sound–sound interaction. The equations for this field behind the shock front have the form

$$\frac{\partial \delta'' v_{2x}}{\partial t} + v_2 \frac{\partial \delta'' v_{2x}}{\partial x} = -\frac{1}{\rho_2}\frac{\partial \delta'' P_2}{\partial x}, \tag{19}$$

$$\frac{\partial \delta'' v_{2y}}{\partial t} + v_2 \frac{\partial \delta'' v_{2y}}{\partial x} = -\frac{1}{\rho_2}\frac{\partial \delta'' P_2}{\partial y}, \tag{20}$$

$$\frac{\partial \delta'' \rho_2}{\partial t} + v_2 \frac{\partial \delta'' \rho_2}{\partial x} + \rho_2 \left(\frac{\partial \delta'' v_{2x}}{\partial x} + \frac{\partial \delta'' v_{2y}}{\partial y} \right) = -2i\left(\omega - v_2 k_x\right) \frac{\left(\delta\rho_2\right)^2}{\rho_2}, \tag{21}$$

$$\frac{\partial \delta'' \zeta_2}{\partial t} + v_2 \frac{\partial \delta'' \zeta_2}{\partial x} - \frac{\Gamma_1 \zeta_2}{\rho_2}\left(\frac{\partial \delta'' \rho_2}{\partial t} + v_2 \frac{\partial \delta'' \rho_2}{\partial x} \right) =$$
$$i\Gamma_1 \zeta_2 \left(\omega - v_2 k_x\right)\left(1 - \Gamma_1\right)\left(\frac{\delta\rho_2}{\rho_2}\right)^2, \tag{22}$$

where $\delta''\mathbf{v}_2$, $\delta''\rho_2$, and $\delta''\zeta_2$ are the second-order corrections to the velocity, density, and specific internal energy, respectively. On the right-hand sides of Eqs. (21) and (22), it is assumed that perturbations of hydrodynamic quantities in the linear approximation are plane waves [14].

According to Eqs. (19)–(22), the closed equation for the second-order correction to the pressure $\delta'' P_2$ has the form

$$\hat{F}_1 \delta'' P_2 = \psi \left(\delta\rho_2/\rho_2\right)^2, \tag{23}$$

where $\delta\rho_2 = \delta P_2 / c_2^2$,

$$\psi = -4\left(\omega - k_x v_2\right)^2 \left[\frac{n+1}{2} + \frac{\Gamma\zeta_2}{2c_2^2}(\Gamma+1)(\Gamma+1-n)\right]\rho_2,$$

$$\hat{F}_1 = \frac{1}{c_2^2}\frac{\partial^2}{\partial t^2} - \left(1 - \left(\frac{v_2}{c_2}\right)^2\right)\frac{\partial^2}{\partial x^2} + 2\frac{v_2}{c_2^2}\frac{\partial^2}{\partial x \partial t} - \frac{\partial^2}{\partial y^2}.$$

The density ρ_2 is indirectly related to the pressure P_2 through the equation of state (7) and the right-hand side of Eq. (23) coincides in order of magnitude with the ratio $\left(\delta P_2 / P_2\right)^2$.

If, according to standard perturbation theory [14], the solution of Eq. (23) is sought in the form of a plane wave, secular (growing in time) terms appear in $\delta'' P_2$ [14]. A similar problem arises when standard perturbation theory is applied to nonlinear ordinary differential equations [15]. A method that makes it possible to avoid the appearance of secular terms in such equations has been known for a long time (see, e.g., [15]). We generalize this method for solving Eq. (23).

We represent the perturbation of the pressure behind the shock front ΔP_2 in the form of the sum

$$\Delta P_2 = \delta \tilde{P}_2 + \delta'' \tilde{P}_2. \tag{24}$$

The first term on the right-hand side of Eq. (24) differs from the first-order correction to the pressure

$$\delta P_2 = A_{\delta P_2} \exp\left(\varphi(x,y,t)\right), \; \varphi(x,y,t) = ik_x x + ik_{0y} y - i\omega t \; \left(\hat{F}_1 \delta P_2 = 0\right) \tag{25}$$

in the phase factor

$$\delta \tilde{P}_2 = \delta P_2 \exp\left(g(x,y,t)\right). \tag{26}$$

The frequency ω in Eq. (25) is given by dispersion relation (15) and the amplitude $A_{\delta P_2}$ is determined from the solution of the linearized system of equations of hydrodynamics (1), (2), and (4)– (6) [13]. The quantity g is chosen such that the function $\delta \tilde{P}_2$ satisfies inhomogeneous equation (23); i.e.,

$$\hat{F}_1 \delta \tilde{P}_2 = \psi \left(\delta \rho_2 / \rho_2\right)^2. \tag{27}$$

The function $\delta'' \tilde{P}_2$ in this case satisfies the homogeneous wave equation

$$\hat{F}_1 \delta'' \tilde{P}_2 = 0, \tag{28}$$

and secular terms in the second-order correction to the pressure do not appear.

Following [15], we assume that the additional phase g is much smaller than the quantity φ (26), $|g| \ll |\varphi|$. As a result, according to Eqs. (26) and (27), the function g in the first approximation satisfies the linear equation

$$\left(\hat{F}_1 + \hat{F}_2\right) g(x,y,t) = \frac{\psi}{\rho_2 c_2^2} \frac{\delta P_2}{\rho_2}, \tag{29}$$

where

$$\hat{F}_2 = -2i \left\{ \frac{\omega}{c_2^2} \frac{\partial}{\partial t} + k_x \left(1 - \left(\frac{v_2}{c_2}\right)^2\right) \frac{\partial}{\partial x} + \frac{v_2}{c_2^2} \left(\omega \frac{\partial}{\partial x} - k_x \frac{\partial}{\partial t}\right) + k_{0y} \frac{\partial}{\partial y} \right\}.$$

The boundary condition for Eq. (29) at the shock front has the form

$$g(d_1 t, y, t) = 0. \tag{30}$$

To determine the second-order corrections to the velocity, density, and specific energy (as well as the quantity $\delta'' \tilde{P}_2$), the system of Eqs. (19)–(22) should be supplemented by equations taking into account the interaction of entropy–vortex waves with each other and with acoustic waves. The solutions of these inhomogeneous equations do not contain secular terms. For this reason, the desired hydrodynamic quantities can be calculated within standard perturbation theory [14].

3.2. Additional phase near the threshold of appearance of surface waves

We now study the spatial structure of the pressure perturbation field behind the shock front near the threshold of appearance of surface waves, $\theta_0^* - \theta_0 \ll \theta_0^*$. This range is of particular interest because the projection of the speed of sound c_{2x} at $\theta_0^* - \theta_0 \ll \theta_0^*$ is close to the velocity of the shock wave

$$c_{2x} = d_1 + \frac{\delta\kappa\, c_2 k_{0y}^2}{\left(\kappa^2 + k_{0y}^2\right)^{3/2}}, \tag{31}$$

where

$$\kappa \equiv k_x(\theta_0^*) = \frac{(d_1 - v_2)\, k_{0y}}{\sqrt{c_2^2 - (d_1 - v_2)^2}},$$

$$\delta\kappa = k_x - \kappa = \frac{c_2^2(v_1 - d_1)k_{0y}}{\left(c_2^2 - (v_2 - d_1)^2\right)^{3/2}} \sqrt{2\cot\theta_0^*\,(\theta_0^* - \theta_0)}. \tag{32}$$

In Eqs. (31) and (32), it is assumed that the angle θ_0 varies only because of an increase in the x projection of the initial wave vector k_{0x} ($k_{0y} = \text{const}$).

According (31), for the angles θ_0 in the immediate vicinity of the angle θ_0^*, the front of the acoustic wave $x = c_{2x}t$ almost coincides with the shock front, $|d_1 - c_{2x}|t \ll |d_1|t$.

We introduce the function

$$f(x, y, t) = \nu(x - d_1 t) + \left(\eta\,(x - d_1 t) + \beta\,(x - d_1 t)^2\right) \delta P_2. \tag{33}$$

The action of the operator $\hat{F}_1 + \hat{F}_2$ on the function f (33) gives the result

$$\left(\hat{F}_1 + \hat{F}_2\right) f = \frac{2}{c_2^2}\left[c_2^2 - (d_1 - v_2)^2\right]\left(i\nu\delta\kappa - \delta P_2\left[2i\eta\delta\kappa - \beta\left(1 + i\delta\kappa(x - d_1 t)\right)\right]\right) \tag{34}$$

We assume that the quantity ν is small,

$$|\nu| \ll |\kappa|, \tag{35}$$

and the coefficients η and β are given by the expressions

$$\eta = 0, \quad \beta = -\frac{\psi}{2\rho_2^2 c_2^2\left[c_2^2 - (d_1 - v_2)^2\right]}. \tag{36}$$

In this case, the function f in neglect of quantities of the order of $|\nu\delta\kappa/\kappa^2| \ll 1$ satisfies Eq. (29) with boundary condition (30) and coincides with the function g, $f \approx g$. The condition of applicability of the perturbation theory

$$|g| \ll |\varphi|. \tag{37}$$

is also satisfied.

Thus, the additional phase g in the angular range $\theta_0^* - \theta_0 \ll \theta_0^*$ can be approximated by the expression

$$g \approx \nu(x - d_1 t) + \beta(x - d_1 t)^2 \delta P_2. \tag{38}$$

The coefficient ν is generally complex:

$$\nu = \nu_1 + i\nu_2. \tag{39}$$

The real and imaginary parts of ν should be negative ($\nu_1 < 0$) and positive ($\nu_2 > 0$), respectively. The first condition is due to the requirement of damping of the acoustic wave behind the shock

IOP Conf. Series: Journal of Physics: Conf. Series **788** (2017) 012041 doi:10.1088/1742-6596/788/1/012041

front. The second condition appears because the absolute value of the projection of the speed of sound c_{2x} at $\theta_0 = \theta_0^*$

$$c_{2x}^* = d_1 + \frac{\nu_2 c_2 k_{0y}^2}{\left(\kappa^2 + k_{0y}^2\right)^{3/2}} \tag{40}$$

cannot exceed the velocity of the shock wave ($d_1 < 0$).

The quantities ν_1 and ν_2 depend on the conditions of appearance of the shock wave (see below) and cannot be determined in the used approximation.

At relatively small times

$$t \ll \left(\frac{1}{\kappa c_2}\right)\left(\frac{\rho_2}{\delta\rho_2}\right), \tag{41}$$

the front of the acoustic perturbation is weakly behind the shock wave and the second term in Eq. (38) can be neglected. As a result, the additional phase is given by the simple formula

$$g \approx (\nu_1 + i\nu_2)(x - d_1 t). \tag{42}$$

Formula (42) is the main result of this section. According to this formula, at angles of incidence $\theta_0^* - \theta_0 \ll \theta_0^*$ of the shock wave on the anisotropic density perturbation field, the acoustic field following the shock wave is formed behind the shock front because of the nonlinear sound–sound interaction. The amplitude of this field decreases with an increase in the distance from the front $x = d_1 t$ ($\mathrm{Re}\, g = -|\nu_1|(x - d_1 t) < 0$).

4. Results of the numerical calculations

Although formula (42) for the additional phase g was obtained under the assumption of the smallness of the amplitude of initial-density perturbation (see (41)), it indicates the behavior of the acoustic field for perturbations of a finite amplitude [14]. In order to confirm the conclusion made in the preceding section concerning the existence of the damping acoustic field behind the shock front, the equations of hydrodynamics (1)-(5) were numerically solved for the case of collision of the iron (impactor) and aluminum (target) plates (see Fig. 1). Figure 2 shows the numerically calculated spatial profiles of the perturbation of the pressure behind the shock front for the angles θ_0 in the range $\theta_0^{**} < \theta_0 \leq \theta_0^*$ ($\theta_0^{**} = 0.83\theta_0^*$) (the calculations were performed with the code from [19]). According to Fig. 2, the decay of the acoustic field behind the shock front is observed not only at $\theta_0 = \theta_0^*$, but also at smaller angles of incidence. The width of the region of acoustic perturbations increases with the time.

The damping ratio depends on time because Eq. (42) is valid at times $t \ll 300$ ns and, therefore, is inapplicable for the times $t = 250$ and 750 ns used in the numerical calculation. However, using the results of the preceding section, we can write an approximate formula providing a qualitatively appropriate description of the time dependence of the quantity $\nu_1(t)$. This formula can be most simply obtained for the angles $\theta_0 < \theta_0^*$ (Figs. 4 c – 4 f). In this case, $\nu_1(t)$ can be estimated as

$$|\nu_1| \simeq \frac{1}{|d_1 - c_{2x}|t}. \tag{43}$$

In view of Eq. (31), Eq. (43) can be represented in the form

$$|\nu_1| \simeq \frac{\left(\kappa^2 + k_{0y}^2\right)^{3/2}}{\delta\kappa\, c_2\, k_{0y}^2 t}. \tag{44}$$

According to the table ($\theta_0 = 53°$, $\delta\kappa = 15.2$ mm^{-1}; $\theta_0 = 48°$, $\delta\kappa = 39.7$ mm^{-1}), the results of calculations of the damping ratio by Eq. (44) are in qualitative agreement with the data obtained from the analysis of the numerical solution. The best agreement is observed at $\theta_0 = 0.89\theta_0^*$.

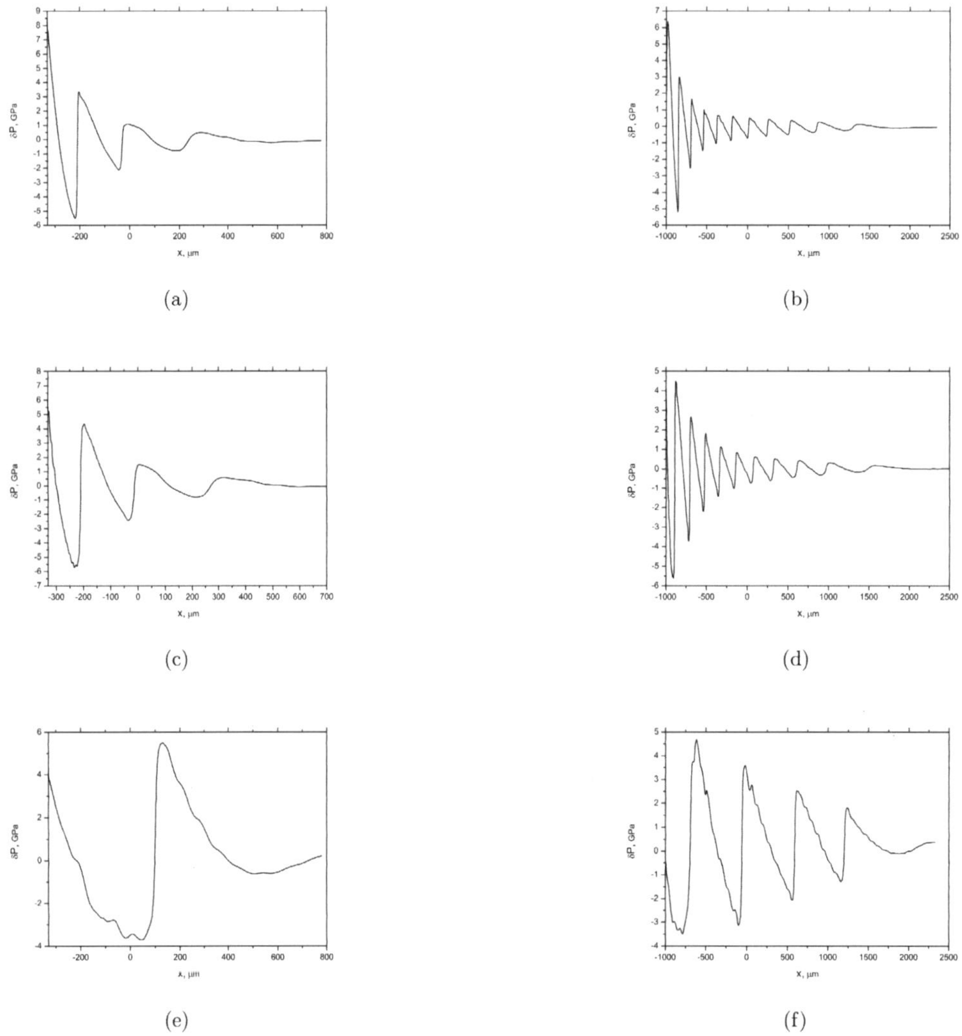

Figure 2. Perturbation of the pressure δP_2 versus the distance from the shock front in the impactor at (a) $\theta_0 = \theta_0^*$, $t = 250$ ns; (b) $\theta_0 = \theta_0^*$, $t = 750$ ns; (c) $\theta_0 = 0.98\theta^*$, $t = 250$ ns; (d) $\theta_0 = 0.98\theta_0^*$, $t = 750$ ns; (e) $\theta_0 = 0.89\theta_0^*$, $t = 250$ ns; and (f) $\theta_0 = 0.89\theta_0^*$, $t = 750$ ns. The velocity of the impactor before the collision is $v_1 = 5$ km/s, the amplitude of the initial perturbation of the density is 5% of the initial density, $\theta_0^* = 53.9°$, and $\theta_0^{**} = 44.5°$. The minimum value of the coordinate x corresponds to the position of the shock front.

The damping ratio was calculated by Eq. (44) with the value $c_2 = 6.44$ km/s for the speed of sound behind the front of the acoustic wave. This value was obtained from the numerical solution of Eqs. (1)–(3) and (7). The projection of the wave vector is $k_{0y} = 51.1mm^{-1}$.

At $\theta_0 = \theta_0^* = 53.9°$, the speed of sound becomes equal to the velocity of the shock wave and estimate (43) is inapplicable. However, the analysis of the results of the numerical calculations shows that the product $t|\nu_1(t)|$ weakly depends on time (see table), $t|\nu_1(t)| \approx$ const, and the

Table

θ_0	53.9° (θ_0^*)		53° ($0.98\theta_0^*$)		48° ($0.89\theta_0^*$)	
t, ns	250	750	250	750	250	750
$\vert\nu_1\vert$, mm^{-1} (theory)	-	-	7.38	2.46	2.5	0.83
$\vert\nu_1\vert$, mm^{-1} (numerical solution)	4.25	1.39	4.78	1.57	2.94	0.78

damping ratio decreases with an increase in the time t as $|\nu_1| \sim 1/t$.

5. Conclusions

The interaction of the shock wave with the anisotropic density perturbation field has been studied in the quadratic approximation. To solve inhomogeneous equations appearing in this approximation, a new analytical method has been proposed to avoid the appearance of secular (growing in time) terms. This method generalizes the approach used in the theory of nonlinear ordinary differential equations [15] and involves the introduction of an additional phase into a solution obtained in the linear approximation. Using this method, we have shown that a damping acoustic field, which is not described by surface waves, appears behind the shock front incident at certain angles on the spatially anisotropic density perturbation field. The width of the region where acoustic perturbations are observed increases with the time. The problem of the propagation of the shock wave in the incident sample, where the spatially anisotropic density perturbation field initially exists, has been numerically solved in application to the collision of two plates. The results of the numerical solution of the equations of hydrodynamics confirm qualitative conclusions made in the quadratic approximation concerning the spatial structure of the acoustic field behind the shock front.

Acknowledgments

We are grateful to P.P. Zakharov for assistance in the numerical calculations.

References

[1] Diakov S P 1954 *Zhur. Exp. i Teor. Fiz.* **27** 288 [in Russian]
[2] Kontorovich V M 1959 Sov. Phys. Acoustics **5** 320
[3] Kuznetsov N M 1989 *Sov. Phys. Usp.* **159** 993
[4] Keller J and Merzkirch W 1990 *Experiments in Fluids* **8** 241
[5] Yang Y, Zhang Q and Sharp D H 1994 *Phys. Fluids* **6** 1856
[6] Wouchuk J G, Huete Ruiz de Lira C and Velikovich A L 2009 *Phys.Rev.E* **79** 066315
[7] Wouchuk J G 2001 *Phys.Rev.E.* **63** 056303
[8] Wouchuk J G, Huete Ruiz de Lira C and Velikovich A L 2009 *Phys.Rev E* **79** 066315
[9] Huete Ruiz de Lira C, Velikovich A L and Wouchuk J G 2011 *Phys.Rev.E* **83** 056320
[10] Nishihara K, Wouchuk J G, Matsuoka C, Ishizaki R and Zhakhovsky V V 2010 *Phil. Trans. R. Soc. A* **368** 1769
[11] Huete Ruiz de Lira C, Wouchuk J G and Velikovich A L 2012 *Phys.Rev E* **85** 026312
[12] Huete Ruiz de Lira C, Wouchuk J G, Canaud B and Velikovich A L 2012 *Journal of Fluid Mechanics* **700** 214
[13] Gorodnichev K E and Kuratov S E 2013 *VANT ser. Mat. Mod. Fiz. Prots.* **2** 37 [in Russian]
[14] Isakovich M A 1973 *General Acoustics* (Moscow: Nauka) [in Russian]
[15] Naife A H 2004 *Perturbation Methods* (Weinheim: WILEY-VCH Verlag GmbH and CO. KGaA)
[16] Zeĺdovich Ya B, Raizer Yu P 2002 *Physics of Shock Waves and High-Temperature Hydrodynamic Phenomena* (New York: Dover Publication Inc. Mineola)
[17] Landau L D and Lifshitz E M 1987 *Fluid Mechanics* (New York: Pergamon)
[18] Zababakhin E I 1997 *Some issues of gasdynamics of explosion* (Snezhinsk) [in Russian]
[19] Menshov I S and Zakharov P P 2014 *Int.J. for Num. Meth. in Fluids* **76** 109

IOP Conf. Series: Journal of Physics: Conf. Series **788** (2017) 012042 doi:10.1088/1742-6596/788/1/012042

Magnetized plasma structures in laser-irradiated curved targets

Philipp Korneev

National Research Nuclear University "MEPhI" (Moscow Engineering Physics Institute), 31 Kashirskoe sh., Moscow, 115409 Russian Federation

E-mail: korneev@theor.mephi.ru

Abstract. This work presents an extensive theoretical study of magnetic field generation effect in hollow targets with curved internal surface, known as "snail" or "escargot" targets [1]. They were recently proposed as a robust setup for generating intense spontaneous quasi-stationary magnetic fields frozen in laser-produced plasmas. The results of such studies are indispensable for planing experimental investigations and for possible applications of the curved targets, including, but not limited to, laboratory astrophysics, fast ignition, and particle acceleration.

1. Introduction

Laboratory-produced magnetic fields attract particular interest both as a special physical phenomenon and as a necessary part of astrophysical studies in laboratory, magnetized solutions of Inertial Confinement Fusion (ICF), particle guiding, atomic physics in magnetic fields, and many other applications. Laser-based (optical) approaches to strong magnetic field generation are compact, precise, and thus looking very promising. By now, the highest quasi-static magnetic fields produced with the use of lasers, reach the values of 10^7 Gauss [2, 3, 4, 5]. More intense magnetic fields have been measured in laser plasma interaction with high-intensity lasers (see, i.e. [6]), but the time of life of those fields is strictly limited by the laser pulse duration. One way to increase the magnetic field value was proposed in 2015 [1]. It was shown on the base of large-scale 2D particle-in-cell (PIC) simulations that an intense laser pulse, caught in the internal volume of a hollow snail-shaped target, may generate a very strong magnetic field, decoupled from the pulse and frozen in the ablated plasma. The generated magnetized plasma then carries the field over the hydrodynamic time scale, so that the magnetized structures may exist for tens of picoseconds and more, becoming extremely interesting for a vast variety of applications.

Most of the conventional laboratory generators of quasi-static magnetic fields, both coil-based, and laser-based (optical) are working within a discharging scenario, where strong currents are released either from a capacitor bank or from a laser-irradiated capacitor-like target. The currents excite and support magnetic fields in a surrounding space. Because of their generation process, these are vacuum magnetic fields. They may be used to magnetize secondary targets or guide particles. Usually, the vacuum magnetic field is a convenient laboratory instrument, though magnetization process becomes sometimes challenging, i.e. for laboratory astrophysical studies, where plasmas usually are both magnetized and collisionless.

The novel approach [1], based on special geometrical design of the targets, allow to produce rather magnetized plasmas than vacuum magnetic fields, so that finally not currents of conducting

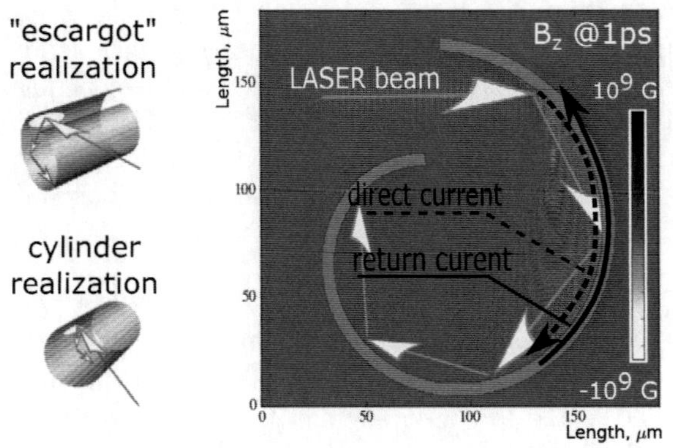

Figure 1. Left panel: Targets proposed for the magnetic field generation, using multiple reflection in the target hollow. Red-yellow arrows indicate laser pulse propagation. Right panel: Results of 2D PIC simulation of the early-stage laser pulse interaction with a snail-type target (Bz component), which presents the generation mechanism. Laser pulse (1.5 ps, 5×10^{19}W/cm^2) propagates from the left along the internal surface of the target. Black arrows show currents near the target internal surface, bright blue contour shows the target.

electrons but the laser plasmas themselves carry the magnetic field. This scheme possesses the advantages of intense laser pulses, which may produce very strong magnetic fields, and the conservation of spontaneous magnetic fields by means of plasma magnetization during the ablation process. Compactness of the target internal volume allows saving the high values of generated fields in magnetized structures over the hydrodynamic time scales. The astrophysical applications of this scheme are straightforward, since many interesting phenomena take place in the target itself. The magnetization of plasmas (one of the most important astrophysical parameters, defined as a relation between the magnetic and the plasma pressure) observed in PIC simulations, reaches the order of unity, and may be controlled by the target size and laser parameters. Many more applications that demand strong magnetic fields as particle guiding or laser-driven magnetized ICF [7] may also follow.

To remind briefly the considered scheme of magnetic field generation, a special target geometry is presented in Fig. 1. On the left panel there are two simplest realizations, a "snail", or "escargot" geometry, and a cylinder geometry. In both cases, the laser pulse reflects multiple times on the internal target volume. To make this possible, in the case of the cylinder, the pulse should enter through a side open edge. Strictly speaking, the reflection geometry is rather complicated and the interaction process may be fully reproduced only by 3D simulations. But for the incidence angles, as shown in Fig. 1, which are close to the right angle with the cylinder axis, interaction process is expected to be similar to that in the "snail" target. The latter allows description by means of 2D PIC simulations, since it has an entrance in the interaction plane. In the presented work this shape is generally used.

The physics of the magnetic field generation in the "snail" target is defined mainly by electron currents excited by laser radiation, shown in Fig. 1 (right panel) with black arrows. The surface current (dashed arrow) is generated by the directly laser-accelerated electrons, which go along the the internal curved target surface because of surface electric and magnetic fields and is supported by multiple laser pulse reflections. The return discharge current raises on the femtosecond time scale in order to neutralize the surface charge in the region of laser-surface interaction. Its propagation is also defined by the target internal surface geometry. During the action of the laser pulse, these currents create the strong magnetic fields in the internal target volume, which is filled at the same time with the plasma. When the pulse ends, the value of these currents drop down, and only the magnetic field frozen in the ablated plasmas survives. The magnetized structure inside the target hollow evolves then on the hydrodynamic time scale. It comes to equilibrium between the plasma pressure and the magnetic field pressure. At this stage, the magnetic field may be increased once more by the compression with a flow of hot plasma, coming from the target internal walls [1]. Later in time, the hot target walls expand, thus leading

IOP Conf. Series: Journal of Physics: Conf. Series **788** (2017) 012042 doi:10.1088/1742-6596/788/1/012042

to the loss of confinement. Because of the large spatial scale and high plasma density in PIC simulations, the simulation time was limited to 30-40 ps, when the magnetic field intensity drops down to values approximately half of the maximum values. This lifetime scaling is suitable for many applications and diagnostics. In this work in the simulations the laser wavelength was chosen to be 1 μm, target material was aluminum with the ion density $\approx 10^{22}$ cm^{-3}. Earlier it was found [1], that the target density is not a critical parameter if it is much higher than the critical density. Simulations were presented with the code PICLS [8], the simulation box was 9496×9216 cells, spatial resolution was ≈ 20 nm, temporal resolution was ≈ 0.07 fs.

2. Intensity, polarization and size dependence

As theoretically proved in this work, the described process of magnetic field generation by intense laser pulse interaction with curved-surface targets is an effective and robust mechanism in a wide range of parameters. The first example, presented in [1], correspond to a target diameter ≈ 100 μm, a 1.5 ps laser pulse of intensity 5×10^{19} W/cm^2 and demonstrate the characteristic values of the generated fields of the order of 2×10^8 Gauss. To understand the scaling of these values with laser pulse intensity, simulations for the same target but various laser intensities were carried out, see Fig. 2. The use of relativistic laser intensities allows to increase strongly the magnetic field amplitude by generation of a direct current of laser-accelerated electrons. While further increase of the laser intensity, as shown in Fig. 2 (laser intensity 10^{22} W/cm^2), the relativistic transparency and hole-boring effects inhibits the process of current generation and only a relatively small part of the laser energy may be converted to the magnetic field.

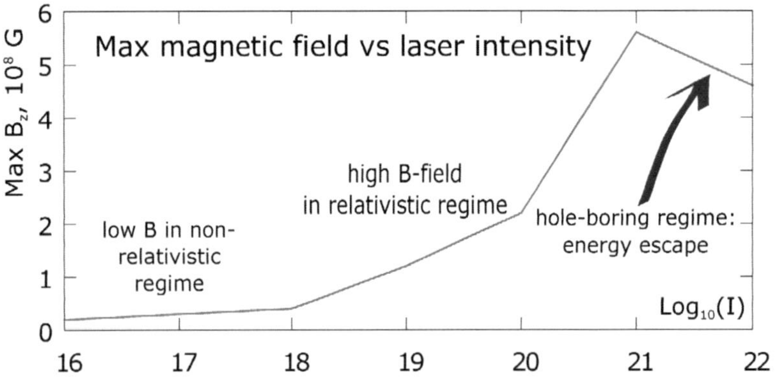

Figure 2. Scaling of the magnetic field amplitude with the laser pulse intensity for a 1.5 ps laser pulse and 200 μm diameter "snail" target with a wall thickness of 10 μm, based on 2D PIC simulations.

Other important parameter is the target size. For the same laser parameters, but different target sizes, the maximum magnetic field values are increasing for smaller targets. Several results of the simulations are presented in Fig. 3, where three target diameters 50, 100, and 200 μm were examined at the fixed laser intensity 10^{19} W/cm^2. The highest magnetic field value was obtained for the smallest diameter of 100 μm ($\sim 3 \times 10^8$ Gauss). For this case the low density magnetized plasma is compressed by high density unmagnetized ablated plasma, imploding from the walls [1]. For less energetic laser pulses, operating in mild relativistic regime, like that on PHELIX facility at GSI, with intensity $I \sim 10^{18}$W/cm^2, the scaling in Fig. 2 predicts magnetic fields of the order of $\sim 2 - 3 \times 10^7$Gauss. In this case decreasing of the target size becomes even more important, since an available for the ablated plasma volume is much less. In Fig. 4 simulation results are presented for the laser pulse intensity $I = 10^{18}$ W/cm^2 and the snail target diameters ≈ 50 μm and ≈ 100 μm. The maximum magnetic field at 6.1 ps is $\sim 3\times 10^7$ Gauss and $\sim 1.5 \times 10^7$ Gauss respectively. Note, that the geometry of the magnetized structure becomes more regular for the smaller target sizes. For the smallest targets, the magnetized structure was almost homogeneous inside the target volume, see Fig. 3 and Fig. 4.

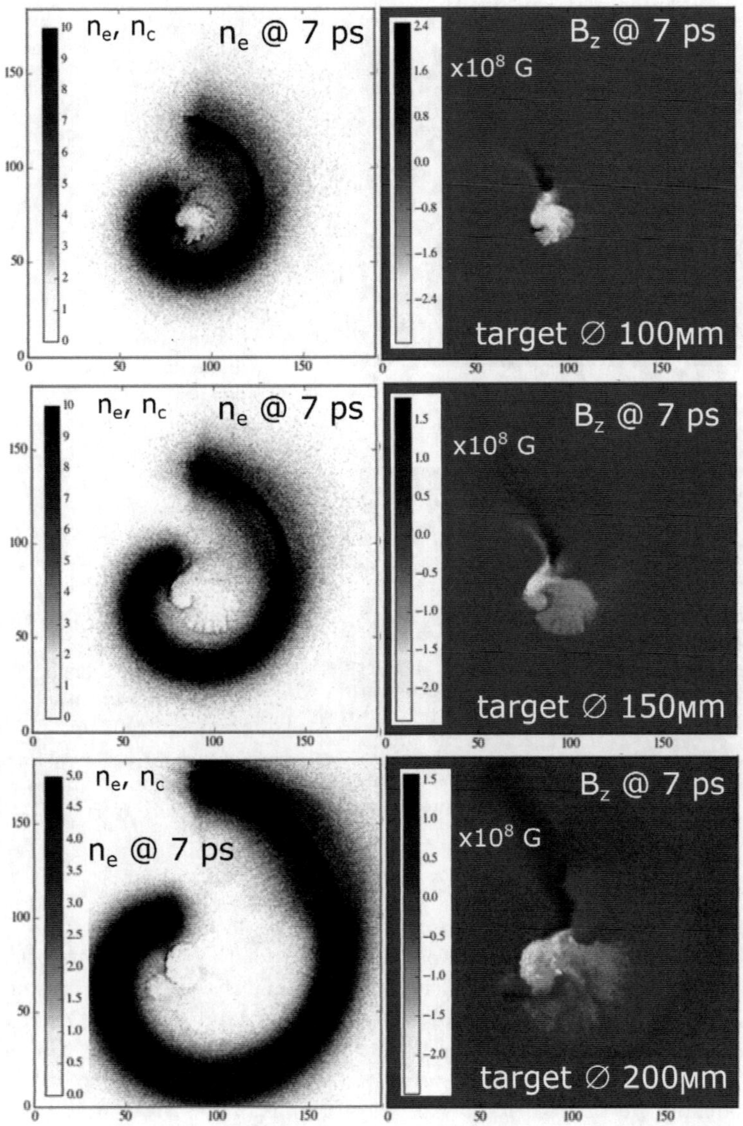

Figure 3. Results of 2D PIC simulations for different diameters of the "snail" targets, for a 1.5 ps laser pulse with the intensity 1×10^{19} W/cm^2, at 7 ps after the beginning of interaction. On the left column – electron density, in the critical densities $n_c = 10^{21}$ cm^{-3}, on the right column B_z component of the magnetic field. One of the main effects which contribute to the magnetic field value is compression of the produced low-density magnetized plasma by high-density ablated plasma at late times. On the presented results for the smallest target diameter 100 μm compression plays the most important role, so that the magnetic field value reaches $|B_z| \sim 3 \times 10^8$ Gauss. For the diameter 150 μm and 200 μm magnetic field has similar values of $|B_z| \sim 2...2.5 \times 10^8$ Gauss.

As the main driver for the magnetic field generation are electron currents, excited during the laser interaction with the internal target surface, the interaction mechanism should put an imprint on the magnetic field value and structure. Indeed, for the case of s- and p- polarizations of the incident laser pulse, the magnetic field structures are quite different, see Fig. 5. The analysis of energy redistribution and the effectiveness of the current generation is not evident because of the changing interaction condition in time and during the propagation. It may up to some point correspond to the Brunel-type mechanism, with the complications because of magnetic fields, stochasticity, plasma resonances etc. The corresponded analysis and the more detailed studies of the interaction mechanisms and the dependence on polarization and incident angles are to be presented elsewhere.

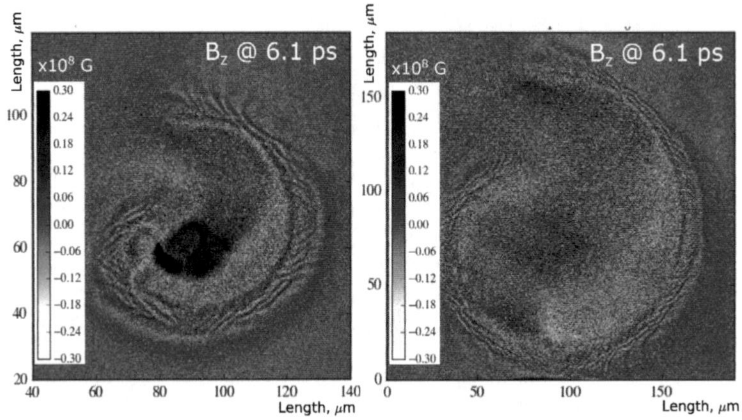

Figure 4. Results of 2D PIC simulations, showing B_z component of the magnetic field, for the laser pulse intensity 10^{18} W/cm^2, and two targets: $\sim$ 50 μm in diameter (left panel) and $\sim$ 100 μm in diameter (right panel).

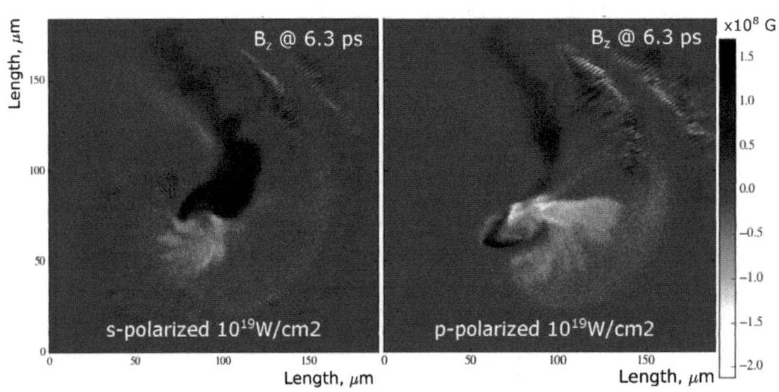

Figure 5. Dependence of the magnetized structure in the snail target on polarization of the laser pulse. For 1.5 ps, 5×10^{19} W/cm^2, s- (left) and p- (right) polarized laser pulses, and for 200 μm in diameter snail target. The showed time moment is 6 ps.

3. Deviations from "snail" geometry

3.1. Non-ideal reflection

The simulation results presented in Fig. 3 and Fig. 4 were carried out for the ideal targets with sharp idealized surfaces, and the geometrical shape defined by an analytical expression from the work [1]. After multiple simulations it was generally found, that a small preplasma of few tens of microns width does not considerably change neither in the magnetic field value, nor in its structure, and by this reason these results are not presented here. However, the important point may be the target production technique. It may be possible, that the surface would not be ideal. In this case, the laser pulse reflection may become less effective, and the current values may drop down. To classify the effect of the non-ideal surface, a test simulation was done with an artificial noise (saw-shaped) in the main reflection region, see Fig. 6. A saw on the surface was composed from seven equilateral triangles with 5 μm side. Surprisingly, the magnetic field structure remained almost the same, but the magnetic field value appeared to be about two times less. This results proves again the robustness of the proposed scheme for the magnetic field generation.

3.2. Modified curved targets

The geometrical properties of the magnetized structure obtained as a result of the interaction process is an important parameter for possible applications. Magnetic field lines are closed, either inside the structure near the target or at infinity. When the lines are closed at far distances from the target area, plasma is promoted to expand in the direction along the lines and locked in

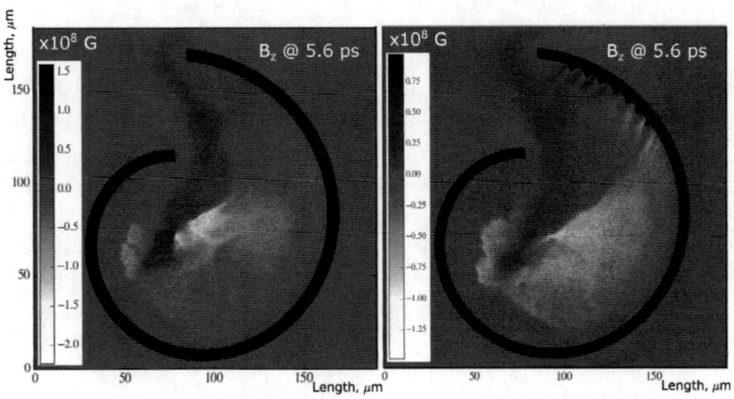

Figure 6. Simulation results for interaction of 1.5 ps 10^{19} W/cm^2 laser pulse with 200 μm "snail" target in case of the ideal (left panel) and artificially degraded surface (right panel). While the magnetic field structure is similar on both panels, the magnetic field value is less for degraded surface. Target shape is shown in solid black.

the transverse direction. Closing of magnetic lines in the target internal volume may help to create a confined magnetized plasma, expanding on slow hydrodynamical time scales. For many magnetized applications, such a confined magnetic structure may be a very interesting solution. Probably, the simplest way is to follow toroidal-like geometry, which may be realized with a spherical target with an entrance hole and a conical tip on the opposite side. The tip may have a different geometry, its mission is to reflect the laser beam and not allow it to escape. The cross-section of an example of such a target is shown in Fig. 7 (left panel) with black color. Laser pulse on both upper and lower sides of the target propagates according to the target shape, rising up correspondent currents. In case of spherical symmetry, the geometry of the currents acquires quasi-toroidal shape, and the magnetic lines close in the target internal volume. On the right panel of Fig. 7 the tip is replaced by the diffusive degraded surface, similar to the saw in Fig. 6 (right panel). In this case directly accelerated by the laser pulse electrons do not propagate along the target surface, and only the return currents survive (solid black line in Fig. 1). As a result, the magnetic field value is less pronounced, though it is still clearly visible. This may be useful for understanding of the roles of two main electron currents during laser-target interaction and may become a good point for experimental "yes or no" confirmation.

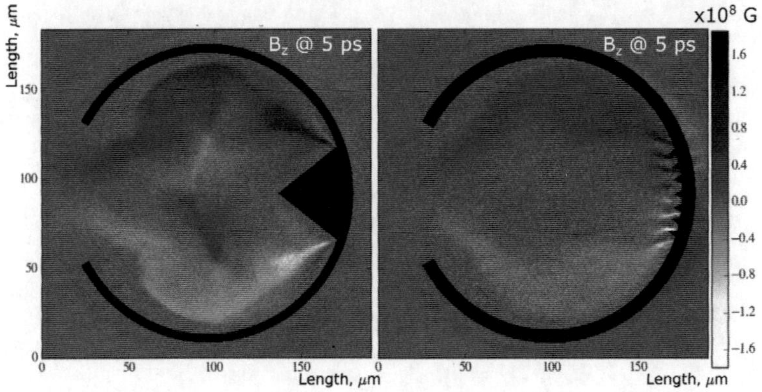

Figure 7. Results of PIC simulations, showing a structure of the magnetic field in case of a tipped cylinder (left panel) and a diffused interaction surface (right panel). While the magnetic field distribution has the similar structures, in the tipped cylinder case the magnetic field value is ~ 2 times higher. Target shape is shown in solid black.

4. Conclusion

The presented extensive numerical studies show a very robust character of the magnetic field generation, which works in a wide range of laser pulse intensities and in different curved targets.

The predicted amplitudes of the magnetic fields reach values of the order of $\sim 1 - 3 \times 10^7$ Gauss ($\sim 1 - 3$ kT), even for moderate mild relativistic intensities. These values are already greater than the modern achievements for the highest optically generated magnetic fields with the capacitor-coil targets [1, 5]. Even higher magnetic fields, up to giga-gauss level, are expected to be generated with more intense laser pulses, according to the scaling in Fig. 2. The limits are defined by the relativistic transparency and the hole-boring effects, preventing further reflection of intense pulse inside the target volume. This limitation may probably be removed by the use of targets with special internal geometries [9].

In conclusion, a novel promising setup for manipulation of spontaneous laser-produced magnetic fields of the order of several kT and higher is numerically studied. This approach may become a flexible and convenient tool for a wide range of applications, from laboratory astrophysical studies to particle acceleration schemes. The proposed study opens a new page in the engineering of laser-driven devices for strong B-field production, operating with the target geometry to direct laser propagation and matter response.

5. Acknowledgments

The work was granted access to the HPC resources of CINES under allocations 2015-056129 and 2016-056129 made by GENCI (Grand Equipement National de Calcul Intensif). A part of the numerical calculations was presented at Joint Supercomputer Center of the Russian Academy of Sciences and resources of NRNU MEPhI high-performance computing center. This work was supported by the Russian Foundation for Basic Research (16-52-50019 ЯФ).

[1] Korneev P, d'Humières E and Tikhonchuk V 2015 *Physical Review E* **91** 043107

[2] Chang P Y, Fiksel G, Hohenberger M, Knauer J P, Betti R, Marshall F J, Meyerhofer D D, Séguin F H and Petrasso R D 2011 *Phys. Rev. Lett.* **107**(3) 035006 URL http://link.aps.org/doi/10.1103/PhysRevLett.107.035006

[3] Yoneda H, Namiki T, Nishida A, Kodama R, Sakawa Y, Kuramitsu Y, Morita T, Nishio K and Ide T 2012 *Phys. Rev. Lett.* **109**(12) 125004 URL http://link.aps.org/doi/10.1103/PhysRevLett.109.125004

[4] Fujioka S, Zhang Z, Ishihara K, Shigemori K, Hironaka Y, Johzaki T, Sunahara A, Yamamoto N, Nakashima H, Watanabe T, Shiraga H, Nishimura H and Azechi H 2013 *Scientific Reports* **3** – URL http://dx.doi.org/10.1038/srep01170

[5] Santos J J, Bailly-Grandvaux M, Giuffrida L, Forestier-Colleoni P, Fujioka S, Zhang Z, Korneev P, Bouillaud R, Dorard S, Batani D, Chevrot M, Cross J E, Crowston R, Dubois J L, Gazave J, Gregori G, d'Humiéres E, Hulin S, Ishihara K, Kojima S, Loyez E, Marquš J R, Morace A, NicolaïP, Peyrusse O, Poyé A, Raffestin D, Ribolzi J, Roth M, Schaumann G, Serres F, Tikhonchuk V T, Vacar P and Woolsey N 2015 *New Journal of Physics* **17** 083051 URL http://stacks.iop.org/1367-2630/17/i=8/a=083051

[6] Sefcik J, Perry M D, Lasinski B F, Langdon A B, Cowan T, Hammer J, Hatchett S, Hunt A, Key M H, Moran M, Pennington D, Snavely R, Trebes J and Wilks S C 2004 *Megagauss Magnetic Field Generation, Its Application to Science and Ultra-High Pulsed-Power Technology* ed Schneider-Muntau H J pp 161–166

[7] Gus'kov S Y and Korneev F A 2016 *JETP Letters* **104** 1–5 ISSN 1090-6487 URL http://dx.doi.org/10.1134/S0021364016130117

[8] Sentoku Y and Kemp A J 2008 *J. Comput. Phys.* **227** 6846–6861 ISSN 0021-9991 URL http://dx.doi.org/10.1016/j.jcp.2008.03.043

[9] Korneev P, Tikhonchuk V and d'Humières E 2016 *arXiv preprint arXiv:1602.02626*

On a Physical Field Theory of Micropolar Thermoelasticity

Vladimir Kovalev[1], **Evgenii Murashkin**[2,3,4a] **and Yuri Radayev**[3b]

[1]Samara State Technical University, 244 Molodogvardeyskaya Str., Samara, 443100, Russia
[2]National Research Nuclear University MEPhI (Moscow Engineering Physics Institute), 31 Kashirskoe shosse, 115409 Moscow, Russia
[3]Institute for Problems in Mechanics of the Russian Academy of Sciences,Vernadski Ave. 101 Bldg. 1, Moscow, 119526, Russia
[4]Bauman Moscow State Technical University, 2nd Baumanskaya Str., 5/1, Moscow, 105005, Russia

E-mail: [a]`evmurashkin@gmail.com`, [b]`radayev@ipmnet.ru`

Abstract. A non-linear mathematical model of thermoelastic micropolar continuum is developed. The model is presented in terms of 4-covariant field theoretical formalism. Lagrangian density for thermoelastic continuum with three micropolar directors is given and the least action principle is formulated. Corresponding field equations of micropolar thermoelasticity are obtained. Variational symmetries of the thermoelastic action are used to formulate covariant conservation laws. Following the usual procedure, micropolar thermoelastic Lagrangians are represented as functions of independent rotationally invariant arguments. The latter constitutes a complete system of objective finite strain measures of micropolar thermoelasticity. Constitutive equations of micropolar thermoelasticity are obtained and discussed.

1. Introduction

In recent years, metamaterials manufacturing and processing have attracted significant attention from the engineers community and researchers. These artificial metamaterials exhibit abnormal physical properties not usually found in nature. Examples include negative Poisson's ratio (auxetic materials), negative thermal expansion, negative electric permittivity and the magnetic permeability. These physical phenomena can not be described in terms of the theory of classical continuum mechanics. In this case, microstructure continuum theories based on the necessity of additional (extra) freedom degrees are required. The physically infinitesimal volume in such continua is not a material point, but a much more sophisticated object, with its intrinsic additional freedom degrees (microrotations, microoscillations). Therefore, the derivations of non-linear Lagrangian, Hamiltonian, extra-stress and extra-strain, valid in the most general case of finite deformations and microrotations, to continua with microstructure [1,2,3]. It stands today as one of the most important problems of the continuum mechanics. That is why a development of complex continuum theories seems to be actual.

The thermomechanical behavior of metamaterials can be described in terms of field formalism and frameworks of micropolar thermoelasticity. In the present study the Green–Naghdi (GN-theory) [4] thermoelastic model is employed. Application of the field formalism principles to continuum mechanics leads to a natural forms of constitutive equations [5,6]. Rotationally invariant Lagrangian forms along with the requirement of the Galilean translational invariance provide determination of a complete system of strain measures.

After the preliminary remarks of Sec. 1 a brief outline of basic principles and requisite field equations are given (Sec. 2). It includes variational formulation of the principle of least action, Euler–Lagrange equations and divergent forms of conservation laws.

In Sec. 3 the thermoelastic Langrangian and least action integral are defined. The differential field equations corresponding to the action integral are derived. The energy-momentum tensor for micropolar thermoelastic continuum is obtained and the corresponding conservation laws are formulated. Its components allow to derive the Eshelby stress tensor, Umov–Poynting vector, pseudomomentum vector and Hamiltonian (total energy) for thermoelastic micropolar media.

In Sec. 4 the objective strain measures for micropolar thermoelastic continuum are obtained from the action invariance under coordinate frame rotations, translations of spatial coordinates and time translations.

The relative strain measures are discussed in Sec.5. The rotationally invariant form of the Helmholtz free energy is given.

In Sec. 6 the objective forms of constitutive equations of micropolar thermoelastic continuum are obtained.

2. Covariant Fields. Field Equations. Action. Conservation Laws

Theory of micropolar thermoelasticity can be developed in terms of the field theory formalism by using the action integral. Following considerations shall be restricted to a plane spacetime. Thus a general form of action furnishes

$$\mathscr{I} = \int \mathscr{L}(X^\beta, \varphi^k, \partial_\alpha \varphi^k) d^4X, \tag{1}$$

where φ^k are covariant physical fields; $\mathscr{L}$ is the Lagrangian density; X^β are the referential coordinates; $d^4X = dX^1dX^2dX^3dX^4$ is spacetime elementary volume.

The least action principle states that the actual field is realized in the spacetime in a way that the action integral (1) is minimum, i.e. for any admissible variations of physical fields $\delta\varphi^k$ and $\delta X^\beta = 0$ the following equation is valid

$$\delta \mathscr{I} = 0. \tag{2}$$

The Euler–Lagrange field equations corresponding to action (1) are

$$\mathscr{E}_k(\mathscr{L}) = \frac{\partial \mathscr{L}}{\partial \varphi^k} - \partial_\alpha \frac{\partial \mathscr{L}}{\partial(\partial_\alpha \varphi^k)} = 0.$$

In field theory a conservation law can be formulated as

$$\partial_\beta J^\beta = 0, \tag{3}$$

where the vector $J^\beta = J^\beta(X^\beta, \varphi^k, \partial_\alpha \varphi^k)$ is called the 4-current. The 4-current can be obtained in the form

$$J^\beta = \frac{\partial \mathscr{L}}{\partial(\partial_\beta \varphi^k)} \delta^\nabla \varphi^k + \left(\mathscr{L} \delta^\beta_\alpha - (\partial_\alpha \varphi^k) \frac{\partial \mathscr{L}}{\partial(\partial_\beta \varphi^k)} \right) \delta^\nabla X^\alpha, \tag{4}$$

wherein $\delta^\nabla = \delta/\varepsilon$ denote finite variations corresponding to symmetry transformations of X^β and φ^k.

In fact the current (4) can be determined if variational symmetries of action (1) are known.

3. Thermoelastic Langrangian. Energy-Momentum Tensor for Micropolar Thermoelastic Continuum

Thermoelastic action for micropolar continuum is taken in the form

$$\mathscr{I} = \int \mathscr{L}(X^\alpha, x^j, \underset{\mathfrak{a}}{d}^j, \vartheta, \partial_4 x^j, \partial_4 \underset{\mathfrak{a}}{d}^j, \partial_4\vartheta, \partial_\beta x^j, \partial_\beta \underset{\mathfrak{a}}{d}^j, \partial_\beta\vartheta) dX^1 dX^2 dX^3 dX^4. \tag{5}$$

In (5)X^α ($\alpha = 1,2,3$) are referential coordinates; x^j ($j = 1,2,3$) are spatial coordinates; $\underset{\mathfrak{a}}{d}^j$ are micropolar directors; ϑ is temperature displacement. Equations $x^j = x^j(X^\alpha)$ determine the deformation of continuum.

We assume the action density (5) in following form

$$\mathscr{L} = \frac{1}{2}(\partial_4 x^k)\rho_{kj}(\partial_4 x^j) + \frac{1}{2}(\partial_4 \underset{\mathfrak{a}}{d}^i)\overset{\mathfrak{ab}}{\mathfrak{J}}_{ij}(\partial_4 \underset{\mathfrak{a}}{d}^j) - \psi(X^\alpha, x^j, \underset{\mathfrak{a}}{d}^j, \vartheta, \partial_4\vartheta, \partial_\alpha x^j, \partial_\alpha \underset{\mathfrak{a}}{d}^j, \partial_\alpha\vartheta). \tag{6}$$

Hereafter $\overset{\mathfrak{ab}}{\mathfrak{J}}_{ij}$ denotes the microinertion tensor; ρ_{ij} is mass density tensor; ψ is referential volume density of the Helmholtz's free energy. For tensors $\overset{\mathfrak{ab}}{\mathfrak{J}}_{ij}$ and ρ_{ij} the symmetry conditions $\rho_{ij} = \rho_{ji}$, $\overset{\mathfrak{ab}}{\mathfrak{J}}_{ij} = \overset{\mathfrak{ab}}{\mathfrak{J}}_{ji}$ are to be valid.

The differential field equations corresponding to action integral (5) read

$$\begin{aligned}
&\partial_\alpha S^{\alpha\cdot}_{\cdot j} - \partial_4 P_j = -\frac{\partial\mathscr{L}}{\partial x^j} && (\alpha = 1,2,3;\ j = 1,2,3),\\
&\partial_\alpha \mathscr{M}^{\alpha\cdot}_{\cdot j} + \mathscr{A}_j - \partial_4(\mathscr{Q}_j) = 0 && (\alpha = 1,2,3;\ j = 1,2,3),\\
&\partial_\alpha j^\alpha_{\mathrm{R}} + \partial_4 s = \frac{\partial\mathscr{L}}{\partial\vartheta} && (\alpha = 1,2,3),
\end{aligned} \tag{7}$$

and supplemented by the following equations:

$$\begin{aligned}
&S^{\alpha\cdot}_{\cdot j} = -\frac{\partial\mathscr{L}}{\partial(\partial_\alpha x^j)}, && \mathscr{M}^{\alpha\cdot}_{\cdot j} = -\frac{\partial\mathscr{L}}{\partial(\partial_\alpha d^j)}, && \mathscr{A}_j = \frac{\partial\mathscr{L}}{\partial d^j},\\
&P_j = \frac{\partial\mathscr{L}}{\partial(\partial_4 x^j)}, && \mathscr{Q}_j = \frac{\partial\mathscr{L}}{\partial(\partial_4 d^j)}, &&\\
&s = \frac{\partial\mathscr{L}}{\partial(\partial_4\vartheta)}, && j^\alpha_{\mathrm{R}} = \frac{\partial\mathscr{L}}{\partial(\partial_\alpha\vartheta)}, &&
\end{aligned} \tag{8}$$

where $S^{\alpha\cdot}_{\cdot j}$ is the first Piola–Kirchhoff tensor; $\mathscr{M}^{\alpha\cdot}_{\cdot j}$ is the extra-stress tensor; $\mathscr{A}_j$ are generalized moments; P_j and $\mathscr{Q}_j$ are generalized momenta; s is entropy density; j^α_{R} is the referential entropy flux. 1st eq. of (7) is the motion balance equation. 2nd eq. of (7) is the momentum balance equation. If $\dfrac{\partial\mathscr{L}}{\partial(\partial_\alpha\vartheta)} = 0$ then 3rd eq. of (7) constitutes the entropy balance equation.

We proceed to conservation laws for the system of field eqs. (7). It is known that action integral $\mathscr{L}$ is invariant under translations of all coordinates X^α if $\mathscr{L}$ does not depends on X^α explicitly. Therefore, the 4-covariant energy-momentum tensor can be easily obtained and the corresponding conservation laws are formulated. There are four groups of equations for the components of the canonical energy-momentum tensor $T^{\mu\cdot}_{\cdot\lambda}$ (λ, $\mu = 1,2,3,4$):

$$T^{\mu\cdot}_{\cdot\lambda} = \mathscr{L}\delta^\mu_\lambda + S^{\mu\cdot}_{\cdot l}(\partial_\lambda x^l) + \overset{\mathfrak{a}}{\mathscr{M}}{}^{\mu\cdot}_{\cdot l}(\partial_\lambda \underset{\mathfrak{a}}{d}^l) - j^\mu_{\mathrm{R}}(\partial_\lambda\vartheta) \qquad (\lambda,\ \mu = 1,2,3), \tag{9}$$

$$T^{\mu\cdot}_{\cdot 4} = S^{\mu\cdot}_{\cdot l}\dot{x}^l + \overset{\mathfrak{a}}{\mathscr{M}}{}^{\mu\cdot}_{\cdot l}\underset{\mathfrak{a}}{\dot{d}}^l - j^\mu_{\mathrm{R}}\dot{\vartheta} \qquad (\mu = 1,2,3), \tag{10}$$

$$T^{4\cdot}_{\cdot\lambda} = -(\partial_\lambda x^l)P_l - (\partial_\lambda \underset{\mathfrak{a}}{d}^l)\overset{\mathfrak{a}}{\mathscr{Q}}_l - s(\partial_\lambda \vartheta) \qquad (\lambda = 1,2,3), \tag{11}$$

$$T^{4\cdot}_{\cdot 4} = \mathscr{L} - \dot{x}^l P_l - \underset{\mathfrak{a}}{\dot{d}}^l \overset{\mathfrak{a}}{\mathscr{Q}}_l - s\dot{\vartheta}. \tag{12}$$

It is clearly seen that the above components of the energy-momentum tensor for micropolar thermoelastic field can be treated as:
$T^{4\cdot}_{\cdot 4} = \mathscr{H}$ is the Hamiltonian,
$T^{4\cdot}_{\cdot\lambda} = \mathscr{P}_\lambda$ is Umov–Poynting vector,
$T^{\mu\cdot}_{\cdot 4} = \Gamma^\mu$ is pseudomomentum vector,
$T^{\mu\cdot}_{\cdot\lambda} = P^{\mu\cdot}_{\cdot\lambda}$ is the Eshelby stress tensor.
Conservation laws corresponding to the translational symmetries of action read

$$\partial_\mu T^{\mu\cdot}_{\cdot\lambda} = 0 \qquad (\lambda,\, \mu = 1,2,3,4). \tag{13}$$

Eqs. (13) are splitted into the following symmetric canonical equations

$$-\dot{\mathscr{H}} + \partial_\mu \Gamma^\mu = 0, \qquad (\mu = 1,2,3) \tag{14}$$

$$-\dot{\mathscr{P}}_\lambda + \partial_\mu P^{\mu\cdot}_{\cdot\lambda} = 0. \qquad (\lambda,\, \mu = 1,2,3) \tag{15}$$

Eqs. (14) and (15) are the energy balance equation and the pseudomomentum balance equation respectively.

4. Objective Strain Measures for Micropolar Thermoelastic Continuum

Action and action density should be satisfied some conditions of invariance with respect to arbitrary rotations of spatial coordinate system and time translations. Since the choice of spatial coordinates is rather arbitrary it should not in any way affect formulations of physical laws. For this reason the action invariance under rotations and translations of spatial coordinates and time translations is presumed.

The action, in particular, should be invariant under translations and rotations of the observer's coordinate system and time translations:

$$\begin{aligned} \widetilde{x}^i &= R^i_{\cdot j} x^j + C^i, \\ \underset{\mathfrak{a}}{\widetilde{d}}^i &= R^i_{\cdot j} \underset{\mathfrak{a}}{d}^j, \\ \widetilde{t} &= t + C, \end{aligned} \tag{16}$$

wherein C^i, C are constants; $R^i_{\cdot j}$ is an arbitrary orthogonal tensor.

The action invariance under spatial coordinates translations is known as the Galilean relativity principle. This principle is supplemented by the temperature displacement translations (C' is an arbitrary constant):

$$\widetilde{\vartheta} = \vartheta + C', \tag{17}$$

that is provided by the following condition

$$\frac{\partial \mathscr{L}}{\partial \vartheta} = 0. \tag{18}$$

By the above reasons the Helmholtz free energy is a function of the variables

$$X^\beta, \;\; \dot{\vartheta}, \;\; \partial_\alpha \vartheta, \tag{19}$$

and following independent and invariant with respect to the rotations of the spatial coordinate system arguments [3]:

$$\begin{aligned} g_{\alpha\beta} &= g_{ij}(\partial_\alpha x^i)(\partial_\beta x^j), \\ \underset{\mathfrak{a}}{\mathscr{R}}_\alpha &= g_{ij}(\partial_\alpha x^i)\underset{\mathfrak{a}}{d}^j, \\ \underset{\mathfrak{a}}{\mathscr{T}}_{\alpha\beta} &= g_{ij}(\partial_\alpha x^i)(\partial_\beta \underset{\mathfrak{a}}{d}^j). \end{aligned} \tag{20}$$

Here g_{ij} is spatial metric tensor; $g_{\alpha\beta}$ is convective metric tensor.

It can be shown that (19) and (20) constitute a complete system of independent rotationally invariant argument of ψ.

The completeness of the rotationally invariant arguments system given by (20) can be proved by means of the algebraic invariants theory considering the contravariant vectors

$$\partial_\alpha x^i,\ \underset{\mathfrak{a}}{d}^j,\ \partial_\beta \underset{\mathfrak{a}}{d}^j. \tag{21}$$

A complete system of vectors invariants (21) consists of pairwise inner products, which leads to the Euler invariants (20).

This invariants system also has all kind of 3×3-determinants, which are located in columns of various components of the Euler vector triples (21). It is clear that the determinants must contain at least one column of Euler component of strain gradient $\partial_\alpha x^i$. The determinants calculation can be carried out with the Gram–Schmidt process, i.e. through the determinants, which are all kinds of internal products of Euler vectors arranged in columns basic determinants, and the metric coefficients ${}^\backprime g_{\alpha\beta}$.

Assuming that the continuum is homogeneous, i.e.,

$$\partial_\beta^{\mathrm{expl}} \mathscr{L} = 0 \qquad (\beta = 1, 2, 3), \tag{22}$$

and, consequently, all Lagrangian variables X^β are cyclic (neglected). We get the following rotationally-invariant form of the Helmholtz free energy, which satisfies the from indifference principle: ($\mathfrak{a}$=1,2,3; α, β=1,2,3)

$$\psi = \psi(g_{\alpha\beta}, \underset{\mathfrak{a}}{\mathscr{R}}_\alpha, \underset{\mathfrak{a}}{\mathscr{T}}_{\alpha\beta}, \dot{\vartheta}, \partial_\alpha\vartheta). \tag{23}$$

We implicitly assume that the reduced form (23) should also depend on the reference metrics and referential position d-vectors ${}^\backprime\underset{\mathfrak{a}}{d}^j$ ($\mathfrak{a}$=1,2,3).

5. Relative Strain Measures

In the equation (23) rotationally invariant argument $g_{\alpha\beta}$ can be replaced by the following relative strain tensor

$$\epsilon_{\alpha\beta} = \frac{1}{2}(g_{\alpha\beta} - {}^\backprime g_{\alpha\beta}). \tag{24}$$

The tensor $\epsilon_{\alpha\beta}$ is called the Green's strain tensor.

Instead of a vector measure of extra strain $\underset{\mathfrak{a}}{\mathscr{R}}_\alpha$ a relative extra strain vector is defined as

$$-\underset{\mathfrak{a}}{\gamma}_\alpha = \underset{\mathfrak{a}}{\mathscr{R}}_\alpha - g_{\alpha\beta}{}^\backprime\underset{\mathfrak{a}}{d}^\beta. \tag{25}$$

Hereafter vectors ${}^\backprime\underset{\mathfrak{a}}{d}^\beta$ denote the referential position of the d-vectors. ${}^\backprime x^i$ are the spatial positions in the reference configuration. Note the following equation:

$${}^\backprime\underset{\mathfrak{a}}{d}^i = \frac{\partial{}^\backprime x^i}{\partial X^\alpha}{}^\backprime\underset{\mathfrak{a}}{d}^\alpha.$$

Vector $\underset{\mathfrak{a}}{\gamma}_\alpha$ vanishes only if each of d-vectors rotates and elongates exactly so as prescribed by the deformation $X^\alpha \to x^i$. If this in effect then $^\backprime d$-vectors and d-vectors are related by

$$\underset{\mathfrak{a}}{d}^i - (\partial_\alpha x^i)^\backprime \underset{\mathfrak{a}}{d}^\alpha = 0.$$

Multiplying this equation by the deformation gradient $\partial_\beta x^j$ and contracting with the metric tensor g_{ij} lead to equation

$$\underset{\mathfrak{a}}{\mathscr{R}}_\beta - g_{\beta\alpha}{}^\backprime \underset{\mathfrak{a}}{d}^\alpha = 0,$$

i.e. relative vector of extra-strain equals to zero:

$$\underset{\mathfrak{a}}{\gamma}_\beta = 0.$$

Finally rotationally invariant form of the Helmholtz free energy is represented by

$$\psi = \psi(\epsilon_{\alpha\beta}, \underset{\mathfrak{a}}{\gamma}_\alpha, \underset{\mathfrak{a}}{\mathscr{T}}_{\alpha\beta}, \dot{\vartheta}, \partial_\alpha \vartheta) \quad (\mathfrak{a} = 1, 2, 3;\ \alpha, \beta = 1, 2, 3). \tag{26}$$

The Helmholtz free energy depends on one scalar argument $\dot{\vartheta}$; four referential vector arguments $\partial_\alpha \vartheta$, $\underset{\mathfrak{a}}{\gamma}_\alpha$ ($\mathfrak{a} = 1, 2, 3; \alpha = 1, 2, 3$); and four referential tensor arguments $\epsilon_{\alpha\beta}$, $\underset{\mathfrak{a}}{\mathscr{T}}_{\alpha\beta}$ ($\mathfrak{a} = 1, 2, 3;\ \alpha, \beta = 1, 2, 3$); $\epsilon_{\alpha\beta}$ symmetric and $\underset{\mathfrak{a}}{\mathscr{T}}$ asymmetric second rank tensors.

6. Constitutive Equations of Micropolar Thermoelasticity

As shown in the previous sections, the objective basis for the thermodynamic micropolar thermoelastic continuum, in which the heat transport is not accompanied by the entropy production consists of the following functional independent variables

$$\epsilon_{\alpha\beta},\ \underset{\mathfrak{a}}{\gamma}_\alpha,\ \underset{\mathfrak{a}}{\mathscr{T}}_{\alpha\beta},\ \dot{\vartheta},\ \partial_\alpha \vartheta \qquad (\mathfrak{a} = 1, 2, 3; \quad \alpha, \beta = 1, 2, 3). \tag{27}$$

The following objective form of the constitutive equations in terms of the rotationally invariant basis can be deduced from (8):

$$\begin{aligned} S^{\alpha\cdot}_{\cdot j} = \frac{\partial \mathscr{L}}{\partial(\partial_\alpha x^j)} - \frac{\partial \mathscr{L}}{\partial \epsilon_{\mu\alpha}} g_{jk} \partial_\mu x^k - \frac{\partial \mathscr{L}}{\partial \underset{\mathfrak{a}}{\gamma}_\beta} g_{jk} \underset{\mathfrak{a}}{d}^k \delta^\alpha_\beta + \\ + \frac{\partial \mathscr{L}}{\partial \underset{\mathfrak{a}}{\gamma}_\beta} g_{jk} (\partial_\beta x^{k\backprime} \underset{\mathfrak{a}}{d}^\alpha + \partial_\sigma x^k \delta^{\alpha\backprime}_\beta \underset{\mathfrak{a}}{d}^\sigma) + \frac{\partial \mathscr{L}}{\partial \underset{\mathfrak{a}}{\mathscr{T}}_{\mu\nu}} g_{jk} \partial_\nu \underset{\mathfrak{a}}{d}^k, \end{aligned} \tag{28}$$

$$-\overset{\mathfrak{c}}{\mathscr{M}}{}^{\alpha\cdot}_{\cdot j} = \frac{\partial \mathscr{L}}{\partial(\partial_\alpha \underset{\mathfrak{c}}{d}^j)} = \frac{\partial \mathscr{L}}{\partial \underset{\mathfrak{c}}{\mathscr{T}}_{\mu\nu}} g_{jk} \partial_\mu x^k \delta^\alpha_\nu, \tag{29}$$

$$\overset{\mathfrak{c}}{\mathscr{A}}_j = \frac{\partial \mathscr{L}}{\partial \underset{\mathfrak{a}}{d}^j} = \frac{\partial \mathscr{L}}{\partial \underset{\mathfrak{c}}{\gamma}_\beta} g_{jk} \partial_\beta x^k. \tag{30}$$

The constitutive equation (28) after obvious transformations can be finally represented as:

$$\begin{aligned} -S^{\alpha\cdot}_{\cdot j} = \frac{\partial \mathscr{L}}{\partial \epsilon_{\mu\alpha}} g_{jk} \partial_\mu x^k + \frac{\partial \mathscr{L}}{\partial \underset{\mathfrak{a}}{\gamma}_\alpha} g_{jk} (\partial_\sigma x^{k\backprime} \underset{\mathfrak{a}}{d}^\sigma - \underset{\mathfrak{a}}{d}^k) + \\ + \frac{\partial \mathscr{L}}{\partial \underset{\mathfrak{a}}{\gamma}_\beta} g_{jk} \partial_\beta x^{k\backprime} \underset{\mathfrak{a}}{d}^\alpha + \frac{\partial \mathscr{L}}{\partial \underset{\mathfrak{a}}{\mathscr{T}}_{\mu\nu}} g_{jk} \partial_\nu \underset{\mathfrak{a}}{d}^k. \end{aligned} \tag{31}$$

7. Acknowledgments

The research was supported in part by the Russian Foundation for Basic Research (project No. 15-31-21111).

8. Conclusions

The study of thermoelastic continuum by methods of field theory allows us to conclude. A non-linear mathematical model of thermoelastic micropolar continuum has been developed. The model has been presented in terms of 4-covariant field theoretical formalism. Thermal field has been incorporated in the theory by the two constitutive constants. The differential field equations corresponding to the action integral have been derived. The energy-momentum tensor for micropolar thermoelastic continuum has been obtained and the corresponding conservation laws have been formulated. Following the usual procedure, micropolar thermoelastic Lagrangians have been represented as functions of independent rotationally invariant arguments. The completeness of the rotationally invariant arguments system has been proved by means of the algebraic invariants theory. Relative strain measures have been discussed. The rotationally invariant form of the Helmholtz free energy has been given. Constitutive equations of micropolar thermoelasticity with three directors have been obtained and discussed.

9. References

[1] Cosserat E and Cosserat F 1909 *Theories of Deformable Bodies* (Paris: Scientific Library A. Hermann et Fils)
[2] Nowacki W 1986 *Theory of Asymmetric Elasticity* (Oxford: Pergamon Press)
[3] Toupin R A 1964 *Arch. Rat. Mech. Analysis* **11** 85
[4] Green A E and Naghdi P M 1992 *J. Therm. Stress* **15** 253
[5] Murashkin E V and Radayev Yu N 2015 *Materials Physics and Mechanics* **23** 10
[6] Kovalev V A and Radayev Yu N 2010 *Wave Problems of Field Theory and Thermomechanics* (Saratov: Saratov. Univ. Pub.)

IOP Conf. Series: Journal of Physics: Conf. Series **788** (2017) 012044 doi:10.1088/1742-6596/788/1/012044

Coulomb Problem for Graphene with Supercritical Impurity

V M Kuleshov, V D Mur and N B Narozhny

National Research Nuclear University MEPhI, Kashirskoe Shosse 31, 115409 Moscow, Russia

E-mail: kuleshov.val@gmail.com

Abstract. The properties of charge carriers in graphene with dopants with charge Z within two-dimensional effective Dirac equation with non-zero gap between upper and lower continuums are analyzed. The closed set of explicit equations determining the spectrum of charge carriers are obtained for the case of the Coulomb potential modified at small distances. The critical values Z_{cr} of the dopant charge at which the energy level with the given quantum numbers crosses the lower continuum boundary are determined. For $Z < Z_{cr}$ for several values of the orbital angular momentum, which may take half-integer values in 2D case, dependence of the position of the energy level as a function of charge Z is obtained. For $Z > Z_{cr}$, the position ε_0 and width γ of the lowest quasidiscrete state are calculated. For such values of Z screening of the impurity charge and resonance scattering of holes are also considered.

1. Introduction

In the presence of multiply charged impurity the electronic characteristics of graphene are described by the effective two-dimensional Dirac equation with the gap in electronic spectrum [1–11]. In this case the energy spectrum of the two-dimensional heterostructure is absolutely similar to one for the three-dimensional Coulomb problem, including the case with the nuclear charge $Z > 137$, e.g. [12, 13]. The screening of the charge for the Coulomb impurity in the gapped graphene was studied in [4,6–11]. In the last paper, the authors discuss the screening by electrons created together with holes from the Dirac sea (by analogy with the three-dimensional case [13]). In the present work, we demonstrate that such a mechanism of screening cannot be put into effect, see also [14, 15], but resonances in the scattering of the holes by overcritical impurities occur.

According to [4, 5, 11], the electronic properties of graphene doped by atomic nuclei or ions with charge Z can be described by the effective two-dimensional Dirac equation

$$\hbar v_F \left(-i\boldsymbol{\sigma}\frac{\partial}{\partial \mathbf{x}} - \frac{q}{\mathbf{x}} + \frac{m_* v_F}{\hbar}\sigma_z \right) \Psi_E(\mathbf{x}) = E\Psi_E(\mathbf{x}) \quad (1)$$

Here v_F is the velocity at the Fermi surface, $\mathbf{x} = (x_1, x_2)$, $\boldsymbol{\sigma} = (\sigma_x, \sigma_y, \sigma_z)$ are the Pauli matrices, $q = Z\alpha_F$, $\alpha_F = e^2/\hbar v_F$ is the effective fine structure constant for graphene, $\Psi_E(\mathbf{x})$ - is the two-component wavefunction for an electron with the effective mass m_* and energy E.

Accounting for [3, 11] the gap width $\Delta = 2m_* v_F^2 = 0.26\, eV$, the effective constant $\alpha_F = 0.4$ and the distance between carbon nuclei $a_{cc} = 1.42$ Å for graphene deposited onto SiC substrate we have

$$v_F = 5.5 \cdot 10^8\ cm/s, \quad m_* = 7.6 \cdot 10^{-4} m_e, \quad a_{CC}/l_F = 5.5 \cdot 10^{-3}, \quad (2)$$

where m_e is the mass of the electron, $l_F = \hbar/m_* v_F$ is the "Compton length" in graphene.

Owing to the axial symmetry, the conserving quantum number [16] is the total angular momentum $J = M + 1/2$ (in $\hbar$ units), i.e., the eigenvalue of the generator for the two-dimensional rotations $-i\partial/\partial\varphi + 1/2\sigma_3$. The orbital angular momentum M is the eigenvalue of the self-adjoint operator $-i\partial/\partial\varphi$ acting in the Hilbert space of square integrable functions on the circle $0 < \varphi < 2\pi$, which eigenfunctions are [18, 19][1]:

$$\Phi_M(\varphi) = \frac{1}{2\pi} e^{\imath M\varphi}, \quad M = \delta + m; \ \delta = 0, \frac{1}{2}, \quad m = 0, \pm 1, \pm 2, \ldots \tag{3}$$

In terms of

$$E = m_* v_F^2 \varepsilon, \quad \mathbf{x} = l_F \boldsymbol{\rho}, \quad \boldsymbol{\rho} = (\rho\cos\varphi, \rho\sin\varphi),$$

the wavefunction with fixed total angular momentum is represented as:

$$\Psi_{\varepsilon,J}(\boldsymbol{\rho}) = \frac{1}{\sqrt{2\pi\rho}} e^{iJ\varphi} \begin{pmatrix} e^{-i\varphi/2} F(\rho) \\ ie^{i\varphi/2} G(\rho) \end{pmatrix} \tag{4}$$

2. Boundary conditions

The functions $F(\rho)$ and $G(\rho)$ satisfy the two-dimensional Dirac equation:

$$H_D \Psi_{\varepsilon,J}(\rho) = \varepsilon \Psi_{\varepsilon,J}(\rho), \ H_D = \begin{pmatrix} 1 - \frac{q}{\rho} & \frac{J}{\rho} + \frac{d}{d\rho} \\ \frac{J}{\rho} - \frac{d}{d\rho} & -1 - \frac{q}{\rho} \end{pmatrix}, \ \Psi_{\varepsilon,J} = \begin{pmatrix} F(\rho) \\ G(\rho) \end{pmatrix}, \ J = M + \frac{1}{2}, \tag{5}$$

which up to the notation coincides with the set of equations for the radial functions for the three-dimensional problem [12]. In the two-dimensional case, however, the value $J = 0$, i.e. $(M = -1/2)$, which corresponds the ground state, and half-integer values $J = m + 1/2, \ m = 0, \pm 1, \pm 2, \ldots$ are possible.

The problem is posed correctly if (5) is supplemented by physically acceptable boundary conditions for the wavefunctions. In the present case, it is equivalent to the condition that the differential expression H_D is the self-adjoined one and corresponds to the Hamiltonian acting in the Hilbert space of the square integrable functions with the defined Hermitian scalar product

$$(\Psi_1, \Psi_2) = \int_0^\infty \{F_1^*(\rho) F_2(\rho) + G_1^*(\rho) G_2(\rho)\} d\rho.$$

Since the Coulomb potential $V_C(\rho)$ vanishes at large distances, so, as in the three-dimensional problem, the values $\varepsilon \geq 1$ and $\varepsilon \leq -1$ correspond to the upper and lower continua of the solutions of the Dirac equation, respectively, whereas the range $-1 < \varepsilon < 1$ – corresponds to the discrete spectrum.

Thus, to ensure the self-adjointness of the Hamiltonian it is essential (as it is shown in [14]) to consider the asymptotic behaviour of the solutions of system (5) at short distances. This behaviour is determined by the value of single parameter $\sigma = \sqrt{J^2 - q^2}$. According to the von Neumann theory of unbounded operators we arrive at conditions [14] which determine a single-parameter family (associated with the differential operator H_D) of self-adjoint operators

[1] The half-integer quantization of the orbital angular momentum comes into effect for two-dimensional quantum dots with an odd number of electrons [17]

in the Hilbert space of wavefunctions[2]:

$$\begin{aligned}
\frac{u_\sigma}{u_{-\sigma}} &= \left(\frac{u_\sigma}{u_{-\sigma}}\right)^* = \tan\theta_\sigma(J), \ -\frac{\pi}{2} \le \theta_\sigma(J) \le \frac{\pi}{2}, \quad 0 < \sigma < 1/2 \\
\frac{u_\tau}{u_{-\tau}} &= \left(\frac{u_{-\tau}}{u_\tau}\right)^* = e^{2i\theta_\tau(J)}, \ \Im\theta_\tau(J) = 0, \ \tau = \sqrt{q^2 - J^2} > 0, \ \ q > |J|, \\
u_{\pm\sigma} &= \frac{\Gamma(\mp 2\sigma)(2\lambda)^{\pm\sigma}}{\Gamma\left(1 \mp \sigma - \frac{\varepsilon}{\lambda} q\right)}[q\sqrt{1-\varepsilon} - (J \pm \sigma)\sqrt{1+\varepsilon}]
\end{aligned} \tag{6}$$

The parameters $\theta_\sigma(J)$ and $\theta_\tau(J)$ define one-parameter families of self-adjoint radial Dirac hamiltonians, which were obtained in a different way in [21–23] in three- and two-dimensional cases respectively. Here $\tau = \sqrt{q^2 - J^2}$, and amplitudes $u_{\pm\tau}$ are obtained from $u_{\pm\sigma}$ by replacement $\sigma \to i\tau$. Hence, we get, in particular, the equation

$$\arg\Gamma\left(2i\sqrt{\left(q_{cr}^{(n)}\right)^2 - J^2}\right) = \sqrt{\left(q_{cr}^{(n)}\right)^2 - J^2} \cdot \ln\left(2q_{cr}^{(n)}\right) - \theta_\tau(J) + \pi n, \ \ n = 0, 1, 2, \ldots \tag{7}$$

for the critical charge $q_{cr}^{(n)} = Z_{cr}\alpha_F$, at which the n-th level with the given quantum number J reaches the boundary of the lower continuum of the solutions of Eq. (5).

3. Short-range Coulomb problem

To fix the parameters $\theta_\sigma(J)$ and $\theta_\tau(J)$ we need additional physical assumptions. According to Pomeranchuk and Smorodinsky [27] we should consider a Coulomb potential modified at small distances $\rho \le R_0 \ll 1$:

$$V_R(\rho) = -\frac{q}{R}\begin{cases} R/\rho, & \rho \ge R \\ f(\rho/R), & \rho \le R \end{cases} \tag{8}$$

Equation (5) with the replacement $V_C(\rho) \to V_R(\rho)$ at $J \ne 0$ has an analytical solution for $f(\rho/R) \equiv 1$. Within the framework of this model we have

$$\tan\theta_\sigma(J;R) = -\frac{(J+\sigma)[qJ_{\mp 1/2 \pm J}(q) \mp (J-\sigma)J_{\pm 1/2 \pm J}(q)]R^{-\sigma}}{(J-\sigma)[qJ_{\mp 1/2 \pm J}(q) \mp (J+\sigma)J_{\pm 1/2 \pm J}(q)]R^{\sigma}}, \tag{9}$$

where $J_\nu(q)$ is the Bessel functions with upper (lower) signs corresponding to $J > 0$ $(J < 0)$ respectively.

If $q > |J| > 0$, assuming in (9) $\sigma = i\tau$, we obtain the equality

$$\exp(2i\theta_\tau(J;R)) = \tan\theta_{i\tau}(J;R), \ \ \tau = \sqrt{q^2 - J^2} > 0, \ \ J \ne 0 \tag{10}$$

In the case of $J = 0$ the analytical solution is possible for any type of modification of the Coulomb potential

$$\exp(2i\theta_\tau(R_0)) = \exp\left(2iq\ln\frac{1}{R_0}\right), \tag{11}$$

[2] At $1/2 < \sigma < |J|$ we arrive at so-called "built-in" boundary condition [20] and the energy is given by a two-dimensional version of the Sommerfeld formula

Table 1. Critical charge values $q_{cr}^{(n)}(J) = Z_{cr}^{(n)}(J)\alpha_F$ at which the lowest $n = 0$ and first excited $n = 1$ levels with $J = 0$ reach the boundary of the lower continuum for $R = 1/20$ and different cutoffs of the Coulomb potential.

$R = 1/20$	$f_0 = 1$	$f_0 = 4/3$	$f_0 = 3/2$	$f_0 = 2$
$q_{cr}^{(0)}(J = 0)$	0.46	0.42	0.40	0.35
$q_{cr}^{(1)}(J = 0)$	1.33	1.23	1.18	1.06

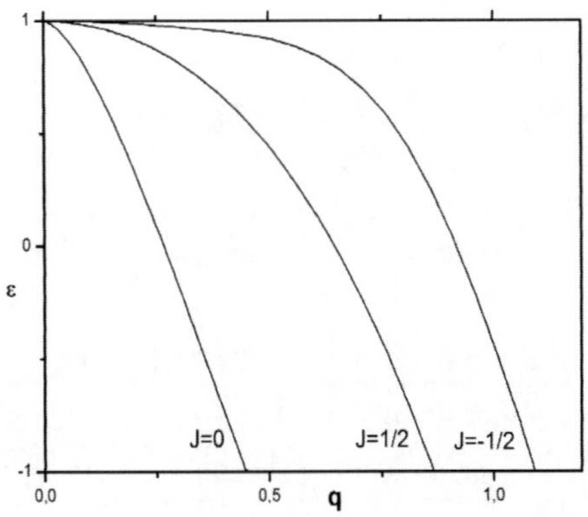

Figure 1. Energy of the ground state $\varepsilon(q, J; R)$ versus the charge $q = Z\alpha_F$ for the Coulomb potential cutoff radius $R = 1/20$ for different values of quantum number J.

where

$$R_0 = R\,e^{-f_0}, \;\; f_0 = \int_0^1 f(\xi)d\xi$$

The value $f_0 = 1$ corresponds to rectangular modification of the Coulomb potential and $f_0 = 4/3$ – to the uniform distribution of the charge over the impurity volume. The charge dependence of the energy $\varepsilon_{J=0}(q)$ is determined by the equation:

$$\frac{(1-\varepsilon-i\lambda)(2\lambda)^{iq}\Gamma(-2iq)\Gamma\left(1+iq-\frac{\varepsilon}{\lambda}q\right)}{(1-\varepsilon+i\lambda)(2\lambda)^{-iq}\Gamma(2iq)\Gamma\left(1-iq-\frac{\varepsilon}{\lambda}q\right)} = \exp\left(2iq\ln\frac{1}{R_0}\right) \tag{12}$$

within the whole range $0 < q \leq q_{cr}$.

Equation (7) along with (9)-(11) determines the critical charge $q_{cr}(J; R)$ as a function of the quantum number J and the cutoff radius R, see Table 1, whereas conditions (6) with (9) and (10) give the charge dependence of the energy $\varepsilon_J(q, R)$ at the fixed cutoff radius (see Fig. 1).

The parameters $\theta_\sigma(J; R)$ and $\theta_\tau(J; R)$ completely determine the energy spectrum and stationary wavefunctions of (5) at any values of the charge $q = Z\alpha_F$.

For example, at $\varepsilon < -1$ we have, see [14]:

$$u_\tau = (2k)^{i\tau}\Gamma(-2i\tau)[Aa - Bb], \;\; u_{-\tau} = (2k)^{-i\tau}\Gamma(2i\tau)[e^{-\pi\tau}Ab^* - e^{\pi\tau}Ba^*], \tag{13}$$

IOP Conf. Series: Journal of Physics: Conf. Series **788** (2017) 012044 doi:10.1088/1742-6596/788/1/012044

where $k = \sqrt{\varepsilon^2 - 1}$, and

$$a = \frac{q\sqrt{-\varepsilon+1} + (iJ-\tau)\sqrt{-\varepsilon-1}}{\Gamma\left(1 - i\tau - i\frac{\varepsilon}{k}q\right)}, \quad b = \frac{q\sqrt{-\varepsilon+1} - (iJ-\tau)\sqrt{-\varepsilon-1}}{\Gamma\left(1 - i\tau + i\frac{\varepsilon}{k}q\right)}.$$

Taking into account second condition from (6) and the wavefunction normalization to $\delta(k-k')$ we obtain

$$e^{2i\delta_J} = \frac{f_J^*(k;q)}{f_J(k;q)}, \quad B = e^{-\pi\tau} e^{\frac{\pi q(-\varepsilon)}{2k}} \frac{e^{-i\delta_J}}{\sqrt{2\pi(-\varepsilon)}}, \tag{14}$$

where $\delta_J(k;q)$ is the scattering phase,

$$f_J(k;q) = -i(e^{\frac{\pi\tau}{2} - i\varphi_\tau} a - e^{\frac{-\pi\tau}{2} + i\varphi_\tau} b^*),$$

is the Jost function, where we introduced the notation

$$e^{2i\varphi_\tau} = \frac{(2k)^{-i\tau}\Gamma(2i\tau)}{(2k)^{i\tau}\Gamma(-2i\tau)} e^{2i\theta_\tau(J)}.$$

The condition $B = 0$ leads to the equation for the spectrum of complex energies of quasistationary states of holes corresponding to the poles of the scattering matrix $S_J(k;q) = \exp(2i\delta_J(k;q))$,

$$\frac{(-2ik)^{i\tau}[q\sqrt{-\varepsilon+1} + (iJ-\tau)\sqrt{-\varepsilon-1}]\Gamma(-2i\tau)\Gamma\left(1 + i\tau - i\frac{\varepsilon}{k}q\right)}{(-2ik)^{-i\tau}[q\sqrt{-\varepsilon+1} + (iJ+\tau)\sqrt{-\varepsilon-1}]\Gamma(2i\tau)\Gamma\left(1 - i\tau - i\frac{\varepsilon}{k}q\right)} = e^{2i\theta_J(q;R)}. \tag{15}$$

The solutions of this equation determine both the position of a quasidiscrete level ε_0 and its width γ:

$$\varepsilon_{qs} = -\varepsilon_p = \varepsilon_0 - \frac{i}{2}\gamma, \quad \varepsilon_0 > 0, \quad \gamma > 0, \tag{16}$$

Due to the Coulomb barrier in the lower continuum the width $\gamma \ll 1$

$$k = k_0' - ik_0'', \quad k_0' = \sqrt{\varepsilon_0^2 - 1} > 0, \quad k_0'' = \frac{\gamma}{2k_0'} > 0,$$

and these poles are located on the second (unphyisical) sheet [28], in accordance with the unitarity of the partial scattering matrix.

The scattering phase $\delta_J(k;q)$ as a function of hole energy $\varepsilon_{hole} = -\varepsilon > 1$ for $J = 0$ and $q > q_{cr}$ is shown in Fig. 3. For $J = 1/2$ and $J = -1/2$ the behaviour of the phases is the same. The poles of scattering matrix determine the complex energies ε_{qs} of quasistationary states of holes (16). Figure 2 shows the position of the lowest quasidiscrete level ε_0 and its width γ as a function of $q - q_{cr}$. If the hole energy ε_{hole} is in the range of dramatic change of scattering phase, such a states are evident as resonances in the hole-impurity scattering and the partial cross-section $\sigma_J = \sin^2\delta_J(k;q)$ have the resonant form, see Fig. 4.

4. Conclusions

Thus, the "cutoff" i.e., the regularization of the Coulomb potential at small distances, ensures that the Dirac Hamiltonian is self-adjoint both for two-dimensional and for three-dimensional problems. This implies that the one-particle approximation for the Dirac equiation is consistent at any values of the Coulomb charge, including the overcritical range. It means that there is no spontaneous creation of electron-hole pairs. At the same time, owing to the presence of low permeable effective Coulomb barrier, the continuum states near the boundary of lower continuum form quasistationary states which may appear as resonances in the scattering of holes on supercritical impurity.

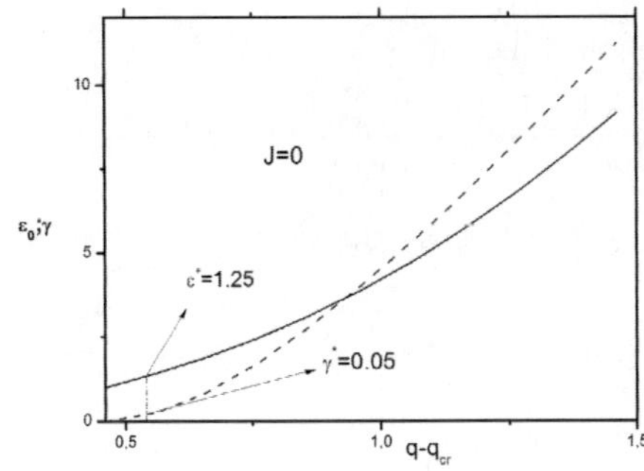

Figure 2. Position of the lowest quasidiscrete level ε_0 and its width γ versus the difference $q - q_{cr}$ at the cutoff radius $R = 1/20$. ε^* γ^* are the position and width of the pole of S-matrix corresponding to the resonance for $q = 0.52$.

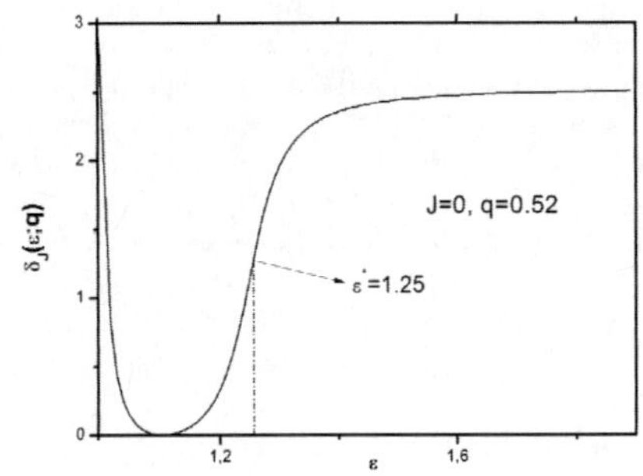

Figure 3. Scattering phase δ_J for $J = 0$ as a function of the hole energy for $q = 0.52$ which is close to q_{cr} for $R = 1/20$, the hole energy $\varepsilon_{hole} \simeq \varepsilon^*$ is in the range of dramatic change of scattering phase with width γ^*.

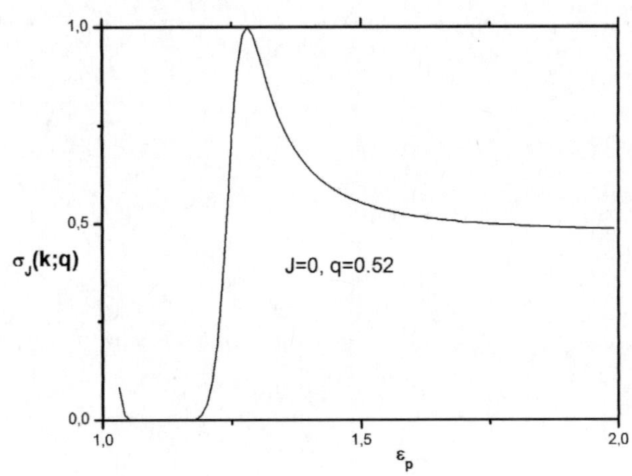

Figure 4. The partial scattering cross-section σ_J.

References

[1] Kane C L and Mele E J 2005 *Phys. Rev. Lett.* **95** 226801
[2] Giovannetti G, Khomyakov P A, Brocks G, Kelly P G and van den Brink J 2007 *Phys. Rev.* B **76** 073103
[3] Zhou S Y, Gweon G-H, Fedorov A V, First P N, de Heer W A, Lee D-H, Guinea F, Castro Neto A H and Lanzara A 2007 *Nature Materials* **6** 770.
[4] Biswas R R and Son D T 2007 *Phys. Rev.* B **76** 205122
[5] Novikov D S 2007 *Phys. Rev.* B **76** 245435
[6] Pereira V M, Nilsson J and Castro Neto A H 2007 *Phys. Rev. Lett.* **99** 236801
[7] Fogler M M, Novikov D S and Shklovskii B I 2007 *Phys. Rev.* B **76** 233402
[8] Shytov A V, Katsnelson M I and Levitov L I 2007 *Phys. Rev. Lett.* **99** 236801
[9] Terekhov I S, Milstein A I, Kotov V N and Sushkov O P 2008 *Phys. Rev. Lett.* **100** 076803

[10] Wei Zhu, Zhengfei Wang, Qinwei Shi, Szeto K Y, Jie Che and Hou J G 2009 *Phys. Rev.* B **79** 155430
[11] Pereira V M, Kotov V N and Castro Neto A H 2008 *Phys. Rev.* B **78** 085101
[12] Akhiezer A I, Berestetskii 1965 *Quantum electrodynamics* (New York: Wiley)
[13] Zel'dovich Ya B, Popov V S 1971 *Sov. Phys. Usp.* **14** 673
[14] Kuleshov V M, Mur V D, Narozhny N B, Fedotov A M, Lozovik Yu E 2015 *JETP Letters* **101** 264
[15] Kuleshov V M, Mur V D, Narozhny N B, Fedotov A M, Lozovik Yu E 2015 *Physics-Uspekhi* **58** 785
[16] DiVincenzo D P and Mele E J 1984 *Phys. Rev.* B **29** 1685
[17] Mur V D, Narozhny N B, Petrosyan A N, Lozovik Yu E 2008 *JETP Letters* **88** 688
[18] Pauli W 1939 *Helv. Phys. Acta* **12** 147
[19] van Vinter C 1968 *Ann. Phys.* **47** 232
[20] Richtmyer R D 1978 *Principle of Advanced Mathematical Physics* (New York, Berlin, Heidelberg: Springer-Verlag)
[21] Voronov B L, Gitman D M, Tyutin I V 2007 *Teor. Math. Phys.* **150** 41
[22] Gitman D M, Levin A D, Tyutin I V, Voronov B L 2012 Electronic Structure of Superheavy Atoms *preprint* http://arxiv.org/abs/1112.2648
[23] Khalilov V R, Lee K E 2011 *Theor. Math. Phys.* **169** 1683
[24] Popov V S 1970 *Yad. Phys.* **12** 429
[25] Popov V S 1971 *Sov. Phys. JETP* **32** 526
[26] Popov V S 1971 *Sov. Phys. JETP* **33** 665
[27] Pomeranchuk I, Smorodinsky Ya 1945 *J. Phys. USSR* **9** 97
[28] Mur V D, Popov V S 1976 *Teor. Math. Phys.* **27** 204

Effect of the single-scattering phase function on light transmission through disordered media with large inhomogeneities

V V Marinyuk[1] and S V Sheberstov[2]

[1] National Research Nuclear University MEPhI, Kashirskoe Shosse 31, Moscow 115409, Russia
[2] P.P. Shirshov Institute of Oceanology, Nahimovski prospect 36, Moscow 117997, Russia

E-mail: vmarinyuk@mail.ru

Abstract. We calculate the total transmission coefficient (transmittance) of a disordered medium with large (compared to the light wavelength) inhomogeneities. To model highly forward scattering in the medium we take advantage of the Gegenbauer kernel phase function. In a subdiffusion thickness range, the transmittance is shown to be sensitive to the specific form of the single-scattering phase function. The effect reveals itself at grazing angles of incidence and originates from small-angle multiple scattering of light. Our results are in a good agreement with numerical solutions to the radiative transfer equation.

1. Introduction

Multiple light scattering in disordered media is a fundamental problem with many applications [1, 2]. An interrelation between characteristics of multiply scattered light and parameters of scattering inhomogeneities is of considerable interest both for understanding of the physical process and for practical use. This interrelation is a basis for numerous techniques for characterization of turbid media (see, e.g., [3, 4, 5, 6]).

In media with large-scale inhomogeneities, the transport mean free path l_{tr} appears to be much greater than the mean free path l, $l_{\text{tr}} \gg l$, and two regimes of multiple light scattering can be distinguished. The diffusive regime develops over spatial scales exceeding l_{tr}. Diffusely scattered light retains no information about the single-scattering phase function (l_{tr} is the only retrievable parameter). The small-angle regime of multiple scattering is more informative. The angular and spatial distributions of multiply scattered light [7, 8, 9] relate directly to the single-scattering law. As is shown in [10], if the medium thickness is less than the transport mean free path, the total transmittance also appears to be sensitive to the single-scattering phase function. The effect reveals itself at grazing angles of incidence ($\zeta_0 \ll 1$, ζ_0 is the angle between the direction of light incidence and the medium boundary) and originates from small-angle multiple scattering of light.

In this paper, we provide a detailed description of the asymptotic method of the total transmittance calculations outlined in [10]. An analytic formula for the total transmittance is derived for an arbitrary relation between the angle ζ_0 and the characteristic angle of single scattering.

2. General relations

Consider transmission of a broad collimated beam of light through a nonabsorbing plane-parallel medium with large-scale inhomogeneities. The intensity of light inside the medium $I(z,\mathbf{\Omega})$ obeys the radiative transfer equation

$$\left(\cos\theta\frac{\partial}{\partial z}+\sigma\right)I(z,\mathbf{\Omega})=\sigma\int d\mathbf{\Omega}'p(\mathbf{\Omega}\mathbf{\Omega}')I(z,\mathbf{\Omega}') \tag{1}$$

where $\sigma = l^{-1}$ is the scattering coefficient, $\mathbf{\Omega}$ is the unit vector in the direction of light propagation, $p(\cos\vartheta)$ is the single-scattering phase function normalized by the condition $\int_0^\pi 2\pi\sin\vartheta p(\cos\vartheta)d\vartheta = 1$. The z-axis is perpendicular to the medium boundaries, $\cos\theta = \Omega_z$.

At grazing angles of incidence, $\zeta_0 \ll 1$, the radiative transfer equation (1) can be simplified using the small-angle approximation (see, e.g., [8])

$$\left(\zeta\frac{\partial}{\partial z}+\sigma\right)I(z,\zeta,\varphi)=\sigma\int\int_{-\infty}^{\infty}d\zeta'd\varphi'p(\vartheta)I(z,\zeta',\varphi') \tag{2}$$

In deriving Eq.(2), we put $\cos\theta = \sin\zeta \approx \zeta$ and $2(1-\mathbf{\Omega}\mathbf{\Omega}') \approx \vartheta^2 = (\zeta-\zeta')^2+(\varphi-\varphi')^2$ where φ is the azimuthal angle of the direction $\mathbf{\Omega}$, $\zeta = \pi/2-\theta$ is the angle between the direction $\mathbf{\Omega}$ and the medium boundary. Angles $\zeta > 0$ and $\zeta < 0$ correspond to light propagating in the forward and backward directions, respectively. As is usual in the small-angle approximation, the angles ζ and φ are formally considered to vary within infinite limits provided that light propagates predominantly at small angles to the direction of incidence $\mathbf{\Omega}_0$ ($\cos\theta_0 \approx \zeta_0$, $\varphi_0 = 0$), and the intensity decreases rapidly for $|\zeta|, |\varphi| > 1$.

To model highly forward scattering in the medium we take advantage of the two-parameter phase function proposed in [11]. The small-angle form of this phase function can be written as

$$p(\vartheta)=\frac{\alpha-2}{2\pi}\frac{\vartheta_0^{\alpha-2}}{(\vartheta_0^2+\vartheta^2)^{\alpha/2}},\quad(\alpha>2) \tag{3}$$

where ϑ_0 is the characteristic angle of single scattering, $\vartheta_0 \ll 1$. The phase function (3) decreases with angle ϑ as $p(\vartheta) \sim 1/\vartheta^\alpha$, and unifies a number of scattering models (see, e.g [7, 12]). For $\alpha = 3$, Eq. (3) corresponds to the Henyey-Greenstein phase function which is used very widely to model scattering of light in natural media [7].

Equation (2) integrated over the azimuthal angle can be presented in the following form

$$\zeta\frac{\partial}{\partial z}I(z,\zeta)=\frac{(\alpha-2)\sigma\vartheta_0^{\alpha-2}}{2\pi}\int\int_{-\infty}^{\infty}\frac{d\varphi'd\zeta'[I(z,\zeta')-I(z,\zeta)]}{(\vartheta_0^2+(\zeta-\zeta')^2+\varphi'^2)^{\alpha/2}} \tag{4}$$

where $I(z,\zeta)$ is the azimuth-integrated intensity of light, $I(z,\zeta) = \int_{-\infty}^{\infty} I(z,\zeta,\varphi)d\varphi$. Boundary conditions for Eq. (4),

$$I(z=0,\zeta>0)=\delta(\zeta-\zeta_0);\quad I(z=L,\zeta<0)=0, \tag{5}$$

determine the intensity within the half-ranges of the angles that correspond to light incident upon the medium (we do not consider the effects caused by Fresnel reflections at the boundaries). Quantities $I(z=0,\zeta<0)$ and $I(z=L,\zeta>0)$ (sometimes referred to as the reflection and the transmission functions, respectively [1]) describe the angular distribution of reflected and transmitted light and should be determined in the course of solving Eqs.(4)-(5). A closed system of integral equations for these functions can be formulated using the eigenfunction method [13]. However, an analytical solution of this system is possible only in the diffusion approximation where the integral term in Eq.(4) is replaced by the differential operator.

A solution to Eq.(4) is expressed in terms of a function of dimensionless variables and parameter $\sigma L/\vartheta_0$,

$$I(z,\zeta) = \frac{1}{\vartheta_0} f\left(\frac{z}{L}, \frac{\zeta}{\vartheta_0}, \frac{\zeta_0}{\vartheta_0}, \frac{\sigma L}{\vartheta_0}\right) \tag{6}$$

If the characteristic angular spread of multiply scattered light exceeds the single-scattering angle ϑ_0, the quantity ϑ_0 can be neglected in the denominator of the phase function in Eq.(4). For $\alpha < 4$, no divergency appears in Eq.(4) because the singularity in the denominator at $\zeta = \zeta'$ is canceled by the difference of the intensities in the numerator. Within this approximation, a solution to Eq.(4) can be presented in the self-similar form as a function of dimensionless variables

$$I(z,\zeta) = \frac{1}{\zeta_L} \tilde{f}\left(\frac{z}{L}, \frac{\zeta}{\zeta_L}, \frac{\zeta_0}{\zeta_L}\right), \quad \zeta_L = (\sigma_{\text{tr}} L)^{1/(\alpha-1)}, \tag{7}$$

where the quantity ζ_L is the characteristic multiple-scattering angle for the slab of thickness L. The transport scattering coefficient $\sigma_{\text{tr}} = l_{\text{tr}}^{-1}$ is equal to $\sigma_{\text{tr}} = [2^{3-\alpha}(\alpha-2)/(4-\alpha)]\sigma\vartheta_0^{\alpha-2}$, $(2 < \alpha < 4)$. Physically, the angle ζ_L separates the transmission through the medium from the reflection by it. Light is mainly reflected from the medium if the angle of incidence ζ_0 is less than ζ_L; otherwise, $\zeta_0 > \zeta_L$, light is transmitted through the medium.

3. Asymptotic expressions for the total transmittance

The total transmittance (total transmission coefficient) is a ratio of the transmitted flux $\int_0^\infty \zeta I(z = L,\zeta)d\zeta$ to the incident flux (the latter equals ζ_0). In a nonabsorbing medium, the condition of the total flux conservation, $\int_0^\infty \zeta I(z = L,\zeta)d\zeta + \int_0^\infty |\zeta| I(z = 0, -|\zeta|)d|\zeta| = \zeta_0$, enables us to express the total transmittance in terms of the intensity of reflected light

$$T = 1 - \zeta_0^{-1} \int_0^\infty d|\zeta| R\left(|\zeta|, \zeta_0\right) \tag{8}$$

where we define the reflection function as $R\left(|\zeta|, \zeta_0\right) = |\zeta| I(z = 0, -|\zeta|)$. According to the reciprocity principle [1], $R(|\zeta|, \zeta_0) = R(\zeta_0, |\zeta|)$.

Consider first a situation where the angle of incidence is greater than the characteristic single-scattering angle, $\zeta_0 > \vartheta_0$. In this case, the reflected flux is governed by light scattered through angles exceeding the value ϑ_0, and ϑ_0 can be neglected in the denominator of the phase function in Eq.(4). As follows from Eqs. (7) and (8), the L-dependence of the total transmittance has a single scale $l_{\text{tr}}\zeta_0^{\alpha-1}$ (T depends on the ratio ζ_0/ζ_L).

To calculate the total transmittance for $L > l_{\text{tr}}\zeta_0^{\alpha-1}$, we take advantage of the solution to Eq.(4) in the semi-infinite medium. The explicit form of the reflection function (for $L \to \infty$),

$$R_\infty(|\zeta|, \zeta_0) = \frac{2(\alpha-1)}{\pi\alpha} \frac{\sin(2\pi/\alpha)}{\left(\frac{|\zeta|}{\zeta_0}\right)^{2(\alpha-1)/\alpha} + \left(\frac{\zeta_0}{|\zeta|}\right)^{2(\alpha-1)/\alpha} - 2\cos(2\pi/\alpha)}, \quad (2 < \alpha < 4), \tag{9}$$

is found in [8] just under assumption that the characteristic multiple-scattering angle exceeds the value ϑ_0 (see Appendix). For given values of the angles ζ and ζ_0, the intensity of reflected light obviously increases with increasing the medium thickness L. Therefore, the solution for the semi-infinite medium, $R_\infty(|\zeta|, \zeta_0) \equiv R_\infty(|\zeta|/\zeta_0)$, is a majorant for the reflection function $R(|\zeta|, \zeta_0)$, that is, with allowance for scaling (7),

$$R(|\zeta|, \zeta_0) = R_\infty(|\zeta|/\zeta_0) F\left(\frac{|\zeta|}{\zeta_L}, \frac{\zeta_0}{\zeta_L}\right) \tag{10}$$

where $0 \le F(x, x_0) \le 1$ for all values x and x_0, $F(x, x_0) = F(x_0, x)$. In the limit $x, x_0 \to 0$ (i.e., $L \to \infty$), $F(0,0) = 1$. Taking into account the flux conservation in a semi-finite medium, $\int_0^\infty d|\zeta| R_\infty(|\zeta|/\zeta_0) = \zeta_0$, we can present the total transmittance in the following form

$$T = x_0^{-1} \int_0^\infty dx\, R_\infty(x/x_0)[1 - F(x, x_0)], \quad x_0 = \zeta_0/\zeta_L \tag{11}$$

For $x_0 \ll 1$, Eqs. (9) and (11) give a simple asymptotic formula $T = c_\alpha x_0^{1-2/\alpha}$ or, in the initial notations,

$$T = c_\alpha \left(\frac{\zeta_0^{\alpha-1}}{\sigma_{\rm tr} L} \right)^{\frac{\alpha-2}{\alpha(\alpha-1)}}, \tag{12}$$

where the numeric constant $c_\alpha = 2(\alpha-1)(\pi\alpha)^{-1} \sin(2\pi/\alpha) \int_0^\infty x^{2/\alpha-2}[1 - F(x,0)]dx$ depends only on the parameter α. The value c_α can be determined only by solving the radiative transfer equation with the relevant boundary conditions. For the Henyey-Greenstein phase function ($\alpha = 3$), analysis of the results of our numerical calculations gives $c_3 \approx 0.7$ [10].

Equation (12) reveals the effect of the phase function on the total transmittance at grazing angles of incidence within the thickness range $l_{\rm tr}\zeta_0^{\alpha-1} < L < l_{\rm tr}$. Surprisingly, the L-dependence (12) appears nearly "universal" for $3 \le \alpha < 4$. The difference of the exponent $(\alpha-2)/\alpha(\alpha-1)$ from 1/6 is less than 3% [10]. The ζ_0-dependence, $T \sim \zeta_0^{1-2/\alpha}$, is more sensitive to the specific form of the medium phase function.

If the phase function decreases more rapidly than $1/\vartheta^4$ ($\alpha > 4$), the radiative transfer equation can be solved analytically within the small-angle diffusion approximation. In this case, the explicit form of the F-function is found in [13], and the total transmittance is equal to $T = 0.84...(\zeta_0^3/\sigma_{\rm tr} L)^{1/6}$ [13] for $L \gg l_{\rm tr}\zeta_0^3$ (where $\sigma_{\rm tr} = \sigma\vartheta_0^2/4$). For $\alpha = 4$, Eq.(12) matches the result derived within the diffusion approximation [13].

For an arbitrary ratio ζ_0/ϑ_0, the relation (10) between the reflection functions changes. According to the scaling relation (6), we can present the reflection function $R(|\zeta|, \zeta_0)$ in the following form

$$R(|\zeta|, \zeta_0) = R_\infty \left(\frac{|\zeta|}{\vartheta_0}, \frac{\zeta_0}{\vartheta_0} \right) \tilde{F} \left(\frac{|\zeta|}{\vartheta_0}, \frac{\zeta_0}{\vartheta_0}, \frac{\vartheta_0}{\zeta_L} \right), \tag{13}$$

where we take into account that $\sigma L/\vartheta_0 \sim (\zeta_L/\vartheta_0)^{\alpha-1}$ and write explicitly the arguments of the R_∞-function, $R_\infty(|\zeta|, \zeta_0) \equiv R_\infty\left(\frac{|\zeta|}{\vartheta_0}, \frac{\zeta_0}{\vartheta_0}\right)$. The reflection function of the semi-infinite medium R_∞ obeys Eq.(A.1) (see Appendix). Equation (8) for the total transmittance takes the form

$$T = \zeta_L \zeta_0^{-1} \int_0^\infty dx R_\infty \left(x\frac{\zeta_L}{\vartheta_0}, \frac{\zeta_0}{\vartheta_0} \right) \left[1 - \tilde{F} \left(\frac{x\zeta_L}{\vartheta_0}, \frac{\zeta_0}{\vartheta_0}, \frac{\vartheta_0}{\zeta_L} \right) \right] \tag{14}$$

For $|\zeta|, \zeta_0, \zeta_L \gg \vartheta_0$, Eqs.(13) and (14) transforms into Eqs.(10) and (11), correspondingly. As follows from Eq.(14), the $L-$dependence of the total transmittance at $L \gg \max\{l\vartheta_0, l_{\rm tr}\zeta_0^{\alpha-1}\}$ (i.e. $\zeta_L \gg \max\{\vartheta_0, \zeta_0\}$) is governed by the asymptotic form of the reflection function R_∞ at relatively large angles. Substituting Eq.(A.8) into Eq.(14), we find an asymptotic dependence of the total transmittance on the medium thickness, $T \sim (\vartheta_0/\zeta_L)^{1-2/\alpha}$, or, with allowance for the definition of ζ_L,

$$T = C_\alpha(\zeta_0/\vartheta_0) \left(\frac{\vartheta_0}{\sigma L} \right)^{\frac{\alpha-2}{\alpha(\alpha-1)}} \tag{15}$$

Equation (15) generalizes the result (12) to the case of an arbitrary relation between the incidence angle ζ_0 and the characteristic single-scattering angle ϑ_0. The explicit form of the function

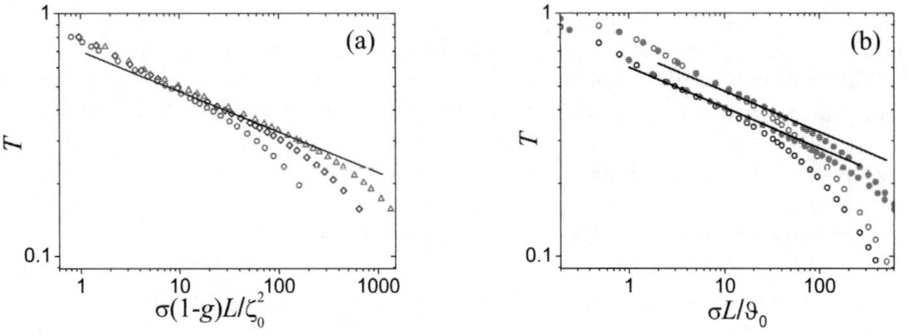

Figure 1. Total transmittance as a function of the medium thickness. Symbols are the results of numerical integration of the radiative transfer equation (1) for the Henyey-Greenstein phase-function. The total number of discrete ordinates is chosen equal to 200. **(a)** The mean cosine of the single scattering angle $g = 0.96$. The incidence angle $\zeta_0 = 0.03$ (triangles), 0.05 (squares) and 0.1 (circles). The solid line is Eq.(12) with $c_3 = 0.7$. **(b)** Upper graphs: $\zeta_0 = \vartheta_0 = 0.1$ (empty) and $\zeta_0 = \vartheta_0 = 0.05$ (filled), lower graphs: $\zeta_0 = \vartheta_0/2 = 0.05$ (empty) and $\zeta_0 = \vartheta_0/2 = 0.02$ (filled). The solid lines are Eq.(15) with $C_3(1) = 0.7$ and $C_3(0.5) = 0.6$, respectively

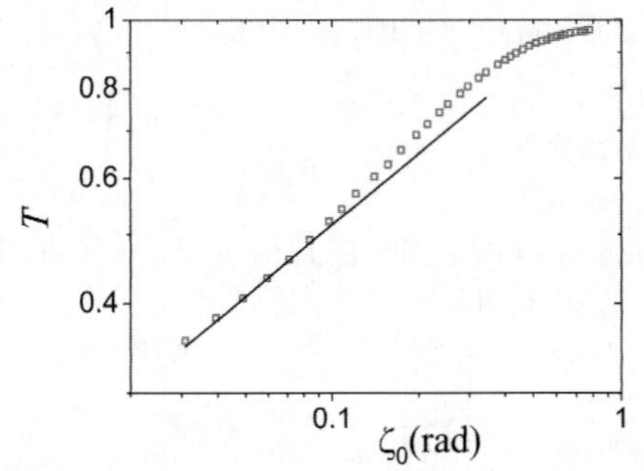

Figure 2. Total transmittance as a function of the incidence angle. The medium optical thickness $\sigma L = 1$. Symbols are the numerical data [1] (sec.13.2, Fig.13.6) for the Henyey-Greenstein phase function with $g = 15/16$. The solid line is Eq.(12) with $c_3 = 0.7$.

$C_\alpha(\zeta_0/\vartheta_0)$ depends on the $\tilde{F}-$function which (as well as F) can be determined only by solving the radiative transfer equation.

Equations (12) and (15) are in a good agreement with the results of numerical integration of the radiative transfer equation (1) for the Henyey-Greenstein phase function (for which $\alpha = 3$ and $\vartheta_0 = 1 - g$, g is the mean cosine of the single-scattering angle). Figure 1 shows the results of our calculations by the discrete-ordinate method [14, 15]. The data [1] are plotted in Fig.2.

For angles $\zeta_0 > \vartheta_0$ the total transmittance is proved to be nearly a universal function of the single parameter $\sigma_{\rm tr}L/\zeta_0^2$ (see Fig.1a) in agreement with scaling relation (7). If the incidence angle ζ_0 is comparable with ϑ_0, the total transmittance appears to be a function of the two parameters ζ_0/ϑ_0 and $\sigma L/\vartheta_0$ (see Fig.1b). In both cases, the self-similarity fails in the crossover to the diffusive regime, $L \geq l_{\rm tr}$, where the small-angle approximation becomes inapplicable, and the numerical solutions falls off more rapidly.

4. Conclusions

In conclusion, we have presented a theoretical study of light transmission through highly forward scattering media. An influence of the single-scattering phase function on the total transmittance has been found within a subdiffusion thickness range. The effect reveals itself at grazing angles of incidence and is governed by small-angle multiple scattering of light. The derived analytic results are in a good agreement with numerical solutions to the radiative transfer equation.

The results obtained above present a way of characterizing inhomogeneities of turbid media from the total transmittance (or reflectance) measurements. This may be useful in studies of highly forward scattering media such as biological tissues.

5. Appendix. Reflection function of a semi-infinite scattering medium

For a semi-infinite medium with the phase function (3), a closed equation for the reflection function $R_\infty(|\zeta|, \zeta_0) = |\zeta| I(z = 0, -|\zeta|)$ is derived in [8]:

$$\zeta_0 \Phi_\lambda(-\zeta_0) = \int_0^\infty d|\zeta| R_\infty(|\zeta|, \zeta_0) \Phi_\lambda(|\zeta|), \quad 0 < \lambda < \infty \tag{A.1}$$

where $\Phi_\lambda(\zeta)$ is the angular eigenfunction of Eq.(4),

$$\Phi_\lambda(\zeta) = \frac{\sqrt{3}}{\pi|\lambda|} \int_0^\infty d\omega \cos\left\{\omega\zeta - \frac{1}{\lambda^3} \int_0^\omega d\omega'(1 - \hat{p}(\omega'))\right\}, \tag{A.2}$$

$\hat{p}(\omega)$ is the Fourier transform of the single-scattering phase function,

$$\hat{p}(\omega) = \int_0^\infty 2\pi\vartheta J_0(\omega\vartheta) p(\vartheta) d\vartheta = \frac{2}{\Gamma(\alpha/2 - 1)} \left(\frac{\omega\vartheta_0}{2}\right)^{\alpha/2-1} \mathrm{K}_{\alpha/2-1}(\omega\vartheta_0), \tag{A.3}$$

$\Gamma(x)$ is the Gamma function, and $\mathrm{K}_\nu(x)$ is the Macdonald function. Equation (A.1) should be satisfied for all eigenvalues λ in the interval $0 < \lambda < \infty$.

If the angles $|\zeta|$ and ζ_0 exceed the value ϑ_0, the $\hat{p}$-function (A.3) can be expanded in the powers of $\omega\vartheta_0$,

$$1 - \hat{p}(\omega) \approx \frac{\Gamma(2 - \alpha/2)}{\Gamma(\alpha/2)} \left(\frac{\omega\vartheta_0}{2}\right)^{\alpha-2}, \tag{A.4}$$

and the angular eigenfunctions take the form

$$\Phi_\lambda(\zeta) \approx \frac{\sqrt{3}}{\pi|\lambda|} \int_0^\infty d\omega \cos\left(\omega\zeta - \frac{a_\alpha \vartheta_0^{\alpha-2}}{\lambda^3} \omega^{\alpha-1}\right) \tag{A.5}$$

where $a_\alpha = \Gamma(2-\alpha/2)/[2^{\alpha-2}(\alpha-1)\Gamma(\alpha/2)]$. The reflection function (9) is a solution to Eq.(A.1) in which the angular eigenfunctions are approximated by Eq.(A.5).

For $|\zeta| \gg \vartheta_0$ and an arbitrary ratio ζ_0/ϑ_0, the approximation (A.5) can be used only in the kernel of Eq.(A.1). The eigenfunction in the inhomogeneous term should be calculated with the exact equations (A.2)-(A.3). Multiplying Eq.(A.1) by $\lambda^{-3s/(\alpha-1)}$ and integrating over all λ, we find the Mellin transform, $\hat{R}_\infty(s, \zeta_0) = \int_0^\infty |\zeta|^{s-1} R_\infty(|\zeta|, \zeta_0) d|\zeta|$, of the reflection function

$$\hat{R}_\infty(s, \zeta_0) = \frac{\zeta_0}{\vartheta_0} \left[\Gamma(1 - s) \sin \frac{\pi\alpha s}{2(\alpha - 1)}\right]^{-1} \int_0^\infty dx \times \tag{A.6}$$

$$\left[a_\alpha^{-1} \int_0^x (1 - \hat{p}(\omega')) d(\omega'\vartheta_0)\right]^{-s/(\alpha-1)} \cos\left(\frac{\zeta_0}{\vartheta_0} x + \frac{\pi s}{2(\alpha - 1)}\right)$$

The inverse transform gives the desired reflection function

$$R_\infty(|\zeta|, \zeta_0) = \int_{\delta-i\infty}^{\delta+i\infty} \frac{ds}{2\pi i} \left(\frac{|\zeta|}{\vartheta_0}\right)^{-s} \hat{R}_\infty(s, \zeta_0) \qquad (A.7)$$

The integration in Eq.(A.7) is performed in the band $0 < \text{Re } s < 1$.

Equations (A.6) and (A.7) enables us to draw a conclusion regarding the asymptotic behavior of the reflection function at relatively large angles, $|\zeta| \gg \max\{\vartheta_0, \zeta_0\}$. The asymptotic form of $R_\infty(|\zeta|, \zeta_0)$ is governed by poles of $\hat{R}_\infty(s, \zeta_0)$ in the complex $s-$plane. As analysis of Eq.(A.6) shows, the pole with the smallest positive real part is located at $s = 2(\alpha-1)/\alpha$. Therefore, the asymptotic expression for the reflection function has the following form

$$R_\infty(|\zeta|, \zeta_0) = A_\alpha(\zeta_0/\vartheta_0)\left(\frac{\vartheta_0}{|\zeta|}\right)^{2(\alpha-1)/\alpha}. \qquad (A.8)$$

The explicit form of the $A_\alpha-$function can be found by calculating the residue of $\hat{R}_\infty(s, \zeta_0)$ at $s = 2(\alpha-1)/\alpha$.

References

[1] van de Hulst H C 1980 *Multiple Light Scattering: Tables, Formulas, and Applications.* (Academic Press, New York) Vols. 1 and 2
[2] Mishchenko M I, Travis L D and Lacis A A 2006 *Multiple Scattering of Light by Particles* (Cambridge University Press, Cambridge)
[3] Prahl S A, van Gemert M J C and Welch A J 1993 *Appl. Opt.* **32** 559-568
[4] Kienle A, Forster F K and Hibst R 2001 *Opt. Lett.* **26** 1571-1573
[5] Li J, Dietsche G, Iftime D, Skipetrov S E, Maret G, Elbert Th, Rockstroh B and Gisler Th 2005 *J. Biomed. Opt.* **10** 044002
[6] Desjardins A E, Vakoc B J, Bilenca A, Tearney G J and Bouma B E 2007 *Opt. Lett.* **32** 1560-1562
[7] Ishimaru A 1999 *Wave Propagation and Scattering in Random Media,* (IEEE Press Series on Electromagnetic Wave Theory, Wiley)
[8] Remizovich V S 1984 *Sov. Phys. JETP* **60** 290-299
[9] Gorodnichev E E and Rogozkin D B 1995 *JETP* **80** 112-126
[10] Marinyuk V V and Sheberstov S V 2016 *Opt. Lett.* **41** 922-925
[11] Reynolds L O and McCormick N J 1980 *J. Opt. Soc. Am.* **70** 1206-1212
[12] Sheppard C J R 2007 *Opt. Lett.* **32** 142-144
[13] Kuzovlev A I and Remizovich V S 1984 *Izv. Acad. Sci. USSR Atmos. Oceanic Phys.* **20** 806-812
[14] Stamnes K, Tsay S C, Wiscombe W and Jayaweera K 1988 *Appl. Opt.* **27** 2502-2509
[15] Marinyuk V V, Rogozkin D B and Sheberstov S V 2014 *Optics and Spectroscopy* **117** 102-110

The effect of the unpaired nucleons on the β-decay properties of the neutron-rich nuclei

E O Sushenok[1,2] **and A P Severyukhin**[1,2]
[1]Bogoliubov Laboratory of Theoretical Physics, Joint Institute for Nuclear Research, 141980 Dubna, Moscow region, Russia
[2]Dubna State University, 141982 Dubna, Moscow Region, Russia

Abstract. Starting from the T45 Skyrme interaction with tensor terms the properties of the β-decay of $^{72-80}$Ni are studied. We take into account the effect of unpaired neutron and proton on the ground state properties of odd-odd and even-odd nuclei. It is shown that the calculated Q_β-values and the β-decay half-lives are in a reasonable agreement with experimental data.

A study of the β-decay properties is an interesting problem not only from the nuclear structure point of view but it is very important for the nuclear astrophysics applications. It is desirable to have theoretical models which can describe the data wherever they can be measured, and predict the properties related to spin-isospin modes in the nuclei with extreme N/Z ratio to allow for experimental studies. One of the successful tools for studying charge-exchange nuclear modes is the quasiparticle random phase approximation (QRPA) with the self-consistent mean-field derived from a Skyrme energy-density functional (EDF), see e.g., [1, 2, 3, 4, 5]. These QRPA calculations enable one to describe the properties of the ground states and excited charge-exchange states using the same EDF. Our tool is the QRPA with Skyrme interactions in the finite rank separable approximation (FRSA) [6, 7, 8, 9], allowing one to perform calculations in large configuration spaces. The FRSA model for the charge-exchange excitations and the β-decay was already introduced in Refs. [10, 11] and in Refs. [12, 13], respectively.

The correct description of the Q_β-values is the important ingredient for the reliable prediction of the half-life of the β-decay. To calculate the binding energy of the odd-odd and even-odd nuclei we take into account the effect of the unpaired neutron and proton on the superfluid properties of nuclei, the well-known blocking effect [14, 15]. As an example, the β-decay properties of neutron-rich nuclei ^{72,74,76,80}Ni and the most neutron-rich ($(N-Z)/A = 0.28$) doubly-magic nucleus ^{78}Ni are studied. The β-decay properties of r-process "waiting-point nucleus" ^{78}Ni have attracted a lot of experimental efforts, see e.g., [16, 17, 18].

We use the EDF T45 which takes into account the tensor force [19]. The T45 set is one of 36 parametrizations, covering a wide range of the parameter space of the isoscalar and isovector tensor term added with refitting the parameters of the central interaction, where a fit protocol is very similar to that of the successful SLy parametrizations. This choice of the Skyrme EDF has been selected to reproduce the experimental Q_β value of ^{78}Ni (see Fig. 1) and enough positive value of the spin-isospin Landau parameter ($G_0' = 0.10$ for T45). It is worth mentioning that the first study of the strong impact of the tensor correlations on the β-decay half-life has been done in [4, 5]. The pairing correlations are generated by a zero-range volume force with a strength of -270 MeVfm^3 and a smooth cut-off at 10 MeV above the Fermi energies [9]. This value of the

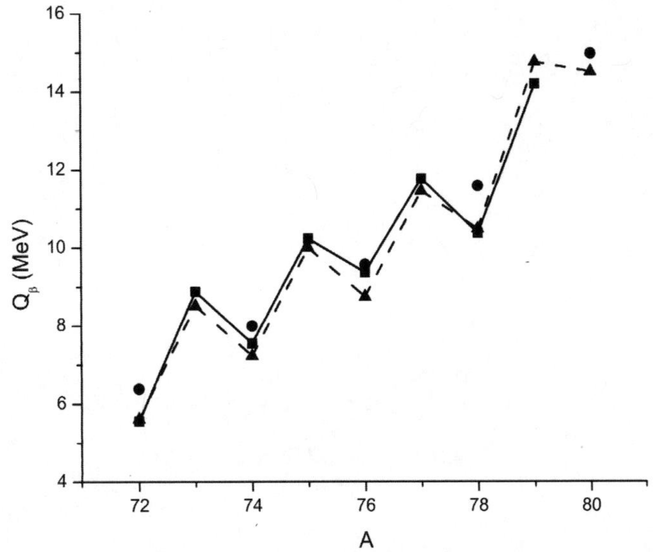

Figure 1. The quasiparticle blocking effect on Q_β-values of $^{72-80}$Ni. Results of the HF-BCS calculations with the blocking effect (triangles) and without the blocking effect (circles) are shown. Experimental data (squares) are from Ref. [26].

pairing strength has been fitted to reproduce the odd-even mass difference in the studied region of nuclei [12, 20].

Assuming the spherical symmetry for the nuclei considered here, the starting point of the method is the self-consistent HF-BCS calculation [21] for the ground state properties of the even-even parent nucleus (N,Z). The continuous part of the single-particle spectrum is discretized by diagonalizing the HF hamiltonian on a harmonic oscillator basis. In the particle-hole channel we use the Skyrme interaction with the tensor components and their inclusion leads to the modification of the spin-orbit potential [19, 22].

The ground state of the odd-odd daughter nucleus (N-1, Z+1) can be obtained as the neutron-quasiparticle proton-quasiparticle state. The neutron and proton quasiparticles can be simultaneously blocked [23]. Using the blocking effect for unpaired nucleons [14, 15, 21] we get the following secular equations,

$$\Delta_j = \frac{1}{2}\sum_{j' \neq j_2} V_{jj'} \frac{(2j'+1)\Delta_{j'}}{\sqrt{\Delta_{j'}^2 + (E_{j'} - \lambda)^2}} + \frac{1}{2} V_{jj_2} \frac{(2j_2 - 1)\Delta_{j_2}}{\sqrt{\Delta_{j_2}^2 + (E_{j_2} - \lambda)^2}}, \qquad (1)$$

where the indexes j denote the quantum numbers nlj. The indexes j_2 emphasize the blocked neutron subshell and the blocked proton subshell near the Fermi energies. For ^{72,74,76,78}Cu the neutron quasiparticle blocking is based on filling the $1g_{9/2}$ subshell and the $2d_{5/2}$ subshell should be blocked for ^{80}Cu. The proton $2p_{3/2}$ and $1f_{5/2}$ subshells are chosen to be blocked in the cases of ^{72,74,76}Cu and ^{78,80}Cu, respectively. It is worth pointing out that there is the closeness of the proton single-particle energies $2p_{3/2}$, $1f_{5/2}$ for ^{76}Cu. The quasiparticle blocking calculations are discussed in more detail in Ref. [24].

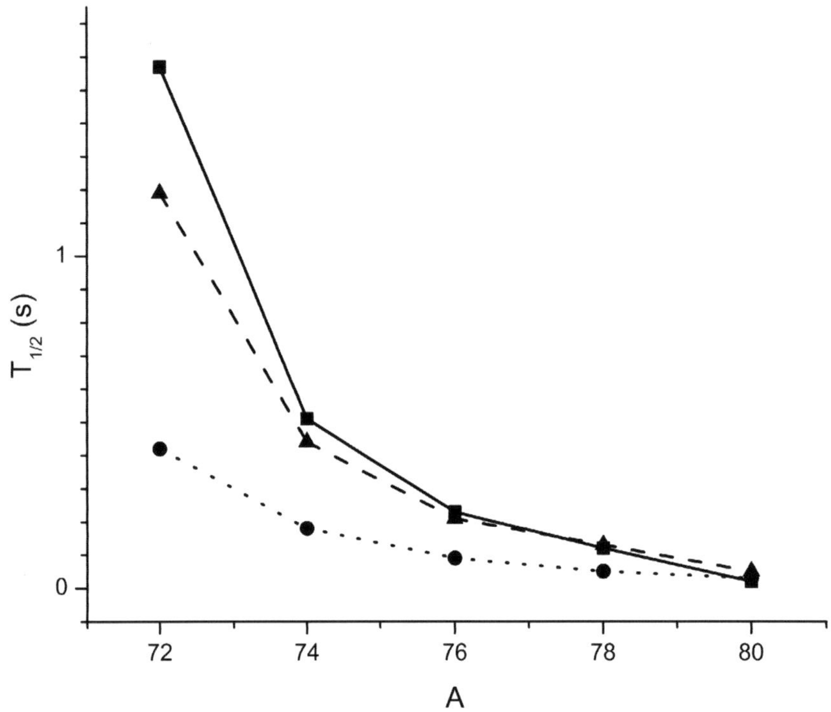

Figure 2. Same as Fig. 1, for the half-lives of the β-decay of ^{72,74,76,78,80}Ni. Experimental data are taken from Refs. [16, 29].

It is interesting to study the blocking effect on Q_β value. The Q_β value can be obtained by the binding-energy difference between the daughter and parent nuclei,

$$Q_\beta = \Delta M_{n-H} + B(Z+1, N-1) - B(Z, N). \tag{2}$$

$\Delta M_{n-H} = 0.782$ MeV is the mass difference between the neutron and the hydrogen atom. As proposed in Ref. [25], the Q_β value of the even-even nucleus can be calculated without the blocking effect,

$$Q_\beta \approx \Delta M_{n-H} + \mu_n - \mu_p - E_{2qp,lowest}, \tag{3}$$

where $E_{2qp,lowest}$ corresponds the lowest two-quasiparticle energy, μ_n and μ_p are the neutron and proton chemical potentials. The calculated Q_β values in the neutron-rich Ni isotones are compared with existing experimental data [26] in Fig.1. The results of the HF-BCS calculation with the blocking effect are in a reasonable agreement with the experimental data. There is a remarkable the odd-even isotope staggering. This feature observed in the calculation occurs because for an odd-neutron nucleus the odd neutron contributes strongly to the β decay. This contribution is absent in an even-even nucleus. For even-even nuclei, the Q_β analysis within the approximation (3) can help to clarify the blocking effect. We find that the blocking effect induces a reduction of the Q_β values and it results in a improvement of the Q_β description, see Fig.1.

To build the QRPA equations on the basis of HF-BCS quasiparticle states of the parent nucleus is the standard procedure [27]. Using the FRSA model the QRPA eigenvalues (E_k) are obtained as the roots of the relatively simple secular equation [10, 11] and we carry out QRPA calculations in very large two-quasiparticle spaces. The cut-off of the discretized continuous part of the single-particle spectra is at the energy of 100 MeV. This is sufficient to exhaust the Ikeda sum rule $3(N-Z)$ for the Gamow-Teller (GT) transitions.

In the allowed GT approximation, the β^{-}-decay half-life is expressed by summing the probabilities (in units of $G_A^2/4\pi$) of the energetically allowed transitions ($E_k^{\mathrm{GT}} \leq Q_\beta$) weighted with the integrated Fermi function

$$T_{1/2}^{-1} = D^{-1}\left(\frac{G_A}{G_V}\right)^2 \sum_k f_0(Z+1, A, E_k^{\mathrm{GT}}) B(GT)_k, \tag{4}$$

$$E_k^{\mathrm{GT}} = Q_\beta - E_{1_k^+}, \tag{5}$$

where G_A/G_V=1.25 and D=6147 s [28]. $E_{1_k^+}$ denotes the excitation energy of the 1_k^+ state of the daughter nucleus. As proposed in Ref. [25], this energy can be estimated by the following expression:

$$E_{1_k^+} \approx E_k - E_{2qp,lowest}. \tag{6}$$

It is worth mentioning that the spin-parity of the lowest two-quasiparticle state is, in general, different from 1^+.

The properties of the low-lying 1^+ states in the daughter nuclei ^{72,74,76,78,80}Cu are studied. There is the gradual reduction of β-decay half-lives with increasing neutron number [16, 29], see Fig.2. One can see that our results calculated with the blocking effect reproduce this behaviour. As expected, the largest contribution (>60%) in the calculated half-life comes from the 1_1^+ state. QRPA results indicate that the dominant configuration of the 1_1^+ wave function is $\{\pi 2p\frac{3}{2}\nu 2p\frac{1}{2}\}$ whose contribution is about 99 % in all five nuclei. The inclusion of the blocking effect for the Q_β calculation reduces the transition energies (5) and this energy shift produces a sizable impact on the β-decay half-life.

In summary, by starting from the Skyrme HF-BCS calculations the Q_β-window has been studied within the model including the blocking effect of unpaired neutron and proton in cases of the even-odd and odd-odd nuclei. Using the EDF T45 containing the tensor terms, we analyze the effect on the β-transition rates of the neutron-rich nuclei $^{72-80}$Ni. It is shown that the inclusion of this effect definitely improves the description of the Q_β-value and the β-decay half-life. A further systematic study of the blocking effect on the β-decay properties is clearly necessary and is in progress.

Acknowledgments

We thank N. N. Arsenyev and I. N. Borzov for useful discussions. This work is partly supported by from the Russian Science Foundation (Grant No. RSF-16-12-1016).

References

[1] Bender M, Dobaczewski J, Engel J and Nazarewicz W 2002 *Phys. Rev.* C **65** 054322
[2] Fracasso S and Colò G 2007 *Phys. Rev.* C **76** 044307
[3] Bai C L, Zhang H Q, Sagawa H, Zhang X Z, Colò G and Xu F R 2011 *Phys. Rev.* C **83** 054316
[4] Minato F and Bai C L 2013 *Phys. Rev. Lett.* **110** 122501
[5] Minato F and Bai C L 2016 *Phys. Rev. Lett.* **116** 089902
[6] Nguyen Van Giai, Stoyanov Ch and Voronov V V 1998 *Phys. Rev.* C **57** 1204
[7] Severyukhin A P, Stoyanov Ch, Voronov V V and Nguyen Van Giai 2002 *Phys. Rev.* C **66** 034304
[8] Severyukhin A P, Voronov V V and Nguyen Van Giai 2004 *Eur. Phys. J.* A **22** 397
[9] Severyukhin A P, Voronov V V and Nguyen Van Giai 2008 *Phys. Rev.* C **77** 024322
[10] Severyukhin A P, Voronov V V and Nguyen Van Giai 2012 *Prog. Theor. Phys.* **128** 489
[11] Severyukhin A P and Sagawa H 2013 *Prog. Theor. Exp. Phys.* **2013** 103D03

[12] Severyukhin A P, Voronov V V, Borzov I N, Arsenyev N N and Nguyen Van Giai 2014 *Phys. Rev.* C **90** 044320
[13] Etilé A, Verney D, Arsenyev N N, Bettane J, Borzov I N, Cheikh Mhamed M, Cuong P V, Delafosse C, Didierjean F, Gaulard C, Nguyen Van Giai, Goasduff A, Ibrahim F, Kolos K, Lau C, Niikura M, Roccia S, Severyukhin A P, Testov D, Tusseau-Nenez S and Voronov V V 2015 *Phys. Rev.* C **91** 064317
[14] Soloviev V G 1961 *Kgl. Dan. Vid. Selsk. Mat. Fys. Skr.* **1** 235
[15] Soloviev V G 1976 *Theory of Complex Nuclei* (Pergamon Press, Oxford)
[16] Xu Z Y, Nishimura S, Lorusso G, Browne F, Doornenbal P, Gey G, Jung H-S, Li Z, Niikura M, Söderström P-A, Sumikama T, Taprogge J, Vajta Zs, Watanabe H, Wu J, Yagi A, Yoshinaga K, Baba H, Franchoo S, Isobe T, John P R, Kojouharov I, Kubono S, Kurz N, Matea I, Matsui K, Mengoni D, Morfouace P, Napoli D R, Naqvi F, Nishibata H, Odahara A, Şahin E, Sakurai H, Schaffner H, Stefan I G, Suzuki D, Taniuchi R and Werner V 2014 *Phys. Rev. Lett.* **113** 032505
[17] Alshudifat M F, Grzywacz R, Madurga M, Gross C J, Rykaczewski K P, Batchelder J C, Bingham C, Borzov I N, Brewer N T, Cartegni L, Fijałkowska A, Hamilton J H, Hwang J K, Ilyushkin S V, Jost C, Karny M, Korgul A, Królas W, Liu S H, Mazzocchi C, Mendez A J, Miernik K, Miller D, Padgett S W, Paulauskas S V, Ramayya A V, Stracener D W, Surman R, Winger J A, Wolińska-Cichocka M and Zganjar E F 2016 *Phys. Rev.* C **93** 044325
[18] Madurga M, Paulauskas S V, Grzywacz R, Miller D, Bardayan D W, Batchelder J C, Brewer N T, Cizewski J A, Fijałkowska A, Gross C J, Howard M E, Ilyushkin S V, Manning B, Matoš, Mendez A J, Miernik K, Padgett S W, Peters W A, Rasco B C, Ratkiewicz A, Rykaczewski K P, Stracener D W, Wang E H, Wolińska-Cichocka M and Zganjar E F 2016 *Phys. Rev. Lett.* **117** 092502
[19] Lesinski T, Bender M, Bennaceur K, Duguet T and Meyer J 2007 *Phys. Rev.* C **76** 014312
[20] Severyukhin A P, Arsenyev N N and Pietralla N 2012 *Phys. Rev.* C **86** 024311
[21] Ring P and Schuck P 1980 *The Nuclear Many Body Problem* (Springer, Berlin)
[22] Stancu F, Brink D M and Flocard H 1977 *Phys. Lett.* **B68** 108
[23] Dobaczewski J, Satuła W, Carlsson B G, Engel J, Olbratowski P, Powałowski P, Sadziak M, Sarich J, Schunck N, Staszczak A, Stoitsov M, Zalewski M and Zduńczuk H 2009 *Comp. Phys. Comm.* **180** 2361
[24] Sushenok E O and Severyukhin A P submitted to *Phys. Atom. Nucl.*
[25] Engel J, Bender M, Dobaczewski J, Nazarewicz W and Surman R 1999 *Phys. Rev.* C **60** 014302
[26] Wang M, Audi G, Wapstra A H, Kondev F G, MacCormick M, Xu X and Pfeiffer B 2012 *Chin. Phys.* C **36** 1603
[27] Terasaki J, Engel J, Bender M, Dobaczewski J, Nazarewicz W and Stoitsov M 2005 *Phys. Rev.* C **71** 034310
[28] Suhonen J 2007 *From Nucleons to Nucleus* (Berlin: Springer-Verlag)
[29] National Nuclear Data Center, http://www.nndc.bnl.gov

About orthogonality one system of functions

A N Barmenkov

National Research Nuclear University MEPHI
(Moscow Engineering Physics Institute),
31 Kashirskoe Shosse, 115409, Moscow, Russia

E-mail: anbarmenkov@mephi.ru

Abstract. For the modeling of physical processes is proposed to use representation of functions by a special trigonometric system. We present a method for constructing full orthogonal sequences of this type depending on the parameter. This parameter can be selected optimally based on specification of modeling task.

1. Introduction

The results of this article were announced in the report [1].

It is known the role of following trigonometric system of functions

$$\{\cos nt; \sin nt\}_{n=0}^{\infty}$$

in mathematical modelling solving a wide class of boundary task of mathematical physics (for example, the Fourier method.).

Consider generalization of trigonometric sequence, namely

$$\{\cos(nt+\alpha(t)); \sin(nt+\alpha(t))\}_{n=0}^{\infty}, \qquad (1)$$

where $\alpha(t)$ - a real function, $t \in [0;2\pi]$.

In 1953 K. Shaidukov [2] by methods of theory of functions of a real variable proved the completeness in $L_2[0;2\pi]$ sequence of functions

$$\{\cos(n+\alpha)t; \sin(n+\alpha)t\}_{n=0}^{\infty}$$

for real α only when $\alpha \le \frac{2}{3}$, initiated a whole line of research of its generalizations.

For the first time in [3],[4] the method of study of the approximation properties (i.e. completeness, minimality and the basis property) of this type of mixing systems by reduction to the investigation of the solvability of boundary tasks of the theory of functions of complex variables.

We got the conditions of completeness, minimality and the basis of such sequences in a fairly general assumptions about the function $\alpha(t)$ depending on the values of its differences (see. [3], p.43), ([4],p.67) points of discontinuity of the first kind $[0;2\pi]$. In particular, for use in modeling functions conveniently to formulate: the sequence (1) is complete and minimal in $L_2[0;2\pi]$, for continuously differentiable $\alpha(t), \alpha(t) \ne const, t \in [0;2\pi]$ where $\frac{1}{4} < \frac{\alpha(2\pi)-\alpha(0)}{2\pi} \le \frac{3}{4}$. That is, the system is considered by K. Shaidukov really, complete when $\alpha \le \frac{3}{4}$ and incomplete if $\alpha > \frac{3}{4}$ in $L_2[0;2\pi]$, since in this case $\alpha(t) = \alpha \cdot t$.

Note that for smooth functions completeness, minimality and the basis of the sequence (1) depend only on the difference of values $\alpha(t)$ on the ends of $[0;2\pi]$, which gives a certain "freedom" in the selection of functions $\alpha(t)$ for use in the modelling.

Biorthogonal system $\{\varphi_n(t); g_n(t)\}_{n=0}^{\infty}$ for (1), expressed in the solutions of boundary tasks in the theory of analytic functions is quite difficult. The most convenient systems (1) for modeling, when these sequences are orthogonal. The article describes the complete and orthogonal sequences of the form (1).

2. Main results

To formulate the theorem we introduce the function $B_a(z)=\frac{|a|}{a}\frac{a-z}{1-\bar{a}z}$,
a conformal mapping of the unit circle on the unit circle centered at the origin and the point of transforming $a\in\{|z|<1\}, a\neq 0$ in the zero point. If $a=0$, so we suppose that $B_a(z)=z$.

Theorem. Let $\alpha(t)$ - a real smooth function different from a constant ($\alpha(t)\neq const, \forall t\in[0;2\pi]$). To ensure that the system functions

$$\{\cos(nt+\alpha(t));\sin(nt+\alpha(t))\}_{n=0}^{\infty}$$

to be complete and orthogonal in space $L_2[0;2\pi]$, it is necessary and sufficient the following conditions to be satisfied $\alpha(t)=\frac{1}{2}\arg B_a(e^{it})+r$, where $B_a(z)=\frac{|a|}{a}\frac{a-z}{1-\bar{a}z}$, $|z|<1$, where $a\in C, |a|<1, a\neq 0$; if $a=0$, than $B_a(z)=z$, ($B_a(e^{it})=e^{it}$), r -arbitrary real constant.

Corollary. General view of the complete and orthogonal sequences of the form (1) $L_2[0;2\pi]$ defined by function ($\alpha(t)\neq const$)

$$\alpha(t)=\frac{1}{2}\arg\frac{|a|}{a}\frac{a-e^{it}}{1-\bar{a}e^{it}}+r \quad , t\in[0;2\pi] \tag{2}$$

when $a\in C, |a|<1, a\neq 0$ and $\alpha(t)=\frac{t}{2}+r$, if $a=0$, where r - arbitrary real constant.

Obviously, the sequence $\{\cos(nt+\frac{t}{2});\sin(nt+\frac{t}{2})\}_{n=0}^{\infty}$ is complete and orthogonal in $L_2[0;2\pi]$.

Proof of the theorem. Orthogonality of complete and minimal sequence (1) in $L_2[0;2\pi]$, for $n=1,2,3,...$ to function $\cos\alpha(t)$

$$\int_0^{2\pi}\cos(nt+\alpha(t))\cos\alpha(t)dt=0$$

$$\int_0^{2\pi}\sin(nt+\alpha(t))\cos\alpha(t)dt=0$$

hereof

$$\int_0^{2\pi}e^{i(nt+\alpha(t))}\cos\alpha(t)dt=0, n=1,2,3,...$$

Similarly, for function $\sin\alpha(t)$

$$\int_0^{2\pi}e^{i(nt+\alpha(t))}\sin\alpha(t)dt=0, n=1,2,3,...$$

So

$$\int_0^{2\pi} e^{\text{int}} e^{2i\alpha(t)} dt = 0, n = 1,2,... \quad (3)$$

$$\int_0^{2\pi} \cos\alpha(t)\sin\alpha(t)dt = 0.$$

The last equality means the orthogonality of functions $\cos\alpha(t)$ and $\sin\alpha(t)$.

Note that equation (1) is performed when we replace pair of functions $\cos\alpha(t);\sin\alpha(t)$ on any other $\cos(kt+\alpha(t));\sin(kt+\alpha(t))$.

Equation (3) means the existence of a function $\phi(z)$ Hardy class H^1 (see. [5], p.102), for angular boundary values which have the equality

$$e^{2i\alpha(t)} = \phi(e^{it}), \ t \in [0;2\pi]. \quad (4)$$

As $\alpha(t),\beta(t)$ - continuous (even the Hölder) function of bounded variation, then according to the theorem of F. and M. Riesz (see., e.g., [6], p. 103) the execution of relations (3) means that actually analytical in $\{|z|<1\}$ function $\phi(z)$ in fact, it is a continuous in the circle $\{|z|\le 1\}$ and absolutely continuous on the unit circumference (i.e. the point z = 1, no jump). This means that the change in the function argument $\phi(z)$ on the unit circumference is an integral multiple of 2π. As it is shown in ([4], p.84) and ([3], p.33) for Hölder functions conditions of completeness and minimality of sequence (1) in the space $L_2[0;2\pi]$ were given in form

$$\frac{1}{4} < \frac{\alpha(2\pi)-\alpha(0)}{2\pi} \le \frac{3}{4} \ (\alpha(t) \ne const, t \in [0;2\pi]).$$

So, change of the argument of function $\phi(z)$ on the unit circumference with (4) and the last double inequality satisfies following equation

$$Var_{[0;2\pi]} \frac{1}{2\pi} \arg\phi(e^{it}) = \frac{2(\alpha(2\pi)-\alpha(0))}{2\pi} = n,$$

where n - necessarily an integer. In view of this inequality is possible only for $n=1$. Therefore, according to the argument principle the function $\phi(z)$ analytical in $\{|z|<1\}$ and continuous in $\{|z|\le 1\}$ has a single zero at some point $a \in C, |a|<1$, $\phi(a)=0$.

Then for previously introduced function $B_a(z) = \frac{|a|}{a}\frac{a-z}{1-\bar{a}z}$ with $a \ne 0$ and $B_a(z) = z$ for $a = 0$ is performed $B_a(a)=0$ and $|B_a(e^{it})| = 1$, that means that, it is exist analytical in $\{|z|<1\}$ and continuous in $\{|z|\le 1\}$ function $\psi(z)$, $\psi(z) \ne 0$ in $\{|z|<1\}$, for which (4) takes the form

$$\frac{e^{2i\alpha(t)}}{B_a(e^{it})} = \psi(e^{it})$$

Taking the logarithm of both sides in view of the fact that

$$\left|\frac{e^{2i\alpha(t)}}{B_a(e^{it})}\right| = 1, t \in [0;2\pi]$$

we get

$$\ln\psi(e^{it}) = i\arg(\frac{e^{2i\alpha(t)}}{B_a(e^{it})})$$

That is analytical in the unit circle and continuous in its closure, function $\ln\psi(z)$ takes on the unit circumference of only imaginary values. It follows immediately, that $\ln\psi(z) \equiv ir$ in $\{|z| \leq 1\}$, where r - arbitrary real constant. It means we get function $\alpha(t)$, which is equivalent of orthogonality of complete system (1). Namely $\alpha(t) = \frac{1}{2}\arg B_a(e^{it}) + r$, which completes the proof of the theorem.

Note that rather arbitrary selection of parameter $a \in C, |a| < 1$, allows us to select a sequence of the form (1) optimally for a particular task.

References

[1] Barmenkov A N 2016.*About orthogonality of the system of functions.*
V international conference, *Problems of mathematical and theoretical physics and mathematical modeling (collection of articles), pp 171-172.*

[2] Shaidukov K 1953, *About the completeness of the trigonometric system.* Success of math Sciences, No. 6, pp 143-153

[3] Barmenkov A N, Kazmin J A 1982 *The completeness of systems of functions of a special form.* Book: Theory of mapping, its generalizations and applications. *Collection of scientific articles, г. Kiev, Naukova Dumka, p 23-43.*

[4] Barmenkov A N 1983 *About approximation properties of certain systems of functions*
Diss..kand.fiz.mat. sciences, Moscow, p 154.

[5] Danyluk I I 1975 *Irregular boundary value problems on the plane*, M. «Science», p 256.

[6] Privalov I I 1950, *Boundary properties of analytic functions*, M. Gostekhizdat, p 336.

IOP Conf. Series: Journal of Physics: Conf. Series **788** (2017) 012048 doi:10.1088/1742-6596/788/1/012048

Model of spontaneous evaporating droplet on solid horizontal substrate

S Z Dunin, O V Nagornov and V P Trifonenkov

National Research Nuclear University MEPhI (Moscow Engineering Physics Institute), 31 Kashirskoeshosse, 115409 Moscow, Russia

E-mail: nagornov@yandex.ru

Abstract. Free evaporation of sessile liquid non-isothermal drop on solid substrate is analyzed. Exact formulae for temperature and concentration fields are found out as functions of dimensionless parameters. The non-uniform temperature distribution at the drop surface creates the thermocapillar Marangonni forces that change their direction in the vicinity of stagnation points. Direction of the forces and disposition of the stagnation points are derived as function of contact angle and thermodynamic parameters of model. Conditions for the stagnation points to appear are found out. Moreover, maximal value of contact angle corresponding to presence of stagnation points in droplet is calculated as a function of the thermal conductivity ratio.

1. Introduction

Evaporation of the drops is the central issue of some complex technological processes. Drying drops used in material processing, DNA/RNA researches, bio-chemical transformations. There are a number of experimental and theoretical researches of flow configuration inside the drop [1-13]. Due to complexity of physical processes inside drop and their mathematical descriptions, researchers were forced to simplify their mathematical models or use numerical methods to solve them. Some works use finite elements method (FEM) to study non-isothermal evaporation [4-8].

In this paper we use derived analytical solution for the temperature and concentration fields to investigate evolution of drop geometry and total mass, to study influence of the thermal conductivity ratio on evaporation process and on distribution of Marangonni forces direction.

2. Problem formulation and solution

We consider a drop of partially wetting liquid resting on flat solid layer. Initial contact angle is $\theta_c < 90^{\circ}$. We assume drop with pinned contact line, so during evaporation process the contact angle decreases to keep the same wetted area. The initial parameters of the drop are radius R, height h and contact angle θ_c. In small drops due to Bond number $Bo = \rho gRh/\sigma = 0.03 \div 0.4$ the surface tension effects dominate gravitational effects and drops can be considered as spherical cap shape (Figure 1). We also consider the Peclet number to be small $Pe << 1$, so the thermal convection is negligible and the thermal diffusivity time to be much less than life time of the drop: $t << R^2 k_{L,s}^{-1} = t_{L,s}$, where R is radius of the drop, k_s and k_L are thermal conductivities of substrate and liquid, respectively.

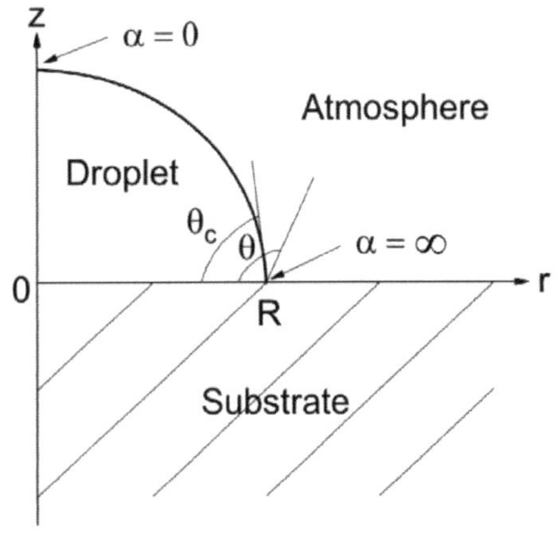

Figure 1. General view of spherical droplet on solid substrate

So, temperature field in substrate and liquid satisfies Laplace's equation:

$$\begin{cases} \Delta T_L = 0 \\ \Delta T_s = 0 \end{cases} \tag{1}$$

In gas phase we consider no thermal convection (i.e. $Pe << 1$), so concentration field also satisfies Laplace's equation:

$$\Delta c = 0 \tag{2}$$

Thermal properties of liquid and substrate are taken into account by boundary conditions. We assume the heat flux from substrate is the main reason of evaporation. The heat inside the drop reaches its surface, where the phase transition occurs. Such problem formulation is closer to problems of the current static [20].

For this geometry the problem can be solved in toroidal coordinates (α, θ), where $\theta = 0$ and $\theta = \theta_c$ represent the drop base and surface respectively. We consider no normal concentration and temperature flows through the substrate-gas boundary and the temperature continuity on the liquid-substrate boundary. The general solution of Laplace's equation for temperature and concentration can be expressed as a result of Mehler-Fock inverse transform:

$$T_L(\alpha, O) - T_\infty + \{2(ch\alpha + cosO)\}^{\frac{1}{2}} \int_0^\infty d\tau \, P_{-\frac{1}{2}+i\tau}(ch\alpha) \frac{ch\tau\theta_c}{ch\tau\pi} A_s(\tau) \frac{ch\tau\pi \, ch\tau\theta + k_R sh\tau\pi \, sh\tau\theta}{ch\tau\pi \, sh\tau\theta_c + k_R sh\tau\pi \, ch\tau\theta_c},$$

$0 < \theta < \theta_c$

$$c_V(\alpha, \theta) = c_\infty + \{2(ch\alpha + cos\theta)\}^{\frac{1}{2}} \int_0^\infty d\tau \, P_{-\frac{1}{2}+i\tau}(ch\alpha) \frac{ch\tau\theta_c}{ch\tau\pi} B_D(\tau) \frac{ch\tau(\pi - \theta)}{ch\tau(\pi - \theta_c)}, \; \theta_c < \theta < \pi \tag{3}$$

$$T_s(\alpha, \theta) = T_\infty + \{2(ch\alpha + cos\theta)\}^{\frac{1}{2}} \int_0^\infty d\tau \, P_{-\frac{1}{2}+i\tau}(ch\alpha) \frac{ch\tau\theta_c}{ch\tau\pi} A_s(\tau) \frac{ch\tau(\pi - \theta)}{ch\tau\pi \, ch\tau\theta_c + k_R sh\tau\pi \, sh\tau\theta_c}$$

$\pi < \theta < 2\pi$

where T_L and T_s are temperature in liquid and substrate, respectively, c_V is a vapor concentration, $k_R = k_s / k_L$, T_∞ and c_∞ are ambient temperature and vapor concentration, respectively, θ_c is the contact angle and $P_{-\frac{1}{2}+i\tau}(ch\alpha)$ is the Legendre function of the first kind.

IOP Conf. Series: Journal of Physics: Conf. Series **788** (2017) 012048 doi:10.1088/1742-6596/788/1/012048

Expressions for kernels $A_s(\tau)$ and $B_D(\tau)$ follow from the boundary conditions and their dependence on the thermal properties of liquid and substrate (also, we neglect the heat flux from gas phase) [13] :

$$A_s(\tau)=A_s(\tau,\theta_c,k_R,\Gamma)=-LD\,B_D(\tau)\frac{tan\tau(\pi-\theta_c)}{k_L(ch\tau\pi\,sh\tau\theta_c+k_R sh\tau\pi\,ch\tau\theta_c)}$$

$$B_D(\tau)=B_D(\tau,\theta_c,k_R,\Gamma)=(c_{sat\infty}-c_\infty)\hat{B}_D(\tau) \tag{4}$$

$$\hat{B}_D(\tau)=\left\{1+\Gamma\,tan\tau(\pi-\theta_c)\frac{ch\tau\pi\,ch\tau\theta_c+k_R sh\tau\pi\,sh\tau\theta_c}{ch\tau\pi\,sh\tau\theta_c+k_R sh\tau\pi\,ch\tau\theta_c}\right\}^{-1},\ \Gamma=c'_{satT}\frac{LD}{k_L}\frac{p_{atm}}{p}.$$

where D is the diffusivity, L is the specific vaporization heat, $c_{sat\infty}$ is the saturated vapor concentration at ambient temperature and p is the pressure.

Note that for non-isothermal case $\Gamma\neq 0$ and the kernel is a function of contact angle and thermal properties of liquid, gas and substrate. The total mass flux and the contact angle can be written as:

$$-\dot{M}(t)=\frac{Q_L}{L}=J_D(\theta_c)=4\pi RD(c_{sat\infty}-c_\infty)f_D(\theta_c)$$

$$\dot{\theta}_c=-\frac{\pi}{\rho R^2}(1+cos^2\theta_c)f_D(\theta_c) \tag{5}$$

$$f_D(\theta_c)=\int_0^\infty d\tau\,\hat{B}_D(\tau)\frac{ch\tau\theta_c}{ch\tau\pi\,ch\tau(\pi-\theta_c)}.$$

For $\Gamma=0$ case the results are in a good agreement with widely used equation for total mass flux, noted by Popov [14]. Equation 5 provides mass flux dependence on properties of liquid and gas phases as well as on thermal properties of substrate.

The Marangonni force in points of the surface can be calculated by temperature gradient:

$$F_M(\alpha,\theta_c)=\nabla_\tau\sigma(\alpha,\theta_c)=\sigma'_T\nabla_\tau T_L(\alpha,\theta_c)=\frac{\sigma'_T}{R}(ch\alpha+cos\theta_c)\frac{d}{d\alpha}T_L(\alpha,\theta_c). \tag{6}$$

3. Results and Analysis

We compared our results with experimental ones, acquired in conditions close to our assumptions. Our results show not only correct trend, but also a good quantitative agreement. Our calculations indicate that for evaporating sessile drop under some conditions the temperature on the top of the drop can be higher, than near the edge of the drop [13].

Some results of our investigation of dependence of Marangonni forces direction distribution on thermal conductivity ratio parameter k_R are presented on figures 2 and 3.

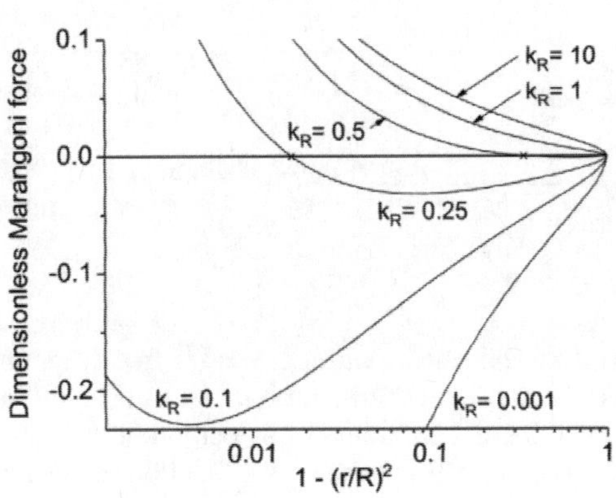

Figure 2. Dimensionless Marangonni force for different substrates

The stagnation points, where temperature reaches minimum, move towards edge of the drop with the contact angle decreasing. Near these points the direction of Marangonni force changes and one can expect existence of two opposite flows inside the drop. Direction of the forces and disposition of the stagnation points are derived as function of contact angle and thermodynamic parameters of model. Conditions for the stagnation points to appear are found out.

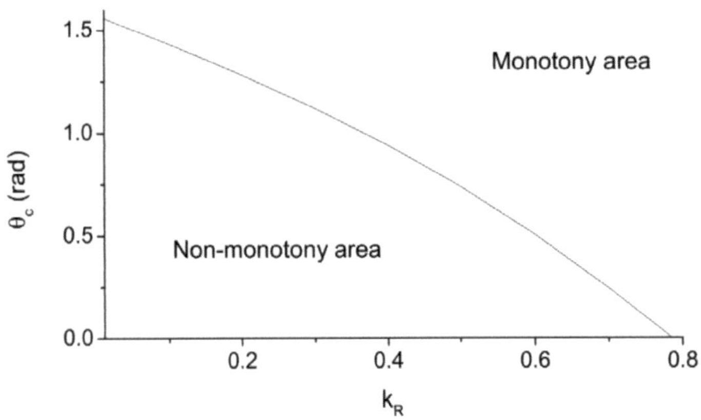

Figure 3. The line separating areas of monotony (the absence of critical points) and non-monotony of temperature distribution on the surface of droplet for water in air under normal environmental conditions

The results of our investigation of influence of substrate thermal parameter k_S on evolution of drop geometry and total mass are presented on figure 3 (dependencies calculated according to model by Popov [14] (i.e. $\Gamma = 0$ case) are denoted as "classics").

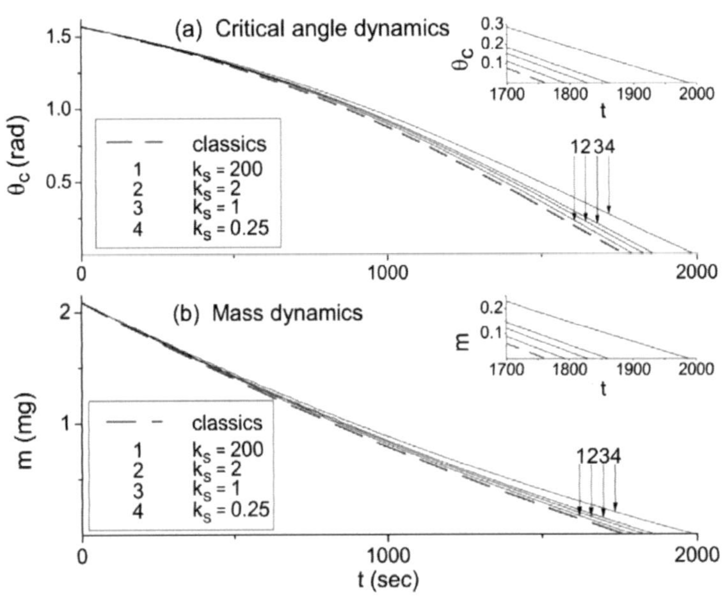

Figure 3. Evaporating sessile drop critical angle (a) and mass (b) dynamics (dependence on substrate thermal parameter Ks)

Gamma ref = 0.11 (water in air) R = 1 mm

4. Conclusion

We derived the analytical equations for temperature and concentration fields for free evaporation of sessile drop with pinned contact line which take into account thermal properties of substrate and liquid. Investigated influence of contact angle and thermal parameters on position of the stagnation points, studied dependence of the Marangonni force direction changes as a function of the edge angle and thermal conductivity ratio parameter. Investigated influence of substrate thermal parameter k_S on evolution of drop geometry and total mass of evaporating sessile drop.

References

[1] Deegan R D, Bakajin O, Dupont T F, Huber G, Nagel S R and Witten T A 1997 Capillary flow as the cause of ring stains from dried liquid drops *Nature* **389(6653)** 827–829
[2] Deegan R D, Bakajin O, Dupont T F, Huber G, Nagel S R and Witten T A 2000 Contact line deposits in an evaporating drop *Phys. Rev. E* **62** 756–765
[3] Deegan R D 1998 Pattern formation in drying drops *Phys. Rev. E* **61** 475–485
[4] Ristenpart W D, Kim P G, Domingues C, Wan J and Stone H A 2007 Influence of Substrate Conductivity on Circulation Reversal in Evaporating Drops *Phys.Rev. Lett.* **99** 234502
[5] Hu H and Larson R G 2005 Analysis of the Effects of Marangoni Stresses on the Microflow in an Evaporating Sessile Droplet *Langmuir* **21** 3972–3980
[6] Hu H and Larson R G 2002 Evaporation of a sessile droplet on a substrate *J. Phys. Chem. B* **106** 1334–1344
[7] Hu H and Larson R G 2005 Analysis of the micro fluid flow in an evaporating sessile droplet *Langmuir* **21** 3963–3971
[8] Hu H and Larson R G 2006 Marangoni effect reverses coffee-ring depositions *J. Phys. Chem. B* **110** 7090–7094
[9] Watanabe T 2011 Simulation of surface flows on a droplet in an oscillating pressure field *International Journal Of Mathematical Models And Methods In Applied Sciences* Issue 8 **5** 1398–1405
[10] Torfi S and Nejad S M H 2011 Droplet Simulation, A Case Study: Evaporation *WSEAS Transactions on Fluid Mechanics* Issue 3 **6** 192–197
[11] Nagornov O V and Starostin N V 2014 Influence of substrate properties on evaporation of the sessile drop *Advances in Engineering Mechanics and Materials* 98–100
[12] Nagornov O V and Starostin N V 2014 Influence of properties of liquid and substrate on evaporation of the sessile drop *Wseas Transactions on Heat and Mass Transfer* 221–225
[13] Dunin S Z, Nagornov O V, Starostin N V and Trifonenkov V P 2014 Analytical Solution for Evaporating Sessile Drops on Solid Substrates *Recent Advances in Applied Mathematics, Modelling and Simulation* 252–255
[14] Popov Y O 2005 Evaporative deposition patterns: spatial dimensions of the deposit *Phys. Rev. E* **71** 036313
[15] Dunn G J, Wilson S K, Duffy B R, David S and Sefiane K 2009 The strong influence of substrate conductivity on droplet evaporation *J. Fluid Mech.* **623** 329–351
[16] Sefiane K, Wilson S K, David S, Dunn G J and Duffy B R 2009 On the Effect of the Atmosphere on the Evaporation of Sessile droplet of water *Phys. Fluids* **21** 062101
[17] David S, Sefiane K and Tadrist L 2007 Experimental investigation of the effect of thermal properties of the substrate in the wetting and evaporation of sessile drops *Colloids and Surfaces A: Physicochem. Eng. Aspects* **298** 108–114
[18] Semenov S, Starov V M, Rubio R G, Agogo H and Velarde M G 2011 Evaporation of sessile water droplets: Universal behaviour in presence of contact angle hysteresis *Colloids Surf. Aspects* **391** 135–1444
[19] Saada M A, Chikh S and Tadrist L 2013 Evaporation of a sessile drop with pinned or receding contact line on a substrate with different thermophysical properties *Int. J. Heat Mass Transf.* **58** 197–208
[20] Lebedev N N 1965 *Special functions and their applications* (Prentice-Hall: Englewood Cliffs, NJ 227-230)

Comparison of two methods for the ODF modeling for magnesium alloy with low symmetry

T M Ivanova[1] **and V N Serebryany**[2]

[1]National Research Nuclear University, 31 Kashirskoe Shosse, Moscow 115409, Russia
[2]Laboratory of Crystal Structural Research, Institute of Metallurgy and Material Science, RAS, 49 Leninsky Prospect, Moscow 119991, Russia

E-mail: ivatatiana@gmail.com, vns@imet.ac.ru

Abstract. In this work we compare two methods for the reconstruction of orientation distribution function (ODF) from experimental pole figures. The magnesium alloy sample Mg-4.5%Nd is used for the investigation. The component fit method with the central normal distributions on SO(3) and the orientation grid method demonstrate similar reconstructed ODF. The component fit method is more robust to experimental errors, however it relies on non-linear optimization with a set of initial approximations provided by a complicated selection problem. The orientation grid method is more sensitive, but easier to implement.

1. Introduction

In quantitative texture analysis, the orientation distribution function provides researches with the most complete information about the texture of the polycrystalline materials.There is a lot of literature on the problem of the ODF modeling from a set of the experimental pole figures (PFs) (for example, [1]–[9]). The problem is known to be ill-posed [5],[6] due to the non-uniqueness of the solution and its instability. The purpose of the paper is to discuss the advantages and disadvantages of the two methods for the ODF modeling: the component fit method with central normal distribution on $SO(3)$ [6] and the orientation grid method [6]. With these methods, the solution is searched in a certain classes of correctness, so the uniqueness is achieved.

2. Fundamentals

Suppose that an ODF can be represented as a finite sum of some standard functions depending on unknown parameters on the rotation group $SO(3)$. This ODF produces pole figures (PFs) as combinations of corresponding functions on the unit sphere S^2 with the same parameters. Orientation distribution function $f(g)$ and pole $P_{\vec{h}}(\vec{y})$ are related by the following integral equation

$$P_{\vec{h}}(\vec{y}) = \frac{1}{2\pi}\int_0^{2\pi}\left\{f\left(\left\{\vec{h},\phi\right\}^{-1}\{\vec{y},0\}\right) + f\left(\left\{-\vec{h},\phi\right\}^{-1}\{\vec{y},0\}\right)\right\}d\phi. \qquad (1)$$

These are the PFs which are usually measured in the experiment, so we can estimate unknown parameters mostly by fitting the set of measured PFs. Let $\{\vec{h}_\lambda, \lambda = 1,...,\Lambda\}$ and $\{\vec{y}_{ij}, i = 1,...,I, j = 1,...,J\}$ be directions in the crystallite and in the sample correspondingly, λ is an index of pole figure and ij are indices of measured points, then

$$P_{\vec{h}_\lambda}(\vec{y}_{ij}) = P^{model}\left(\vec{c}, \vec{t}, \vec{h}_\lambda, \vec{y}_{ij}\right) + \text{err}\left(\lambda, i, j\right). \tag{2}$$

In equation (2) vector $\vec{c}$ is a set of linear parameters of a model, $\vec{t}$ is a vector of nonlinear parameters, and $\text{err}\left(\lambda, i, j\right)$ are the measurement errors which are influenced by many factors.

The maximum likelihood estimators for unknown parameters of the model are the values which minimize the sum of weighted squared residuals

$$r^2 = \sum_{\lambda=1}^{\Lambda}\sum_{i=1}^{I}\sum_{j=1}^{J} W\left(\lambda, i, j\right)\left\{N_\lambda P_{\vec{h}_\lambda}(\vec{y}_{ij}) - P^{model}\left(\vec{c}, \vec{t}, \vec{h}_\lambda, \vec{y}_{ij}\right)\right\}^2 \to \min. \tag{3}$$

Here the statistical weight $W\left(\lambda, i, j\right)$ reflects the degree of the significance of the measurement at point $\vec{y}_{ij}$ on the pole figure $P_{\vec{h}_\lambda}(\vec{y})$. The parameter N_λ should be chosen to normalize the model (and, of course, experimental pole figures too). Let $S = \Lambda IJ$ be the total number of the experimental data. We will refer to a particular set of $\{\lambda, i, j\}$ by single number s, where $s = 1,..., S$.

First, consider the linear problem with model function in (3) depending on $\vec{c} = \left(c_1,...,c_K\right), \vec{h}_\lambda$ and $\vec{y}_{ij}$. Then the number of parameters to be fitted is K. Let $A = \left(A_{sk}\right)$ be the design matrix [10] of this problem, vector $\vec{b}$ be the representative of experimental PF data, that is

$$A_{sk} = W\left(s\right)\frac{\partial P^{model}}{\partial c_k}, \qquad \vec{b} = W\left(s\right)\left(N_\lambda P_{\vec{h}_\lambda}(\vec{y}_{ij})\right)_s. \tag{4}$$

Then the minimization of r^2 yields normal equations of the least square problem (3):

$$\left(A^T A\right)\vec{c}^T = A^T\vec{b}^T. \tag{5}$$

In many cases, normal equations are very close to singularity, so a regularization step for solution of problem (5) is necessary. The regularization can be achieved by applying an orthogonal decomposition to the design matrix A, such as SVD [10],[11].

In the case with nonlinear parameters $\vec{t}$ in the problem (3), the minimization has to be employed iteratively. Starting from $\vec{t}_{cur} = \vec{t}_0$ in each step of the iterations, we update the values $\vec{t}_{cur} \to \vec{t}_{next}$. The initial value set $\vec{t}_0$ is usually selected by a researcher (i.e. human factor). Consider the quadratic approximation of the common residual

$$r^2\left(\vec{t}_{next}\right) \approx r^2\left(\vec{t}_{cur}\right) + \left(\vec{t}_{next} - \vec{t}_{cur}\right)\nabla r^2 + 0.5\left(\vec{t}_{next} - \vec{t}_{cur}\right)D\left(\vec{t}_{next} - \vec{t}_{cur}\right)^T, \tag{6}$$

where $D = \left(D_{pq}\right)$ is Hessian matrix $D_{pq} = \dfrac{\partial^2 r^2}{\partial t_p \partial t_q}$. In Levenberg–Marquard method [12] for solution of the problem (3) Hessian D is replaced by $\tilde{D}$: $\tilde{D}_{pq} = \dfrac{\partial r^2}{\partial t_p}\dfrac{\partial r^2}{\partial t_q}\left(1 + \mu \cdot \delta_{pq}\right)$ with some heuristic parameter μ, and updating procedure looks as follows

$$\vec{t}_{next}^T = \vec{t}_{cur}^T - \tilde{D}^{-1}\nabla r^2\left(\vec{t}_{cur}\right). \tag{7}$$

If $r^2\left(\vec{t}_{next}\right) < r^2\left(\vec{t}_{cur}\right)$, then $\vec{t}$ gets the new value $\vec{t}_{next}$ and μ decreases by some factor, if it is not the case, then $\vec{t}$ remains the old value and μ increases by some factor. Iterations stop when common residual value becomes stable. It should be noticed that matrix $\tilde{D}$ may degenerate if the dependence on some parameter t_p is rather weak, and in some other cases when μ is very small. That is why the regularization of the same type is desirable in the nonlinear case too.

3. The component fit method with central normal distributions

The component fit method is well known [2],[4],[6]. It's heavily based on the idea of ideal orientations for the texture interpretation and can be treated as smoothing and high compressing of the experimental data. The ODF model with modest number of standard functions describes the main features of the texture. Various analogs of normal distributions for mathematical modeling of ODF and PFs are used in quantitative texture analysis. The canonical normal distributions obeying the central limit theorem on the rotation group SO(3) are suggested in [2],[3]. The modified method of components with the central normal distributions is considered in [6]. Unlike linear methods, the component fit method involves a moderate number of parameters (about 50). The texture is supposed to consist of K normal components (axial with an axis $\vec{n}_{0k}$ or peak with center) with volume fractions c_k , parameters of dispersion ε_k , centers g_{0k} , and a volume fraction of the nonstructural component c_0. This model ODF generates model pole density as follows

$$P^{model}\left(\vec{c},\vec{t},\vec{h},\vec{y}\right) = c_0 + \sum_{k=1}^{K} c_k P^{cnd}\left(\vec{h},\vec{y},g_{0k},\vec{n}_k,\varepsilon_k\right), \tag{8}$$

where P^{cnd} can be either the peak (9) or the axial (10) component of reduced CND represented by the restricted series over Legendre polynomials

$$P^{p}\left(\vec{h}_\lambda,\vec{y},g_0,\varepsilon^2\right) = \sum_{l=0(2)}^{L} (2l+1)\exp\left(-l(l+1)\varepsilon^2\right) P_l\left(\left(\vec{h}_\lambda g_0 \vec{y}\right)\right), \tag{9}$$

$$P^{a}\left(\vec{h}_\lambda,\vec{y},g_0,\varepsilon^2,\vec{n}\right) = \sum_{l=0(2)}^{L} (2l+1)\exp\left(-l(l+1)\varepsilon^2\right) P_l\left(\vec{y}\vec{n}\right) P_l\left(\vec{h}_\lambda g_0 \vec{n}\right). \tag{10}$$

For unknown adjustable parameters $\left\{c=\left(c_0,\dots,c_k\right), t=\left(g_{01},\varepsilon_1,\dots,g_{0K},n_K,\varepsilon_K\right)\right\}$ and the set of measured PFs $\left\{P_{\vec{h}_\lambda}(\vec{y}_{ij}), \lambda=1,\dots,\Lambda\right\}$, we get nonlinear problem (3). The statistical weights $W\left(\lambda,i,j\right)$ are supposed to be proportional to the element of area of S^2.

The minimization of common residual r^2 in this case refers to the problems of nonlinear optimization (6)-(7) and is solved by Levenberg–Marquard method [12]. The singular value decomposition [10],[11] for ill–conditioned matrix $\tilde{D}$ in (7) is applied. Regularization boils down to zeroing of small singular numbers. This causes condition number of $\tilde{D}$ to decrease and improves statistical properties of the solution: it becomes smoother and robust to the experimental errors [10]-[12]. This method recovers the main texture features even in case when experimental errors are considerable. For estimating the adequacy of the model $RP\left(\Delta\right)-$ factor [1]-[4] is used in this work.

4. The ODF modelling by the grid of orientations

Another possible representation of $f(g)$ is a large set of standard distributions with identical small dispersion ε and centers $\left\{g_{0k}\right\}$, forming a grid of orientations on SO(3) [7]-[9]. The model ODF is represented as a superposition of these distributions. The central normal distribution [2],[3] $f^G(g, g_0, \varepsilon)$ as a standard function is offered. For narrow peaks ($\varepsilon < 0.3$) Gaussian distribution has a good approximation [2]:

$$f^s(g, g_0, \varepsilon) \equiv f(\omega, \varepsilon) = \left[\frac{\sqrt{\pi}}{\varepsilon^3}\exp(\varepsilon^2/4)\,\mathrm{erfc}(\varepsilon/2) + 1/\varepsilon^2\right] \times \frac{\omega/2}{\sin(\omega/2)}\exp\left(-\omega^2/4\varepsilon^2\right) \quad (11)$$

with $\omega_k(g_0, g) = \arccos(0.5(Tr(g_0^{-1}g) - 1))$. The distribution (11) generates the reduced standard pole density

$$P^s(\vec{h}, \vec{y}, g_0, \varepsilon) \equiv P(\theta, \varepsilon) = \frac{1}{2\varepsilon^2}\left[\exp\left(-\theta^2/4\varepsilon^2\right) + \exp\left(-(\pi - \theta)^2/4\varepsilon^2\right)\right], \quad (12)$$

where $\cos\theta$ is a scalar product of vectors $\vec{h}$ and $g_0\vec{y}$: $\cos\theta = \left(\vec{h}g_0\vec{y}\right)$.

Since the model ODF is a superposition of K standard functions (11) with volume fractions $c_k > 0$, the corresponding model PFs is a superposition of (12) with the same parameters

$$P(\vec{h}, \vec{y}) = \sum_{k=1}^{K} c_k P^s(\vec{h}, \vec{y}, g_{0k}, \varepsilon), \quad (13)$$

Here, the centers g_{0k} are not adjustable parameters. They form some regular grid in $SO(3)$. The choice of the dispersion parameter ε depends on the angular distance ψ between grid nodes. An initial value of ε can be based on the relation $\psi \approx 4\varepsilon\sqrt{\ln 2}$. The true value minimizes common residual (3).

The design matrix of the problem has elements $A_{sk} = P^G\left(\left\{\vec{h}, \vec{y}\right\}_s, g_{0k}, \varepsilon\right)$, and the solution of the problem (3) is obtained by the iterative method of projections [7]. The condition of the volume fraction positivity and normalization are taken into account during iteration. The result of this method always converges to the estimations of the least square method [10]. The solution with minimum norm ($\sum c_k^2 \to \min$) is obtained. This method produces complete nonnegative ODF.

5. Results and discussion

The methods discussed were applied for the sample of magnesium alloy Mg-4,5%Nd. Experimental pole figures $[0004], [11\bar{2}0], [10\bar{1}0], [10\bar{1}1], [10\bar{1}2], [10\bar{1}3]$ are produced by x–ray scattering method. It is obtained by the component fit method that the texture can be adequately described (the value $RP(0.5)$ for the recalculated PFs is about 14%) by seven axial components with following parameters:

$c_1 = 29.8$, $\varepsilon_1 = 0.010$, $\vec{n}_1 = \{259.2^o, 23.0^o\}$, $\rho_1 = 0.94^o$, $c_2 = 18.7$, $\varepsilon_2 = 0.012$, $\vec{n}_2 = \{295.2^o, 30.1^o\}$, $\rho_2 = 2.17^o$,

$c_3 = 13.3$, $\varepsilon_3 = 0.007$, $\vec{n}_3 = \{223.2^o, 32.6^o\}$, $\rho_3 = 6.86^o$, $c_4 = 8.7$, $\varepsilon_4 = 0.024$, $\vec{n}_4 = \{259.2^o, 43.7^o\}$, $\rho_4 = 6.62^o$,

$c_5 = 6.3$, $\varepsilon_5 = 0.014$, $\vec{n}_5 = \{74.7^o, 36.5^o\}$, $\rho_5 = 6.84^o$, $c_6 = 4.9$, $\varepsilon_6 = 0.010$, $\vec{n}_6 = \{114.3^o, 37.3^o\}$, $\rho_6 = 6.81^o$,

$c_7 = 3.4$, $\varepsilon_7 = 0.008$, $\vec{n}_7 = \{90.0^o, 32.7^o\}$, $\rho_7 = 3.20^o$, $c_0 = 14.9$.

In the orientation grid method the ODF model is represented by a superposition of 3000 standard peaks with $\psi = 8^o$. The value $RP(0.5)$ for the recalculated PFs is about 11%.

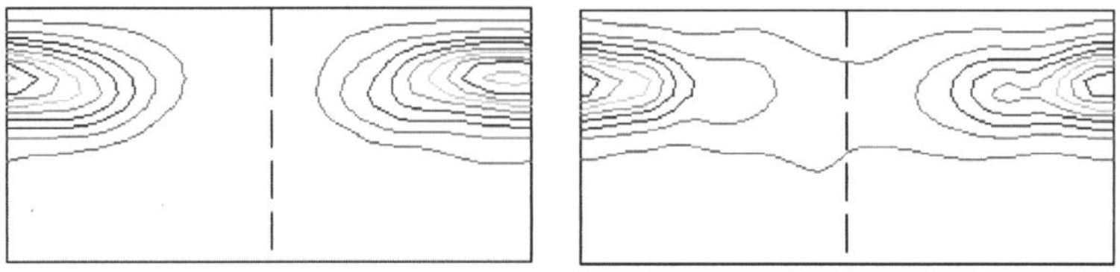

levels: 1.0 2.0 3.0 ... 9.0
a) $f_{\min} = 0.15$, $f_{\max} = 9.20$
b) $f_{\min} = 0.04$, $f_{\max} = 8.70$

Figure 1. The sections of model ODF a) in the component fit method, b) in the orientation grid method. Maximum and minimum values of the model ODF in both methods and levels are given.

The Figure **1** represents zero section of the ODF function obtained by the aforementioned methods. Both methods lead to similar results, but non-texturized component's value in the orientation grid method is much smaller than the corresponding value in the component fit method, and the max value of ODF in grid method is less than the one in component fit method. This is due to the fact that the non-texturized component is approximated by a large number of standard functions. The following method development requires taking into account the adjustable constant c_0 , which is equal to the volume fraction of the non-texturized component. The component fit method is more stable, but definitely requires human assistance in order to select initial values of parameters and also the number and type of components. The orientation grid method is fully automatic, but unfortunately less stable. Moreover, it does not provide the researcher with information needed to investigate the texture development process.

References

[1] Bunge H J 1982 *Texture Analysis in Material Science* (Butterworths:London)
[2] Bucharova T I and Savyolova T I 1993 Application of normal distribution on SO(3) and S2 for orientation distribution function approximation *Text. Microstruct.* **23** 161–176
[3] Nikolayev D I and Savyolova T I 1997 Normal distribution on the rotation group SO(3) , *Text. Microstruct.* **29** 201–233
[4] Helming K and Eschner T 1990 A new approach to texture analysis of multiphase materials using a texture component *Model Cryst. Res. Technol.* **25** K203–K208
[5] Matthies S 1979 On the reproducibility of the orientation distribution function from pole figures (ghost phenomena) *phys. stat. sol.* (b) **92** N2 K135–137
[6] Ivanova T M and Savelova T I 2006 Robust method of approximating the orientation distribution function by canonical normal distributions *Phys. Met. Metallog.* **101** N2 114–118
[7] Kourtasov S F 2007 *Indust. Lab.* **73** N7 41-44
[8] Savyolova T I and Kourtasov S F 2005 *Proc. Int. Conf. on Textures of Materials (Louven)* (Switzerland: Trans Tech Publication) pp 301–306
[9] Serebryany V N, Kourtasov S F and Savyolova T I 2005 *Proc. Int. Conf. on Textures of Materials (Louven)* (Switzerland: Trans Tech Publication) pp 1693–1697
[10] Kahaner D, Moler C, Nash S 1989 *Numerical Methods and Software* (Englewood Cliffs, NJ: Prentice Hall)
[11] Golub G and Van Loan Ch 1996 *Matrix Computations* (Johns Hopkins University Press, Baltimore)
[12] Dennis J E and Schnabel R B 1983 *Numerical Methods for Unconstrained Optimization and Nonlinear Equations* (Englewood Cliffs, NJ: Prentice Hall)

On stability of the solutions of inverse problem for determining the right-hand side of a degenerate parabolic equation with two independent variables

V L Kamynin and T I Bukharova

National Research Nuclear University MEPhI (Moscow Engineering Physics Institute), 31 Kashirskoe shosse, 115409 Moscow, Russia

E-mail: VLKamynin@mephi.ru, TIBukharova@mephi.ru

Abstract. We prove the estimates of stability with respect to perturbations of input data for the solutions of inverse problems for degenerate parabolic equations with unbounded coefficients. An important feature of these estimates is that the constants in these estimates are written out explicitly by the input data of the problem.

1. Introduction. Unique solvability of the inverse problem

In the paper we obtain the estimate of stability with respect to perturbations of input data for the solution $\{u(t,x), p(t)\}$ of the inverse problem for nonuniformly parabolic equation

$$u_t - a(t,x)u_{xx} + b(t,x)u_x + d(t,x)u = p(t)g(t,x) + r(t,x), \quad (t,x) \in Q; \tag{1}$$

with initial and boundary conditions

$$u(0,x) = u_0(x), \ x \in [0,l], \ u(t,0) = u(t,l) = 0, \ t \in [0,T]; \tag{2}$$

and the additional condition of integral observation

$$\int_0^l u(t,x)\omega(x)\,dx = \varphi(t), \ t \in [0,T]; \tag{3}$$

here $Q = [0,T] \times [0,l]$, where T and l are some numbers.

In the setting of the inverse problem (1)–(3) one allowed the unboundedness of all the coefficients of the equation (1) and the functions $g(t,x)$, $r(t,x)$ in the right hand side of this equation as well as degeneration of the leading coefficient $a(t,x)$.

The questions of unique solvability of the problem (1)–(3) were considered in [1] (see also [2]).

Note that inverse problems for degenerate parabolic equation are important in applications, in particular, they arise in the study of models of price formation for options in financial markets (see, for example, [3-5], etc.).

In this paper we use Lebesgue and Sobolev spaces with corresponding norms in usual sense, and all equalities and inequalities are satisfied almost everywhere.

We also need the well-known the Poincare'–Steklov inequality which for n=1 is of the form

$$\|z\|^2_{L_2(0,l)} \le \frac{l^2}{2}\|z_x\|^2_{L_2(0,l)}, \quad \forall z \in \overset{0}{W}{}^1_2(0,l). \tag{4}$$

We assume that the functions occuring in the input data of the problem (1)–(3) are measurable and satisfy the following conditions:

$$0 \le a(t,x),\ (t,x)\in Q;\ a(t,x), \frac{1}{a(t,x)} \in L_q(Q),\ q>1,\ \|a\|_{L_q(Q)} \le a_1,\ \left\|\frac{1}{a}\right\|_{L_q(Q)} \le a_2; \tag{A}$$

$$\begin{gathered} b(t,x),\ d(t,x)\in L_\infty\left(0,T;L_2(0,l)\right),\ \frac{b^2(t,x)}{a(t,x)},\ \frac{d^2(t,x)}{a(t,x)} \in L_\infty(Q), \\ \|b\|_{L_\infty(0,T;L_2(0,l))} \le K_b,\ \|d\|_{L_\infty(0,T;L_2(0,l))} \le K_d,\ \left\|\frac{b^2}{a}\right\|_{L_\infty(Q)} \le K_b^*,\ \left\|\frac{d^2}{a}\right\|_{L_\infty(Q)} \le K_d^*; \end{gathered} \tag{B}$$

$$\begin{gathered} g(t,x),\ r(t,x)\in L_\infty\left(0,T;L_1(0,l)\right),\ \|r\|_{L_\infty(0,T;L_1(0,l))} \le K_r,\ \frac{g^2(t,x)}{a(t,x)} \in L_\infty\left(0,T;L_1(0,l)\right), \\ \left\|\frac{g^2}{a}\right\|_{L_\infty(0,T;L_1(0,l))} \le K_g^*,\ \frac{r^2(t,x)}{a(t,x)} \in L_1(Q),\ \left\|\frac{r^2}{a}\right\|_{L_1(Q)} \le K_r^*; \end{gathered} \tag{C}$$

$$u_0(x)\in \overset{0}{W}{}_2^1(0,l); \tag{D}$$

$$\varphi(t),\ \varphi'(t)\in L_\infty(0,T),\ \|\varphi'(t)\|_{L_\infty(0,T)} \le K_\varphi^*; \tag{E}$$

$$\begin{gathered} \omega(x)\in L_\infty(0,l),\ (a\omega)_x \in L_\infty\left(0,T;L_2(0,l)\right),\ a(t,x)\omega(x)\big|_{x=0,l} = 0, \\ \|\omega\|_{L_\infty(0,l)} \le K_\omega,\ \|(a\omega)_x\|_{L_\infty(0,T;L_2(0,l))} \le K_a; \end{gathered} \tag{F}$$

$$\left|\int_0^l g(t,x)\omega(x)\,\mathrm{d}x\right| \ge g_0 > 0; \tag{G}$$

$$\varphi(0) = \int_0^l u_0(x)\omega(x)\,\mathrm{d}x. \tag{H}$$

Here $a_1,\ a_2,\ K_g^*,\ K_\omega,\ K_a,\ g_0 = \text{const}>0,\ \ K_b,\ K_d, K_b^*,\ K_d^*, K_r,\ K_\varphi^* = \text{const} \ge 0.$

Definition 1. By generalized solution of the inverse problem (1)–(3) we mean a pair of functions $\{u(t,x), p(t)\}$ such that

1) $u(t,x)\in W_s^{1,2}(Q)\cap \overset{0}{W}{}_2^1(0,l),\ s>1,\ a(t,x)u_{xx}^2,\ \dfrac{u_t^2}{a(t,x)} \in L_1(Q),\ p(t)\in L_\infty(0,T);$

2) this pair satisfies equation (1) almost everywhere in Q;

3) $\lim\limits_{t\to 0^+} \int_0^l |u(t,x) - u_0(x)|\ \mathrm{d}x = 0;$

4) equality (3) is satisfied at each point $t\in[0,T]$.

Under the assumptions (A)–(H) in the paper [1] (see also [2]) we proved the following theorem of existence and uniqueness of the generalized solution of the problem (1)–(3).

Theorem 1. Let conditions (A)–(H) are satisfied. Set

$$q^* = \frac{2q}{q+1},\ \ \lambda = \frac{3}{2}\left(K_b^* + \frac{l^2}{2}K_d^*\right),\ \ \beta = \frac{\sqrt{3K_g^*}}{g_0}\mathrm{e}^{\lambda T}\left(K_a + K_b K_\omega + K_d K_\omega \frac{l}{\sqrt{2}}\right),\ \ \gamma = 4\beta^2. \tag{5}$$

Then there exists a unique generalized solution $\{\hat{u}(t,x),\hat{p}(t)\}$ of the inverse problem (1)–(3), $\hat{u}(t,x)\in W^{1,2}_{q^*}(Q)$ and the following estimates hold:

$$\|\hat{p}\|_{L_\infty(0,T)} \le \frac{2e^{\gamma T/2}}{g_0}\left\{ e^{\lambda T}\left(K_a + K_b K_\omega + K_d K_\omega \frac{l}{\sqrt{2}}\right)\times\left(\|u_0'\|^2_{L_2(0,l)} + 3K_r^*\right)^{1/2} + K_\varphi^* + K_\omega K_r\right\}, \quad (6)$$

$$\sup_{0\le t\le T}\|\hat{u}_x(t,\cdot)\|^2_{L_2(0,l)} \le e^{2\lambda T}\left(\|u_0'\|^2_{L_2(0,l)} + 6K_g^* T\|\hat{p}\|^2_{L_\infty(0,T)} + 6K_r^*\right), \quad (7)$$

$$\|\hat{u}_{xx}\|^2_{L_{q^*}(Q)} + \|\hat{u}_t\|^2_{L_{q^*}(Q)} \le e^{2\lambda T}(a_1+a_2)\left(\|u_0'\|^2_{L_2(0,l)} + 6K_g^* T\|\hat{p}\|^2_{L_\infty(0,T)} + 6K_r^*\right). \quad (8)$$

2. Main result

Now let us obtain the estimates of stability.

Theorem 2. Consider the following two inverse problems in Q:

$$u_t - a(t,x)u_{xx} + b(t,x)u_x + d(t,x)u = p(t)g^{(j)}(t,x) + r^{(j)}(t,x), \quad (9)$$

$$u(0,x) = u_0^{(j)}(x), \quad u(t,0) = u(t,l) = 0, \quad (10)$$

$$\int_0^l u(t,x)\omega(x)\,\mathrm{d}x = \varphi^{(j)}(t), \quad (11)$$

$j=1,2.$

Suppose that for these problems conditions (A)–(H) hold. Let $\{u^{(j)}(t,x),p^{(j)}(t)\}, j=1,2$ be the corresponding solutions of these problems. (These solutions exist and are unique by theorem 1.) Then the following estimates hold:

$$\|p^{(1)} - p^{(2)}\|_{L_\infty(0,T)} \le \frac{2e^{\gamma T/2}}{g_0}\left\{ e^{\lambda T}\left(K_a + K_b K_\omega + K_d K_\omega \frac{l}{\sqrt{2}}\right)\times\right.$$

$$\times\left(\|u_{0x}^{(1)} - u_{0x}^{(2)}\|^2_{L_2(0,l)} + 6T\,\|p^{(2)}\|^2_{L_\infty(0,T)} \times \left\|\frac{\left(g^{(1)}-g^{(2)}\right)^2}{a}\right\|_{L_\infty(0,T;L_1(0,l))} + \left\|\frac{\left(h^{(1)}-h^{(2)}\right)^2}{a}\right\|_{L_1(Q)}\right)^{1/2} + \quad (12)$$

$$\left. + \|\varphi_t^{(1)} - \varphi_t^{(2)}\|_{L_\infty(0,T)} + K_\omega \|p^{(2)}\|_{L_\infty(0,T)}\|g^{(1)} - g^{(2)}\|_{L_\infty(0,T;L_1(0,l))} + K_\omega\,\|r^{(1)} - r^{(2)}\|_{L_\infty(0,T;L_1(0,l))}\right\},$$

$$\sup_{0\le t\le T}\|u_x^{(1)}(t,\cdot) - u_x^{(2)}(t,\cdot)\|^2_{L_2(0,l)} + \|u_{xx}^{(1)} - u_{xx}^{(2)}\|^2_{L_{q^*}(Q)} + \|u_t^{(1)} - u_t^{(2)}\|^2_{L_{q^*}(Q)} \le$$

$$\le e^{2\lambda T}(1+a_1+a_2)\left(\|u_{0x}^{(1)} - u_{0x}^{(2)}\|^2_{L_2(0,l)} + 6K_g^* T\|p^{(1)} - p^{(2)}\|^2_{L_\infty(0,T)} + 12T\|p^{(2)}\|^2_{L_\infty(0,T)} \times\right. \quad (13)$$

$$\left.\times\left\|\frac{\left(g^{(1)}-g^{(2)}\right)^2}{a}\right\|_{L_\infty(0,T;L_1(0,l))} + 12\left\|\frac{\left(r^{(1)}-r^{(2)}\right)^2}{a}\right\|_{L_1(Q)}\right).$$

Proof. Set

$$v^*(t,x)=u^{(1)}(t,x)-u^{(2)}(t,x),\ \delta^*(t)=p^{(1)}(t)-p^{(2)}(t),\ v_0(x)=u_0^{(1)}(x)-u_0^{(2)}(x),$$
$$\psi(t)=\varphi^{(1)}(t)-\varphi^{(2)}(t),\ \rho(t,x)=g^{(1)}(t,x)-g^{(2)}(t,x),\ \sigma(t,x)=r^{(1)}(t,x)-r^{(2)}(t,x),$$
$$G^{(j)}(t)=\int_0^l g^{(j)}(t,x)\omega(x)\,\mathrm{d}x,\qquad j=1,2.$$

Then the pair $\{v^*(t,x),\delta^*(t)\}$ is a solution in Q of the inverse problem

$$v_t-a(t,x)v_{xx}+b(t,x)v_x+d(t,x)v=\delta(t)g^{(1)}(t,x)+p^{(2)}(t)\rho(t,x)+\sigma(t,x),\tag{14}$$
$$v(0,x)=v_0(x),\quad v(t,0)=v(t,l)=0,\tag{15}$$
$$\int_0^l v(t,x)\omega(x)\,\mathrm{d}x=\psi(t).\tag{16}$$

Let us multiply equation (14) by $\omega(x)$ and integrate the resulting relation over the segment $[0,l]$. Taking into account conditions (15), (16) and integrating by parts using the assumptions (E)–(D) we obtain the relation

$$\delta^*(t)=\frac{1}{G^{(1)}(t)}\left[\int_0^l\left((a\omega)_x+b\omega\right)v_x^*\,\mathrm{d}x+\int_0^l d\omega v^*\mathrm{d}x+\psi'(t)-p^{(2)}(t)\int_0^l\rho\omega\,\mathrm{d}x-\int_0^l\sigma\omega\,\mathrm{d}x\right].\tag{17}$$

Let us introduce the operator $B:L_\infty(0,T)\to L_\infty(0,T)$ by the formula

$$B(\delta)=\frac{1}{G^{(1)}(t)}\left[\int_0^l\left((a\omega)_x+b\omega\right)v_x\,\mathrm{d}x+\int_0^l d\omega v\,\mathrm{d}x\right],\tag{18}$$

where $v(t,x)$ is a solution of the direct problem (14), (15) with a given function $\delta(t)$ in the right–hand side of equation (14). Then we derive from (18) that $\delta^*(t)$ is a solution of operator equation

$$\delta(t)=B(\delta)(t)+\frac{1}{G^{(1)}(t)}\left[\psi'(t)-\int_0^l\left(p^{(2)}(t)\rho+\sigma\right)\omega\,\mathrm{d}x\right].\tag{19}$$

Let us introduce in space $L_\infty(0,T)$ the norm

$$\|z\|_\gamma=\left\|\mathrm{e}^{-\gamma t/2}z\right\|_{L_\infty(0,T)},\quad z\in L_\infty(0,T),$$

where γ is defined in (5).

Then by analogy with the proof of theorem 3.1 from [2] we obtain that the operator B is a contraction operator in $L_\infty(0,T)$ with chosen norm $\|\cdot\|_\gamma$ and the following estimate

$$\left\|B\left(\delta^{(1)}\right)(t)-B\left(\delta^{(2)}\right)(t)\right\|_\gamma\le\frac{1}{2}\left\|\delta^{(1)}(t)-\delta^{(2)}(t)\right\|_\gamma,\tag{20}$$

$\delta^{(1)}(t)$, $\delta^{(2)}(t)\in L_\infty(0,T)$, hold.

Denote by $w_0(t,x)$ the solution of direct problem (14), (15) with $\delta(t)\equiv 0$ in the right-hand side of (14). Then for $w_0(t,x)$ we have the estimate of the form (7) which in our case looks as follows

$$\sup_{0\le t\le T}\left\|w_{0x}(t,\cdot)\right\|^2_{L_2(0,l)}\le\mathrm{e}^{2\lambda T}\left(\left\|v_0'\right\|^2_{L_2(0,l)}+6T\left\|p^{(2)}\right\|^2_{L_\infty(0,T)}\left\|\frac{\rho^2}{a}\right\|_{L_\infty(0,T;L_1(0,l))}+\left\|\frac{\sigma^2}{a}\right\|_{L_1(Q)}\right).\tag{21}$$

Let $\delta(t)\equiv 0$ be the zeroth approximation of the solution of equation (19) when applying the method of iterations. Then from formula (18) in view of conditions (B), (F), (G) and inequality (4) we obtain

$$\left|B(\delta)(t)\right|\leq\frac{1}{g_0}\left(K_a+K_bK_\omega+K_dK_\omega\frac{l}{\sqrt{2}}\right)\left\|w_{0x}(t,\cdot)\right\|_{L_2(0,l)}. \tag{22}$$

Let $\delta_1(t)$ be the first approximation of the solution of equation (19) when applying the method of iterations. Then using the well-known estimate of the n-th approximation of the solution in the iterative process (see, for example, [6], p.43) and also (20) we have

$$\left\|\delta_1(t)-\delta^*(t)\right\|_\gamma\leq\left\|B(0)(t)+\frac{\psi'}{G^{(1)}}+\frac{1}{G^{(1)}}\int_0^l\left(p^{(2)}(t)\rho+\sigma\right)\omega\,\mathrm{d}x\right\|_\gamma,$$

whence in view of (22) and (21) we obtain the inequality

$$\left\|\delta^*\right\|_\gamma\leq\frac{2}{g_0}\left\{e^{\lambda T}\left(K_a+K_bK_\omega+K_dK_\omega\frac{l}{\sqrt{2}}\right)\times\right.$$

$$\times\left(\left\|v_0'\right\|^2_{L_2(0,l)}+6T\left\|p^{(2)}\right\|^2_{L_\infty(0,T)}\times\left\|\frac{\rho^2}{a}\right\|_{L_\infty(0,T;L_1(0,l))}+6\left\|\frac{\sigma^2}{a}\right\|_{L_1(Q)}\right)^{1/2}+$$

$$\left.+\left\|\psi'\right\|_{L_\infty(0,T)}+\left\|p^{(2)}\right\|^2_{L_\infty(0,T)}K_\omega\left\|\rho\right\|_{L_\infty(0,T;L_1(0,l))}+K_\omega\left\|\sigma\right\|_{L_\infty(0,T;L_1(0,l))}\right\}.$$

This inequality obviously implies estimate (12). Note that the equation (14) is similar to the equation (1) with replacement of the function $r(t,x)$ on $p^{(2)}(t)\rho(t,x)+\sigma(t,x)$ (there are three terms in the right-hand side of (14), not two as in the equation (1)). So with this in mind, estimate (13) follows from estimates (7) and (8).

References

[1] Kamynin V L 2015 On correct solvability of the inverse problem for degenerate parabolic equation with condition of integral observation *Abstracts of the International workshop on inverse and ill-posed problems* (Moscow) (Moscow: RUDN) 90–91

[2] Kamynin V L 2015 On the solvability of the inverse problem for determining the right-hand side of a degenerate parabolic equation with integral observation *Math. Notes* **98** N5 765–777

[3] Bouchouev I and Isakov V 1999 Uniqueness, stability and numerical methods for the inverse problem that arises in financial markets *Inverse problems* **15** N3 95–116

[4] Jiang Lishang, Chen Qihong, Wang Lijun and Zhang J E 2003 A new well-posed algorithm to recover implied local volatility *Quantitative Finance* **3** N6 451–457

[5] Jiang Lishang and Tao Yourshan 2001 Identifying the volatility of underlying assets from option prices *Inverse problems* **17** N1 137–155

[6] Lyusternik L A and Sobolev V I 1982 *Kratkii kurs Functional'nogo Analiza* (Moskva: Vysh. Shkola)

Two-dimensional prognostic experiments for fast-flowing ice streams from the Academy of Sciences Ice Cap

Y V Konovalov, O V Nagornov

National Research Nuclear University MEPhI (Moscow Engineering Physics Institute), Kashirskoe shosse, 31, 115409, Moscow, Russian Federation

Abstract. The prognostic experiments for fast-flowing ice streams on the southern side of the Academy of Sciences Ice Cap in the Komsomolets Island, Severnaya Zemlya archipelago, are implemented in this study. These experiments are based on inversions of basal friction coefficients using a two-dimensional flow-line thermo-coupled model and the Tikhonov's regularization method. The modeled ice temperature distributions in the cross-sections were obtained using the ice surface temperature histories that were inverted previously from the borehole temperature profile derived at the summit of the Academy of Sciences Ice Cap and employing elevational gradient of ice surface temperature changes, which is equal to about 6.5 ^{0}C km^{-1}. Input data included InSAR ice surface velocities, ice surface elevations, and ice thicknesses obtained from airborne measurements and the surface mass balance, were adopted from previous investigations for the implementation of both the forward and inverse problems. The prognostic experiments reveal that both ice mass and ice stream extents decline for the reference time-independent surface mass balance. Specifically, the grounding line retreats (a) along the B–B′ flow line from ~40 km to ~30 km (the distance from the summit), (b) along the C–C′ flow line from ~43 km to ~37 km, and (c) along the D–D′ flow line from ~41 km to ~32 km considering a time period of 500 years and assuming time-independent surface mass balance. Ice flow velocities in the ice streams decrease with time and this trend results in the overall decline of the outgoing ice flux. Generally, the modeled evolution is in agreement with observations of deglaciation of Severnaya Zemlya archipelago.

1. Introduction

There are relevant diagnostic observations of glaciers such as digital Landsat imagery and satellite synthetic aperture radar interferometry (InSAR), airborne measurements, borehole ice temperature and ice surface mass balance measurements. These observations provide data for prognostic experiments that allow prediction of future glacier conditions for different climatic scenarios in the future. These experiments can be performed employing the mathematical modeling and in this study a two-dimensional ice flow model is applied for prediction of the future conditions of fast-flowing ice streams on the southern side of the Academy of Sciences Ice Cap in the Komsomolets Island, Severnaya Zemlya archipelago (figure 1 of [1]).

The observations were based on digital Landsat imagery and satellite synthetic aperture radar interferometry (InSAR) and revealed four drainage basins and four fast-flowing ice streams on the southern side of the Academy of Sciences Ice Cap in the Komsomolets Island, Severnaya Zemlya archipelago (figure 10 of [1]). The four ice streams are 17–37 km long and 4–8 km wide [1]. Bedrock elevations of these areas are below the sea level, and the ice flow velocities attain a value of 70–140 m/a (figure 10 of [1]). Such fast flow-line features are typical for outlet glaciers and ice streams in both the Arctic and the Antarctic. These ice streams are the major locations of iceberg calving from the Academy of Sciences Ice Cap [1].

The flow-line profiles of the three ice streams on the southern side of the Academy of Sciences Ice Cap are shown in figure 8 of [1]. Ice flow in these ice streams is simulated with a two-dimensional flow-line higher-order finite-difference model, e.g., [2]-[4]. This model describes an ice flow along a flow line [3], [4]. The results of the diagnostic experiments obtained in [5], for instance, for the C–C′ flow-line profile show that the ice surface velocity along the flow line attains a value of 100 m/a assuming that ice is sliding. However, the observed surface velocity distribution along the C–C′ flow-line profile [1] is not similar to that obtained by the model experiments for constant values of friction coefficient and for both linear and nonlinear friction laws [5]. Similarly, the diagnostic experiments carried out for the B–B′ and D–D′ profile data show the same results for the ice flow velocities. The deviation between the observed and modeled surface velocities suggests that the friction coefficient should be a spatially variable parameter. Therefore, to achieve a better agreement between the observed and simulated velocities, the spatial distribution of the friction coefficients requires to be optimized and an inverse problem needs to be solved, e.g., [6]-[14].

The inversion of friction coefficients is based on the minimization of the deviation between the observed and modeled surface velocities. A series of test experiments [5], in which modeled surface velocities are used as observations in the inverse problem, have shown that the inverse problem for the full 2D ice flow-line model is ill posed. More precisely, the surface velocity is weakly sensitive to small perturbations in friction coefficients, and as a result the perturbations appear in the inverted friction coefficients [5].

Herein, in the prognostic experiments we use the friction coefficients inversions obtained by applying the Tikhonov's regularization method, in which Tikhonov's stabilizing functional is added to the main discrepancy functional [15].

The inversions of friction coefficient are used in the prognostic experiments for the fast-flowing ice streams. The considered 2D prognostic experiments are the numerical simulations with the ice thickness distribution changes performed by the 2D flow-line thermo-coupled model, which includes diagnostic equations as the heat-transfer equation and the mass-balance equation [3], [4]. In this study, we present the results of the prognostic experiments performed for the B–B′, C–C′, and D–D′ profiles (figure 8 of [1]). Specifically, the prognostic experiments are carried out for the three ice streams (figure 10 of [1]) that are the main sources of the ice flux from the ice cap to the ocean. The results of the prognostic experiments include future modeled histories of ice thickness distributions along the flow lines of grounding line locations and outgoing ice fluxes. The surface mass balance in the performed experiments is considered as time-independent, so the prognostic experiments show the assessment of the minimal ice mass loss in the ice streams in the future, because the obtained forecasts don't include for future global warming. Nevertheless, the results of the prognostic experiments are in agreement with the observations of ice mass loss on the Severnaya Zemlya archipelago [16].

2. Field equations

2.1. Forward problem: Diagnostic equations

The 2D flow-line higher-order model includes the continuity equation for incompressible medium, the mechanical equilibrium equation in terms of stress deviator components [3], [4], and the rheological Glen law [17]:

$$\begin{cases} \int\limits_{h_b}^{z} \frac{\partial u}{\partial x} dz' + \frac{1}{b} \frac{db}{dx} \int\limits_{h_b}^{z} u\, dz' + w - w_b = 0, \\ 2\frac{\partial \sigma'_{xx}}{\partial x} + \frac{\partial \sigma'_{yy}}{\partial x} + \frac{\partial^2}{\partial x^2} \int\limits_{z}^{h_s} \sigma'_{xz}\, dz + \frac{\partial \sigma'_{xz}}{\partial z} = \rho g \frac{\partial h_s}{\partial x}, \\ \sigma'_{ik} = 2\eta \dot{\varepsilon}_{ik}; \quad \eta = \frac{1}{2}(m\, A(T))^{-\frac{1}{n}} \dot{\varepsilon}^{\frac{1-n}{n}}, \\ 0 < x < L; \; h_b(x) < z < h_s(x), \end{cases} \tag{1}$$

where (x,z) is a rectangular coordinate system with the x-axis along the flow line and the z-axis pointing vertically upward; u, w are the horizontal and vertical ice flow velocities, respectively; b is the width along the flow-line, σ'_{ik} is the stress deviator; $\dot{\varepsilon}_{ik}$ is the strain-rate tensor; $\dot{\varepsilon}$ is the second invariant of the strain-rate tensor; ρ is the ice density; g is the gravitational acceleration; η is the ice effective viscosity; $A(T)$ is the flow-law rate factor; T is the ice temperature; $h_b(x)$, $h_s(x)$ are the ice bed and ice surface elevations, respectively; and L is the glacier length.

The boundary conditions and some complementary experiments that were carried out applying this model, were considered in [5]. In particular, the technique, when the boundary conditions are included in the momentum equations [5], was applied in the considered here prognostic experiments.

2.2. *Inverse problem for the friction coefficient*

The inversion of friction coefficient has been carried out using the gradient minimization procedure for the "smoothing" functional [15].

The details of the gradient minimization procedure and the problem of the regularization parameter choice are discussed in [18], [19], [5]. In this manuscript the inversions have been obtained for the linear (viscous) friction law, implying the experiments implemented in [5] with the inversions for the C-C' profile, that have shown a good agreement between the observed (u_{obs}) and the calculated (u_{mod}) surface velocities for the linear friction law.

2.3. *Prognostic equations*

The thermo-coupled prognostic experiments imply that the 2D flow-line model includes the heat-transfer equation [3], [4]:

$$\frac{\partial T}{\partial t} = \chi \left(\frac{\partial^2 T}{\partial x^2} + \frac{1}{b} \frac{db}{dx} \frac{\partial T}{\partial x} + \frac{\partial^2 T}{\partial z^2} \right) - \left(u \frac{\partial T}{\partial x} + w \frac{\partial T}{\partial z} \right) + \frac{2 A^{-\frac{1}{n}} \dot{\varepsilon}^{\frac{1+n}{n}}}{\rho C}, \tag{2}$$

where χ and C are the thermal diffusivity and the specific heat capacity, respectively. The terms in the first and in the second brackets respectively define the heat transfer due to heat diffusion and due to ice advection. The last term is associated with strain heating.

In this model it is suggested that the ice surface temperature at the Academy of Sciences Ice Cap varies with an elevational gradient of temperature changes, which is equal to about $6.5\ ^0\mathrm{C/km}$. Hence, the ice surface temperature distribution along the flow line is defined by the temperature history at the summit $T_{s0}(t)$ and by the elevational changes, and it is expressed as

$$T_s(x,t) = T_{s0}(t) + \theta_T (h_s(0) - h_s(x)), \tag{3}$$

where θ_T is the elevational gradient. Therefore, equation (3) provides the boundary condition on the ice surface. However, it should be noted that equation (3) does not account firn warming through refreezing meltwater.

The boundary condition at the ice base is defined by the geothermal heat flux and by the heating due to the basal friction, and it is expressed as [3], [4]

$$\frac{\partial T}{\partial z}=-\frac{1}{k}\left(Q+(\sigma'_{xz})_b u_b\right), \tag{4}$$

where k is the thermal conductivity.

The boundary conditions at the ice (ice-shelf) terminus and at the ice-shelf base are defined by sea water temperature, which is considered as $-2\,°\mathrm{C}$ in this study.

The ice thickness temporal changes along the flow line are described by the mass-balance equation [3], [4]:

$$\frac{\partial H}{\partial t}=M_s-M_b-\frac{1}{b}\frac{\partial(\bar{u}\,b\,H)}{\partial x}, \tag{5}$$

where $\bar{u}$ is the depth-averaged horizontal velocity, M_s is the annual surface mass balance, and M_b is the melting rate at the ice base.

The mass-balance equation requires two boundary conditions at the summit and at the ice terminus. The first condition at the ice cap summit implies that $\partial h_s/\partial x = 0$. The second condition applied in the ice terminus originates from the fact that the ice thicknesses in the ice shelf along the flow line attain a constant value at the terminus.

2.4. Grounding line evolution

In the model the grounding line position is defined from the hydrostatic equilibrium [20]-[22]. That is, since sea water flow under ice shelf is not considered in the model and, hence, the pressure in equations (10)-(11) from [21] is equal to hydrostatic pressure, and the grounding line position is at the location where

$$-\rho_w\, h_r = H \tag{6}$$

and h_r is the bedrock elevation, ρ_w is the water density.

3. Results of the prognostic experiments

The main input data along with flow-line profiles for the prognostic experiments, namely, the surface mass balance, are adopted from [23]. For instance, figure 7 of [24] shows the elevational mass-balance distribution along the C–C′ flow line, i.e., it shows how the surface mass balance changes with elevation in the C–C′ direction [23]. For the B–B′ and D–D′ flow lines, the elevational mass-balance distributions are qualitatively the same [23]. In the prognostic experiments that have been carried out, the mass balance is considered as time-independent. That is, the elevational mass-balance distributions are kept unchanged for the considered time period in the future. Thus, we intend to assess the maximum ice thickness in the ice streams in the future, because the forecasts implemented with the time-independent surface mass balance, don't imply a future global warming and, so, they don't suggest a future decreasing of the surface mass balance M_s in equation (5). Similarly, the ice surface temperature is suggested to be time-independent but dependent on elevation, i.e., according to equation (3), it is changed with elevation with a constant value of $T_{s0}(t)$. From the borehole temperature measurements, the present ice surface temperature at the summit is about $-7.2\,°\mathrm{C}$. The initial ice temperatures applied in the prognostic experiments are shown in figure 5 of [24].

Despite that future warming scenarios are not included into the prognostic experiments, the modeled ice cap response to the present environmental impact, which is reflected in the elevational mass-balance distribution [23], reveals that the ice thicknesses gradually diminish along all the three flow lines. Figure 1 shows the modeled successive ice surfaces divided into 50-year time intervals for the B–B′, C–C′, and D–D′ profiles, respectively. The prognostic experiments are performed by applying a rectangular ice-shelf geometry (figure 1). The cumulative impact of sea water, surface mass balance, and ice flow changes in the glacier has produced the future modeled ice shelve geometries (figure 1; figures 8-10 of [24]).

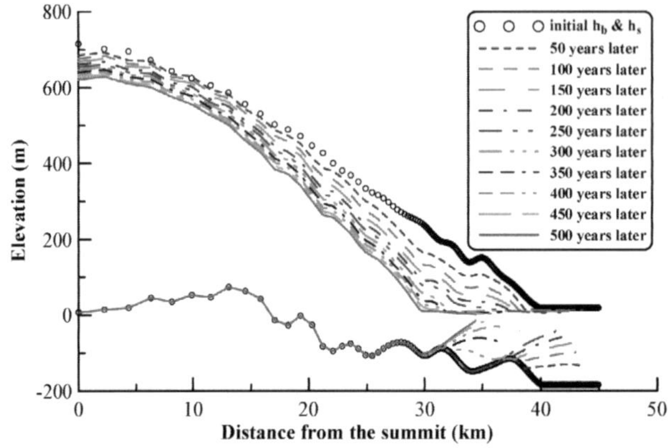

Figure 1 (a)

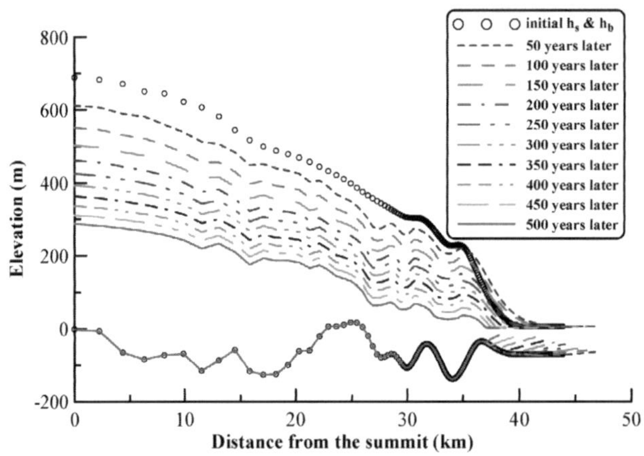

Figure 1 (b)

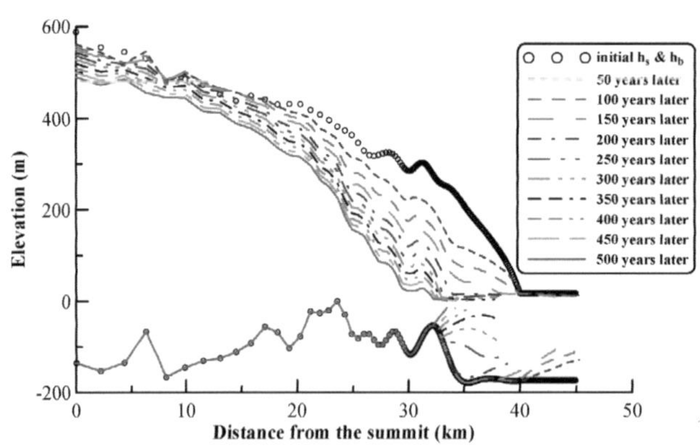

Figure 1 (c)

Figure 1. (a) The modeled successive B-B' cross-section geometries separated by 50-year intervals from the present to the future 500 years later. **(b)** The modeled successive C-C' cross-section geometries separated by 50-year intervals from the present to the future 500 years later. **(c)** The modeled successive D-D' cross-section geometries separated by 50-year intervals from the present to the future 500 years later.

The ancillary black circles in figure 1 are aligned with the grid nodes and, thus, they show the spatial resolution, at which the prognostic experiments have been implemented. The spatial resolution is irregular and it decreases from about $2 \cdot 10^3$ m at the summit to about 10^2 m in the grounding line vicinity and in the ice shelf. The spatial grid is considered unchangeable throughout the period of the modeling.

The grounding line history, i.e., grounding line retreat or advance, specifically reflects the growing or diminishing ice mass, i.e., the history is an indicator of the glacier evolution. The grounding line retreats (a) along the B–B′ flow line from ~40 km to ~30 km (figure 11 (a) of [24]), (b) along the C–C′ flow line from ~43 km to ~37 km (figure 11 (b) of [24]), and (c) along the D–D′ flow line from ~41 km to ~32 km (figure 11 (c) of [24]) considering a time period of 500 years.

Furthermore, the results of the prognostic experiments can be likewise treated suggesting a changes in the friction coefficient. The glacier terminus, currently fast flowing and therefore at pressure melting, becomes eventually frozen to the ground – ice thickness insufficient to insulate from cold atmosphere and reduced driving stress and strain heating. So basal friction coefficients could change drastically, given the simulated changes in glacier geometry.

The ice flow velocities in the ice streams decrease with time and this trend diminishes the outgoing ice fluxes in the future (figure 12 of [24]). Figure 13 of [24] shows the future history of the overall outgoing ice flux, i.e., it is the sum of the three future modeled outgoing ice fluxes obtained for B–B′, C–C′ and D–D′ ice streams.

There are small peaks that periodically disturb main historical trends of the three outgoing ice fluxes (figure 12 of [24]). Every peak reflects ice calving at the ice-shelf terminus. Similarly, the ice calving provides a sudden change in the value of the outgoing ice flux ($\bar{u}\,H\,b$) due to a sudden change in the ice thickness (H) at the terminus. Considering a complex environmental impact on ice shelves [25], from the mathematical point of view it can be suggested that the calving processes are described by a stochastic model. Hence, the overall annual (or decadal) sizes of the anticipated ice debris can be described by a frequency distribution function. In the model In the model the periodic calving of equal-size debris is considered, i.e., δ-function is considered as the frequency distribution function.

In this model the both ice-shelf length and ice-shelf thickness at the terminus are considered as the variables that should satisfy a certain conditions. If the ice-shelf length exceeds a value l_{cr} (the parameter of the model) or the ice-shelf thickness beside the terminus becomes smaller than a value H_{cr}, then the calving of the appropriate part of ice occurs in the model. To investigate the impact of the parameters on the results of the modeling, the parameters were varied in a series of the experiments. However, the simulation reveals that the mass balance, friction coefficient, ice temperature have the main impact to the assessment of the grounding line retreat derived by the modeling.

4. Conclusions

The modeled present ice temperatures (figure 5 of [24]) are qualitatively the same in the three cross-sections. There are resembling zones of relatively cold ice that can be distinguished in the modeled temperatures in the middle of the cross-sections. These cold ice zones reflected the surface temperature minimum about 150–200 years ago in the inverted past temperature history [18], [19]. This surface temperature minimum corresponds to an event that is known as Little Ice Age.

The inversions of the friction coefficient performed for the three cross-sections can be interpreted as follows. The two levels that are evidently distinguished in the inverted friction coefficient distributions (figure 6 of [24]) can be explained by changing the physical properties of the bedrock along the flow lines. Similarly, the large values of the friction coefficient at 0 km $< x <$ 20 km justify the rock-type bottom where ice is frozen to the bed (the ice temperature at 0 km $< x <$20 km is lower than the melting point). The lower values of the friction coefficient at 25 km $< x <$40 km presumably indicate the existence of the till layer (or the sandy layer) at the bottom. Specifically, the till layer provides the basal ice sliding [26]-[33].

The prognostic experiments carried out with the reference mass balance [23] show that the grounding line has been retreated at about 10 km in the three ice streams considering a time period of 500 years. Similarly, the grounding line retreats (a) along the C–C′ flow line from ~43 km to ~37 km (the distance from the summit), (b) along the B–B′ flow line from ~40 km to ~30 km, and (c) along the D–D′ flow line from ~41 km to ~32 km considering a time period of 500 years and assuming time-independent mass balance. In the experiments, the ice flow velocities in the ice streams decrease with time due to (a) diminishing of the ice thicknesses and (b) retreating of the grounding lines from the sliding zones toward the zones where ice is frozen to the bed. Thus, the maxima of the ice flow velocities in the ice streams decrease from ~80–120 m/a to ~20–30 m/a. These trends in the ice flow velocities diminish the outgoing ice fluxes and as a result diminish the overall ice flux (figure 13 of [24]). The modeled evolution of the ice streams is in agreement with observations of ice mass loss on Severnaya Zemlya archipelago [16].

Acknowledgements. Authors are grateful to Prof. J.A. Dowdeswell et al. for the data that have been used in the manuscript. Authors are grateful to Prof. F. Pattyn for the useful comments to the manuscript. Authors are grateful to Dr. T. Dunse for the reviewing the manuscript.

References

[1] Dowdeswell J A, Bassford R P, Gorman M R, Williams M, Glazovsky A F, Macheret Y Y, Shepherd A P, Vasilenko Y V, Savatyugin L M, Hubberten H W and Miller H 2002 Form and flow of the Academy of Sciences Ice Cap, Severnaya Zemlya, Russian High Arctic *J. Geophys. Res.* **107** 1-15

[2] Colinge J, Blatter H 1998 Stress and velocity fields in glaciers: Part I. Finite difference schemes for higher-order glacier models *J. Glaciol.* **44** 448–456

[3] Pattyn F 2000 Ice-sheet modeling at different spatial resolutions: focus on the grounding zone *Ann. Glaciol.* **31** 211-216

[4] Pattyn F 2002 Transient glacier response with a higher-order numerical ice-flow model *J. Glaciol.* **48** 467-477

[5] Konovalov Y V 2012 Inversion for basal friction coefficients with a two-dimensional flow line model using Tikhonov regularization *Research in Geophysics* **2:e11** 82-89

[6] MacAyeal D R 1992 The basal stress-distribution of Ice Stream-E, Antarctica, inferred by control methods *J. Geophys. Res. Solid Earth* **97** 595-603

[7] Sergienko O V, Bindschadler R A, Vornberger P L, MacAyeal D R 2008 Ice stream basal conditions from block-wise surface data inversion and simple regression models of ice stream flow: Application to Bindschadler Ice Stream *J Geophys. Res.* **113** 1-11

[8] Arthern R and Gudmundsson H 2010 Initialization of ice-sheet forecasts viewed as an inverse Robin problem *J. Glaciol.* **56** 527-533

[9] Gagliardini O, Jay-Allemand M, Gillet-Chaulet F 2010 Friction distribution at the base of a surging glacier inferred from an inverse method *AGU Fall Meeting* C13A-0540 (San Francisco CA USA Dec. 13-17).

[10] Habermann M, Maxwell D A, Truffer M 2010 A principled stopping criterion for the reconstruction of basal properties in ice sheets *AGU Fall Meeting* C21C-0556 (San Francisco CA USA Dec. 13-17)

[11] Morlighem M, Rignot E, Seroussi H, Larour E, Ben Dhia H and Aubry D 2010 Spatial patterns of basal drag inferred using control methods from a full-Stokes and simpler models for Pine Island Glacier, West Antarctica *Geophys. Res. Lett.* **37** L14502

[12] Jay-Allemand M, Gillet-Chaulet F, Gagliardini O, Nodet M 2011 Investigating changes in basal conditions of Variegated Glacier prior to and during its 1982-1983 surge *The Cryosphere* **5** 659-672

[13] Larour E, Seroussi H, Morlighem M, Rignot E 2012 Continental scale, high order, high spatial resolution, ice sheet modeling using the Ice Sheet System Model (ISSM) *J. Geophys. Res.*

117 1-20
[14] Sergienko O V and Hindmarsh R C A 2013 Regular patterns in frictional resistance of ice-stream beds seen by surface data inversion *Science* **342(6162)** 1086–1089
[15] Tikhonov A N, Arsenin V Ia 1977 *Solutions of ill posed problems* (Washington: Winston & Sons)
[16] Moholdt G, Wouters B, Gardner A S 2012 Recent mass changes of glaciers in the Russian High Arctic *Geophys. Res. Lett.* **39** L10502
[17] Cuffey K and Paterson W S B 2010 *The physics of glaciers* 4th ed (Butterworth-Heineman, Elsevier).
[18] Nagornov O V, Konovalov Y V, Tchijov V 2005 Reconstruction of past temperatures for Arctic glaciers subjected to intense subsurface melting *Ann. Glaciol.* **40** 61-66
[19] Nagornov O V, Konovalov Y V, Tchijov V 2006 Temperature reconstruction for Arctic glaciers *Palaeogeogr. Palaeoclimatol. Palaeoecol.* **236** 125-134
[20] Schoof C 2007 Ice sheet grounding line dynamics: Steady states, stability, and hysteresis *J. Geophys. Res.* **112** 1–19
[21] Pattyn F, Schoof C, Perichon L, Hindmarsh R C A, Bueler E, de Fleurian B, Durand G, Gagliardini O, Gladstone R, Goldberg D, Gudmundsson G H, Huybrechts P, Lee V, Nick F M, Payne A J, Pollard D, Rybak O, Saito F and Vieli A 2012 Results of theMarine Ice Sheet Model Intercomparison Project, MISMIP *The Cryosphere* **6(3)** 573–588
[22] Seroussi H, Morlighem M, Larour E, Rignot E, Khazendar A 2014 Hydrostatic grounding line parameterization in ice sheet models *The Cryosphere* **8** 2075–2087
[23] Bassford R P, Siegert M J, Dowdeswell J A, Oerlemans J, Glazovsky A F and Macheret Y Y 2006 Quantifying the Mass Balance of Ice Caps on Severnaya Zemlya, Russian High Arctic I: Climate and Mass Balance of the Vavilov Ice Cap *Arctic, Antarctic, and Alpine Research* **38 (1)** 1-12
[24] Konovalov Y V, Nagornov O V 2015 Two-dimensional prognostic experiments for fast-flowing ice streams from the Academy of Sciences Ice Cap: future modeled histories obtained for the reference surface mass balance *Earth Syst. Dynam. Discuss.* **6** 2211–2242
[25] Bassis J N, Fricker H A, Coleman R, Minster J-B 2008 An investigation into the forces that drive ice-shelf rift propagation on the Amery Ice Shelf, East Antarctica *J. Glaciol.* **54** 17-27
[26] Engelhardt H F, Harrison W D, and Kamb B 1978 Basal sliding and conditions at the glacier bed as revealed by bore-hole photography *J. Glaciol.* **20(84)** 469-508
[27] Engelhardt H F, Kamb B, Raymond C F, and Harrison W D 1979 Observation of basal sliding of Variegated Glacier, Alaska *J. Glaciol.* **23(89)** 406-407
[28] Boulton G S 1979 Processes of glacier erosion on different substrata *J. Glaciol.* **23(89)** 15-38
[29] Boulton G S and Jones A S 1979 Stability of temperate ice caps and ice sheets resting on beds of deformable sediment *J. Glaciol.* **24(90)** 29-43
[30] MacAyeal D R 1989 Large-scale ice flow over a viscous basal sediment: theory and application to ice stream B, Antarctica *J. Geophys. Res.* **94** 4071–4088
[31] Engelhardt H and Kamb B 1998 Basal sliding of Ice Stream B, West Antarctica *J. Glaciol.* **44(147)** 223-230
[32] Iverson N R, Hooyer T S, Baker R W 1998 Ring-shear studies of till deformation: Coulomb-plastic behavior and distributed strain in glacier beds *J. Glaciol.* **44** 634–642
[33] Tulaczyk S, Kamb W B, Engelhardt H F 2000 Basal mechanics of Ice Stream B, West Antarctica I. Till mechanics *J. Geophys. Res.* **105** 463–482

On complex roots of an equation arising in the oblique derivative problem

A B Kostin and V B Sherstyukov

National Research Nuclear University MEPhI (Moscow Engineering Physics Institute), 31 Kashirskoe shosse, 115409 Moscow, Russia

E-mail: abkostin@yandex.ru, shervb73@gmail.com

Abstract. The paper is concerned with the eigenvalue problem for the Laplace operator in a disc under the condition that the oblique derivative vanishes on the disc boundary. In a famous article by V. A. Il'in and E. I. Moiseev (Differential equations, 1994) it was found, in particular, that the root of any equation of the form

$$\mu\, J_n'(\mu) + i\, n \tan\alpha\, J_n(\mu) \;=\; 0, \qquad n \in \mathbb{Z},$$

with the Bessel function $J_n(\mu)$ determines the eigenvalue $\lambda = \mu^2$ of the problem. In our work we correct the information about the location of eigenvalues. It is specified explicit view of the corner, containing all the eigenvalues. It is shown that all the nonzero roots of the equation are simple and given a refined description of the set of their localization on the complex plane. To prove these facts we use the partial differential equations methods and also methods of entire functions theory.

1. Statement of the problem

In the unit disc $D = \big\{(x,\, y) \mid x^2 + y^2 < 1\big\}$ we consider the problem

$$\Delta\, w \,+\, \lambda\, w \,=\, 0 \quad \text{in} \quad D, \qquad \frac{\partial w}{\partial l} \,=\, 0 \quad \text{on} \quad \partial D. \tag{1}$$

Here, Δ is the Laplace operator, $\lambda \in \mathbb{C}$ is the spectral parameter, $\boldsymbol{l}$ is a direction at a fixed angle $\alpha \in (-\pi/2, \pi/2)$ with the outer normal vector $\boldsymbol{n}$ to the boundary $\partial D = \big\{(x, y) \mid x^2 + y^2 = 1\big\}$.

The problem with oblique derivative was extensively studied [1]–[5]. Our paper is closest to the papers [1], [5] and has been strongly influenced by these papers. Of great interest in problem (1) is to find nontrivial solutions $w = u + iv$ and the corresponding eigenvalues $\lambda = \mu^2 = \xi + i\eta$. It is known (see, for example, [1], [6, Ch. 8]), that any weak solution to problem (1) from the Sobolev class $W_2^2(D)$ is a classical solution. It is also known [7, Ch. 4, § 1, Theorem 3], [8, Ch. 5, § 19], that in the case $\alpha = 0$ (that is, when $\partial w/\partial l \equiv \partial w/\partial n$) all the eigenvalues of problem (1) are real and nonnegative, the eigenfunctions forming an orthogonal basis for $L_2(D)$. In [1] it was proved that for any $\varepsilon > 0$ outside the set $\{\lambda \in \mathbb{C} \mid |\arg\lambda| < \varepsilon\}$ there is at most finite number of eigenvalues of problem (1), and, hence, they all located in some angle on the complex plane. Besides, the system of eigenfunctions and associated functions was shown to be complete in L_2 in [1] for general elliptic problems (in particular, including problem (1)). The main result of [5] is that the system of root functions of problem (1) fails to have the basis property both in the space $L_2(D)$ and in the space $L_p(D)$ for any $p > 1$. In [5] it was also shown that any root of an equation of the form

$$\mu J_n'(\mu) + i\, n \tan\alpha\, J_n(\mu) = 0, \qquad n \in \mathbb{Z}, \tag{2}$$

with the Bessel function $J_n(\mu)$ defines an eigenvalue $\lambda = \mu^2$ of problem (1) with the corresponding eigenfunction $J_n(\mu r)\exp(in\varphi)$. Besides, there exists a subsequence of eigenvalues $\lambda = \mu^2$ for which the set $|\mathrm{Im}\,\mu|$ is unbounded. In other words, this subsequence does not lie "inside" the so-called Carleman parabola [9], [10]. In [11] an asymptotically exact set on the complex plane was found to contain all the eigenvalues of an elliptic operator with classical conditions on the boundary. Even though we failed to directly apply these result to problem (1), nevertheless the approaches of [11] were found to be instrumental in treating problem (1). The results of our paper extend the results of [1], [5] on the location of eigenvalues of problem (1). Our arguments depend on the methods of partial differential equations and methods of special considerations from the theory of entire functions.

2. Statement of the main results

In what follows, the phrase "μ is a root of equation of the form (2)" will mean that μ satisfies at least one equation from the family of equations (2).

We have the following results.

Theorem 1. *Let $\lambda = \xi + i\eta \in \mathbb{C}$ be an arbitrary eigenvalue of problem* (1)*. Then*

$$|\eta| \leqslant \xi\,|\tan\alpha|\,, \tag{3}$$

which shows that all the eigenvalues lie in the right half-plane.

Thus, for the angle appearing in [1] we give an explicit description by (3).

Theorem 2. *Each root $\mu \neq 0$ of an arbitrarily chosen equation from* (2) *is simple,* $\mathrm{Im}\,\mu \neq 0$ *with $n \neq 0$.*

Theorem 3. *For a fixed $n \in \mathbb{Z}$, the corresponding equation from* (2) *has no roots $\mu \neq 0$ on the set $\left\{\mu \in \mathbb{C} \mid \mathrm{Re}\,\mu^2 \leqslant n^2\right\}$.*

Theorem 4. *$\lambda = \mu^2$ is an eigenvalue of problem* (1) *if and only if when μ is a root of an equation of the form* (2)*.*

Theorem 5. *All the eigenfunctions of problem* (1) *are given in polar coordinates by formulas of the form*

$$w(r,\varphi) = J_n(\mu r)\exp(i\,n\varphi)\,, \qquad n \in \mathbb{Z}, \tag{4}$$

where μ assumes all such values from the set of roots of equation of the form (2) *with given n for which $w(r,\varphi) \not\equiv 0$. There are no associated functions in problem* (1)*.*

Remark 2.1. The assertion $\mathrm{Im}\,\mu \neq 0$ with $n \neq 0$ in Theorem 2, the sufficiency in Theorem 4, as well as the fact from Theorem 5 that there are no associated functions in problem (1) were proved in [5]; these results are shown for the sake of completeness of exposition. We also note that Theorem 5 is an immediate consequence of Theorem 4 and the results of [5].

Remark 2.2. For each $n \in \mathbb{Z}$ a fixed equation from (2) has an infinite countable set of roots.

For later purposes, it will be convenient to split the family of equations (2) into two subfamilies:

$$\mu J_n'(\mu) + i\, n \tan\alpha\, J_n(\mu) = 0, \qquad n \in \mathbb{N}, \tag{5}$$

$$\mu J_n'(\mu) - i\, n \tan\alpha\, J_n(\mu) = 0, \qquad n \in \mathbb{N} \cup \{0\}. \tag{6}$$

The following properties of subfamiles (5), (6) are immediate.

(i) If some $\mu = \mu(n)$ is a root of an equation from (5), then its complex conjugate $\overline{\mu}$ is a root of an equation in (6) with the same n, and vice versa.

(ii) For each $n \neq 0$ the change α by $-\alpha$ carries over an equation from (5) into an equation from (6), and vice versa.

(iii) If μ is a root of equation of the form (2), then $-\mu$ is a root of the same equation.

Thus, in the study of complex roots of the family of equations (2) it suffices to consider only equations (5) and only in the right half-plane, assuming that $\alpha \in (0, \pi/2)$, because the case $\alpha = 0$ has received a good deal of attention (see, for example, [8, Ch. 5, § 19]).

3. Proofs of the main results

We give only sketches of the proofs; detailed arguments will be given in a forthcoming paper.

3.1. Proof of Theorem 1.

Let $\boldsymbol{n}$, $\boldsymbol{\tau}$ be, respectively, the unit normal vector and the unit tangent vector at a point $(x, y) \in \partial D$. We assume that the shortest rotation from $\boldsymbol{\tau}$ to $\boldsymbol{n}$ forms a counterclockwise rotation. We fix an eigenvalue $\lambda = \xi + i\eta$ and the corresponding eigenfunction $w = u + iv$, $w \not\equiv 0$. On ∂D we have

$$\frac{\partial w}{\partial l} = (\boldsymbol{n}\cos\alpha + \boldsymbol{\tau}\sin\alpha, \nabla w) \equiv \cos\alpha\frac{\partial w}{\partial n} + \sin\alpha\frac{\partial w}{\partial \tau} \equiv (\boldsymbol{n}\cos\alpha + \boldsymbol{\tau}\sin\alpha)\,\nabla w^T\,,$$

where $\nabla w = \{w_x; w_y\}$ is the vector of the gradient, ∇w^T is the corresponding column. The vectors $\boldsymbol{n}$ and $\boldsymbol{\tau}$ depend on a point (x, y), while the angle $\alpha \in (-\pi/2, \pi/2)$ between $\boldsymbol{l}$ and the outer normal vector $\boldsymbol{n}$ is fixed. A rotation for α in a clockwise direction is assumed to be positive. In polar coordinates, for $(x, y) \in \partial D$ we write $\boldsymbol{n} = \{\cos\varphi; \sin\varphi\}$, $\boldsymbol{\tau} = \{\sin\varphi; -\cos\varphi\}$. Consider the matrix

$$A = \begin{pmatrix} \cos\alpha & -\sin\alpha \\ \sin\alpha & \cos\alpha \end{pmatrix},$$

and calculate

$$\operatorname{div}\left(\nabla w\, A^T\right) = \operatorname{div}\left\{w_x\cos\alpha - w_y\sin\alpha; w_x\sin\alpha + w_y\cos\alpha\right\} \equiv \cos\alpha\,\Delta w.$$

This enables us to rewrite equation from the statement of problem (1) in the equivalent form

$$\operatorname{div}\left(\nabla w\, A^T\right) + w\,\lambda\cos\alpha = 0.$$

Multiplying the last equality in $L_2(D)$ by w, we find that

$$\int_D \operatorname{div}\left(\nabla w\, A^T\right)\overline{w}\, dxdy + \lambda\cos\alpha\int_D |w|^2\, dxdy = 0. \tag{7}$$

Transforming the first integral in (7) we find that

$$\int_D \operatorname{div}\left(\nabla w\, A^T\right)\overline{w}\, dx\, dy = -\int_D \left[\cos\alpha|\nabla w|^2 + \sin\alpha\left(w_x\overline{w}_y - w_y\overline{w}_x\right)\right]\, dx\, dy\,.$$

We shall assume that the eigenfunction is normalized; that is, $\|w\|^2 \equiv \int_D |w|^2\, dx\, dy = 1$. Equating to zero in succession the real and imaginary parts of (7), this gives

$$\xi = \int_D |\nabla w|^2\, dx\, dy\,, \qquad \eta\cos\alpha = \sin\alpha\,\mathrm{Im}\int_D \left(w_x\overline{w}_y - w_y\overline{w}_x\right)\, dx\, dy\,.$$

Hence,

$$|\eta| \cos\alpha = |\sin\alpha|\,|\operatorname{Im} \int_D (w_x\,\overline{w}_y - w_y\,\overline{w}_x)\;dx\,dy| \leqslant |\sin\alpha| \int_D |w_x\,\overline{w}_y - w_y\,\overline{w}_x|\;dx\,dy \leqslant$$
$$\leqslant |\sin\alpha| \int_D |\nabla w|\;|\nabla \overline{w}|\;dx\,dy = |\sin\alpha| \int_D |\nabla w|^2\;dx\,dy = |\sin\alpha|\,\xi\,.$$

This proves inequality (3). The proof of Theorem 1 is complete. □

3.2. Proof of Theorem 2.
Let us prove that each root $\mu \neq 0$ of equation (2) with fixed $n \in \mathbb{Z}$ is simple. That for $n = 0$ the corresponding equation has simple roots is well-known (see, for example, [12, Ch. 15]). For definiteness, let us consider equation (5) with parameter $\alpha \in (0, \pi/2)$. Assume that $\mu \neq 0$ is a multiple root of (5) with fixed $n \in \mathbb{N}$. Then the following equalities hold simultaneously:

$$\begin{cases} \mu\,J_n'(\mu) + i\,n\,\tan\alpha\,J_n(\mu) & = 0, \\ i\,n\,\mu\,\tan\alpha\,J_n'(\mu) - (\mu^2 - n^2)J_n(\mu) & = 0. \end{cases} \tag{8}$$

System (8) is linear and homogeneous in $J_n'(\mu)$, $J_n(\mu)$. Hence, it has either only the zero solution or its determinant is zero. We claim that none of such cases is realized. If $J_n'(\mu) = J_n(\mu) = 0$, then the classical Bessel's equation with $\mu \neq 0$ shows that the derivatives $J_n^{(m)}(\mu)$ are zero for all $m \in \mathbb{N}$, which is impossible. If the determinant of system (8) is zero, then $\mu^2 = n^2(1 + \tan^2\alpha) > 0$, and therefore, $\mu \in \mathbb{R} \setminus \{0\}$, which contradicts the first equation in (8). The property $\operatorname{Im}\mu \neq 0$ for each $\mu \neq 0$ was established in [5]. This proves Theorem 2. □

3.3. Proof of Theorem 3.
Let us consider equation (5) with fixed $n \in \mathbb{N}$ and $\alpha \in (0, \pi/2)$. Let $\mu \in \mathbb{C} \setminus \{0\}$ be a root of this equation. By Theorem 2 we have $\operatorname{Im}\mu \neq 0$. The complex conjugate $\overline{\mu}$ satisfies the equality

$$\overline{\mu}\,J_n'(\overline{\mu}) - i\,n\tan\alpha\,J_n(\overline{\mu}) = 0. \tag{9}$$

Consider the real function

$$\Phi(x) = |J_n(\mu x)|^2 \equiv J_n(\mu x) J_n(\overline{\mu} x), \qquad x \geqslant 0. \tag{10}$$

The first and second derivatives of Φ are as follows:

$$\Phi'(x) = (J_n(\mu x)J_n(\overline{\mu}x))' = \mu J_n'(\mu x)J_n(\overline{\mu}x) + \overline{\mu}J_n'(\overline{\mu}x)J_n(\mu x),$$

$$\Phi''(x) = -\frac{1}{x}\,\Phi'(x) - 2\left(\operatorname{Re}\mu^2 - \frac{n^2}{x^2}\right)\Phi(x) + 2|\mu|^2\left|J_n'(\mu x)\right|^2, \qquad x > 0.$$

We have $\Phi(0) = \Phi'(0) = 0$. Besides, using (5), (9) this gives $\Phi'(1) = 0$. So, the function $\Phi(x)$, as defined by (10), is a solution to the problem

$$\Phi''(x) + \frac{1}{x}\,\Phi'(x) + 2\left(\operatorname{Re}\mu^2 - \frac{n^2}{x^2}\right)\Phi(x) = 2|\mu|^2\left|J_n'(\mu x)\right|^2, \quad x > 0, \tag{11}$$

$$\Phi(0) = \Phi'(0) = 0, \qquad \Phi'(1) = 0. \tag{12}$$

Since $\Phi(x)$ is continuous and nonnegative on $[0,1]$, there exists a point $x_0 \in (0,1]$ such that

$$\max_{x\in[0,1]} \Phi(x) = \Phi(x_0) \geqslant 0.$$

Assume that the so-chosen root $\mu \neq 0$ satisfies the condition $\operatorname{Re}\mu^2 \leqslant n^2$. Then in equation (11) the coefficient multiplying $\Phi(x)$ is nonpositive:

$$c(x) \equiv 2\left(\operatorname{Re}\mu^2 - \frac{n^2}{x^2}\right) \leqslant 0, \quad x \in (0,1].$$

We first consider the case $x_0 < 1$. Taking into account equation (11), we see that subelliptic function $\Phi(x)$ attains its nonnegative maximum inside the interval $[0,1]$. Hence, this function is constant. But in this case $\Phi(x) \equiv 0$, which leads to a contradiction with (10). Assume now that $x_0 = 1$. By the normal derivative lemma (see, for example, [13, Ch. 1, § 1, Theorem 4]) we have $\Phi'(1) > 0$, contradicting (12). So, equation (5) cannot have roots $\mu \neq 0$ that satisfy the condition $\operatorname{Re}\mu^2 \leqslant n^2$. This proves Theorem 3. □

3.4. Proof of Theorem 4.
If μ is a root of an equation of the form (2), then a direct calculation shows that the function $J_n(\mu r)\exp(in\varphi) \not\equiv 0$ is a solution to problem (1); that is, $J_n(\mu r)\exp(in\varphi)$ is an eigenfunction corresponding to the eigenvalue $\lambda = \mu^2$. Why problem (1) has no other eigenvalues? Such results are usually verified using the completeness and orthogonality (the basis property) of eigenfunctions (see [8, Ch. 5, § 19], [14, Lecture 27]). But problem (1) has no such properties!

We claim that if $\lambda = \mu^2$ is an eigenvalue of problem (1), then μ is a root of at least one of the equations from family (2). Let $w = w(r,\varphi)$ be the eigenfunction corresponding to this eigenvalue. We write the equation for $w(r,\varphi)$:

$$\frac{1}{r}\frac{\partial}{\partial r}\left(r\frac{\partial w}{\partial r}\right) + \frac{1}{r^2}\frac{\partial^2 w}{\partial \varphi^2} + \mu^2 w = 0. \tag{13}$$

For each fixed r the function $w(r,\varphi)$ is φ-smooth and 2π–periodic, and hence, it can be expanded into a Fourier series.

$$w(r,\varphi) = \sum_{m\in\mathbb{Z}} R_m(r)\, e^{im\varphi}, \qquad R_m(r) = \frac{1}{2\pi}\int_0^{2\pi} w(r,\varphi)\, e^{-im\varphi}\, d\varphi,\ m \in \mathbb{Z}. \tag{14}$$

Here, at least one of the Fourier coefficients does not vanish identically; that is, there exists an $n \in \mathbb{Z}$ such that $R_n(r) \not\equiv 0$.

Multiplying equation (13) by $\exp(-im\varphi)$ and integrating in $\varphi \in [0,2\pi]$, we see that for each $m \in \mathbb{Z}$

$$\frac{1}{r}\frac{d}{dr}\left(r\frac{dR_m(r)}{dr}\right) + \left(\mu^2 - \frac{n^2}{r^2}\right) R_m(r) = 0, \quad r \in (0,1).$$

Since the functions $R_m(r)$ are bounded, it follows that $R_m(r) = C_m\, J_m(\mu r)$ for each Fourier coefficients with some constant C_m. Substituting the expression for $R_m(r)$ into the Fourier series in (14), we have

$$w(r,\varphi) = \sum_{m\in\mathbb{Z}} C_m\, J_m(\mu r)\, e^{im\varphi}.$$

The boundary condition of problem (1) implies that

$$\sum_{m\in\mathbb{Z}} C_m\, e^{im\varphi}\left[\mu\, J_m'(\mu) + i\, m \tan\alpha\, J_m(\mu)\right] = 0\,.$$

But in this case

$$C_m\left[\mu J'_m(\mu) + i\, m \tan\alpha\, J_m(\mu)\right] = 0, \qquad m \in \mathbb{Z}.$$

We have $R_n(r) \not\equiv 0$, and hence $C_n \neq 0$. Therefore, for such $n \in \mathbb{Z}$ the number μ is a root of an equation of the form (2). This proves Theorem 4. □

4. Concluding remarks

Consider an equation of the form (5) with fixed $\alpha \in (0, \pi/2)$, $n \in \mathbb{N}$. We expand the entire function $J_n(\mu)$ as an infinite product

$$J_n(\mu) = \frac{1}{n!}\left(\frac{\mu}{2}\right)^n \prod_{k=1}^{\infty}\left(1 - \frac{\mu^2}{j_{n,k}^2}\right), \quad \mu \in \mathbb{C},$$

where $j_{n,k}$ are its positive roots (see [12, Ch. 15, p. 498]). Taking the logarithmic derivative we conclude that, for the above α, n with $\mu \neq 0$ equation (5) is equivalent to the equation

$$\sum_{k=1}^{\infty} \frac{1}{\mu^2 - j_{n,k}^2} + \frac{n}{2\mu^2}(1 + i\tan\alpha) = 0.$$

Setting $\mu^2 = \xi + i\eta$, we have the following system for eigenvalues:

$$\begin{cases} \displaystyle\sum_{k=1}^{\infty} \frac{\xi - j_{n,k}^2}{(\xi - j_{n,k}^2)^2 + \eta^2} + \frac{n}{2}\,\frac{\xi + \eta\tan\alpha}{\xi^2 + \eta^2} = 0, \\ \displaystyle\sum_{k=1}^{\infty} \frac{\eta}{(\xi - j_{n,k}^2)^2 + \eta^2} - \frac{n}{2}\,\frac{\xi\tan\alpha - \eta}{\xi^2 + \eta^2} = 0. \end{cases} \tag{15}$$

System (15) provides additional tools for the study of the eigenvalues of problem (1). In particular, in this way we get a different proof of Theorem 1. Indeed, (15) may only have solutions that simultaneously satisfy the inequalities $\eta > 0$, $\xi\tan\alpha - \eta > 0$. Thus, for the roots of equation (5) with $\alpha \in (0, \pi/2)$ the eigenvalues $\lambda = \mu^2$ lie in the angle $0 < \eta < \xi\tan\alpha$. Note that this fact cannot be 'felt' by the methods of partial differential equations.

References

[1] Agmon S 1962 On the eigenfunctions and on the eigenvalues of general elliptic boundary value problems *Comm. on pure and appl. math.* **15** 119–147
[2] Egorov Ju V and Kondrat'ev V A 1969 The oblique derivative problem *Mathematics of the USSR-Sbornik* **7** 139–169
[3] Bitsadze A V 1988 *Some classes of partial differential equations* (CRC Press)
[4] Moiseev E I 1976 Completeness of eigenfunctions of some boundary-value problems *Doklady Akademii Nauk SSSR* **226** 1012–14
[5] Il'in V A and Moiseev E I 1994 On the Absence of the Basis Property of a System of Root Functions of a Problem with an Oblique Derivative *Differ. Uravn.* **30** 128–143
[6] Gilbarg D and Trudinger N S 1984 *Elliptic Partial Differential Equations of Second Order* (Berlin: Springer)
[7] Mikhailov V P 1978 *Partial differential equations* translated from the Russian by P C Sinha (Moscow: Mir)
[8] Vladimirov V S 1971 *Equations of mathematical physics* (New York: Marcel Dekker)
[9] Geppert H 1927 Über Randwertprobleme bei linearen eliptischen Differentialgleichungen *Mathematische Annalen* **98** 264–272
[10] Carleman T 1936 Über die asymptotische Verteilung der Eigenwerte partieller Differentialgleichungen *Berichte der Sächsischen Akademie der Wissenschaften zu Leipzig. Math.-Phys. Klasse* **88** 119–132
[11] Kostin A B 2013 Complex eigenvalues of elliptic operators and an example of inverse problem with nonunique solution *Doklady Mathematics* **88** 651–654
[12] Watson G N 1944 *A Treatise on the Theory of Bessel Functions* (Cambridge: Cambridge University Press)
[13] Protter M H and Weinberger H F 1984 *Maximum Principles in Differential Equations* (New York: Springer-Verlag)
[14] Sobolev S L 1964 *Partial Differential Equations of Mathematical Physics* (Oxford: Pergamon Press)

Closed solutions for model problems in generalized thermoelasticity

S. A. Lychev[1,2] **and V.V. Klindukhov**[2]
[1]Ishlinsky Institute for Problems in Mechanics of the Russian Academy of Sciences, prosp. Vernadskogo 101-1, Moscow, 119526 Russian Federation
[2]National Research Nuclear University MEPhI (Moscow Engineering Physics Institute), 31 Kashirskoe shosse, 115409 Moscow, Russia
E-mail: `lychev@ipmnet.ru`

Abstract. Closed solutions for model problems in non-dissipative thermoelasticity are obtained. The solutions are in the form of spectral expansion over biorthogonal system of eigenfunctions corresponded to mutual conjugate pair of operator pencils.

1. Introduction

Thermal disturbances in some materials (in particular, liquid and crystalline helium) are known to propagate at low temperatures as progressive waves (Peshkov [1]). This phenomenon is usually referred to as "second sound." The wave nature of heat propagation substantially changes the medium heat conductivity, and this change should be taken into account when designing cryogenic devices. Various theories have been devised to explain this phenomenon (Tisza–Landau [2], Atkin et al. [3], etc.). Among them are Green–Naghdi theories of thermoelasticity [4], in which the time primitive of the temperature is chosen as an independent constitutive variable. Such theories results in very special nonself-adjoint boundary value problems, for which classical approaches apparently fail. Closed solution for the initial boundary value problems in the framework of Green-Naghdi type II thermoelasticity were obtained in [9]. The aim of this paper is to obtain closed solutions for more general problems arising in Green-Naghdi type III thermoelasticity. The solutions are obtained in the form of spectral expansions in complete biorthogonal systems of root functions (eigenfunctions and associated functions) of the corresponding nonself–adjoint operator pencils.

2. Differential operators

Consider an isotropic homogenous body $\mathscr{B}$. The equations of motion and generalized heat conduction (Green-Naghdi type III) have the form:

$$\mu\nabla^2\boldsymbol{u} + (\lambda+\mu)\nabla\nabla\cdot\boldsymbol{u} - \alpha\nabla\theta - \rho\ddot{\boldsymbol{u}} + \rho\boldsymbol{b} = \boldsymbol{0},$$
$$\Lambda\nabla^2\theta + \Lambda_*\nabla^2\dot{\theta} - \kappa\ddot{\theta} - \varpi\nabla\cdot\dot{\boldsymbol{u}} + \rho s = 0,$$

where $\boldsymbol{u}$ is the displacement field, θ is the temperature, $\boldsymbol{b}$ is the density of mass force, s is the power of heat supply, λ and μ are the Lamé elastic moduli and Λ, Λ_* κ, and ϖ are the thermomechanical moduli. Parameters Λ, Λ_* are specific for Green-Naghdi type III

IOP Conf. Series: Journal of Physics: Conf. Series **788** (2017) 012053 doi:10.1088/1742-6596/788/1/012053

thermoelasticity. The limit case $\Lambda \to 0$ corresponds to conventional (Fourier) thermoelasticity, whereas the limit case $\Lambda_* \to 0$ corresponds to Green-Naghdi type II thermoelasticity. These equations, together with the prescribed boundary and initial conditions

$$\boldsymbol{u}\Big|_{t=0} = \boldsymbol{u}_0, \quad \theta\Big|_{t=0} = \Theta_0, \quad \dot{\boldsymbol{u}}\Big|_{t=0} = \dot{\boldsymbol{u}}_0, \quad \dot{\theta}\Big|_{t=0} = \dot{\theta}_0,$$

form a well-posed statement of the initial–boundary value problem.

Consider the space of square integrable four-component complex-valued vector functions $(\boldsymbol{u}, \theta)$ with the inner product

$$\langle(\boldsymbol{u}_1, \theta_1), (\boldsymbol{u}_2, \theta_2)\rangle = \int\limits_{\mathscr{B}} \left(\boldsymbol{u}_1 \cdot \overline{\boldsymbol{u}}_2 + \theta_1 \overline{\theta}_2\right) dV.$$

The initial–boundary value problem can be stated as the following Cauchy problem with operator coefficients:

$$\mathcal{L}(\boldsymbol{u}, \theta) = \boldsymbol{f}, \quad \mathcal{L}(\boldsymbol{u}, \theta) = \mathcal{A}_0 (\boldsymbol{u}, \theta) + \mathcal{A}_1 \frac{\partial}{\partial t} (\boldsymbol{u}, \theta) + \mathcal{A}_2 \frac{\partial^2}{\partial t^2} (\boldsymbol{u}, \theta), \quad \boldsymbol{f} = -(\rho \boldsymbol{b}, \rho s),$$

$$\mathcal{A}_0 = \begin{pmatrix} \mu\nabla^2 + (\lambda+\mu)\nabla\nabla\cdot & -\alpha\nabla \\ 0 & \Lambda\nabla^2 \end{pmatrix}, \quad \mathcal{A}_1 = \begin{pmatrix} 0 & 0 \\ 0 & \Lambda_*\nabla^2 \end{pmatrix}, \quad \mathcal{A}_2 = -\begin{pmatrix} \rho & 0 \\ \varpi\nabla & \kappa \end{pmatrix}.$$

The problem in question gives rise to the polynomial operator pencil

$$\mathcal{L}_\nu = \mathcal{A}_0 + \nu \mathcal{A}_1 + \nu^2 \mathcal{A}_2,$$

The domain $\mathcal{D}$ of the pencil $\mathcal{L}_\nu$ is determined by the boundary operator $\mathcal{B}$ as follows:

$$\mathcal{D} = \left\{ (\boldsymbol{u}, \theta) \Big| \mathcal{B}(\boldsymbol{u}, \theta)\big|_{\partial\mathscr{P}} = \boldsymbol{0} \right\},$$

The adjoint pencil $\mathcal{L}_\nu^*$ can be obtained directly from the definition:

$$\forall (\boldsymbol{u}_1, \theta_1) \in \mathcal{D},\ (\boldsymbol{u}_2, \theta_2) \in \mathcal{D}^* \quad \langle \mathcal{L}_\nu (\boldsymbol{u}_1, \theta_1), (\boldsymbol{u}_2, \theta_2)\rangle = \langle (\boldsymbol{u}_1, \theta_1), \mathcal{L}_\nu^* (\boldsymbol{u}_2, \theta_2)\rangle.$$

provided that

$$\mathcal{L}_\nu^* = \mathcal{A}_0^* + \overline{\nu} \mathcal{A}_1 + \overline{\nu}^2 \mathcal{A}_2^*, \quad \mathcal{A}_0^* = \begin{pmatrix} \mu\nabla^2 + (\lambda+\mu)\nabla\nabla\cdot & 0 \\ \alpha\nabla\cdot & \Lambda\nabla^2 \end{pmatrix}, \quad \mathcal{A}_2^* = \begin{pmatrix} -\rho & \varpi\nabla \\ 0 & -\beta \end{pmatrix}.$$

We define the domain $\mathcal{D}^*$ of the adjoint pencil $\mathcal{L}_\nu^*$ as the set of vector functions $\boldsymbol{k}^* = (\boldsymbol{u}^*, \theta^*)$ such that the bilinear form

$$\langle \mathcal{L}_\nu (\boldsymbol{u}, \theta), (\boldsymbol{u}^*, \theta^*)\rangle - \langle (\boldsymbol{u}, \theta), \mathcal{L}_\nu^* (\boldsymbol{u}^*, \theta^*)\rangle = [(\boldsymbol{u}, \theta), (\boldsymbol{u}^*, \theta^*)]_{\partial\mathscr{B}}$$

is zero for arbitrary $(\boldsymbol{u}, \theta) \in \mathcal{D}$. Hence we obtain the adjoint boundary operator $\mathcal{B}^*$ in closed form. It is well known that the system of root functions of the pencil $\mathcal{L}_\nu$ is biorthogonal to that of the adjoint pencil. Note that if $\mathcal{D} = \mathcal{D}^*$, then this system is a basis. This provides a theoretical background for the representation of the solution in the form of a spectral expansion.

3. Solution in the general case

The obtained solutions have the form of expansions based on the systems of eigenfunctions $\boldsymbol{k}_i$ and $\boldsymbol{k}_j^*$ that can be found as the solutions of the mutually adjoint boundary value problems

$$\mathcal{L}_\nu \boldsymbol{k}_i = \boldsymbol{0}, \quad \mathcal{L}_\nu^* \boldsymbol{k}_j = \boldsymbol{0}, \quad \boldsymbol{k}_i = (\boldsymbol{u}_i, \theta_i), \quad \boldsymbol{k}_j^* = (\boldsymbol{u}_j^*, \theta_j^*); \quad i = 1, \dots \infty; \quad j = 1, \dots \infty. \tag{1}$$

We assume that the body $\mathscr{B}$ is compact. Hence the set of eigenfunctions and the corresponding set of eigenvalues are countable. According to [8], the following biorthogonal relations hold for $\boldsymbol{k}_i$ and $\boldsymbol{k}_j^*$:

$$\langle \mathcal{A}_1 \boldsymbol{k}_i, \boldsymbol{k}_j^* \rangle + (\nu_i + \nu_j) \langle \mathcal{A}_2 \boldsymbol{k}_i, \boldsymbol{k}_j^* \rangle = 0, \quad \langle \mathcal{A}_0 \boldsymbol{k}_i, \boldsymbol{k}_j^* \rangle - \nu_i \nu_j \langle \mathcal{A}_2 \boldsymbol{k}_i, \boldsymbol{k}_j^* \rangle = 0.$$

These properties permit one to represent the solution in the form [8]

$$\boldsymbol{y} = \sum_{i=1}^{\infty} \left[\left(\langle \mathcal{A}_1^* \boldsymbol{k}_i^* + \nu_i \mathcal{A}_2^* \boldsymbol{k}_i^*, \boldsymbol{y}_0 \rangle + \langle \mathcal{A}_2^* \boldsymbol{k}_i^*, \dot{\boldsymbol{y}}_0 \rangle \right) \exp(\nu_i t) - \int_0^t \langle \boldsymbol{f}(\tau), \boldsymbol{k}_i^* \rangle \exp\big(\nu_i (t-\tau)\big)\, d\tau \right] \boldsymbol{k}_i \mathcal{Q}_i^{-1},$$

where

$$\boldsymbol{y} = (\boldsymbol{u}, \theta), \quad \boldsymbol{y}_0 = (\boldsymbol{u}_0, \theta_0), \quad \dot{\boldsymbol{y}}_0 = \left(\dot{\boldsymbol{u}}_0, \dot{\theta}_0\right), \quad \boldsymbol{f} = (\rho \boldsymbol{b}, \rho s), \quad Q_i = \langle \mathcal{A}_1 \boldsymbol{k}_i, \boldsymbol{k}_i^* \rangle + 2\nu_i \langle \mathcal{A}_2 \boldsymbol{k}_i, \boldsymbol{k}_i^* \rangle.$$

The spectral expansions in complete biorthogonal systems of root functions (eigenfunctions and associated functions) are of the same form and discussed in details in [8].

In the special case $\mathcal{A}_1 = 0$ and $\operatorname{Re} \nu_i = 0$, this representation can be written as follows:

$$\boldsymbol{y} = \sum_{i=1}^{\infty} \left[\left(\langle \mathcal{A}_2^* \boldsymbol{k}_i^*, \boldsymbol{y}_0 \rangle \cos(\lambda_i t) + \frac{1}{\lambda_i} \langle \mathcal{A}_2^* \boldsymbol{k}_i^*, \dot{\boldsymbol{y}}_0 \rangle \right) \sin(\lambda_i t) - \frac{1}{\lambda_i} \int_0^t \langle \boldsymbol{f}(\tau), \boldsymbol{k}_i^* \rangle \sin\big(\lambda_i (t-\tau)\big)\, d\tau \right] \boldsymbol{k}_i \mathrm{N}_i^{-1}.$$

Here $N_i = 2 \langle \mathcal{A}_2 \boldsymbol{k}_i, \boldsymbol{k}_i^* \rangle$ and $\lambda_i = \operatorname{Im} \nu_i$.

4. Solution in a special case

The mutually adjoint systems (1) are reduced to

$$\begin{aligned} &\mu \nabla^2 \boldsymbol{u}_i + (\lambda + \mu) \nabla \nabla \cdot \boldsymbol{u}_i - \alpha \nabla \theta_i - \nu_i^2 \rho \boldsymbol{u}_i = \boldsymbol{0}, \quad (1 + \nu_i \epsilon) \nabla^2 \theta_i - \nu_i^2 \beta \theta_i - \nu_i^2 \gamma \nabla \cdot \boldsymbol{u}_i = 0, \\ &\mu \nabla^2 \boldsymbol{u}_i^* + (\lambda + \mu) \nabla \nabla \cdot \boldsymbol{u}_i^* + \overline{\nu}_i^2 \gamma \nabla \theta_i^* - \overline{\nu}_i^2 \rho \boldsymbol{u}_i^* = \boldsymbol{0}, \quad (1 + \overline{\nu}_i \epsilon) \nabla^2 \theta_i^* - \overline{\nu}_i^2 \beta \theta_i^* + \alpha \nabla \cdot \boldsymbol{u}_i^* = 0 \end{aligned} \tag{2}$$

Here $\beta = \varkappa / \Lambda$, $\gamma = \alpha / \Lambda$ and $\epsilon = \Lambda_* / \Lambda$. Now let us apply the Stokes–Helmholtz resolution

$$\boldsymbol{u}_i = \nabla \phi_i + \nabla \times \boldsymbol{\psi}_i, \quad \nabla \cdot \boldsymbol{\psi}_i = 0, \quad \boldsymbol{u}_i^* = \nabla \phi_i^* + \nabla \times \boldsymbol{\psi}_i^*, \quad \nabla \cdot \boldsymbol{\psi}_i^* = 0. \tag{3}$$

where ϕ_i, ϕ_i^* and $\boldsymbol{\psi}_i$, $\boldsymbol{\psi}_i^*$ are scalar and vector functions, respectively. By substituting (3) into (2), we obtain

$$\begin{aligned} &(\lambda + 2\mu) \nabla^2 \phi_i - \alpha \theta_i - \nu_i^2 \rho \phi_i = 0, \quad (1 + \nu_i \epsilon) \nabla^2 \theta_i - \nu_i^2 \beta \theta_i - \nu_i^2 \gamma \nabla^2 \phi_i = 0, \quad \mu \nabla \boldsymbol{\psi}_i - \nu_i^2 \rho \boldsymbol{\psi}_i = \boldsymbol{0}, \\ &(\lambda + 2\mu) \nabla^2 \phi_i^* + \overline{\nu}_i^2 \gamma \theta_i^* - \overline{\nu}_i^2 \rho \phi_i^* = 0, \quad (1 + \overline{\nu}_i \epsilon) \nabla^2 \theta_i^* - \overline{\nu^2} \beta \theta_i^* + \alpha \nabla^2 \phi_i^* = 0, \quad \mu \nabla \boldsymbol{\psi}_i^* - \overline{\nu}_i^2 \rho \boldsymbol{\psi}_i^* = \boldsymbol{0}. \end{aligned}$$

In this section, we assume that the desired solutions are generated by scalar potentials alone[1] ($\boldsymbol{\psi}_i = \boldsymbol{\psi}_i^* = 0$). It follows in a standard way that

$$\begin{pmatrix} \phi_i \\ \theta_i \end{pmatrix} = \begin{pmatrix} \varrho \\ \varsigma \end{pmatrix} \Xi_i.$$

[1] This assumption, of course, restricts the problem under consideration to potential displacement fields, body forces, and initial data. However, the rotational components of the above-mentioned fields do not interact with the temperature field. Thus, the rotational components can be omitted in the case of purely thermal excitation.

Here ϱ and ς are constants, Ξ_i is a solution of the auxiliary equation

$$\nabla^2\Xi_i = \zeta_i\Xi_i$$

of the Helmholtz type, and ζ_i satisfies the algebraic equation ($\sigma = \lambda + 2\mu$):

$$\begin{vmatrix} \sigma\zeta_i - \rho\nu_i^2 & -\alpha \\ -\gamma\zeta_i\nu_i^2 & (1+\nu_i\epsilon)\zeta_i - \beta\nu_i^2 \end{vmatrix} = 0.$$

By solving the latter for ν_i, we obtain

$$\nu_{i1} = \frac{\zeta\epsilon}{4\beta} - \frac{a_{i1}}{2} - \frac{\sqrt{a_{i2}-a_{i3}}}{2}, \quad \nu_{i2} = \frac{\zeta\epsilon}{4\beta} - \frac{a_{i1}}{2} + \frac{\sqrt{a_{i2}-a_{i3}}}{2},$$

$$\nu_{i3} = \frac{\zeta\epsilon}{4\beta} + \frac{a_{i1}}{2} - \frac{\sqrt{a_{i2}-a_{i3}}}{2}, \quad \nu_{i4} = \frac{\zeta\epsilon}{4\beta} + \frac{a_{i1}}{2} + \frac{\sqrt{a_{i2}-a_{i3}}}{2},$$

where

$$a_{i1} = \sqrt{\frac{p_{i1}}{3\sqrt[3]{2}\beta\rho} + p_{i2} + p_{i3}}, \quad a_{i2} = -\frac{p_{i1}}{3\sqrt[3]{2}\beta\rho} - p_{i2} + 2p_{i3},$$

$$a_{i3} = \left(\frac{\zeta^3\epsilon^3}{\beta^3} - \frac{8\zeta^2\sigma\epsilon}{\beta\rho} - \frac{4\zeta p_{i0}\epsilon}{\beta^2\rho}\right) / (4a_{i1});$$

$$p_{i3} = \frac{\zeta^2\epsilon^2}{4\beta^2} - \frac{2p_{i0}}{3\beta\rho}, \quad p_{i2} = \frac{\sqrt[3]{2}q_{i1}}{3\beta p_{i1}\rho}, \quad p_{i1} = \sqrt[3]{\sqrt{q_{i0}^2 - 4q_{i1}^3} + q_{i0}}, \quad p_{i0} = -\zeta(\alpha\gamma + \beta\sigma + \rho);$$

$$q_{i1} = 3\zeta^2\rho\sigma\left(4\beta + \zeta\epsilon^2\right) + p_{i0}^2, \quad q_{i0} = 27\zeta^4\rho\sigma\epsilon^2(\beta\sigma + \rho) + 2p_{i0}^3 + 9\zeta^2 p_{i0}\rho\sigma\left(\zeta\epsilon^2 - 8\beta\right).$$

In the case of simple spectrum the eigenfunctions can be obtained as

$$\boldsymbol{k}_{i,s} = (\boldsymbol{u}_{i,s}, \theta_{i,s}) = \left(\left[1 - \beta\nu_{i,s}^2/\zeta + \nu_{i,s}\epsilon\right]\nabla\Xi_i, \gamma\nu_{i,s}^2\Xi_i\right),$$
$$\boldsymbol{k}_{i,s}^* = (\boldsymbol{u}_{i,s}^*, \theta_{i,s}^*) = \left(\left[1 - \beta\overline{\nu}_{i,s}^2/\zeta + \overline{\nu}_{i,s}\epsilon\right]\nabla\Xi_i, -\alpha\Xi_i\right).$$

Here $s = 1, \ldots, 4$.

It is clear that the solution thus obtained depends on the boundary conditions for the auxiliary scalar function Ξ. Consider the Neumann type boundary conditions $\boldsymbol{n}\cdot\nabla\Xi\Big|_{\partial\mathscr{B}} = 0$. This corresponds to the following boundary conditions in terms of the variables $\boldsymbol{u}$ and θ (where $\boldsymbol{I}$ is the unit tensor)

$$\boldsymbol{n}\cdot\boldsymbol{u} = 0, \quad \boldsymbol{n}\cdot\boldsymbol{T}\cdot(\boldsymbol{I} - \boldsymbol{n}\otimes\boldsymbol{n}) = \boldsymbol{0}, \quad \boldsymbol{n}\cdot\nabla\theta = 0.$$

Now let $\mathscr{B}$ be a cube with edge length π, and let x, y, z be Cartesian coordinates. Assume that the body force and the entropy production are zero, and so are the initial displacements, velocities, and temperature rate; i.e.,

$$\boldsymbol{b} = \boldsymbol{0}, \quad s = 0, \quad \boldsymbol{u}\Big|_{t=0} = \boldsymbol{0}, \quad \dot{\boldsymbol{u}}\Big|_{t=0} = \boldsymbol{0}, \quad \dot{\theta}\Big|_{t=0} = 0.$$

The solution of the auxiliary problem is

$$\Xi_k = \cos nx \cos my \cos lz, \quad \zeta_k = -n^2 - m^2 - l^2.$$

Let the initial temperature distribution be given by $\theta_0 = H(z) - H(z-h)$, $0 < h < \pi$, where $H(z)$ is the Heaviside step function. In this case only $\zeta_k = -l^2$ may be taken into account. The normalising factors $\mathfrak{Q}_{i,s}$ are of the form

$$\mathfrak{Q}_{i,s} = \pi^3\nu_{i,s}\left(-\alpha\gamma n^4(\nu_{i,s}\epsilon + 2) - 2\rho\left(\beta\nu_{i,s}^2 + n^2(\nu_{i,s}\epsilon + 1)\right)^2\right)/(2n^2),$$

and non-zero expansion coefficients $\langle \mathcal{A}_1^* \boldsymbol{k}_i^* + \nu_i \mathcal{A}_2^* \boldsymbol{k}_i^*, \boldsymbol{y}_0 \rangle$ may be calculate as follows

$$P_{i,s} = \left(\pi^2 \alpha\right) \sin(hn) \left(\beta \nu_{i,s} + n^2 \epsilon\right) / n$$

Thus the solution of the corresponding initial–boundary value problem can be represented in the form of a spectral expansion:

$$\theta = \frac{h}{\pi} + \sum_{i=1}^{\infty} \left(S_{i,1} + S_{i,2} + S_{i,3} + S_{i,4}\right), \quad \boldsymbol{u} = \boldsymbol{k} w, \quad w = \sum_{i=1}^{\infty} \left(W_{i,1} + W_{i,2} + W_{i,3} + W_{i,4}\right),$$

where

$$S_{i,s} = \frac{\gamma \nu_{i,s}^2 P_{i,s} \exp(\nu_{i,s} t) \cos(nz)}{\mathfrak{Q}_{i,s}}, \quad W_{i,s} = \left(\beta \nu_{i,s}^2 / \zeta - \nu_{i,s} \epsilon - 1\right) \frac{n P_{i,s} \exp(\nu_{i,s} t) \sin(nz)}{\mathfrak{Q}_{i,s}}.$$

Obtained solution generalise the solution given in [9].

Acknowledgments

The research was financially supported by the Russian Foundation for Basic Research (under grants Nos. 15-08-06330).

References

[1] Pitaevskii L. P. 1968 *Phys. Uspekhi* **95** 139–44.
[2] Landau L. D. 1967 *Phys. Uspekhi* **11** 495–520.
[3] Atkin R. J., Fox N. and Vasey M. W. 1975 *J. Elasticity* **5** 237–48.
[4] Green A. E. and Naghdi P. M. 1993 *J. Elasticity* **61** 189–208.
[5] Truesdell C. and Toupin R. A. 1960 *Handbuch der Physik. The Classical Field Theories.* Band III/1 pp 226–858.
[6] Kalpakides V. K. and Maugin G. A. 2004 *Rep. Math. Phys.* **53** 371–91.
[7] Lychev S. A. and Semenov D. A. 2008 *Vestnik SSU* **2** 183–217 (in Russian).
[8] Lychev S. A. 2008 *Mech. Solids* **43** 769-784.
[9] Lychev S. A. and Manzhirov A. V. 2009 *JOP: Conference series* **181** 012096.

On Finite Strains in Continua with Complex Rheology

E Murashkin[1,2,3a] **and V Klindukhov**[1]

[1]National Research Nuclear University MEPhI (Moscow Engineering Physics Institute), 31 Kashirskoe shosse, 115409 Moscow, Russia
[2]Institute for Problems in Mechanics of the Russian Academy of Sciences,Vernadski Ave. 101 Bldg. 1, Moscow, 119526, Russia
[3]Bauman Moscow State Technical University, 2nd Baumanskaya Str., 5/1, Moscow, 105005, Russia

E-mail: [a]evmurashkin@gmail.com

Abstract. The present study deals with the multiphysics modelling frameworks for the elastic-creep-plastic material. The proposed theory of large deformations is based on the classical formalism of nonequilibrium thermodynamics. Reversible and irreversible components of total deformations are derived by the constitutive differential equations. The development of this theory involves statements and solutions of the correct boundary value problems. The problem of deformation of the material in the vicinity of microdefect was considered. Resulting integro-differential equations system was derived.

1. Preliminary Remarks

Modern additive manufacturing technology [1, 2] states the problem of creating and developing of new mathematical models of growth and deformation of structural materials. The calculation of stress-strain state parameters is needed during the forming processes of material [3]. In the design of ship hulls and aircraft are widely used panels and profiles of hardly-deformed aluminum alloys [4]. Traditional methods of formation of such structural elements often leads to the appearance of the plastic breaks, cracks and other damages. Thus, the effective way of this fabrication is metal forming under creep and low strain rates. These processes ensures the production of construction with high accuracy, which reduces the complexity of assembly and welding, and increase the residual life and the quality of construction.

The mathematical description of the processes of thermomechanical treatment of contracture materials is faced with the necessity to consider the elastic properties of materials at all stages of the producing and utilization. Consideration of the problems in the classical models of small deformations is impossible when the relative shape of the body is strongly changed. Such problem is the process of deformation metals in vicinity of micropore under the intense pressure action. In this case, we are faced to assumption of large strains. Well known experiments [3, 4] when the long-term strength of metal products after their treatment under hydrostatic pressure is significantly increased. Attempts to simulate such process of the micropores "healing" in the metal have been made repeatedly [4-14]. In [5] such problems are considered due to the model of large elastic-plastic deformation. In this case, the effect of adaptability to periodic loading on the cycle "loading and unloading" was shown in [6].

Since classic work E. Lee [7], lots of plastic flow models with large reversible and irreversible deformations were developed [4-14]. Lot of them use a Lagrangian description [13, 14]. But in this case the results of mathematical modeling is physically difficult interpreted. If we want to build a framework of the flow in Eulerian descriptions, then we are faced with two fundamental problems. The first problem is identification irreversible and reversible components of the total strain tensor. The second problem is constitution of irreversible strain source. The mathematical model which was proposed in [5, 8, 12] is used throughout the paper. The problem of a spherically symmetric compression of the ball with micropore in the center is considered.

2. Constitutive Equation

Throughout the paper we use mathematical model of the finite elastocreep deformations (see details in [11]). The kinematic equation for parts of the Almansi total strain d_{ij} can be written in the Cartesian system (Eulerian coordinates) in the form

$$\begin{aligned}\frac{De_{ij}}{Dt} &= \varepsilon_{ij} - \gamma_{ij} - \frac{1}{2}\left((\varepsilon_{ik} - \gamma_{ik} + z_{ik})e_{kj} + e_{ik}(\gamma_{kj} - \varepsilon_{kj} - z_{kj})\right),\\ \frac{Dp_{ij}}{Dt} &= \gamma_{ij} - p_{ik}\gamma_{kj} - \gamma_{ik}p_{kj},\end{aligned} \tag{1}$$

where $e^e_{ij} = e_{ij} - 0.5e_{ik}e_{kj}$ is the reversible part of the Almansi total strain tensor, p_{ij} is the irreversible part of the Almansi total strain tensor, D/Dt denotes the convective derivative with respect to time, γ_{ij} is the irreversible strain rate tensor, ε_{ij} is the strain rate tensor. The strain rate tensor can be computed by the equation

$$\varepsilon_{ij} = \frac{1}{2}\left(v_{i,j} + v_{j,i}\right), \tag{2}$$

where v_i are the components of velocity vector, the index after comma denotes partial derivative with respect to the corresponding spatial coordinate.

The convective derivative with respect to time in (1) from an arbitrary tensor n_{ij} read:

$$\begin{aligned}&\frac{Dn_{ij}}{Dt} = \frac{dn_{ij}}{dt} - r_{ik}n_{kj} + n_{ik}r_{kj},\\ &r_{ij} = w_{ij} + z_{ij}\left(e_{ij}, \varepsilon_{ij}\right), \quad w_{ij} = \frac{1}{2}\left(v_{i,j} - v_{j,i}\right),\end{aligned} \tag{3}$$

wherein w_{ij} is the angular rate tensor and $z_{ij}(e_{ij}, \varepsilon_{ij})$ is the nonlinear part of the rotation tensor r_{ij} (see in full in [5]). Thus, the components of the Almansi total strains d_{ij} in terms of its parts e_{ij} and p_{ij} taking account of equations (1) and (3) are presented as follows

$$d_{ij} = \frac{1}{2}\left(u_{i,j} + u_{j,i} - u_{i,k}u_{k,j}\right) = e_{ij} + p_{ij} - \frac{1}{2}e_{ik}e_{kj} - e_{ik}p_{kj} - p_{ik}e_{kj} + e_{ik}p_{km}e_{mj}, \tag{4}$$

where u_i are the translational displacements. This assumption allows to obtain the constitutive equation like the Murnaghan equation of state well known in the non-linear elasticity [15]:

$$\sigma_{ij} = -p\delta_{ij} + \frac{\partial Q}{\partial e_{ik}}\left(\delta_{kj} - e_{kj}\right), \tag{5}$$

where Q is the strain-energy function, p denotes the hydrostatic pressure function as a Lagrangian multiplier to enforce the incompressibility constraint. For isotropic hyperelastic materials, the strain-energy function can be expressed in terms of the invariants of the reversible strain tensor. Let us expand Q into the Taylor series in the vicinity of the natural state $e_{ij} = 0$, disregarding the terms of higher order than the second one. The following form of the expansion is obtained for isotropic, homogeneous and incompressible bodies

$$\begin{aligned} &Q = (\alpha - \mu)J_1 + \alpha J_2 + \beta J_1^2 - \kappa J_1 J_2 - \zeta J_1^2, \\ &J_1 = e^e_{jj}, \quad J_2 = e^e_{ij} e^e_{ji}, \end{aligned} \tag{6}$$

wherein $\alpha, \mu, \beta, \kappa, \zeta$ are elastic modules.

During process, anticipating plastic flow, and in the unloading, the irreversible strain rate tensor γ_{ij} is identified by the creep strain rate tensor $\gamma_{ij} = \varepsilon^v_{ij}$. The energy dissipation law is valid for creep stage of deforming. Let accept the dissipation potential in the power form like Norton-Bailey creep law [16, 17]:

$$\begin{aligned} &\varepsilon^v_{ij} = \frac{\partial V(\Sigma)}{\partial \sigma_{ij}}, \\ &V(\Sigma) = B\Sigma^n(\sigma_{ij}), \\ &\Sigma = \sqrt{\tfrac{3}{2}\left((\sigma_1 - \sigma)^2 + (\sigma_2 - \sigma)^2 + (\sigma_3 - \sigma)^2\right)}. \end{aligned} \tag{7}$$

Here $\sigma_1, \sigma_2, \sigma_3$ are principal values of Cauchy stress tensor and B, n are the creep constants.

3. Boundary Value Problem

Let examine the changes in the geometry of a single spherical microdefect (micropore) under hydrostatic compression and stress relaxation process during unloading of the material within the proposed model [5]. We consider the solid ball of the initial radius R_0 with a single spherical defect (micropore) of the initial radius s_0 in the center of the sphere. The deformation is given by the boundary conditions

$$\begin{aligned} &\sigma_{rr}|_{r=R(t)} = -P(t), \\ &\sigma_{rr}|_{r=s(t)} = 0, \end{aligned} \tag{8}$$

where σ_{rr} is the radial component of the stress tensor in the spherical coordinate system (r, θ, φ), $R(t) >> r_0$ is the radius of the spherical surface which is given by the external pressure $P(t)$, $s(t)$ is the current radius of the micropore. Reversible (elastic) e_{ij} and irreversible (creep) p_{ij} parts of the Almansi total strain tensor equation (4) are defined by the differential equations of change (transfer) (1). Stresses with reversible deformations are related by equations (5). The constraints of the incompressibility in the present case of the spherical symmetry leads to the continuity equation

$$(1 - u_{r,r}) \cdot \left(1 - \frac{u_r}{r}\right)^2 = 1, \tag{9}$$

where u_r is the only nonzero displacement. The solution of the equation (9) is obtained in form

$$\begin{aligned} &u_r = r - \left(r^3 + \varphi(t)\right)^{\frac{1}{3}}, \\ &\varphi(t) = s_0^3 - s^3(t) = R_0^3 - R^3(t). \end{aligned} \tag{10}$$

Note that the kinematics is specifies with an accuracy of an unknown function $\varphi(t)$.

The equation of motion in considered case of spherical symmetry can be deduced in form

$$\sigma_{rr,r} + 2\frac{\sigma_{rr} - \sigma_{\theta\theta}}{r} = -\rho_0 \left(\frac{\ddot{\varphi}(t)}{3r^2} + \frac{2}{9}\frac{\dot{\varphi}^2(t)}{r^5}\right), \tag{11}$$

wherein $\sigma_{\theta\theta}$ denotes the angular stress, ρ_0 denotes the mass density.

Equation of motion (11) should be supplemented by equation for components of irreversible strain tensor

$$\begin{aligned} \frac{dp_{rr}}{dt} &= Bn(1-2p_{rr})\Phi^{n-1}(e_{rr}, e_{\theta\theta}), \\ p_{\theta\theta} &= \frac{1}{2}\left(1 - \frac{1}{\sqrt{1-2p_{rr}}}\right), \end{aligned} \tag{12}$$

wherein Φ is derived by formula

$$\begin{aligned} &\Phi(a,b) = 2\mu(a-b) - \lambda_1 a^2 + \lambda_2 b^2 + \lambda_3 a^3 - \lambda_4 b^3 + \\ &+\lambda_5\left(\frac{1}{2}ab^2 - ab\right) + \lambda_6\left(a^2 b - \frac{1}{2}a^2 b^2\right) - \lambda_7 a^4 + \lambda_8 b^4 + \\ &+\lambda_9\left(a^4 b + 2a^3 b^2 - \frac{1}{2}a^4 b^2 - 4a^3 b\right) + \lambda_{10}\left(\frac{1}{2}a^6 - 3b^5\right), \\ &\lambda_1 = \mu + 4\alpha + 4\beta + 2\xi, \\ &\lambda_2 = \mu + 4\alpha + 8\beta + 4\xi, \\ &\lambda_3 = 2(2\alpha + 2\beta + 4\xi + 3\chi), \\ &\lambda_4 = 4(\alpha + 2\beta + 4\xi + 6\chi), \\ &\lambda_5 = 4\beta + 2\xi, \\ &\lambda_6 = 2\beta + 7\xi + 18\chi, \\ &\lambda_7 = \alpha + \beta + \frac{19}{2}\xi + 9\chi, \\ &\lambda_8 = \alpha + 2\beta + 19\xi + 36\chi, \\ &\lambda_9 = \frac{3}{4}(\xi + 3\chi), \\ &\lambda_{10} = \frac{1}{2}(5\xi + 11\chi). \end{aligned} \tag{13}$$

Resulting system of the integro-differential equations after integrating equation (11) under condition (8) is transformed into

$$\begin{aligned} &P(t) = 2\int\limits_{s(t)}^{R(t)} \frac{\Psi(r, p_{rr}(r,t), \varphi(t))}{r} dr - \rho_0\left(\frac{\ddot{\varphi}(t)}{3}\left(\frac{1}{R(t)} - \frac{1}{s(t)}\right) + \frac{\dot{\varphi}^2(t)}{18}\left(\frac{1}{R(t)^4} - \frac{1}{s(t)^4}\right)\right), \\ &\frac{dp_{rr}}{dt} = (1-2p_{rr})Bn\Psi^{n-1}(r, p_{rr}(r,t), \varphi(t)), \end{aligned} \tag{14}$$

wherein

$$\Psi(r, p_{rr}(r,t), \varphi(t)) = \Phi(e_{rr}, e_{\theta\theta}),$$

after substituting $e_{rr}, e_{\theta\theta}$ by the rules

$$\begin{aligned} e_{rr} &= 1 - H^{-\frac{2}{3}}(1-2p_{rr})^{-\frac{1}{2}}, \\ e_{\theta\theta} &= 1 - H^{\frac{1}{3}}(1-2p_{rr}), \\ H &= 1 + r^{-3}\varphi(t). \end{aligned} \tag{15}$$

For the calculations of the strain-stress state parameters, we define the following dimensionless parameters:

$$\begin{gathered} \alpha\mu^{-1} = 0.9, \quad \beta\mu^{-1} = 4, \quad \xi\mu^{-1} = 20, \\ B_0 = nB\rho_0 R_0 \mu^{n-2} = 3.5, \quad \chi\mu^{-1} = 80, \\ n = 3, \quad k\mu^{-1} = 0.003, \quad s_0 R_0^{-1} = 0.03. \end{gathered}$$

The resulting system (14) is numerically analyzed by using symbolic computation algebra *Mathematica*.

4. Acknowledgments
The authors wish to thank Professor A. A. Burenin for the statement of the problem, scientific advice and necessary support. This work was financially supported by the Russian Science Foundation under Grant 14-19-01280.

5. Conclusions
The presented mathematical model of large elastocreep deformations is based on equations (1), (4), (5), (7), (11). One can use the presented approach for mechanical design creep fabricated parts from creep-elastic materials determining the strength and the shape of final products. Solved problems for a ball-shaped body with a single spherical defect are consistent with well-known microdefects welding process [3]. The proposed method of the loading pressure computing for a given displacements can be used to optimize the treatment metal process under creep conditions [18]. Moreover, on the basis of this mechanical analysis one can work out effective recommendations for refinements of the technological process.

6. References
[1] Manzhirov A V and Parshin D A 2015 *Mechanics of Solids* **50:5** 559
[2] Manzhirov A V and Parshin D A 2015 *Mechanics of Solids* **50:6** 657
[3] Gorelov V I 1984 *Journal of Applied Mechanics and Technical Physics* **25** 813
[4] Rogachev E E Polonik M V, Dudko O V and Murashkin E V 2014 *Advanced Materials Research* **1040** 898
[5] Burenin A A and Kovtanyuk L V 2011 *On Elastic Features Under Intensive Irreversible Deformation* (Vladivostok: FEFU)
[6] Burenin A A, Kovtanyuk L V and Polonik M V 2000 *Doklady Physics* **45** 694
[7] Lee E H 1969 *Journal of Applied Mechanics* **36** 1
[8] Shitikov A V 1995 *Journal of Applied Mathematics and Mechanics* **59:1** 147
[9] Burenin A A, Kovtanyuk L V and Murashkin E V 2006 *Journal of Applied Mechanics and Technical Physics* **47** 241
[10] Murashkin E and Polonik M 2013 *Applied Mechanics and Materials* **249-250** 833
[11] Bazhin A A and Murashkin E V 2012 *Doklady Physics* **57(8)** 294
[12] Shitikov A V and Bykovtsev G I 1990 *Dokl. Akad. Nauk SSSR*, **311:1** 32
[13] Levitas V I 1996 *Large deformation of materials with complex rheological properties at normal and high pressure* (NY, Nova Science Publishers)
[14] Bruhns O T 2007 *Key Engineering Materials* **340-341** 787
[15] Murnaghan F D 1937 *American Journal of Mathematics* **59** 235
[16] Norton F H 1929 *Creep of Steel at High Temperatures* (NY: McGraw Hill)
[17] Bailey R W 1929 *Transactions of the World Power Conference* **3** 528
[18] Anop M F, Polonik M V, Mikhalichuk V I and Murashkin E V 2016 *Key Engineering Materials* **685** 83

Comparative analysis of the past glacier surface temperatures based on various probe-functions

O V Nagornov, S A Tyuflin, V G Nikitaev and A N Pronichev

National Research Nuclear University MEPhI (Moscow Engineering Physics Institute), 31 Kashirskoe shosse, 115409 Moscow, Russia

E-mail: nagornov@yandex.ru

Abstract. There have been carried out many reconstructions of the past surface temperatures based on the glacier borehole data and the grid functions. Also there has been derived that reconstructions are unique and stable if they based on the finite segments of the Fourier series. We compare results of glacier surface temperature reconstructions by both these methods.

1. Introduction

The studies of the past temperatures at the Earth surface is important problem for prediction of the climate changes. The systematic instrumental temperature measurements took place no more than two centuries. Thus, indirect estimations of the past temperatures present main information on the past climate. There are two different sources of information about paleotemperatures. These are the rock and ice boreholes [1, 2, 3], the tree rings [4], corals and sclerosponges [5] etc. The most reliable data are kept in the measured borehole temperatures that present response to the surface temperature history. Measured temperature profiles in glacier boreholes can keep information on the past temperature changes at the surface for last centuries and more. There are several methods to inverse the measured borehole temperatures in the past surface temperature. That is why it is needed to compare these methods to understand their advatages.

The known paleotemperature reconstructions are based on the control methods [6], the Monte-Carlo method [7] and the Tikhonov's regularization method [8]. In these approaches the past temperature at the surface was looking for the grid function. The uniqueness and stability properties of the solution of the inverse problem were not established. On the other hand it was found out that the solution of inverse problem in form of the finite series of the Fourier poses is unique and stable [9]. Therefore, these three methods can be used to derive the past surface temperature in form of the finite series of the Fourier. In this paper we compare solutions that derived these three methods.

2. Mathematical model

The mathematical statement of the inverse problem consists of the thermal conductivity equation that takes into account the vertical advection term, the initial condition, the boundary condition at the bottom of glacier and the re-determination condition. The measured-temperature-depth profile is used as the re-determination condition, $\chi(z)$, where z is vertical coordinate. Thus, the following one-dimensional inverse problem for the heat conduction equation has to be solved to find out the past surface temperature:

$$\begin{cases} V_t + w(z)V_z = a^2 V_{zz}, & 0 < t < t_f, 0 < z < H, \\ V(0,t) = \mu(t), & 0 < t \le t_f, \\ V_z(H,t) = 0, & 0 < t \le t_f, \\ V(z,0) = 0, & 0 \le z \le H, \\ V(z,t_f) = \theta(z), & 0 \le z \le H, \end{cases} \tag{1}$$

where H is the glacier thickness, $w(z)$ is the advection rate of glacier layers, a^2 is the thermal diffusivity, $\mu(t)$ is temperature variations on the surface in time from the moment $t = 0$ ($\mu(0) = 0$) to the time of measurements of the borehole temperature profile t_f, $\theta(z) = \chi(z) - U(z)$. $U(z)$ is the steady-state surface temperature in the past:

$$\begin{cases} U_{zz} - w(z)U_z = 0, & 0 < z < H, \\ U(0) = U_s, \\ U_z(H) = -q/k, \end{cases} \tag{2}$$

where U_s is the steady-state surface temperature in the past, k is the thermal conductivity, q is the geothermal heat flux,

3. Methods for inverse problem solutions

3.1. The control method

The control method is an analog of the method of the Lagrange undetermined multipliers. The initial functional minimized to reconstruct the surface temperature is a functional consisting of the difference and integral determining the standard deviation of the desired surface temperature from a certain function $\eta(t)$ [6]:

$$J = \frac{1}{2}\int_0^H [R\{\eta(t)\} - \theta(z)]^2 dz + \frac{\beta}{2}\int_0^{t_f} [\mu(t) - \eta(t)]^2 dt \tag{3}$$

where $R\{\eta(t)\}$ is the solution of the direct problem specified by equations (1) without the re-determination condition for temperature variations $\eta(t)$ and β is the parameter matched with the accuracy of the input data (temperatures measured in a borehole).

Additional conditions determining $R\{\eta(t)\}$ include the equation describing the heat transfer process and the initial and boundary conditions. Taking into account these conditions, the problem of the reconstruction of the surface temperature is reduced to the minimization of the following functional (Lagrange function) with the Lagrange multiplier $\lambda(z,t)$:

$$J' = J + \int_0^H \int_0^{t_f} \lambda(z,t)[V_t - a^2 V_{zz} + w(z)V_z] dt dx + \\ \int_0^{t_f} \lambda(H,t)V_z(H,t)dt + \int_0^H \lambda(z,0)V(z,0)dz \tag{4}$$

If the functional variations are equated to zero $\frac{\partial J'}{\partial V(z,t)} = 0$, $\frac{\partial J'}{\partial V_z(0,t)} = 0$, $\frac{\partial J'}{\partial V(H,t)} = 0$, $\frac{\partial J'}{\partial V(z,t_f)} = 0$, $\frac{\partial J'}{\partial \mu} = 0$, we arrive at the following system of equations for determining the Lagrange multiplier $\lambda(z,t)$:

$$\begin{cases} \lambda_t + a^2\lambda_{zz} + (w\lambda)_z = 0, & 0 < t < t_f, 0 < z < H, \\ \lambda(0,t) = 0, & 0 < t \le t_f, \\ a^2\lambda_z(H,t) + w(H)\lambda(H,t) = 0, & 0 < t \le t_f, \\ \lambda(z,t_f) = \theta(z) - R\{\eta(t)\}, & 0 \le z \le H. \end{cases} \tag{5}$$

Then the equation for the temperature determination is:

$$\mu(t) = \eta(t) + \frac{1}{\beta}(a^2\lambda_z(0,t) + w(0)\lambda(0,t)). \tag{6}$$

Thus, the algorithm for the inverse problem solution by the control method is of the iteration procedure where $\eta(t)$ equals to $\mu(t)$ from the previous iteration step. The past surface temperature at the first step can be given by any value.

Let us write the surface temperature in the form of finite set of the Fourier series:

$$\mu(t) = \frac{a_0}{2} + \sum_{m=1}^{M} a_m \cos(\frac{2\pi lt}{t_f}) + b_m \sin(\frac{2\pi lt}{t_f}) \tag{7}$$

Then the Fourier series coefficients can be found out by the followinf iteration procedure:

$$\begin{aligned} a_0^{n+1} &= a_0^n + \frac{2}{\beta t_f}\int_0^{t_f}(a^2\lambda_z(0,t) + w(0)\lambda(0,t))dt, \\ a_m^{n+1} &= a_m^n + \frac{2}{\beta t_f}\int_0^{t_f}(a^2\lambda_z(0,t) + w(0)\lambda(0,t))\cos(\frac{2\pi lt}{t_f})dt, \\ b_m^{n+1} &= b_m^n + \frac{2}{\beta t_f}\int_0^{t_f}(a^2\lambda_z(0,t) + w(0)\lambda(0,t))\sin(\frac{2\pi lt}{t_f})dt, \quad m = 1,2,...M. \end{aligned} \tag{8}$$

3.2. The Monte-Carlo method

The Monte-Carlo method [7] tests randomly chosen variations in the surface temperature using them as the input data for the direct problem and taking into account the consistency degree between the calculated and measured temperature profiles. The Monte Carlo simulation involves random walks in the multidimensional space of all the possible models of $\vec{m}$. The vector of the model parameters has the form $\vec{m} = (\mu_1, \mu_2, ...\mu_L)$ for a grid function, and $\vec{m} = (a_0, a_1, ...a_M, b_1, ...b_M)$ for a finite set of the Fourier series. The probability function for the vector of the model parameters has the form $L(\vec{m}) = \gamma\exp(-S(\vec{m}))$, where $S(\vec{m})$ is the difference between the calculated temperature in the borehole and measured temperature profile and γ is the normalization constant.

Any point of the model space can be taken as the start point for the random walk $\vec{m}^1_{cur}$. The random walk can be described as follows:

(1) the trial value of the surface temperature $\vec{m}^{n+1}_{trail}$ is taken near the current surface temperature $\vec{m}^n_{cur}$;

(2) $\vec{m}^{n+1}_{trail}$ is tested with the probability $P_{accept} = \min(1, L_{trail}/L_{cur})$;

(3) if $\vec{m}^{n+1}_{trail}$ is accepted, then $\vec{m}^{n+1}_{cur} = \vec{m}^{n+1}_{trail}$, otherwise, $\vec{m}^{n+1}_{cur} = \vec{m}^n_{cur}$.

Thus, the necessary number of allowable model parameters the differences for which are smaller than a certain value S_0 (which is usually matched with the error of the borehole temperature measurements) can be collected. Using this statistics, histograms can be plotted for each parameter in the vector of model parameters. Such distributions usually have the maximum region, i.e., the most probable values of the corresponding parameters, which constitute the solution of the inverse problem.

3.3. The Tikhonov regularization method

The Tikhonov regularization method is the determination of the boundary temperature $\mu(t)$ minimizing a smoothing functional consisting of the difference and stabilizer [8]:

$$\Psi = \frac{1}{2}\int_0^H [R\{\mu(t)\} - \theta(z)]^2 dz + \alpha\Omega\{\mu(t)\} \tag{9}$$

where α is the regularization parameter matched with the accuracy of the input data. The functional $\Omega\{\mu(t)\}$ is called the stabilizing functional or stabilizer

$$\Omega\{\mu(t)\} = \int_0^{t_f} \sum_{j=0}^{r} q_j \left(\frac{d^j\mu(t)}{dt^j}\right)^2 dt \tag{10}$$

where r is the stabilizer order, $q_j \geq 0$, and $q_r > 0$. The procedure of the minimization of the smoothing functional Ψ can be performed by means of the gradient method and is an iteration procedure. The iteration procedure is carried out until the functional Ψ reaches the minimum with a given accuracy, which corresponds to the optimal solution of the inverse problem.

We consider case when the surface temperature is in form of the finite set of the Fourier series (7). The initial Fourier coefficients are given in the first iteration step while the next n-th iterations are determinated by the following equations:

$$\begin{aligned} a_0^{n+1} &= a_0^n - \gamma^n \frac{\partial \Psi^n}{\partial a_0^n}, \\ a_m^{n+1} &= a_m^n - \gamma^n \frac{\partial \Psi^n}{\partial a_m^n}, \\ b_m^{n+1} &= b_m^n - \gamma^n \frac{\partial \Psi^n}{\partial b_m^n}, \quad m = 1, 2, ...M, \end{aligned} \tag{11}$$

where $\gamma^n > 0$ is the gradient step. The derivatives of the functional in (11) with respect to the corresponding Fourier coefficients are given by the expressions:

$$\begin{aligned} \frac{\partial \Psi^n}{\partial a_0^n} &= \int_0^H W_{a_0}(z)[R\{\mu^n(t)\} - \theta(z)]dz + \alpha\frac{\partial \Omega^n}{\partial a_0^n}, \\ \frac{\partial \Psi^n}{\partial a_m^n} &= \int_0^H W_{a_m}(z)[R\{\mu^n(t)\} - \theta(z)]dz + \alpha\frac{\partial \Omega^n}{\partial a_m^n}, \\ \frac{\partial \Psi^n}{\partial b_m^n} &= \int_0^H W_{b_m}(z)[R\{\mu^n(t)\} - \theta(z)]dz + \alpha\frac{\partial \Omega^n}{\partial b_m^n}, \quad m = 1, 2, ...M, \end{aligned} \tag{12}$$

Here, the profiles $W_{a_0}(z)$, $W_{a_m}(z)$, and $W_{b_m}(z)$ are the solutions of the direct problem with the boundary conditions on the surface $\mu(t) = 1/2$, $\mu(t) = cos(2\pi t/T_m)$ and $\mu(t) = sin(2\pi t/T_m)$, respectively. It is easy to show that the term with the stabilizer in (9) when the boundary condition on the surface $\mu(t)$ has the form of the segment of the trigonometric Fourier series has the form: $\alpha\Omega = \alpha_0 \frac{a_0^2}{2} + \sum_{m=1}^{M}(a_m^2 + b_m^2)\xi_m$, where $\xi_m = \alpha_0 + \alpha_1(2\pi/T_m)^2 + ... + \alpha_r(2\pi/T_m)^2 r$, $\alpha_i = \alpha q_i t_f/2$, $i = 0, 1, ...r$. In this case

$$\begin{aligned} \alpha\frac{\partial \Omega^n}{\partial a_0^n} &= \alpha_0 a_0^n, \\ \alpha\frac{\partial \Omega^n}{\partial a_m^n} &= 2a_m^n \xi_m, \\ \alpha\frac{\partial \Omega^n}{\partial b_m^n} &= 2b_m^n \xi_m, \quad m = 1, 2, ...M. \end{aligned} \tag{13}$$

Thus, to determine the Fourier coefficients of the boundary condition given by (7), the iteration procedure specified by (11) is performed with the use of (12) and (13).

4. Comparative analysis of the inverse problem solutions

We carry out comparison for the test problems for the following parameters of glacier: the thickness of glacier is $H = 500\ m$, $a^2 = 10^{-6}\ m^2/s$, the advection rate is linear function decreasing from 0.3 $m/year$ at the surface to 0 at the glacier bottom. The direct problems are solved for two probe functions: $\mu(t) = 2\sin(2\pi t/t_f)$ and $\mu(t) = 2\sin(\pi t/t_f)$. We consider the time interval from 0 to $t_f = 400\ years$. The solutions of these two direct problems are the corresponding temperature-depth profiles $V(z, t_f)$. After that we solve the inverse problems (1) with the determined temperature-depth profiles $\theta(z)= V(z, t_f)$. Moreover, to check the stability of the methods we introduce the random errors in the input data with amplitude of 0.05 °C. The reconstructions are calculated by three methods using the surface temperature in the form of both the grid fuctions (Fig. 1) and the finite series of the Fourier (Fig. 2).

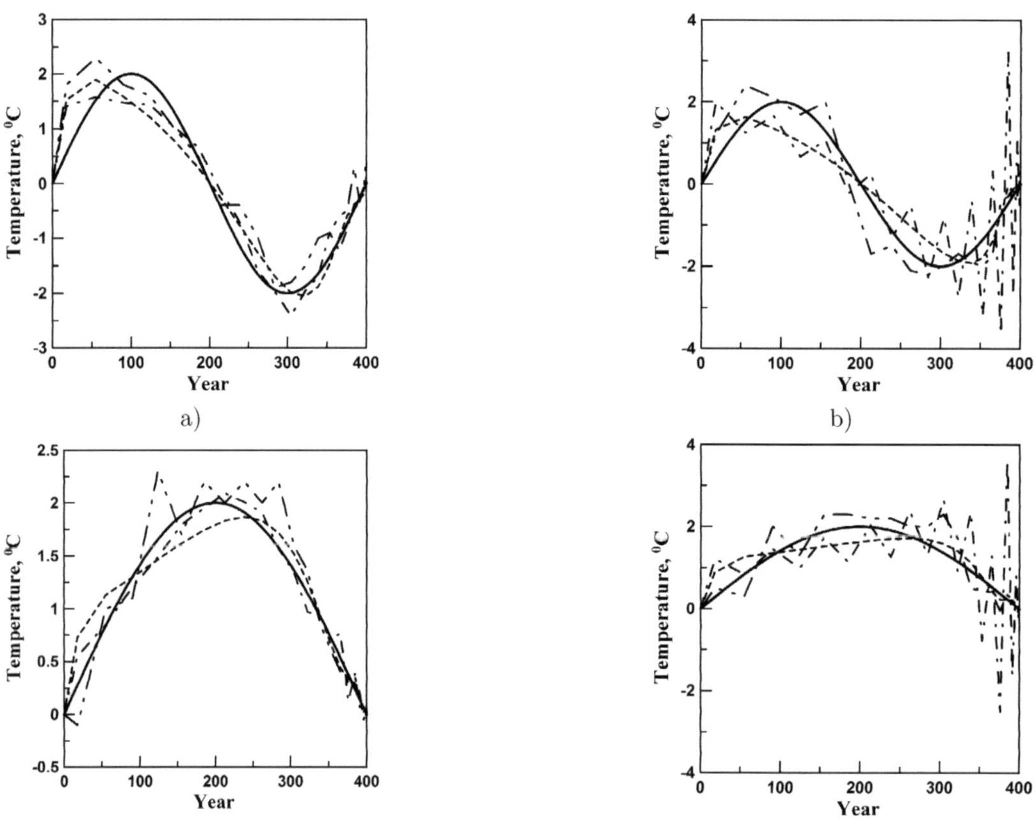

Figure 1. The surface temperature changes at the surface (solid line) and the reconstructed solutions be the Tikhonov regularization method (dashed line), the control method (dash-dot line) and the Monte-Carlo method (dash-dot-dot line). The reconstructions are based on the grid functions. a and c are the input data without errors; b and d are the input data containing the random errors with amplitude of 0.05 °C

Comparison shows that all methods provide similar results in case of the exact input data. However, the errors in the input data result in the instability of the inverse problem solutions. The worst result takes place for the grid function solution obtained by the control method. The best results is provided by the Tikhonov regularization method (Fig. 1).

It is interesting to note that if we look for the inverse problem solution by the finite series of the Fourier then three considered methods are of similar solutions (Fig. 2).

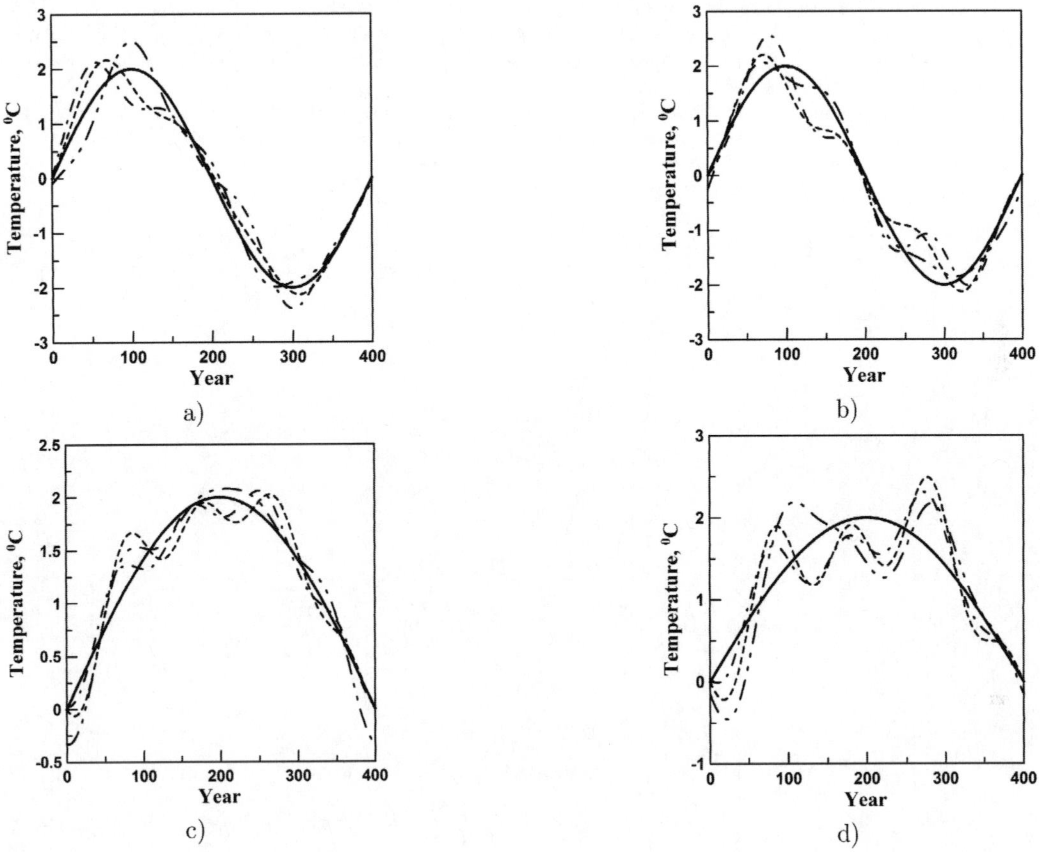

Figure 2. The surface temperature changes at the surface (solid line) and the reconstructed solutions be the Tikhonov regularization method (dashed line), the control method (dash-dot line) and the Monte-Carlo method (dash-dot-dot line). The reconstructions are based on the finite series of the Fourier; a and c are the input data without errors; b and d are the input data containing the random errors with amplitude of 0.05 $^{\circ}C$

References

[1] Shen P Y and Beck A E 1991 Least squares inversion of borehole temperature measurements in functional space *Journal of Geophysical Research* **96** 19965–19979

[2] Nagornov O V, Nikitaev V G, Pronichev A N, Tyuflin S A and Bukharova T I 2016 Analysis of past surface temperature reconstructions based on the tree-ring chronologies and borehole temperature measurements *AIP Conference Proceedings* **1738** 230006

[3] Zagorodnov V, Nagornov O, Scambos T A, Muto A, Mosley-Thompson E, Pettit E C and Tyuflin S 2012

Borehole temperatures reveal details of 20th century warming at Bruce Plateau, Antarctic Peninsula *The Cryosphere* **6** 675–686
[4] Fritts H C 1976 *Tree-Rings and Climate* (London: Academic Press)
[5] Dunbar R B and Cole J E 1999 *Annual Records of Tropical Systems, PAGES Workshop ReportSeries, 99-1* (PAGES, BERN Switzerland)
[6] MacAyeal D R, Firestone J and Waddington E 1991 Paleothermometry by control methods *Journal of Glaciology* **37(127)** 326–338
[7] Dahl-Jensen D, Mosegaard K, Gundestrup N, Clow G D, Johnsen S J, Hansen A W and Balling N 1998 Past temperatures directly from the Greenland Ice Sheet *Science* **282** 268–271
[8] Nagornov O V, Konovalov Yu V, Zagorodnov V S, Thompson L G 2001 Reconstruction of the surface temperature of Arctic glaciers from the data of temperature measurements in wells *Journal of engineering physics and thermophysics* **74** 253–265
[9] Nagornov O V, Tyuflin S A and Bukharova T I 2014 The properties of solutions of the inverse paleotemperature problems *Proceedings of the 1st International Conference on Mathematical Methods and Computational Techniques in Science and Engineering (MMCTSE 2014)* 48–52

Mathematical model of the chromatin structure of the nuclei of blood cells

V G Nikitaev, O V Nagornov, A N Pronichev, V V Dmitrieva, E V Polyakov

National Research Nuclear University MEPhI (Moscow Engineering Physics Institute), 31 Kashirskoe shosse, 115409 Moscow, Russia

E-mail: VGNikitayev@mephi.ru

Abstract. This paper describes the model of images of the nuclei of blood cells for research informative texture features in the diagnostics of acute leukemias on the basis of computer microscopy. The proposed model allows to simulate the structure of chromatin and factors distorting the signal in the formation of image.

1. Introduction

Currently, diagnosis of acute leukemia is based on an objective analysis of the results obtained in laboratory studies [1-3]: blood smears prepared with an general clinical analysis, bone marrow preparations, cytochemical, immunological, cytogenetic.

In practice, the use of modern diagnostic technology does not replace the study of blood smears and bone marrow preparations under the microscope for analysis of cellular composition [4-5].

Microscopic analysis of blood smears and bone marrow is difficult, tedious, time-consuming, requires mental and physical tension, according to some estimates, the error reaches 30 – 40 % depending on experience and qualifications of a physician hematologist.

The automated analysis techniques, based on pattern recognition methods, are used to reduce the complexity and difficulty of visual study of microscopic preparations. The use of medical decision support systems, based on computer microscopy would reduce the subjectivity of the diagnostic process. Unexplored region in automated image processing systems for the diagnosis of acute leukemia is the use of models of blood cell images based on texture analysis. Texture characteristics of nuclei of blood cells can be used for automatic classification of the blood cells. In this regard, the task of modeling texture images of nuclei of blood cells is relevant [6].

The aim of this work is the creation of model of images of the nuclei of blood cells for experimental studies of the information content of texture features in the diagnostics of acute leukemias on the basis of computer microscopy.

Computer graphics systems display information about the processes or objects in the form of synthesized display on the screen or other surface. For computer graphics systems, the source of the input data is not the physical processes, but their mathematical models. Such models generally represent an ordered set of data, numerical characteristics, parameters, mathematical and logical dependency mapping the structure, properties, relationships and the relationship between the elements of the object, as well as between the object and its surroundings.

Models usually are generalized, intended to describe a class of objects. Individual objects described by the input specific values of system parameters. The image that is associated with the simulated analog, known as the original.

2. Formation model

The model of formation of images of the nucleus of blood cells based on the specified distribution of the basic elements – model of chromosomes. In our model, one chromosome is a sequence of circles, whose centers are equidistant from each other and are on the same line. The length of the chromosomes, the radius of the circles and their brightness is a random variable, distributed according to the normal law of mathematical expectation. In two-dimensional space models of chromosomes "attached" to the nodes of the grid covering the entire area of the digital image.

The whole field is covered with a conventional grid, a step which is set as a parameter, and the nodes are circles of random brightness, the center of which is deviated from the node to the random variable, radius, brightness, and distance along the axes OX and OY are normally distributed, their mean value and standard deviation are specified as parameters. Areas of intersection are colored weighted sum of brightness of the circles.

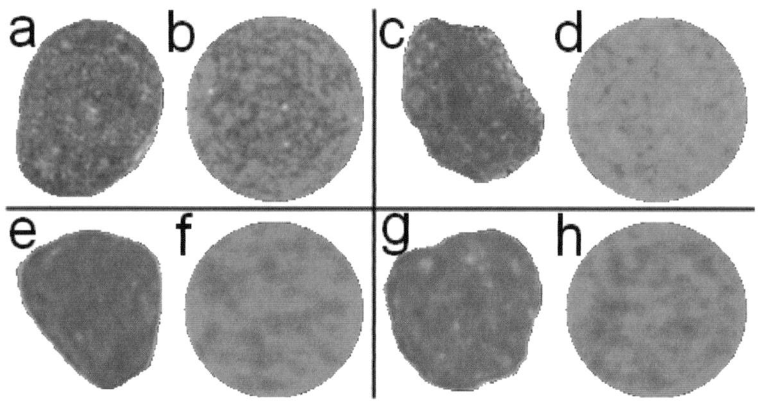

Figure 1. Image of blast cells a), myelocyte c),lymphoid cell e) and lymphocyte g) and the generated structures for each type b), d), f), h).

The advantages of this model include the ability to generate specific textures for all proposed types of blood cells, the corresponding model and the original image is set visually and meaningfully textural features.

Based on the analysis of the structure of blood cell nucleus suggests the following mathematical model of the nucleus image.

$F(x,y) = A(x,y) + B(x,y) + C(x,y)$, где $F(x,y)$ – the optical density of the substance in the nucleus of blood cells at the point with coordinates (x,y), $A(x,y)$ - the optical density of a substance generated by set of chromosomes, $B(x,y)$ - the optical density of the substance generated by the nucleoli, $C(x,y)$ - the optical density of a substance generated by nucleoplasm, $A(x,y) = \sum A_i(x,y)$, where $A_i(x,y)$ – weight of i chromosome, coordinates of the chromosome anchor.

$$\begin{cases} x_1 = n * d + \Delta r * \cos\varphi_1 \\ y_1 = m * d + \Delta r * \cos\varphi_1 \end{cases}$$

where d – grid spacing, φ_1 – uniformly distributed random variable from $0°$ to $360°$, n and m – whole numbers, Δr – normally distributed quantity. The coordinates of the second end of chromosome

$$\begin{cases} x_2 = x_1 + L * \cos\varphi_2 \\ x_2 = y_1 + L * \cos\varphi_2 \end{cases}$$

where φ_2 uniformly distributed random variable from $0°$ to $360°$, L – length of rod, normally distributed quantity, $A_i(x, y) = \sum a_i(x, y) / z$, where $a_i(x, y)$ – weight of elementary circle of radius r, z - the number of intersections of elementary circles of one chromosome at the point (x, y).

$$\begin{cases} x_c = x_1 + (x_1 - x_2) * k * r / l \\ y_c = y_1 + (y_1 - y_2) * k * r / l \end{cases}$$

where (x_c, y_c) – elementary centers of circles that make up the chromosome k – non-negative integer, not exceeding L / r.

In a straight line of a given length will postpone the circles with variable brightness, brightness of each point is determined by the averaged sum of brightness of all circles containing a point with coordinates (x, y).

Formation of the image texture of the nucleus of blood cells represented by a formula: $G(x, y) = (\max - F(x, y)) * 255 / \max$, where $G(x, y)$ – brightness, $\max$ – maximum value $F(x, y)$. Blur: $G_2(x, y) = \sum G_1(x + i, y + j) / (2 * d + 1)^2$, for all integer i and j not exceeding modulo d, where $(2 * d + 1)$ – size of averaging matrix, $G_2(x, y)$ – generated image, $G_1(x, y)$ – original image. Adding noise: $G_3(x, y) = G_2(x, y) + \eta(x, y)$, where $\eta(x, y)$ – additive and independent of the signal noise with Gaussian (or other) distribution density function.

Table 1. Values of the input parameters.

Parameter	1 class	2 class	3 class	4 class
Step size	10	18	25	25
Standard deviation (SD) of Rod	3	3	10	0
The radii length	200	150	200	200
SD of Rod length	100	30	50	100
Radius of the base circle	4	6	17	18
SD of Radius of the base circle	1	3	6	5
Weight of the basic of circles	3	4	7	7
SD of Weight of the basic of circles	7	3	3	2
Number of nucleoli	5	1	1	1
Radius of the nucleoli	6	2	2	2
SD of Radius of the nucleoli	2	0	0	0
Weight of the nucleoli	-120	100	350	350
SD of Weight of the nucleoli	10	0	0	0
Weight of the nucleoplasm	50	50	200	200
Degree of averaging	3	3	3	3

For the simulation of nucleoli used circles whose centers are uniformly distributed in the image area. The radii and brightness of nucleoli - random variables distributed according to the normal law of mathematical expectation, standard deviation, and number of nucleoli are set as initial data of the model. Since nucleoplasm is liquid uniformly distributed in the nucleus, to model it the brightness of the entire working area changes by an amount set as a model parameter. As analyzed images obtained in a microscopic system, the diffraction of light waves is simulated by image blur filter.

3. Experiment

To check the adequacy of the model, we conducted an experiment study during which selected parameters of the model, allowing to obtain characteristics similar to real cells (presents four classes: 1 – blast, 2 – myelocytes, 3 – lymphocyte, 4 – lymphoid cell). The selected values are presented in Table 1.

Visually, the model images like the original on the degree of graininess of the texture of chromatin, although it looks more blurry.

4. Conclusion

The developed computational model allows you to create a texture image corresponding to the nuclei of various types of blast and non-blast cells. Computing model parameters found experimentally to ensure compliance generated image with the original image. The proposed model is designed to generate a texture image characteristic of different types of leukocytes (including lymphocytes, blast cells, etc.) to study the informativeness of features in the diagnostics of acute leukemias.

References

[1] Reyes E Luis, L X Bautista Rozo, F A Rojas Morales 20th Symposium on Signal Processing Images and Computer Vision (STSIVA) (IEEE, Bogota, Colombia, 2015) p1

[2] Amin M, Kermani S, Talebi A, Oghli M 2015 Journal of Medical Signals & Sensors 5(1) 49

[3] Azam B, Qureshi R J, Jan Z, Khattak T A 2014 Research Journal of Recent Sciences 3(4) 34

[4] Nikitaev V G, Nagornov O V, Pronichev A N , Polyakov E V, Sel'chuk V Y, Chistov S K, Blyndar V N , Dmitrieva V V, Zaytsev S M, Gordeev V V 2015 Measurement Techniques 57(10) 1203

[5] Nikitaev V G, Nagornov O V, Pronichev A N, Polyakov E V, Sel'chuk VY, Chistov S K, Blyndar V N, Dmitrieva V V, Gordeev V V 2015 Measurement Techniques 57(5) 560

[6] Nikitaev V G, Nagornov O V, Pronichev AN, Polyakov E V, Dmitrieva V V 2015 WSEAS TRANSACTIONS on BIOLOGY and BIOMEDICINE 12 16

One effective algorithm for finding a minimum cut of any transport network

S V Petrov[1], A N Snezhin[2] and V M Prostokishin[1,2]

[1] National Research Nuclear University MEPhI (Moscow Engineering Physics Institute), 31 Kashirskoe shosse, 115409 Moscow, Russia
[2] Gazprom VNIIGAZ LLC, box 130, 115583 Moscow, Russia

E-mail: VMProstokishin@mephi.ru

Abstract. Finding the minimal cut on the directed graph is an important task that has not yet been solved in a general form. The solution of this problem allows one to find "the pipeline bottleneck" of the transport network and manage changing flows efficiently. In this article, authors discussed a new algorithm for solving this problem, built on the basis of consideration of the corresponding dual problem.

Introduction

Transport networks are one of the bases part of the fuel and energy complexes. An important problem is the localization of "the pipeline bottleneck" of the transport network also. A directed graph with varying flows and possible inflows and outflows of substance at the vertices and at the edges of the graph is the model of the network. This article describes an algorithm for the solution of the generalized problem (assuming constant inflows and outflows in arbitrary network nodes) finding the minimal cut and determine the maximum flow through the network.

1. Problem statement

Suppose to each edge ij of the graph assign the positive number q_{ij}. Any such number q_{ij} is called the capacity constraint on the edge. Consider two vertices s and t over from the graph. Let s be the source, t be the sink. Such graph, with some additional conditions, is called the network.

Assume that to each edge ij of the graph assign the non-negative number x_{ij}; then a set $\{x_{ij}\}$ is called the flow from s into t if the following conditions hold:

$$\sum_i x_{ij} - \sum_k x_{jk} = \begin{cases} -\nu, & \text{if } j = s, \\ 0, & \text{if } j \neq s, t, \\ \nu, & \text{if } j = t, \end{cases} \tag{I}$$

$$\nu \geqslant 0, \tag{II}$$

$$0 \leqslant x_{ij} \leqslant q_{ij}, \text{ for all } i, j. \tag{III}$$

IOP Conf. Series: Journal of Physics: Conf. Series **788** (2017) 012057 doi:10.1088/1742-6596/788/1/012057

By ν denote the maximum flow of the network. Then the maximum flow problem is of the form:

$$\nu \to \max . \qquad \text{(IV)}$$

Dual problem: To find "the pipeline bottleneck" of the network such that the value of a flow in the network can be less than or equal to the value $V = \max \nu$.

The cut is the minimal cut if the capacity constraint of it is minimal over from all them.

Let $S = \{s_1, s_2, \ldots, s_r\}$ be a nonempty set of the numbers of the vertices, where there exist sources, $T = \{t_1, t_2, \ldots, t_u\}$ be a nonempty set of the numbers of the vertices, where there exist sinks.

Note that any vertex can have either a source or a sink only. Also, it can have neither a source nor a sink. By assumption, there exist one source and one sink in the network. It means that $S \neq \emptyset$, $T \neq \emptyset$, $S \cap T = \emptyset$. Then for amount of the sources and the sinks the following conditions hold: $2 \leqslant r + u \leqslant m$, $1 \leqslant r \leqslant m$, $1 \leqslant u \leqslant m$.

Finite inflows and outflows onto the edges (also into the vertices) of the network are added in this work as distinguished from the problems discussed in [1].

Let the transport network be represented with a corresponding graph. Then the problem has the following form:

$$x_{T(sj)} + \sum_i x_{T(ij)} - \sum_k x_{T(jk)} = f_j - \sum_i g_{T(ij)},\ j = s_1, s_2, \ldots, s_r, \qquad (1)$$

$$\sum_i x_{T(ij)} - \sum_k x_{T(jk)} = f_j - \sum_i g_{T(ij)},\ j \neq s_1, s_2, \ldots, s_r, t_1, t_2, \ldots, t_u, \qquad (2)$$

$$\sum_i x_{T(ij)} - \sum_k x_{T(jk)} - x_{T(jt)} = f_j - \sum_i g_{T(ij)},\ j = t_1, t_2, \ldots, t_u, \qquad (3)$$

$$x_{T(ij)} \leqslant q_{T(ij)}, \qquad (4)$$

$$- x_{T(ij)} \leqslant g_{T(ij)}, \qquad (5)$$

$$x_{T(ij)} \geqslant 0,\ x_{T(sj)} \geqslant 0,\ x_{T(jt)} \geqslant 0, \qquad (6)$$

$$\nu = \sum_{i=1}^{r} x_{n+i} = x_{n+1} + x_{n+2} + \cdots + x_{n+r} \to \max, \qquad (7)$$

where m is the number of all vertices of the graph, n is the number of all edges, x_{ij} is a variable of the flow starting from the i-th vertex and ending in the j-th vertex, q_{ij} is the capacity constraint of the network's edge, g_{ij} is the inflow onto the edge, and f_i is the inflow into the i-th vertex.

Here, a transition is made from numbering of the variables and constants with two indeces towards numbering them with one index, preserving a bijection between them [2]. This bijection is denoted by T and it has the following form:

$$T : ij \leftrightarrow l; \quad i, j = 1, 2, \ldots, m;\ l = 1, 2, \ldots, n,$$
$$T : \nu \leftrightarrow x_0, \quad T : x_{ij} \leftrightarrow x_l, \quad T : q_{ij} \leftrightarrow q_l, \quad T : c_{ij} \leftrightarrow c_l,$$
$$T : ss_i \to n + i,\ i = 1, 2, \ldots, r, \quad T : tt_k \to n + r + k,\ k = 1, 2, \ldots, u.$$

where (i, j) is an index of the coordinate of a nonzero element in the adjacency-matrix representation, $l = 1, 2, \ldots, n$, $T(z_{ij}) = z_{T(ij)} = z_l$, and z is either a variable or a given value.

The matrix form of the system of equations and inequalities (1)-(7) is

$$F(x) = \sum_q c_q x_q \to \max, \tag{8}$$

$$A'x = d, \tag{9}$$

$$A''x \leqslant q, \tag{10}$$

$$A'''x \leqslant g, \tag{11}$$

$$x \geqslant 0, \tag{12}$$

where

$$d_j = f_j - \sum_i g_{T(ij)},\ j = 1, 2, \ldots, m. \tag{13}$$

We must substitute two inequalities for the equation (9)

$$A'x \leqslant d, \tag{14}$$

$$A'_{-}x \leqslant d_{-}, \tag{15}$$

where $A'_{-} = -A'$, $d_{-} = -d_{-}$.

Thus, we obtain the standard form of the linear-programming problem. Let a solution our problem be feasible. Here, we take inverse bijection T^{-1} and can receive the values of the flows over all edges.

2. Problem of finding a minimal cut

Dual problem. The aim of this section is to solve the following problem: to have got edges of a minimal cut such that it should separate the sources S and the sinks T. Also, the following conditions hold: the constant inflows and outflows in vertices and over edges are balanced.

Now, we take the primal linear-programming problem having the standard form

$$\max \langle x, c \rangle,$$

$$Ax \leqslant b,$$

$$x \geqslant 0.$$

It follows in the standard way that the dual linear-programming problem is

$$\min \langle b^{tr}, y \rangle, \tag{16}$$

$$A^{tr} y \geqslant c^{tr}, \tag{17}$$

$$y \geqslant 0, \tag{18}$$

where $y = (y_1, y_2, \ldots, y_{2\cdot(m+n)})^{tr}$ are dual variables.

It can easily be checked (see [3]) that the standard form of the problem (16 – 18) is

$$\max \langle -b^{tr}, y \rangle, \tag{19}$$

$$-A^{tr} y \leqslant -c^{tr}, \tag{20}$$

$$y \geqslant 0. \tag{21}$$

Note that, the formula $\min \langle b^{tr}, y \rangle = -\max \langle -b^{tr}, y \rangle$ is correct.

We stress that the components of formulae (19 – 21) are

$$b^{tr} = (d^{tr}, d_-^{tr}, q^{tr}, g^{tr}) = (d_1, \ldots, d_m, -d_1, \ldots, -d_m, q_1, \ldots, q_n, g_1, \ldots, g_n), \quad (22)$$

$$A^{tr} = ((A')^{tr}(A'_-)^{tr}(A'')^{tr}(A''')^{tr}) = ((A')^{tr} - (A')^{tr}(A'')^{tr}(A''')^{tr}), \quad (23)$$

$$-c = (0, \cdots, 0, -1_{n+1}, -1_{n+2}, \cdots, -1_{n+r}, 0, \cdots, 0), \quad (24)$$

where the matrix A^{tr} has a size of $(n+r+u) \times (2 \cdot (m+n))$, the row vector b^{tr} has a dimension $1 \times (2 \cdot (m+n))$, the column vector c^{tr} has a dimension $(n+r+u) \times 1$, and the vector of dual variables $y = (y_1, y_2, \ldots, y_{2\cdot(m+n)})^{tr}$ has a dimension $(m+n) \times 1$.

The coordinates of the column vector b^{tr} is of the form (22). In this way, in our numbering, the set the edges of the network takes the coordinates y_i, $(i = 2 \cdot m+1, \ldots 2 \cdot m+n)$ of the vector y. The other coordinates correspond to the terms of a distribution of the flow at the vertices or the edges of the network.

Let a column vector $\hat{y} = (\hat{y}_1, \hat{y}_2, \ldots, \hat{y}_{2\cdot(m+n)})^{tr}$ be a solution of our problem (16 – 18). Consider the column vector

$$\tilde{y} = (0, \ldots, 0, \tilde{y}_{2\cdot m+1}, \ldots, \tilde{y}_{2\cdot m+n}, 0, \ldots, 0)^{tr} = (0, \ldots, 0, \hat{y}_{2\cdot m+1}, \ldots, \hat{y}_{2\cdot m+n}, 0, \ldots, 0)^{tr}.$$

Lemma. *Any coordinates of the column vector $\tilde{y}$ are equal to either 0 or 1.*

And our main result is the following:

Theorem. *The edge with the number i corresponds the unit value of the coordinates $\tilde{y}_{2\cdot m+i}$, where $i = 1, 2, \ldots, n$, of vector $\tilde{y}$. All these edges are the minimal cut such that this cut separates the sources S and the sinks T.*

This statement is the criterion: if the arc is marked by number 1 then this arc belongs to the minimal cut; if the arc is marked by number 0 then it does not belong to the minimal cut. We find one of "the pipeline bottleneck" of the network for the maximum-flow.

Constant inflows and outflows in vertices and over edges take some part of the flow. Thus the sum of capacity constraints of edges for such a cut can be larger than the minimum of the dual problem. Let us remark that it follows from numerical experiments: the assertion of the theorem is correct for an unconnected graph.

The provided solution to the task of finding the pipeline bottlenecks can be used effectively in the optimization of the strategic planning in the reconstruction and development of network objects for the Gazprom gas-transport system, the Transneft oil pipeline system, etc. Also, this solution can be used for the control of flows upon the whole.

References

[1] Beklemishev D V 2008 *Additional chapters of linear algebra: Tutorial. Second edition* (SPb, Moscow, Krasnodar: Lan Press) pp 450-456

[2] Ribnikov K K 2010 *Introduction to discrete mathematics and the theory of extreme problems* (Moscow: Gelios ARV Press) pp 243-247

[3] Gorelik V A and Ushakov I A 1986 *Operations research. Textbook for college with a degree in Applied Mathematics* (Moscow: Mashinostroenie Press) pp 62-65

[4] Cormen T H, Leiserson C E, Rivest R L and Stein C 2008 *Introduction to Algorithms. Third edition* (Cambridge, Massachusetts / London, England: The MIT Press) pp 843-862, 879-885

The heat equation source determination for the case of non-smooth boundary and initial conditions

V V Solovi'ev and D S Tkachenko

National Research Nuclear University MEPhI (Moscow Engineering Physics Institute), 31 Kashirskoe shosse, 115409 Moscow, Russia

E-mail: DSTkachenko@mephi.ru

Abstract. An inverse problem of reconstructing the source of a special kind for parabolic equations in a bounded region with smooth boundary is considered. Solutions are sought in the Hölder classes. We prove an uniqueness criterion for the solution and sufficient conditions of Fredholm property of the task at hand. As a consequence of the sufficient conditions for existence and uniqueness of solution of the inhomogeneous inverse problems are found.

1. The problem statement

Let $T > 0$, $0 < \alpha < 1$ is a fixed constant, $\Omega \subset \mathbb{R}^n$ is a bounded region with boundary of class $C^{2,\alpha}$. Let us consider in the cylinder $\Omega_T = \Omega \times (0, \mathrm{T}]$ the inverse problem of determining a pair of functions (u, f) satisfying the conditions

$$u_t(x,t) = \Delta u(x,t) + f(x)h(x,t) + g(x,t), \qquad (x,t) \in \Omega_T \tag{1}$$

$$u(x,0) = \varphi(x), \quad x \in \overline{\Omega}, \tag{2}$$

$$u(x,t) = \mu(x,t), \quad (x,t) \in \Gamma_T = \partial\Omega \times [0,T], \tag{3}$$

$$u(x,T) = \chi(x), \quad x \in \overline{\Omega}. \tag{4}$$

The inverse problem (1) – (4) has been studied in various functional spaces. The history of the study of such problems are presented in [1], [2], [3], [4]. The solvability of problem (1) – (4) results have been obtained for the case of Hölder spaces $u \in C^{2+\alpha,1+\frac{\alpha}{2}}(\overline{\Omega}_T), f \in C^{\alpha}(\overline{\Omega})$ (see [5]) or in the Sobolev space $u \in W_2^{1,0}(\Omega_T), f \in L_2(\Omega)$ (see [6]). In all previous results it is assumed that μ, φ are smooth functions. In the first case it is assumed that $\mu \in C^{2+\alpha,1+\frac{\alpha}{2}}(\Gamma_T), \varphi \in C^{2+\alpha}(\overline{\Omega})$, in the second one $\mu \in W_2^{2,1}(\Gamma_T)$, $\varphi \in W_2^1(\Omega)$. For the first case, in addition, matching conditions of the first order for functions φ, χ, μ, h and their derivatives on $\partial\Omega$, at t=0 and t=T are required. This paper is studying the problem (1) – (4) assuming that the function $\mu \in C(\Gamma_T)$, $\varphi \in C(\overline{\Omega})$ and the function μ is derived only if $t \in (\bar{t}, T], 0 < \bar{t} < T$.

The direct problem (1) – (3) is studied on a wider class of functions and does not imply smoothness of its solution up to $t = 0$.

As we'll see, this will allow to refuse requirements for the matching conditions of the first order to hold, and thus significantly expands the class of functions φ, g, μ, χ, for which the problem (1) – (4) has a unique solution. Let us give a precise definition.

2. Main results

Denote as $\Omega_{\bar{t}} = \bar{\Omega} \times (\bar{t}, T]$ and $\Gamma_{\bar{t}} = \partial\Omega \times (\bar{t}, T]$. The solution of the problem (1) – (4) is a pair of functions

$$(u, f) \in \left(C_{x,t}^{2,1}(\Omega_T) \cap C(\bar{\Omega}_T) \cap C^{2+\alpha, 1+\frac{\alpha}{2}}(\Omega_{\bar{t}}) \right) \times C^{\alpha}(\bar{\Omega}) = \mathrm{U}(\Omega_T) \times C^{\alpha}(\bar{\Omega}),$$

satisfying conditions (1) – (4).

Following solvability theorem for the direct problem (1) – (3) in this not quite conventional class of functions $\mathrm{U}(\Omega_T)$ holds.

Theorem 1. Let the functions φ, g, μ, f, h lie in classes $g \in C^{\alpha, \frac{\alpha}{2}}(\bar{\Omega}_T), \mu \in C^{2+\alpha, 1+\frac{\alpha}{2}}(\Gamma_{\bar{t}}) \cap C(\Gamma_T), \varphi \in C(\bar{\Omega}),\ f \in C^{\alpha}(\bar{\Omega})$, the matching condition of order zero holds:

$$\varphi(x) = \mu(x, 0), \quad x \in \partial\Omega.$$

Then there exists a unique solution of problem (1) – (3) a function $u \in \mathrm{U}(\Omega_T)$.

The proof is directly obtained by successive application of the results on the solvability of the direct problem from [7], 469 p., [8], 95, [7] p. 260.

Let us first consider the most important for inverse problems topic of the solution of problem (1) – (4) uniqueness in the considered class of functions.

Theorem 2. Let $h, h_t \in C^{\alpha, \frac{\alpha}{2}}(\bar{\Omega}_T)$ and $h(x,t)h_t(x,t) \geq 0$ for $(x,t) \in \bar{\Omega}_T$.

Then the problem (1) – (4) cannot have two different solutions in the class when, and then only when $\operatorname{supp} h(x, T) = \bar{\Omega}$.

The proof of the theorem is based on some modification of the reasoning made in the proof of the solution uniqueness for the inverse problem of the in [5].

Now consider the question of existence of solutions of problem (1) – (4). Let us assume the following definition: we say that matching condition of order zero holds for the problem (1) – (4) if the functions φ, χ, μ satisfy the conditions

$$\varphi(x) = \mu(x, 0), \ \ \chi(x) = \mu(x, T), \ \ x \in \partial\Omega. \qquad (5)$$

Remark. It is clear that the matching conditions are necessary for the existence of solutions of problem (1) – (4) in the class $u \in C(\bar{\Omega}_T)$.

We will prove the following statement clarifying the nature of solvability of problem (1) – (4).

Theorem 3. Let $h, h_t \in C^{\alpha, \frac{\alpha}{2}}(\bar{\Omega}_T), |h(x,T)| \geq h_T > 0,\ x \in \bar{\Omega}$. Then the Fredholm Alternative holds for the inverse problem (1) – (4), which assumes equivalence of the two statements:

1) problem (1) – (4) when $g \equiv 0,\ \mu \equiv 0,\ \chi \equiv 0$ has only trivial solution, i.e. the solution of problem (1) – (4) is unique;
2) problem (1) – (4) has a unique solution $(u, f) \in \mathrm{U}(\Omega_T) \times C^{\alpha}(\bar{\Omega})$ for given functions $(g, \mu, \chi, \varphi) \in C^{\alpha, \frac{\alpha}{2}}(\bar{\Omega}_T) \times \left(C^{2+\alpha, 1+\frac{\alpha}{2}}(\Gamma_T) \cap C(\Gamma_T) \right) \times C^{2+\alpha}(\bar{\Omega}) \times C(\bar{\Omega})$, that satisfy matching conditions of order zero.

Proof. Show first that if condition 1) is true, then condition 2) holds. Let us build this solution. We seek a pair of functions (u, f) in the form

$$(u, f) = \left(u^{(1)}, 0\right) + \left(u^{(2)}, f\right),$$

where the function $u^{(1)} \in \mathrm{U}(\Omega_T)$ is the solution of the direct problem

$$\left(Lu^{(1)}\right)(x,t) = g(x,t), \qquad (x,t) \in \Omega_T, \qquad (6)$$

$$u^{(1)}(x, 0) = \varphi(x), \quad x \in \bar{\Omega}, u^{(1)}(x,t) = \mu(x,t), \quad (x,t) \in \Gamma_T. \qquad (7)$$

The pair $(u^{(2)}, f) \in \mathrm{U}(\Omega_T) \times C^{\alpha}(\bar{\Omega})$ is a solution of the inverse problem:

$$\left(Lu^{(2)}\right)(x,t) = f(x)h(x,t), \qquad (x,t) \in \Omega_T, \qquad (8)$$

$$u^{(2)}(x, 0) = 0, \quad x \in \bar{\Omega}, \qquad u^{(2)}(x,t) = 0, \quad (x,t) \in \Gamma_T, \qquad (9)$$

$$u^{(2)}(x, T) = \chi(x) - u^{(1)}(x, 0) = \tilde{\chi}(x), \quad x \in \bar{\Omega}. \qquad (10)$$

Because of the matching conditions and known results (see [7], 461, [7], p. 260) on the solvability of the problems for the heat equation the function $u^{(1)}(x,t)$ must lie in the class $\mathrm{U}(\Omega_T)$. Then $\tilde{\chi} \in$

$C^{2+\alpha}(\overline{\Omega})$. Thus to prove that assertion 1) implies assertion 2) it is sufficient to prove the existence of solutions of problem (8) – (10) under the assumption that $\tilde{\chi} \in C^{2+\alpha}(\overline{\Omega})$, $\tilde{\chi}(x) = 0,\ x \in \partial\Omega$. For this, we consider the problem of determining the three functions (u, v, f) from the conditions:

$$(Lu)(x,t) = f(x)h(x,t), \quad (x,t) \in \Omega_T \tag{11}$$

$$u(x,0) = 0, \quad x \in \overline{\Omega}, \qquad u(x,t) = 0, \quad (x,t) \in \Gamma_T, \tag{12}$$

$$(Lv)(x,t) = f(x)h_t(x,t), \quad (x,t) \in \Omega_T \tag{13}$$

$$v(x,0) = f(x)h(x,0), \ x \in \overline{\Omega}, \qquad v(x,t) = 0, \quad (x,t) \in \Gamma_T, \tag{14}$$

$$f(x) = \frac{1}{h(x,T)}v(x,T) - \frac{1}{h(x,T)}\Delta\tilde{\chi}, \quad x \in \overline{\Omega}. \tag{15}$$

Under the conditions (11) – (12) the function u will lie in the class $\mathrm{U}(\Omega)$, v is generalized solution (see [11]) of the direct problem (13) – (14), i.e., v is a bounded function from the class $v \in C^{2,1}_{x,t}(\Omega_T) \cap C(\overline{\Omega}_T \setminus (\partial\Omega \times \{t = 0\}))$. The solution of this problem, as shown in [11], can be produced with the Green's function. Define a linear operator A: $C^\alpha(\Omega) \longmapsto C^\alpha(\Omega)$, by the rule:

$$A(f)(x) = \frac{1}{h(x,T)}v(x,T), \quad x \in \overline{\Omega}. \tag{16}$$

Consider a function $v(x,t)$ in the form:

$$v(x,t) = v^{(1)}(x,t) + v^{(2)}(x,t),$$

where $v^{(1)} \in C^{2+\alpha,1+\alpha/2}_{x,t}(\overline{\Omega}_T)$ is the solution of the problem

$$\left(Lv^{(1)}\right)(x,t) = f(x)h_t(x,t), \quad (x,t) \in \Omega_T \tag{17}$$

$$v^{(1)}(x,0) = 0, \ x \in \overline{\Omega}, \qquad v^{(1)}(x,t) = 0, \quad (x,t) \in \Gamma_T, \tag{18}$$

function $v^{(2)}$ is generalized solution of the problem

$$\left(Lv^{(2)}\right)(x,t) = 0, \quad (x,t) \in \Omega_T, \tag{19}$$

$$v^{(2)}(x,0) = f(x)h(x,0), \ x \in \Omega, v^{(2)}(x,t) = 0, \ (x,t) \in \Gamma_T \setminus (\partial\Omega \times \{t = 0\}). \tag{20}$$

Note that for the function $v^{(1)}(x,t)$ Schauder estimation holds (see [8], p. 88):

$$\left\|v^{(1)}\right\|^{(2+\alpha,1+\alpha/2)}_{\Omega_T} \le c_1 \|f\, h_t\|^{(\alpha,\alpha/2)}_{\Omega_T}. \tag{21}$$

But then, by definition of the Hölder norm for the function $v^{(1)}$ (see, e.g., [7], p. 16), we obtain the estimate:

$$\left\|v^{(1)}_{x_i}(x,T)\right\| \le c_2 \|f\|^{(\alpha)}_{\Omega}, \tag{22}$$

the constant c_2 here does not depend on the function f.

Let us use the evaluation of the green's function (see [7], p. 469) for the function $v^{(2)}$:

$$\left|G_{x_i}(x,\xi;T,0)\right| \le c_3 T^{-1/2} \exp\left\{-c_4 \frac{|x-\xi|^2}{T}\right\}.$$

Then $v^{(2)}(x,T)$ well satisfy the following estimate:

$$\left|v^{(2)}_{x_i}(x,T)\right| \le c_4 \|fh(x,0)\| \le c_5 \|f\|^{(\alpha)}_{\Omega}.$$

From this, operator A can be estimated this way:

$$\left|A(f)_{x_i}(x)\right| \le c_6 \|f\|^{(\alpha)}_{\Omega}.$$

And now we get the final estimate:

$$\left\|A(f)_{x_i}\right\| = \max_{x\in\Omega}\left|A(f)_{x_i}(x)\right| \le c_6 \|f\|^{(\alpha)}_{\Omega}. \tag{23}$$

Since the space $C^1(\Omega)$ is compactly embedded into the space $C^\alpha(\Omega)$ (see, e.g., [11]), the estimate (23) implies that equation (15) is a Fredholm equation of the second kind with a compact operator. But for such equation from the Riesz theorem (see, e.g., [12], p. 212) we get that the uniqueness of the solution of the homogeneous equation (15) implies the existence of solutions of the equation (15) for any function $\frac{L\chi}{h(x,T)}$. But the uniqueness of the solution of the inverse problem (8) – (10) implies the uniqueness of the solution of the homogeneous equation (15). It is clear that the found function f and the solution of the direct problem (11) – (12) satisfy all the equalities that define the inverse problem (1) – (4) (the proof is standard and is given in e.g. [5]). Thus, it is proven that the uniqueness of the solution of the inverse problem (5) – (7) implies that this solution exists.

Prove that, other way, if the inverse problem (1) – (4) has a unique solution for any four functions, then by setting $g \equiv 0,\ \mu \equiv 0,\ \varphi \equiv 0$, we obtain that the inverse problem (8) – (10) has a solution for any function $\chi \in C^{2+\alpha}(\bar{\Omega})$. But then (15) has a solution for any function $\Delta\tilde{\chi}$, as an equation of the second kind with a compact operator (16). From the Riesz theorem for these equations we find that the homogeneous equation (15) has only trivial solution. But then the inverse problem (1) – (4) has only trivial solution either.

Theorem 2 is completely proved.

3. The corollaries

Corollary 1. Let $h, h_t \in C^{\alpha,\frac{\alpha}{2}}(\bar{\Omega}_T), |h(x,T)| \geq h_T > 0,\ x \in \bar{\Omega},\ h(x,t)h_t(x,t) \geq 0, (x,t) \in \Omega_T$. Then for any four functions $g,\ \mu,\ \varphi,\ \chi$, such that $g \in C^{\alpha,\frac{\alpha}{2}}(\bar{\Omega}_T)$, $\mu \in C^{2+\alpha,1+\frac{\alpha}{2}}(\Gamma_{\bar{t}}) \cap C(\Gamma_T)$, $\varphi \in C(\bar{\Omega})$, $\chi \in C^{2+\alpha}(\bar{\Omega})$ satisfying the matching conditions:

$$\varphi(x) = \mu(x,0),\quad \chi(x) = \mu(x,T),\quad x \in \partial\Omega,$$

exists the solution of problem (1) – (4) in the class considered and it is unique.

The proof follows from theorem 1 and theorem 2.

Corollary 2. Let $h, h_t \in C^{\alpha,\frac{\alpha}{2}}(\bar{\Omega}_T), |h(x,T)| \geq h_T > 0,\ x \in \bar{\Omega}, l = \mathrm{diam}\,\Omega$. Then there exists a number l_0 such that for any domains $\Omega, \mathrm{diam}\Omega < l_0$ inverse problem (1) – (4) has a unique solution for any four functions $g,\ \mu,\ \varphi,\ \chi$ from the class considered satisfying the matching conditions.

The proof of uniqueness of the solution of the inverse problem (1) – (4) follows from the estimates for Green function (see [7], p. 169). A detailed proof of some similar propositions see in [5]. The existence follows from theorem 2.

Choose one of the axes $x = x_i,\ i \in \{1,\cdots,n\}$, and define the parameters

$$l_i = \sup_{(x,y)\in\Omega}|x_i - y_i| = \mathrm{diam}_i\Omega.$$

Then the following statement is true.

Corollary 3. Let $h, h_t \in C^{\alpha,\frac{\alpha}{2}}(\bar{\Omega}_T), |h(x,T)| \geq h_T > 0,\ x \in \bar{\Omega}$. Then exists a number l_0^ithat for any Ω of class $\partial\Omega \in C^{2,\alpha}$, $\mathrm{diam}_i\Omega < l_0^i$ inverse problem (1) – (4) has a unique solution for any four functions $g,\ \mu,\ \varphi,\ \chi$ from the class considered satisfying the matching conditions.

The proof of uniqueness of the solution of the inverse problem (1) – (4) in the conditions of the Corollary 3 is in detail carried out in [5].

Existence again follows from theorem 2.

Acknowledgements

In conclusion, the authors express deep gratitude to all participants of the workshop on inverse problems under the guidance of V. A. Sadovnichii and A. I. Prilepko, Mechanics and Mathematics department, MSU for their active and useful discussion the above results.

References

[1] Denisov A M 1994 *Introduction into the inverse problems theory* (Moscow: Moscow State University Press) p 207

[2] Lavrent'ev M M 1973 *Conditionally well-posed problems for differential equations* (Novosibirsk: Novosibirsk State University Press) p 71
[3] Isakov V 2006 *Inverse problems for partial differential equations* (New York: Springer) p 344
[4] Prilepko A I, Orlovsky D G, Vasin I A 2000 *Methods for solving inverse problems in mathematical physics* (New York – Basel: Marcel Dekker Inc.) p 723
[5] Soloviev V V 1989 *Differential equations* On solvability of source restoration problem with final overdetermination for a parabolic equation **25** 1577-83
[6] Prilepko A I, Kostin A B 1993 *Sib. Mathem. Journal* **34** 147 –162
[7] Ladyjenskaya O A, Solonnikov V A, Ural'ceva N N 1967 *Linear and quasilinear equations of parabolic kind* (Moscow: Nauka) p 736
[8] Fridman A 1968 *Equations with partial derivatives of parabolic kind* (Moscow: Mir) p 428
[9] Mihailov V P 1983 *Differential equations in partial derivatives* (Moscow: Nauka) p 391
[10] Gilbarg D, Trudinger N 1989 *Elliptic Differential equations with partial derivatives of the second order* (Moscow: Nauka) p 465
[11] Landis E M 1971 *Equations of the second order of elliptic and parabolic kinds* (Moscow: Nauka) p 288
[12] Trenogin V A 2002 *Functional analysis* (Moscow: Fizmatlit) p 488

A Sufficient Condition for the Harmonicity of Functions of Two Variables That Satisfy a Difference Laplace Equation

D S Telyakovskii

National Research Nuclear University MEPhI (Moscow Engineering Physics Institute), 31 Kashirskoe shosse, 115409 Moscow, Russia

E-mail: dtelyakov@mail.ru

Abstract. In this paper we weaken the sufficient conditions of harmonicity for functions of two variables. I.I. Privalov had shown that a continuous function that satisfies the Laplace equation in each point of the domain is harmonic. For function of two variables the Privalov's condition on continuity can be weakened. G.P. Tolstov replaced the continuity condition by the boundness condition, later the author had shown that summability is sufficient. At the same time summability condition can not be weakened substantially.

In this paper, while we keep the summability condition, we provide the sufficient condition for harmonicity of the functions, that satisfy less restricted condition than the Laplace equation in all points of the domain. We assume that arbitrary close to any point ζ there exists a collection of four nodes for which a difference relation of Schwartz type for the Laplace equation can be made arbitrary small by the absolute value. Nodes are the ends of two mutually perpendicular segments, that intersect at point ζ. We need to impose a certain weakened continuous assumption on function itself, in case of function that satisfy the traditional Laplace condition this continuity condition follows from the existence of the partial derivatives.

Analytical function is called harmonic in the domain of the function, if it satisfies the Laplace equation in each point of the domain. If function satisfies the Laplace equation, then to obtain that it is harmonic we do not need to assume its analyticity, we can impose substantially less restrictive conditions on the function. At the same time there are examples that show that if a function only satisfies the Laplace condition or some of kind its generalization, at all points of the domain, it is not sufficient for the harmonicity of the function. Some additional conditions are required.

I.I. Privalov (1925, [1]) have shown that the continuous functions that satisfy in each point of the domain the Laplace equation are harmonic. In Privalov's theorem the number of variables is arbitrary, only those partial derivatives assumed to exist that are included in the Laplace equation, and the derivatives can be considered in the sense of Schwartz. The continuity condition can be substantially weakened in case of the functions of two variables.

Let function $u(z) = u(x,y)$, $z \in G \subset \mathbb{R}^2$, satisfies the Laplace equation in each point of the domain G. G.P. Tolstov (1951, [2]) have shown that for harmonicity of $u(z)$ in G boundness of $u(z)$ is sufficient; next the author (1986, [3]) have shown that it is sufficient to assume the summability of $u(z)$ with respect to the planar Lebesgue measure.

Function

$$u(z) := \Re\Big(\frac{i}{z^2}\Big) = \frac{2xy}{(x^2+y^2)^2}, \ z \neq 0, \quad u(0) := 0,$$

satisfies the Laplace equation at every point of the unit disk D, it is harmonic inside of D except for $z = 0$ and the absolute value of $|u(z)|$ is summable for all positive powers less than 1. From this example we can see that in the sufficient condition for the harmonicity we can not relax the summability assumption of $|u(z)|$ in any substantial way. That's why while keeping the summability condition for $|u(z)|$, in order to relax the sufficient condition for harmonicity we replace the requirement that the Laplace equation holds at every point of the domain by a less strict condition.

Thus this author (2009, [4]) considered functions that satisfy the generalized Laplace equation in region G, where instead of the traditional second derivatives we take the second derivatives in the sense of Peano. Expression $\Delta^{\times} u$ at point $\zeta \in G$ equals to the sum of the second Peano derivatives of function $u(z)$ in point ζ, that are taken along some mutually perpendicular lines that intersect at point ζ. The value of $\Delta^{\times} u$ in general can depend on choice of two lines along which the derivatives are taken. In [4] it was shown that if the function $u(z)$ is summable in region G and for every point $\zeta \in G$ there exists a pair of perpendicular to each other lines that intersect in ζ for which $\Delta^{\times} u = 0$, then function $u(z)$ is harmonic in G. The condition of mutual perpendicularity of lines in the pair can not be removed, while the directions of lines for different points of the region are not related to each other.

In work [4] first was established the harmonicity of continuous functions that satisfy the equation $\Delta^{\times} u = 0$, next was proved the continuity of summable functions that satisfy this condition. The conditions on derivations that appear in the equation $\Delta^{\times} u = 0$ are only used to prove the harmonicity of the continuous functions. In the proof of continuity of summable functions in G we do not require the existence of derivatives along certain directions for function $u(z)$ in each point $\zeta \in G$, moreover the existence of the second derivatives, but we only need the Lipschitz condition to hold along those direction for point ζ. The fact that the Lipschitz condition holds, which in this case is used as a continuity condition, is clear from the existence of directional derivatives for $u(z)$ at point ζ

In this paper the conditions of continuity and differentiability are imposed separately on the function. The differentiability condition is imposed not on the sum of second derivatives of the function along certain two mutually perpendicular directions, but on a sum of difference relations of Schwartz type for the second derivatives of function $u(z)$, that are taken along such directions in non-equally spaced nodes in general. We do not require the existence of any kind of derivatives. Also we impose some condition on continuity of the function by direction, this condition is less strict than the Lipschitz condition, but more strict than the ordinary continuity.

The second Schwartz partial derivatives of the function

$$u(x,y) := \begin{cases} 1 & \text{for } x > y, \\ 0 & \text{for } x = y, \\ -1 & \text{for } x < y, \end{cases}$$

are equal to zero in every point. This example shows, that if the derivatives appearing in Laplace equation are Schwartz derivatives, then without any continuity condition we can not obtain a sufficient condition for the harmonicity even for the bounded functions.

Let us introduce some definitions.

We define as *a right angle cross* with center at the point ζ the union of two mutually perpendicular intervals $a_1(\zeta)$ and $a_2(\zeta)$ that intersect in the middle at the point ζ.

Now we obtain the expression for difference relation of Schwartz type for the second derivative of the function of one variable for unequally spaced nodes.

Difference relation for the Schwartz second derivative of function $u(x)$ in point x for nodes $x-h$ and $x+h$ is equal to the twice the value of the coefficient in x^2 term in the equation for parabola $y = ax^2+bx+c$ that passes through points $(x-h; u(x-h))$, $(x; u(x))$ and $(x+h; u(x+h))$. So analogously to it we call as a difference relation $\delta^{(2)}u(x_0; x_1, x_2)$ for the second Schwartz derivative of function $u(x)$ at point x_0 with nodes x_1 and x_2 twice the value of coefficient in x^2 term in the equation for parabola that passes through points $(x_1; u(x_1))$, $(x_0; u(x_0))$ and $(x_2; u(x_2))$. We write equation for this parabola using the Lagrange interpolation polynomial. To simplify the notation we denote as $u_1{:=}u(x_1)$, $u_0{:=}u(x_0)$, $u_2{:=}u(x_2)$. We have

$$ax^2+bx+c = \frac{(x-x_0)(x-x_2)}{(x_1-x_0)(x_1-x_2)}u_1 + \frac{(x-x_1)(x-x_2)}{(x_0-x_1)(x_0-x_2)}u_0 + \frac{(x-x_0)(x-x_1)}{(x_2-x_0)(x_2-x_1)}u_2.$$

From the above follows

$$\begin{aligned}\delta^{(2)}u(x_0; x_1, x_2){:=}2a{=}2\left(\frac{u_1}{(x_1-x_0)(x_1-x_2)} + \frac{u_0}{(x_0-x_1)(x_0-x_2)} + \frac{u_2}{(x_2-x_0)(x_2-x_1)}\right){=}\\ = 2\frac{u_1(x_2-x_0) - u_0(x_2-x_1) + u_2(x_0-x_1)}{(x_0-x_1)(x_2-x_1)(x_2-x_0)}.\end{aligned} \quad (1)$$

For equally spaced nodes $x_1 = x_0 - h$, x_0 and $x_2 = x_0 + h$ the expression $\delta^{(2)}u(x_0; x_1, x_2)$ is the difference relation for a regular Schwartz second derivative of the function $u(x)$ at point x_0. In fact,

$$\delta^{(2)}u(x_0;x_1,x_2) = 2\frac{u_1(x_2-x_0)-u_0(x_2-x_1)+u_2(x_0-x_1)}{(x_0-x_1)(x_2-x_1)(x_2-x_0)} = 2\frac{hu_1-2hu_0+hu_2}{2h^3} = \frac{u_1-2u_0+u_2}{h^2}.$$

Equation for the parabola that passes through three arbitrary points does not depend on the order in which these points are enumerated. We enumerate nodes in a more convenient way for us. Since the difference relation $\delta^{(2)}u(x_0; x_1, x_2)$ is considered as a difference relation for the second derivative at point x_0, it is natural to assume that x_0 is located between x_1 and x_2. If $x_1 < x_0 < x_2$, then all differences between coordinates of the nodes in formula (1) are positive and are equal to the distances between the corresponding nodes. So if we denote as $r_1 := |x_1-x_0|$ and $r_2 := |x_2-x_0|$, then the expression for $\delta^{(2)}u(x_0; x_1, x_2)$ can be rewritten as:

$$\delta^{(2)}u(x_0;x_1,x_2) = 2\frac{u_1r_2 - u_0(r_1+r_2) + u_2r_1}{r_1r_2(r_1+r_2)}. \quad (2)$$

It is easy to show that the difference relation $\delta^{(2)}u(x_0; x_1, x_2)$ will give in the limit the second Peano derivative of function $u(x)$ at the point x_0, if that derivative exists.

Now we obtain the expression for the difference Laplace equation that we analyze in this paper. Let function $u(z)$ be defined in region $G \subset \mathbb{R}^2$, take a point $z_0 \in G$ and a set $A_{z_0} = \{z_1, z_2, z_3, z_4\}$ of four points in the region G for which segments $[z_1;z_2]$ and $[z_3;z_4]$ are mutually perpendicular and intersect at the internal point z_0. Denote as $r_j := |z_j - z_0|$, $u_j := u(z_j)$, $j = 1,\ldots,4$. Consider

$$\Delta^{(*)}u(z_0) = \Delta^{(*)}u(z_0;A_{z_0}) := \delta^{(2)}_{z_1z_2}u(z_0; z_1, z_2) + \delta^{(2)}_{z_3z_4}u(z_0; z_3, z_4),$$

where the quantities $\delta^{(2)}_{z_1z_2}u(z_0; z_1, z_2)$ and $\delta^{(2)}_{z_3z_4}u(z_0; z_3, z_4)$ are defined by formula (2). We obtain the expression

$$\Delta^{(*)}u(z_0) = 2\frac{u_1r_2 - u_0(r_1+r_2) + u_2r_1}{r_1r_2(r_1+r_2)} + 2\frac{u_3r_4 - u_0(r_3+r_4) + u_4r_3}{r_3r_4(r_3+r_4)}. \quad (3)$$

IOP Conf. Series: Journal of Physics: Conf. Series **788** (2017) 012059 doi:10.1088/1742-6596/788/1/012059

We say that the function $u(z)$ satisfies *the generalized difference Laplace equation* $\Delta^{(*)}u = 0$ at point z_0, if the limit of $\Delta^{(*)}u(z_0; A_{z_0}^{(n)})$ is equal to zero at least for one sequence of sets of nodes $\{A_{z_0}^{(n)}\}$, that contract to z_0. For different sequences of sets of nodes $\{A_{z_0}^{(n)}\}$, that contract to z_0, the limits of $\Delta^{(*)}u(z_0; A_{z_0}^{(n)})$ could be different or not exist.

Now we define the continuity condition that we impose on function $u(z)$ in points of the region G.

Let $h(t)$, $t \geqslant 0$, be a modulus of continuity type function and let function $u(z)$ be defined on some set B_ζ, $z \in B_\zeta$, where point ζ is the limiting point. If for some value $L_\zeta > 0$ at every point $z \in B_\zeta$ the following inequality holds $|u(z)-u(\zeta)| \leqslant L_\zeta h(|z-\zeta|)$, then we say that function $u(z)$ is *h-regular at point ζ with respect to set B_ζ with coefficient* L_ζ, or simply *h-regular at ζ with respect to B_ζ*, if specific value of L_ζ is not important.

Clearly if function $u(z)$ is differentiable at point ζ along B_ζ as a function of two real arguments, then for some $L_\zeta > 0$ in all points $z \in B_\zeta$ the inequality $|u(z) - u(\zeta)| \leqslant L_\zeta|z - \zeta|$ holds. It means that function $u(z)$ is h-regular at point ζ with respect to B_ζ for a linear function $h(t)$.

In points $\zeta \in G$ as a continuity condition we consider h-regularity condition with respect to some right angle cross T_ζ or some star S_ζ.

Theorem 1. *Let function $u(z)$ be locally summable in region $G \subset \mathbb{R}^2$, while $h(t)$, $t \geqslant 0$ is some function of modulus of continuity type. If for each point $\zeta \in G$ the function $u(z)$ satisfies the generalized difference Laplace equation and is h-regular with respect to some right-angle cross T_ζ, then function $u(z)$ is harmonic in region G.*

There is no connection between the locations of the nodes for different sets $A_\zeta^{(n)}$ for a given point $\zeta \in G$. Similarly we do not assume any connection between sets of nodes for different points of region G. The orientation and the length of the intervals that make up the crosses T_ζ for different points $\zeta \in G$ do not depend from each other.

In this paper we prove of a special case of the Theorem 1.

Theorem 2. *If function $u(z)$ is continuous in a region G and satisfies the generalized difference Laplace equation at each point $z \in G$, then function $u(z)$ is harmonic in the region G.*

Proof of Theorem 2 follows the approach similar to the Theorem 2, I.I. Privalov [1]. Proof of the Privalov's theorem is also provided in the footnote in the work of I.G. Petrovskii [5], pg. 107.

In the proof of the Theorem 2 we use the following Lemma. We do not give its proof since it amounts to the immediate check.

Lemma 1. *Assume that function $u(z)$ has the second Peano differential at point $z_0 = (x_0, y_0)$*

$$u(z) = u(z_0) + du + \frac{1}{2}d^2u + \alpha(z)|z - z_0|^2, \tag{4}$$

where $du = p(x-x_0)+q(y-y_0)$ and $d^2u = r(x-x_0)^2+2s(x-x_0)(y-y_0)+t(y-y_0)^2$, while $\alpha(z) \to 0$ when $z \to z_0$. Then for any sequence of sets of nodes $\{A_{z_0}^{(n)}\}$ that contract to z_0

$$\lim_{n\to\infty} \Delta^{(*)}u(z_0, A_{z_0}^{(n)}) = r + t.$$

Corollary 1.1. *If the function is harmonic in the region, then in every point of the region it satisfies the generalized difference Laplace equation for any sequence of sets of nodes that contract to that point.*

Corollary 1.2. *For the polynomial of the second degree $u(x) = ax^2 + bx + c$ for any collection of nodes $\{x_0, x_1, x_2\}$ the expression $\delta^{(2)}u(x_0; x_1, x_2)$ is exactly equal to $2a$. That why, if function $u(z) = a_{11}x^2+2a_{12}xy+a_{22}y^2+a_1x+a_2y+a_0$ is a polynomial of the second degree in two variables, then for any point z and any set of nodes A_z we have the identity $\Delta^{(*)}u(z, A_z) \equiv 2(a_{11} + a_{22})$.*

IOP Conf. Series: Journal of Physics: Conf. Series **788** (2017) 012059 doi:10.1088/1742-6596/788/1/012059

Proof of the Theorem 2. Proof of the Theorem 2. Assume that the statement is not true and that the function $u(z)$ is not harmonic everywhere in G. We take a disk $D \subset G$ inside of which $u(z)$ is not harmonic everywhere. Let $v(z)$ be a function that is harmonic inside of the disk D and coincides with the continuous function $u(z)$ on ∂D. Then from the Corollary 1.1 to Lemma follows that function $w(z) := u(z) - v(z)$ satisfies the generalized difference Laplace equation everywhere inside of D.

By the definition $w(z) \equiv 0$ on ∂D, but this function is not identically equal to zero inside of D. Assume that $\max_D w(z) > 0$, if not then instead of $w(z)$ we consider the function $-w(z)$. Let $z_0 = (x_0,y_0)$ be a point of maximum $w(z)$ in disk D: $\max_D w(z) = w(z_0) > 0$, and this point z_0 belongs to interior of D. Denote as

$$\varphi(z) := w(z) + \frac{w(z_0)}{2\operatorname{diam}^2 D}\left((x-x_0)^2 + (y-y_0)^2\right), \quad z \in D.$$

Since the function $\varphi(z)$ is equal to the sum of function $w(z)$ and polynomial of the second degree $\dfrac{w(z_0)}{2\operatorname{diam}^2 D}\left((x-x_0)^2+(y-y_0)^2\right)$, it means that from the Corollary 1.2 to Lemma, we obtain that for any point ζ from the interior of disk D and any set of nodes $A_\zeta \subset D$:

$$\Delta^{*}\varphi(\zeta, A_\zeta) = \Delta^{(*)}w(\zeta, A_\zeta) + \frac{w(z_0)}{\operatorname{diam}^2 D}. \tag{5}$$

By the definition of $\varphi(z)$ everywhere inside of D we have $\varphi(z) \geqslant w(z)$, therefore

$$\max_D \varphi(z) \geqslant \max_D w(z) = w(z_0) > 0.$$

Since all the points of ∂D are closer to z_0 (the interior point of D) than the value of $\operatorname{diam} D$ and on ∂D the function $w(z) \equiv 0$, it follows from the definition of $\varphi(z)$ that on ∂D we have

$$\varphi(z) = w(z) + \frac{w(z_0)}{2\operatorname{diam}^2 D}\left((x-x_0)^2+(y-y_0)^2\right) = \frac{w(z_0)}{2\operatorname{diam}^2 D}\left((x-x_0)^2+(y-y_0)^2\right) \leqslant \frac{w(z_0)}{2}.$$

This is the reason why $\varphi(z)$ takes the maximum value in the interior of disk D. Let ζ_0 be the point of maximum: $\max_D \varphi(z) = \varphi(\zeta_0)$.

For each interior point ζ of disk D there exists a set of nodes A_ζ for which the absolute value of $\Delta^{(*)}w(\zeta, A_\zeta)$ can be made arbitrary small. Let $A_{\zeta_0} = \{\zeta_1,\zeta_2,\zeta_3,\zeta_4\} \subset D$ be the set of nodes for which $\left|\Delta^{(*)}w(\zeta_0, A_{\zeta_0})\right| < \dfrac{w(z_0)}{2\operatorname{diam}^2 D}$. According to the expression (5) for this set the following inequality holds

$$\Delta^{(*)}\varphi(\zeta_0, A_{\zeta_0}) = \Delta^{(*)}w(\zeta_0, A_{\zeta_0}) + \frac{w(z_0)}{\operatorname{diam}^2 D} > \frac{w(z_0)}{2\operatorname{diam}^2 D} > 0.$$

Assuming that $\varphi_j := \varphi(\zeta_j)$ and $r_j := |\zeta_j - \zeta_0|$ for $j = 1, 2, 3, 4$, from formula (3) we obtain that

$$\Delta^{(*)}\varphi(\zeta_0, A_{\zeta_0}) = 2\frac{\varphi_1 r_2 - \varphi_0(r_1+r_2) + \varphi_2 r_1}{r_1 r_2(r_1+r_2)} + 2\frac{\varphi_3 r_4 - \varphi_0(r_3+r_4) + \varphi_4 r_3}{r_3 r_4(r_3+r_4)} > 0.$$

At least one of the terms in $\Delta^{(*)}\varphi(\zeta_0)$ is positive. For example, assume that it is the first term. We can rewrite the first term as:

$$\frac{\varphi_1 r_2 - \varphi_0(r_1+r_2) + \varphi_2 r_1}{r_1 r_2(r_1+r_2)} = \frac{r_2(\varphi_1-\varphi_0) + r_1(\varphi_2-\varphi_0)}{r_1 r_2(r_1+r_2)} > 0.$$

Values of r_1 and r_2 are positive, since they are distances. So at least one of the differences $\varphi(\zeta_1) - \varphi(\zeta_0)$ or $\varphi(\zeta_2) - \varphi(\zeta_0)$ is positive, but it contradicts to the statement that at the point ζ_0 the function $\varphi(z)$ reaches the maximum value. Therefore our assumption, that the function $u(z)$ is not harmonic everywhere inside of the region G is not true. It finishes the proof of the Theorem 2.

The author thanks I.V. Tikhonov for discussion and useful advice.

References

[1] Priwaloff I Sur les fonctions harmoniques. — Mat. Sb., 1925, 23:3, 464–471.
[2] Tolstov G P On bounded functions satisfying Laplace's equation, Mat. Sb., 1951, 29:3, 559–564 [in Russian].
[3] Telyakovskii D S Generalization of the Looman–Men'shov theorem. Mathematical Notes, 1986, 39:4, 296–301.
[4] Telyakovskii D S A sufficient condition for the harmonicity of functions of two variables. Mathematical Notes, 2009, 86:4, 591–601.
[5] Petrovskii I G Perron's method for solving the Dirichlet problem, Uspekhi Mat. Nauk, 1941, 8, 107–114 [Selected Works, Gordon and Breach Publishers, Amsterdam, 1996, Vol. 2, 20–29].

On a distinctive feature of problems of calculating time-average characteristics of nuclear reactor optimal control sets

A V Trifonenkov and V P Trifonenkov

National Research Nuclear University MEPhI (Moscow Engineering Physics Institute), 31 Kashirskoe shosse, 115409 Moscow, Russia

E-mail: AVTrifonenkov@mephi.ru

Abstract. This article deals with a feature of problems of calculating time-average characteristics of nuclear reactor optimal control sets. The operation of a nuclear reactor during threatened period is considered. The optimal control search problem is analysed. The xenon poisoning causes limitations on the variety of statements of the problem of calculating time-average characteristics of a set of optimal reactor power off controls. The level of xenon poisoning is limited. There is a problem of choosing an appropriate segment of the time axis to ensure that optimal control problem is consistent. Two procedures of estimation of the duration of this segment are considered. Two estimations as functions of the xenon limitation were plot. Boundaries of the interval of averaging are defined more precisely.

1. Introduction

The present advancement of nuclear power allows building nuclear power plants in areas of unstable climatic conditions or with an increased risk of seismic activity, as well as regions of political instability and terroristic threats [1, 2].

Such power plants may undergo force majeure circumstances, which cause abrupt shutdown of nuclear reactors and their consequent idleness. The consumers of energy produced by the reactor may suffer damages induced by an unexpected shutdown and outage of the chief source of power.

The time when force majeure situation occurs can not be predicted precisely. There is usually a threatened period, which is declared beforehand, given by its duration and probability distribution for the event time. When the threatened event comes, nuclear reactors are shut down. Shutdowns at certain states of a reactor may cause long-term outages due to very high levels of xenon poisoning [2]. Taking the above into consideration, the problem of the search of the best reactor power control set during the threatened period is analyzed in this article.

Consider the time interval between the threatened period's start and the time of the reactor shutdown. The problem is to find the optimal control function, defined on this interval, which satisfies the conditions and minimizes the cost functional.

The cost functional is defined by some parameter, which depends on the state of the system and the control function. For example, the power generation loss before the reactor shutdown, as well as the reactivity margin or the concentration of xenon-135 in the reactor core at the time of the shutdown, etc. may be used as such parameters.

As the reactor control is impossible in case of exceeding the marginal xenon-135 concentration level inside the reactor core [3, 4], the limitation of xenon-135 concentration before the reactor shutdown is introduced. Also, as it was mentioned above, the reactor shutdown at some set of states may cause a long-term reactor outage. The long-term reactor outage due to xenon poisoning may cause damage to the infrastructure and industry. That is why the condition for the values of both iodine-135 and xenon-135 at the end of process is introduced in order to limit the concentration of xenon-135 after the reactor shutdown. Finally, it is stated, that after the reactor shutdown occurs, there is an additional time span, reserved for the elimination of the accident's consequences, when the reactor can't operate.

Thus, the optimization problem statement conforms to the problem of optimal control on a limited interval with the limitation of xenon-135 concentration before shutdown and both iodine-135 and xenon-135 at the time of shutdown.

The threatened event time is random; it defines the duration of the process and the control function, which corresponds to the duration value. In practice, for the problem stated the optimal process can't be determined unless the period of control is given. Nevertheless, if the optimal control search problem is resolved in advance for any process duration value, the functions derived and their characteristics may be useful for applied control modes development.

However, there is a set of process duration values, for which the solution to the problem given doesn't exist.

2. Statement and proof

That is why issues occur, that are related to problems, which require existence of optimal control for any value of process duration. Problems of calculating time-averaged characteristics of optimal control are of that type. In the time axis, originating simultaneously with the start of the reactor control process, there exists the time interval, containing instants, when the reactor shutdown doesn't satisfy the conditions. Further this interval is denoted by Δ_τ. The origin of this interval coincides with the origin of the time axis, the ending is defined by the parameters of the system and the limitations stated.

One-group point model of a nuclear reactor [5] was used in the research. The process of xenon poisoning is described by the system of differential equations [6]

$$\begin{cases} \dfrac{\mathrm{d}x(t)}{\mathrm{d}t} = \lambda_1 i(t) - \lambda_X x(t) - \sigma_X \varphi x(t) \\ \dfrac{\mathrm{d}i(t)}{\mathrm{d}t} = \sigma_X \varphi - \lambda_1 i(t) \end{cases} \quad (1)$$

Then for the case of the nuclear reactor after having been shut down, the system takes the following form

$$\begin{cases} \dfrac{\mathrm{d}x(t)}{\mathrm{d}t} = \lambda_1 i(t) - \lambda_X x(t) \\ \dfrac{\mathrm{d}i(t)}{\mathrm{d}t} = -\lambda_1 i(t) \end{cases} \quad (2)$$

Two ways of estimating the length of the interval were considered. The first way corresponds with the lower estimation.

Let t_0 be the origin of the threatened period and t_1 be the instant of the reactor shutdown. In these terms, if $t > t_1$ then $\varphi(t) \equiv 0$ and the transients are described by (2).

Let Φ be the set of admissible control functions $\varphi(t)$, such that $0 \le \varphi(t) \le \varphi_{\max}$, where $t_0 \le t \le t_1$ is the time span of the control process. The set of states of the model (1) is represented by the quadrant $Q = \{(i, x) \mid i \ge 0, x \ge 0\}$ of the phase plane Oix.

Let A be the set of probable model states at time τ, given $i(t_0) = i_0$, $x(t_0) = x_0$ for any $\varphi(t) \in \Phi$. Let B be the set of acceptable states at $t = t_1$ that is the end of the control process. In these terms, if

$A \cap B = \{\varnothing\}$ at some instant τ, then, if $t_1 = \tau$, the problem is inconsistent. This is to be proved by the following.

To investigate the properties of the set B, consider the function $F(i,x)$ over the quadrant Q; the function is defined by (2) solution properties as follows. Let $i(t)$, $x(t)$ be the solution of (2) given initial conditions $i(t_1) = i_1, x(t_1) = x_1$, where $(i_1, x_1) \in Q$ and $t \geq t_1$. By construction, $F(i_1, x_1) = \max\limits_{t \in [t_1, +\infty)} x(t)$. Consequently, $B = \{(i,x) \in Q \mid F(i,x) \leq x_{\text{lim}}\}$.

The equation describing the phase curve for the system (2) has the form:

$$x = \frac{\lambda_1 i_{\text{m}}}{\lambda_1 - \lambda_{\text{X}}} \left(\left(\frac{i}{i_{\text{m}}} \right)^{\frac{\lambda_{\text{X}}}{\lambda_1}} - \frac{i}{i_{\text{m}}} \right), \text{ with } 0 \leq i \leq i_{\text{m}}.$$

Let $f(i)$ be the right-hand member of this equation; note also that $f(i)$ is convex upwards at $0 < i < i_{\text{m}}$. Then the equiscalar lines of $F(i,x)$ have the form of $x = L(i)$, where

$$L(i) = \begin{cases} f(i), i^* \leq i \leq i_{\text{m}} \\ f(i^*), 0 \leq i < i^* \end{cases} \text{ and } i^* = i_{\text{m}} \left(\frac{\lambda_{\text{X}}}{\lambda_1} \right)^{\frac{\lambda_1}{\lambda_1 - \lambda_{\text{X}}}}$$ - the maximum point of $f(i)$ (see figure 1 below).

It is clear that the value of $F(i,x)$ over the equiscalar line equals the maximal value of $f(i^*)$.

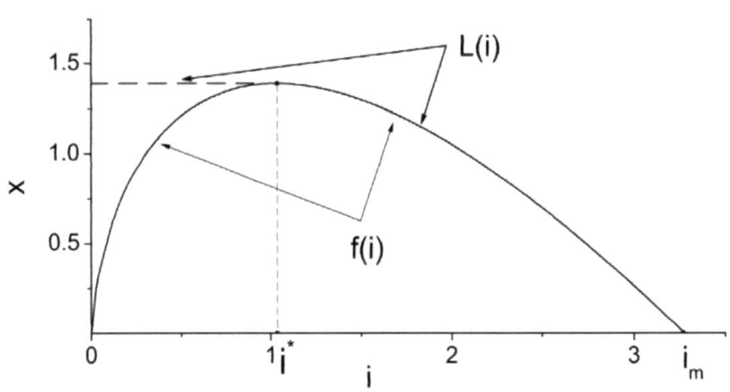

Figure 1. The form of the equiscalar line $x = L(i)$

Consequently, the function $L(i)$ is also convex upwards at $0 < i < i_{\text{m}}$; furthermore, the set B is also convex. The boundary of the set B at $i > 0$ and $x > 0$ is defined by the equiscalar line $x = L(i)$, corresponding to the value $F(i,x) = x_{\text{lim}}$.

The set B is closed, and, if $(i_0, x_0) \notin B$, then there exists a circular area $U_0 = \{(i,x) \mid \sqrt{|i - i_0|^2 + |x - x_0|^2} \leq a\}$ $(a > 0)$, such that $U_0 \cap B = \{\varnothing\}$. So, the estimation [7]

$(i(t), x(t)) \in U_0$ where $t_0 \leq t \leq a/M$ for any $\varphi(t) \in \Phi$, given

$$M = \sup_D \sqrt{(\lambda_1 i - (\lambda_{\text{X}} + \sigma_{\text{X}}\varphi)x)^2 + (\sigma_{\text{X}}\varphi - \lambda_1 i)^2}$$

and $D = \{(i,x,\varphi) \in R^3 \mid (i,x) \in U_0,\ 0 \leq \varphi \leq 1\}$,

which follows from the boundedness of right-hand members of (1) over the set D, results in $\Delta_\tau > a/M > 0$, because for any $t_0 \leq t \leq a/M$ the set $A \subset U_0$ and, consequently, $A \cap B = \{\varnothing\}$.

More precise lower estimation of Δ_τ may be obtained by adjusting the shape of sets A and B. To simplify the computation procedure convex polygons are considered; they are denoted by A_{c} and B_{c}, and contain sets A and B respectively. Thus, if $A_c \cap B_c = \{\varnothing\}$, then sets A and B do not intersect either.

The conditions, considered in this article, implying that the initial state (i_0, x_0) coincides with the equilibrium point of the system (1) given $\varphi = \varphi_{\max}$ and $(i_0, x_0) \notin B$, then the construction of the set B_c may be based on the following reasoning. In the coordinate system $Oixh$ the tangential plane defined by the equation $h = F(i,x)$ at the point $(i_0, x_0, F(i_0, x_0))$ is considered. The Oix phase plane projection of the intersection line between the tangential plane and the plane $h = x_{\lim}$ may be used as a part of the boundary of the set B_c (line Γ in figure 2).

The sides of the polygon A_c, estimating the set A, are formed by K estimations for the solutions of (1) denoted by $i(t), x(t)$ given $i(t_0) = i_0, x(t_0) = x_0$ and any $\varphi(t) \in \Phi$

$s_k(t) \le \alpha_k\, i(t) + \beta_k\, x(t)$, $|\alpha_k| + |\beta_k| > 0$, for $t_0 \le t \le t_1$, $k = 1, \dots, K$.

The estimations of such type are obtained by the integration of differential inequalities, which follow from (1) given the limitation $0 \le \varphi(t) \le \varphi_{\max}$, signs of summands in right-hand members of equations, as well as signs of coefficients α_k and β_k for the specific estimation line. For example,

1) $s_1(t) = i_0 \exp(-\lambda_I (t - t_0))$ при $\alpha_1 = 1, \beta_1 = 0$;

2) $s_2(t) = (x_0 - c_0)\exp(-(\lambda_X + \Lambda)(t - t_0)) + c_0 \exp(-\lambda_1 (t - t_0))$, где $c_0 = \dfrac{\lambda_1 i_0}{\lambda_X + \Lambda - \lambda_1}$,

$\Lambda = \sigma_X \varphi_{\max}$, при $\alpha_2 = 0, \beta_2 = 1$;

3) $s_3(t) = i_0 + x_0 - (\lambda_X + \Lambda)\big((x_0 - d_0 - \Lambda/\lambda_X)(1 - \exp(-\lambda_X (t - t_0)))/\lambda_X +$

$+ \Lambda (t - t_0)/\lambda_X + d_0 (1 - \exp(-\lambda_1 (t - t_0)))/\lambda_1\big)$, где $d_0 = \dfrac{\lambda_1 i_0 - \Lambda}{\lambda_X - \lambda_1}$, при $\alpha_3 = 1, \beta_3 = 1$;

etc. The choice of the count K of estimations and their coefficients α_k and β_k may be varied to adjust the set A_c (see figure 2 below).

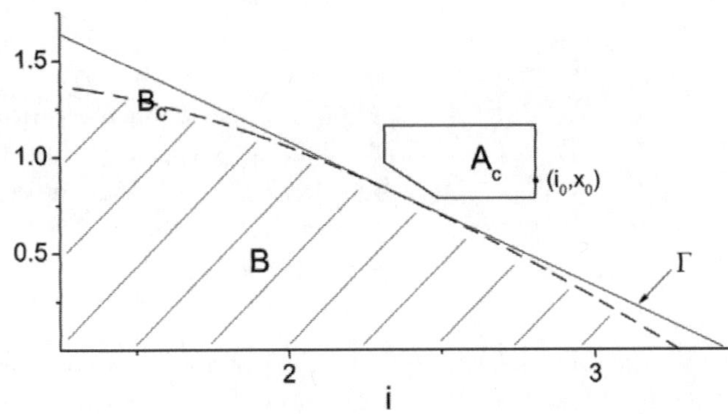

Figure 2. Approximations of sets A and B; the intersection line Γ

Then the relative position of A_c and B_c is analyzed for various values of τ and the point of time is estimated when the condition $A_c \cap B_c = \{\varnothing\}$ is violated. The maximal duration of the time interval $t_0 \le \tau < t_0 + \Delta_\tau$, where each point τ satisfies $A_c \cap B_c = \{\varnothing\}$, defines the lower estimation for Δ_τ.

The upper estimation of the length of the interval was based on the solution of the problem of optimal speed reactor power decrease with limited xenon concentration. The statement of this problem and analytical principles of the solution construction are stated in [8]. The duration of the numerically calculated process is the upper estimation for the interval. The duration of the process as the function of the xenon limitation value was plot.

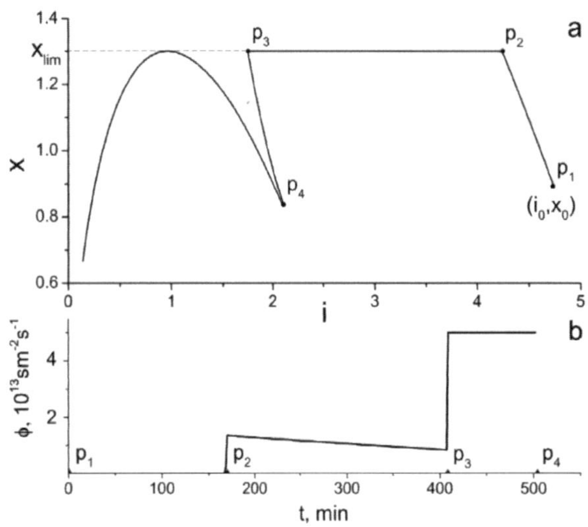

Figure 3. (a) The example of a transient, representing the optimal speed reactor power decrease, over the phase plane, (b) optimal control

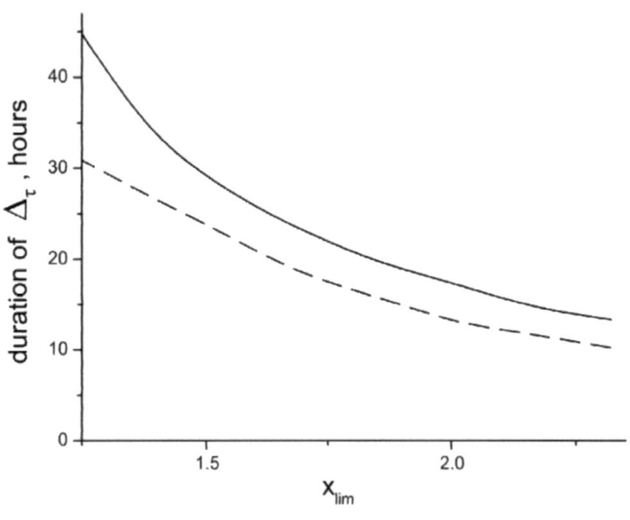

Figure 4. The estimations of the duration of segment Δ_τ as the functions of the limitation value for the concentration of xenon: solid and dashed lines for upper and lower estimations respectively

As results of the research carried two ways of estimation of the time interval observed were developed and applied and the dependencies of the estimations as functions of xenon limitation value were plot. The results obtained may be used in further research of the improvements to the problem statement considered in this article.

In particular, the following researches may include the problems of averaging the reactor control characteristics over the threatened period, which point of origin is known in advance.

Acknowledgments
We would like to express our sincere gratitude to our advisor Ph.D. Prof. Zagrebayev A. M. for sharing his ideas and knowledge, for his guidance, support and interest in our research.

References

[1] Zagrebayev A M, Ovsyannikova N V and Sadchikov S M 2012 Optimization of a mode of change of a nuclear reactor capacity during the threatened period under force-majeur circumstances *Modern problems of science and education* **4** 1–7
[2] Zagrebayev A M, Ovsyannikova N V and Sadchikov S M 2012 The average loss of energy production at a random stop of a reactor with a limited operational reserve of reactivity *Natural and technical sciences* **58** 418–22
[3] Dollezhal' N A and Emel'yanov I Ya 1980 *Channel Nuclear Power Reactor* (Moscow: Atomizdat)
[4] Ovchinnikov F Ya, Golybev L I, Dobrynin V D, Klochkov D I, Semenov V V and Tsybenko V I 1979 *Operating conditions of water-moderated-water-cooled power reactors* (Moscow: Atomizdat)
[5] Zweifel P 1997 *Reactor Physics* (Moscow: Atomizdat)
[6] Rudik A P 1979 *Optimization of reactor physical characteristics* (Moscow: Atomizdat)
[7] Pontryagin L S 1974 *Ordinary differential equations* (Moscow: Nauka)
[8] Rudik A P 1971 *Nuclear reactors and Pontryagin's maximum principle* (Moscow: Atomizdat)

IOP Conf. Series: Journal of Physics: Conf. Series **788** (2017) 012061 doi:10.1088/1742-6596/788/1/012061

Linear inverse problem of the reactor dynamics

N P Volkov

National Research Nuclear University MEPhI (Moscow Engineering Physics Institute), 31 Kashirskoe shosse, 115409 Moscow, Russia

E-mail: NPVolkov@yandex.ru

Abstract. The aim of this work is the study transient processes in nuclear reactors. The mathematical model of the reactor dynamics excluding reverse thermal coupling is investigated. This model is described by a system of integral-differential equations, consisting of a non-stationary anisotropic multispeed kinetic transport equation and a delayed neutron balance equation. An inverse problem was formulated to determine the stationary part of the function source along with the solution of the direct problem. The author obtained sufficient conditions for the existence and uniqueness of a generalized solution of this inverse problem.

1. Introduction.

The need for reliable, manageable and secure nuclear power plants is increased in recent years. This is an additional incentive for the development of the mathematical theory of nuclear reactors, which is relatively new and is constantly evolving. The proof can be found in a number of papers [1-10] devoted to this subject. While developing this theory, scientists together with technicians try to create a model reactor, which reliable work is less likely to be dependent on the human factor. The agenda raises questions of mathematical modeling of processes occurring in nuclear reactors. This is necessary both in the design and in the design and in the management fnd control of the operation of the reactor.

2. Formulation of linear inverse problem of the reactor dynamics.

We consider a mathematical model of the reactor dynamics [5], which describes a system of integral-differential equations:

1) non-stationary multispeed anisotropic kinetic transport equation

$$\frac{\partial u}{\partial t}(\mathbf{x},\mathbf{v},t) + (\mathbf{v},\nabla_x)u(\mathbf{x},\mathbf{v},t) + \Sigma(\mathbf{x},\mathbf{v},t)u(\mathbf{x},\mathbf{v},t) =$$
$$= \int_V J(\mathbf{x},\mathbf{v},\mathbf{v}',t)u(\mathbf{x},\mathbf{v}',t)d\mathbf{v}' + \sum_{k=1}^{N} z_k R_k(\mathbf{x},\mathbf{v},t) + F(\mathbf{x},\mathbf{v},t), \quad (1)$$

2) delayed neutron balance equation

$$\frac{\partial R_k}{\partial t}(\mathbf{x},\mathbf{v},t) = -z_k R_k(\mathbf{x},\mathbf{v},t) + \int_V J_k(\mathbf{x},\mathbf{v},\mathbf{v}',t)u(\mathbf{x},\mathbf{v}',t)d\mathbf{v}', \quad \forall k = \overline{1,N}, \quad (2)$$

$$(\mathbf{x},\mathbf{v},t) \in D = G \times V \times (0,T).$$

This model describes the function $u(\mathbf{x}, \mathbf{v}, t)$ of the density distribution of neutrons that pass through a point $\mathbf{x} \in G$ at a rate $\mathbf{v} \in V$ in time $t \in (0, T)$. Functions $\Sigma(\mathbf{x}, \mathbf{v}, t), J(\mathbf{x}, \mathbf{v}, \mathbf{v}', t), F(\mathbf{x}, \mathbf{v}, t)$ characterize properties of a medium where a mass transfer process occurs. More specifically, $\Sigma(\mathbf{x}, \mathbf{v}, t)$ is an absorption coefficient, $J(\mathbf{x}, \mathbf{v}, \mathbf{v}', t)$ is a scattering phase function, and $F(\mathbf{x}, \mathbf{v}, t)$ is a radiation source density. Here, G is a range of spatial coordinates, which is assumed to be a strictly convex domain, while V is a range of particles velocities $\mathbf{v}$, which is a closed set contained in a spherical layer $\{0 < v_0 \leqslant |\mathbf{v}| \leqslant v_1 < \infty\}$. Kernels $J_k(\mathbf{x}, \mathbf{v}, \mathbf{v}', t)$ of scattering integrals characterize a density distribution of secondary neutrons, and a function $R_k(\mathbf{x}, \mathbf{v}, t)$ is a density distribution of carriers neutron k-th group $\forall k = \overline{1, N}$.

Let us assume that the process under consideration occurs in the absence of external sources of particles; i.e., for example, external radiation does not pass through the reactor walls. This assumption is mathematically expressed by a boundary condition

$$u(\mathbf{x}, \mathbf{v}, t) = 0, \qquad (\mathbf{x}, \mathbf{v}, t) \in \gamma_- \times [0, T], \tag{3}$$

where $\gamma_- = \{(\mathbf{x}, \mathbf{v}) \in \partial G \times V : (\mathbf{v}, n_x < 0)\}$, and n_x is an outward normal to a boundary ∂G of G at a point $\mathbf{x}$.

Additionally, we set initial conditions

$$u(\mathbf{x}, \mathbf{v}, 0) = \varphi(\mathbf{x}, \mathbf{v}), \qquad (\mathbf{x}, \mathbf{v}, t) \in \overline{G} \times V, \tag{4}$$

$$R_k(\mathbf{x}, \mathbf{v}, 0) = R_{k0}(\mathbf{x}, \mathbf{v}), \quad \forall k = \overline{1, N}, \qquad (\mathbf{x}, \mathbf{v}, t) \in \overline{G} \times V. \tag{5}$$

The problem (1) - (5) is a direct problem of the nuclear reactor dynamics.

Let us formulate an inverse problem where together with solving a direct problem, we simultaneously need to find some parameter of the equation (1) by applying an additional condition. For example, we take a final condition, i.e., at $t_1 \in (0, T)$

$$u(\mathbf{x}, \mathbf{v}, \ t_1) = \psi(\mathbf{x}, \mathbf{v}), \qquad (\mathbf{x}, \mathbf{v}, t) \in \overline{G} \times V, \tag{6}$$

a condition which is called an over-determination condition.

From a physical point of view, the task is to find conditions in the input data of this problem under which a considered process can be driven from initial state (4) to terminal state (6) over a fixed time $t_1 \in (0, T)$ by controlling a stationary part $f(\mathbf{x}, \mathbf{v})$ of a source function

$$F(\mathbf{x}, \mathbf{v}, t) = f(\mathbf{x}, \mathbf{v}) g(\mathbf{x}, \mathbf{v}, t), \tag{7}$$

where $f(\mathbf{x}, \mathbf{v})$ is an admissible control function (distributed time-invariant control) and $g(\mathbf{x}, \mathbf{v}, t)$ is a given function (correction function).

Inverse problem. Find functions $u(\mathbf{x}, \mathbf{v}, t), R_k(\mathbf{x}, \mathbf{v}, t), f(\mathbf{x}, \mathbf{v})$, meeting the conditions (1) - (7).

Remark 1. The role of initial conditions in such inverse problems for transient processes can be played by some steady reactor states.

3. Solution of the linear inverse problem of the reactor dynamics.

The inverse problem stated above will be studied in the following function spaces, which were introduced in [7].

1) $H_\infty(D)$ is the space of functions u that belong to a class $L_\infty(D)$ together with their generalized derivatives u_t and $(\mathbf{v}, \nabla_x)u$ on D and have traces $u\,|_{\Gamma_-}$ on $\Gamma_- = \gamma_- \times (0, T)$ from $L_\infty(\Gamma_-)$, i.e.,

$$H_\infty(D) = \left\{ u \in L_\infty(D) : \frac{\partial u}{\partial t}, (\mathbf{v}, \nabla_x)u \in L_\infty(D), u\,|_{\Gamma_-} \in L_\infty(\Gamma_-) \right\};$$

$H_\infty(D)$ is a Banach space with respect to the norm

$$\|u\|_{H_\infty} = [\|u\|_{\infty,D} + \|\frac{\partial u}{\partial t}\|_{\infty,D} + \|(\mathbf{v}, \nabla_x)u\|_{\infty,D} + \|u\,|_{\Gamma_-}\,\|_{\infty,\Gamma_-}] < \infty.$$

2)$L^t_\infty(D)$ is the space of functions R_k that belong to a class $L_\infty(D)$ together with their generalized derivatives $\frac{\partial R_k}{\partial t}$ on D , i.e.,

$$L^t_\infty(D) = \left\{R_k \in L_\infty(D) : \frac{\partial R_k}{\partial t} \in L_\infty(D)\right\}, \quad \forall k = \overline{1, N};$$

$L^t_\infty(D)$ is a Banach space with respect to the norm

$$\|R_k\|_{L^t_\infty} = [\|R_k\|_{\infty,D} + \|\frac{\partial R_k}{\partial t}\|_{\infty,D}] < \infty, \quad \forall k = \overline{1, N}.$$

Where $\| \bullet \|_{\infty,D}$ is a norms in $L_\infty(D)$.

The solution of the linear inverse problem under study can be provided by the following result.

Theorem 1. *Let* $\Sigma, \frac{\partial \Sigma}{\partial t} \in L_\infty(D)$; $J, \frac{\partial J}{\partial t} \in L_\infty(D \times V)$; $\varphi, (\mathbf{v}, \nabla_x)\varphi \in L_\infty(G \times V), \varphi|_{\gamma_-} \in L_\infty(\gamma_-)$; $\psi, (\mathbf{v}, \nabla_x)\psi \in L_\infty(G \times V), \psi\,|_{\gamma_-} \in L_\infty(\gamma_-)$; $g, \frac{\partial g}{\partial t} \in L_\infty(D), |g(\mathbf{x}, \mathbf{v},\ t_1)| \geqslant g_0 > 0$; $J_k, \frac{\partial J_k}{\partial t} \in L_\infty(D \times V)$; $R_{k0} \in L_\infty(D)$, $\forall k = \overline{1, N}$; *and let the compatibility conditions* $\varphi(\mathbf{x}, \mathbf{v}) = 0$ *and* $\psi(\mathbf{x}, \mathbf{v}) = 0$ *for* $(\mathbf{x}, \mathbf{v}) \in \gamma_-$ *are met.*

If $\frac{diamG}{v_0} < b$ *with a constant b depending on the norms* $\|\Sigma\|, \|J\|, \|g\|, \|\varphi\|, \|\psi\|$, $\|J_k\|, \|R_{k0}\|, \forall k = \overline{1, N}$ *in the corresponding spaces and on numbers* t_1, g_0, *and* $m(V)$, *where* $m(V)$ *is a measure of the set* V, *then there exists a unique generalized solution of the inverse problem , i.e., there are functions* $u \in H_\infty(D)$, $R_k \in L^t_\infty(D)$ *and* $f \in L_\infty(G \times V)$, *for which a process under study can transfer from initial state (4) to terminal state (6).*

The idea behind the proof of this theorem is the reduction to a system of integral equations of the second kind for the functions $\{u, R_k, f\}$, considering the constraints as the smallness of $\frac{diamG}{v_0}$. Next, a right-hand side operator of this system is examined for contractibility. It turns out that its second power is a contraction if $\frac{diamG}{v_0} < C$, where C is a constant depending on the norms $\|\Sigma\|, \|J\|, \|g\|, \|\varphi\|, \|\psi\|, \|J_k\|, \|R_{k0}\|, \forall k = \overline{1, N}$ and on the numbers $t_1, g_0, m(V)$. Therefore, the system of integral equations has a unique solution $\{u, R_k, f\} \in H_\infty(D) \times L^t_\infty(D) \times L_\infty(G \times V)$ and, hence, inverse problem is uniquely solvable.

4. Conclusion.

The linear inverse problem is studied in this paper. In terms of transient processes in nuclear reactors, a reactor can be transferred from one state to another by applying a time-invariant control to the source function or to an absorption coefficient of the non-stationary multispeed anisotropic kinetic transport equation. Sufficient conditions for existence and uniqueness of generalized solution of the above inverse problem were obtained. The proof of Theorem 1 implies that a desired generalized solution depends continuously on the input data of a problem and can be found by the method of successive approximations.

Remark 2. Inverse problem discussed in this paper, can be interpreted as a task of managing of transition processes in nuclear reactors. This function $f(\mathbf{x}, \mathbf{v})$ will play a role of a distributed control effect on the neutron sources .

References

[1] Marchuk G I The Methods of Calculation for Nuclear Reactors. Atomizdat, Moscow, Russia,1961 (Russian). 667 p.

[2] Case K M, Zweifel P F Existance and Uniquenes Theorems for the Neutron Transport Equation. – J. Math. Phys., 1963, v. 4, N 11,p.1376.

[3] Akcasu A Z, Lellouche G S and Shotkin L M Mathematical methods in nuclear reactor dynamics. N.-Y.: Academic Press, 1971.

[4] Shikhov S B Problems of the Mathematical Theory of Reactors: Linear Analysis, Atomizdat, Moscow (1973) (Russian). 376 p.

[5] Kryanev A V and Shikhov S B Questions in the Mathematical Theory of Reactors: Nonlinear Analysis, Atomizdat, Moscow (1983) (Russian). 280 p.

[6] Lewins J Nuclear reactor kinetics and control / Jefery Lewins. Pergamon Press, 1978. - 264 c.

[7] Germogenova T A Local Properties of Solutions to Transport Equations. Nauka, Moscow, Russia, 1986 (Russian). 272 p.

[8] Prilepko A I and Volkov N P Inverse problems of determining the parameters of a non-stationary kinetic transport equation for more information about the following desired function. - Differential Equations, 1988, v. 24, N 1, p. 136-146

[9] Volkov N P On Some Inverse Problems for Time-Dependent Transport Equation. //ILL-Posed Problems in Natural Sciences. Moscow: TVP-VSP, 1992, p. 431-438.

[10] Volkov N P Solvability of Certain Inverse Problems for the Nonstationary Kinetic Transport Equation. - Computational Mathematics and Mathematical Physics. Pleiades Publishing, 2016, v. 56, N 9, pp. 1598-1603.